Luigia Sabbatini, Inez Dorothé van der Werf (Eds.)
Chemical Analysis in Cultural Heritage

Also of interest

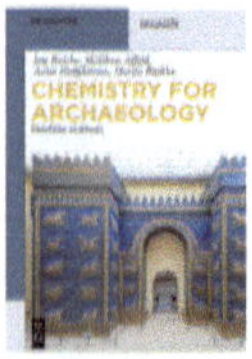

Chemistry for Archaeology – Heritage Sciences
Reiche (Ed.), 2020
ISBN 978-3-11-044214-4, e-ISBN 978-3-11-044216-8

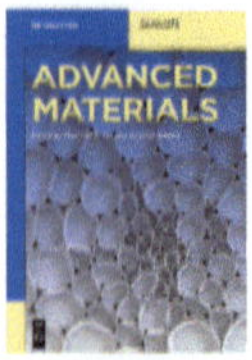

Advanced Materials
Van de Ven, Soldera (Eds.), Gauvin, 2019
ISBN 978-3-11-053765-9, e-ISBN 978-3-11-053773-4

Inorganic Trace Analytics - Trace Element Analysis and Speciation
Matusiewicz, Bulska (Ed.), 2017
ISBN 978-3-11-037194-9, e-ISBN 978-3-11-036673-0

Inorganic Pigments
Pfaff, 2017
ISBN 978-3-11-048450-2, e-ISBN 978-3-11-048451-9

Contents

Editors
Prof. Luigia Sabbatini
University of Bari
Chemistry Department
70125 Bari
Italy
luigia.sabbatini@uniba.it

Dr. Inez Dorothé van der Werf
Cultural Heritage Laboratory
Cultural Heritage Agency of the Netherlands
Hobbemastraat 22
1071 ZC Amsterdam
The Netherlands
i.van.der.Werf@cultureelerfgoed.nl

ISBN 978-3-11-045641-7
e-ISBN (PDF) 978-3-11-045753-7
e-ISBN (EPUB) 978-3-11-045648-6

Library of Congress Control Number: 2019951715

Bibliographic information published by the Deutsche Nationalbibliothek
The Deutsche Nationalbibliothek lists this publication in the Deutsche Nationalbibliografie; detailed bibliographic data are available on the Internet at http://dnb.dnb.de.

Cover image: Dmitrty / iStock / Getty Images Plus
Typesetting: Integra Software Services Pvt. Ltd.
Printing and binding: CPI books GmbH, Leck

www.degruyter.com

Chemical Analysis in Cultural Heritage

Edited by
Luigia Sabbatini and Inez Dorothé van der Werf

DE GRUYTER

List of contributing authors

Maurizio Aceto
Universita degli Studi del Piemonte Orientale Amedeo Avogadro Sede di Alessandria
Dipartimento di Scienze e Innovazione Tecnologica (DISIT)
Viale T. Michel, 11
15121 Alessandria
Italy
maurizio.aceto@uniupo.it

Pasquale Acquafredda
Universita degli Studi di Bari Aldo Moro
Dipartimento di Scienze della Terra e Geoambientali
via Edoardo Orabona 4
70125 Bari
Italy
pasquale.acquafredda@uniba.it

Luís C. Alves
Universidade de Lisboa Instituto Superior Tecnico
C2TN
Bobadela
Lisboa
Portugal
lcalves@ctn.tecnico.ulisboa.pt

Demetrios Anglos
University of Crete - Department of Chemistry
Heraklion
Greece
anglos@iesl.forth.gr

Rita Araújo
NOVA University of Lisbon, Faculty of Sciences and Technology - Department of Conservation and Restoration and LAQV-REQUIMTE
Caparica
Portugal
a.araujo@campus.fct.unl.pt

Giovanni Bartolozzi
National Council of Research
Institute of Applied Physics "Nello Carrara", IFAC-CNR
50019 Sesto Fiorentino
Italy
g.bartolozzi@ifac.cnr.it

Klaas Jan van den Berg
Cultural Heritage Laboratory
Cultural Heritage Agency of the Netherlands
Hobbemastraat 22
1071 ZC Amsterdam
The Netherlands
K.van.den.Berg@cultureelerfgoed.nl

Susanna Bracci
National Council of Research
Institute for the Conservtaion and Valorization of Cultural Heritage
Via Madonna del Piano, 10
50019 Sesto Fiorentino
Italy
s.bracci@icvbc.cnr.it

Maria Cristina Caggiani
Via R. Canudo, 40
70123 Bari
Italy
mariacristinacaggiani@virgilio

Antonio Candeias
Universidade de Evora Escola de Ciencias e Tecnologia
Hercules Centre
Évora
Portugal
candeias@uevora.pt

https://doi.org/10.1515/9783110457537-203

Laura Cartechini
Consiglio Nazionale delle Ricerche
Istituto di Scienze e Tecnologie Molecolari
Via Elce di sotto, 8
06123 Perugia
Italy
laura.cartechini@cnr.it

Pavel Cejnar
University of Chemistry and Technology,
Department of Computing and Control
Engineering
Technicka 5
Prague
166 28
Czech Republic
cejnarp@vscht.cz

Victor J Chen
Indianapolis Museum of Art at Newfields
Conservation Science
Indianapolis
Indiana
United States
vchen@discovernewfields.org

Philippe Colomban
CNRS
MONARIS
Sorbonne Université, c49, 4 Place Jussieu
Paris
75005
France
philippe.colomban@upmc.fr

Inês Coutinho
Universidade Nova de Lisboa Faculdade de
Ciencias e Tecnologia
Department of Conservation and
Restoration
Caparica
Setúbal
Portugal
icoutinho@fct.unl.pt

Ilaria Degano
Università di Pisa
Department of Chemistry and Industrial
Chemistry
via Moruzzi 13
56123 Pisa
Italy
ilaria.degano@unipi.it

Giacomo Eramo
Universita degli Studi di Bari Aldo Moro
Dipartimento di Scienze della Terra e
Geoambientali
Via Orabona, 4
70125 Bari
Italy
giacomo.eramo@uniba.it

Lorena Carla Giannossa
Università degli Studi di Bari Aldo Moro
Dipartimento di Chimica
Via Edoardo Orabona, 4
70125 Bari
Italy
lorenacarla.giannossa@uniba.it

Bernard Gratuze
Institut de Recherche sur les
Archeomateriaux
IRAMAT
Orleans
Centre
France
gratuze@cnrs-orleans.fr

Maria Filomena Guerra
CNRS
UMR 8096
MAE
21 Allée de l'Université
Nanterre
92023
France
maria.guerra@cnrs.fr

Kurt F Hostettler
Indianapolis Museum of Art at Newfields
Conservation Science
Indianapolis
Indiana
United States
khostettler@discovernewfields.org

Radovan Hynek
University of Chemistry and Technology, Department of Biochemistry and Microbiology
Prague
Czech Republic
hynekr@vscht.cz

Henk van Keulen
Cultural Heritage Laboratory
Cultural Heritage Agency of the Netherlands
Hobbemastraat 22
1071 ZC Amsterdam
The Netherlands
H.van.Keulen@cultureelerfgoed.nl

Stepanka Kuckova
University of Chemistry and Technology
Department of Biochemistry and Microbiology
Technicka 3
Prague
16628
Czech Republic
kuckovas@vscht.cz

Francisco Laborda
Universidad de Zaragoza
Analytical Chemistry
Pedro Cerbuna, 12
Zaragoza
50009
Spain
flaborda@unizar.es

Andreia Machado
Universidade Nova de Lisboa Faculdade de Ciencias e Tecnologia
Department of Conservation and Restoration
Caparica
Setúbal
Portugal
machado_andreia@sapo.pt

Annarosa Mangone
Universita degli Studi di Bari Aldo Moro
Dipartimento di Chimica
Via Edoardo Orabona, 4
70125 Bari
Italy
annarosa.mangone@uniba.it

Teresa Medici
Universidade Nova de Lisboa Faculdade de Ciencias e Tecnologia - Research Unit VICARTE
Caparica, Setúbal
Portugal
teresa.medici@gmail.com

Maria Joao Melo
NOVA University of Lisbon, Faculty of Sciences and Technology - Department of Conservation and Restoration and LAQV-REQUIMTE
Caparica
Portugal
mjm@fct.unl.pt

Costanza Miliani
Consiglio Nazionale delle Ricerche
Istituto di Scienze e Tecnologie Molecolari
Via Elce di sotto, 8
06123 Perugia
Italy
costanza.miliani@cnr.it

Paula Nabais
NOVA University of Lisbon - Faculty of Sciences and Technology, Department of Conservation and Restoration and LAQV-REQUIMTE
Caparica, Lisbon
Portugal
p.nabais@campus.fct.unl.pt

Josefina Pérez-Arantegui
Universidad de Zaragoza
Analytical Chemistry
Pedro Cerbuna, 12
Zaragoza
50009
Spain
jparante@unizar.es

Caitlyn E Phipps
Indianapolis Museum of Art at Newfields
Conservation Science
Indianapolis
Indiana
United States
caitlynephipps@gmail.com

Marcello Picollo
IFAC-CNR
SABeC
Via Madonna del Piano
50019 Sesto Fiorentino
Italy
m.picollo@ifac.cnr.it

Francesca Rosi
Consiglio Nazionale delle Ricerche
Istituto di Scienze e Tecnologie Molecolari
Via Elce di sotto, 8
06123 Perugia
Italy
francesca.rosi@cnr.it

Diego Sali
Bruker Italia srl
Viale Vincenzo Lancetti, 43
20158 Milano
Italy
diego.sali@bruker.com

Irina Crina Anca Sandu
Munch Museum
Conservation Department
Tøyengata 53
Oslo
0608
Norway
irina.sandu@munchmuseet.no

Erika Gohde Sandbakken
Munch-museet
Conservation Department
Oslo
Norway
erika.sandbakken@munchmuseet.no

Jiri Santrucek
University of Chemistry and Technology, Department of Biochemistry and Microbiology
Prague
Czech Republic
santrucj@vscht.cz

Gregory Dale Smith
Indianapolis Museum of Art at Newfields
Conservation Science
4000 Michigan Road
Indianapolis
Indiana
46208
United States
gdsmith@discovernewfields.org

Ken Sutherland
Art Institute of Chicago
Conservation and Science
111 South Michigan Ave
Department of Conservation and Science
Chicago
Illinois
60603
United States
ksutherland@artic.edu

Vitorino, Tatiana
NOVA University of Lisbon - Faculty of Sciences and Technology, Department of Conservation and Restoration and LAQV-REQUIMTE
Campus de Caparica, Caparica, Lisbon
2829-516
Portugal
tm.vitorino@campus.fct.unl.pt

Eva Storevik Tveit
Munch-museet
Conservation Department
Oslo
Norway
estveit@munchmuseet.no

Márcia Vilarigues
Universidade Nova de Lisboa Faculdade de Ciencias e Tecnologia Ringgold standard institution - Department of Conservation and Restoration
Caparica, Setúbal
Portugal
mgv@fct.unl.pt

Inez Dorothé van der Werf
Cultural Heritage Laboratory
Cultural Heritage Agency of the Netherlands
Hobbemastraat 22
1071 ZC Amsterdam
The Netherlands
I.van.der.Werf@cultureelerfgoed.nl

Part I: **Analytical Techniques**

Lorena Carla Giannossa and Inez Dorothé van der Werf

1 Overview of materials in Cultural Heritage

Pottery

Raw materials and production

- clay
- water
- firing process

Clay is an earthy, fine-grained material that combines one or more clay minerals with possible traces of quartz (SiO_2), metal oxides (Al_2O_3, MgO, etc.) and organic matter. Geologic clay deposits are mainly composed of phyllosilicate minerals and variable amounts of water trapped in the mineral structure. Surely, it becomes plastic and easily moldable when mixed with limited amount of **water** and rigid when dried and **fired**.

Clay is a rock originated from the alteration of minerals or other rocks. It can be divided in primary and secondary. The formers originate by minerals changing in situ, producing a rock with a chemical composition essentially unaltered with respect to the bedrock. The secondary ones are sedimentary rocks whose deposits were formed in places different from that of alteration, due to natural transport. Skeleton and matrix well mixed compose the clay. The skeleton is made of fragments of crystals of various kinds whose dimensions can reach up to 4 mm; whereas the matrix is made of clay minerals containing crystals smaller than 4 mm. Raw materials to produce ceramic artifacts are the plastic component consisting of clay materials, the inert or temper component consisting principally of silica and the glazing and fluxing component. The latter often concurs because fluxes favor the formation of a vitreous phase. Basically, they are mixtures of carbonates and feldspars. Inert, temper and flux materials are often added to clay to obtain a ceramic object with the desired characteristics.

The list of necessary steps in ceramic manufacturing includes discovering sources of raw materials, selecting those to be used, extracting and transport raw materials to the site of manufacturing (e. g. a workshop), preparing and shaping the ceramic paste, drying and firing the vase. Non-essential actions consist, for example, of adding decorations (engravings or stamps) or applying some kind of slip (glaze or enamel). Following possible crushing, grinding and purification procedures of raw materials, the next step is the mixing of raw materials with an appropriate amount of water to obtain the ceramic paste: an adequate workability is a mandatory pre-firing condition. Plasticity can be increased adding plasticizers or mix of clays owning different characteristics. On the contrary, it can be decreased by non-plastic additives (quartz, calcite, seeds, shell, crushed pottery, etc.).

Modelling

The actions aimed at modifying the shape of the unformed clay is defined forming or modelling. Of course, after the invention of the wheel, everything changed. Wheel function is turning the object incessantly, so working faces change, and pressure can be applied. Such a high technological advance increased hugely vessel production.

Drying and firing

Drying provides cohesion and mechanical resistance to the ceramic mixture, a slow evaporation of water, avoids deformations, breakages and possible explosions during firing step. During firing,

https://doi.org/10.1515/9783110457537-001

process transformations, decomposition reactions, redox reactions and formation of new minerals occur. It involves the transformation of some of the constituent minerals into new ones (neoformation process), influencing the properties of the ceramic product. Clearly pottery characteristics will also depend on firing conditions and raw materials. For example, the reached temperature influences the permeability of the handcraft or depending on firing atmosphere, the object looks a certain color. The main methods of firing can be grouped into two categories: one in which the products are in close contact with the fuel (open fire) and the other in which the fuel and products are physically separated (kilns). The firing cycle is normally divided in three steps: (1) progressive heating of the dried material up to the selected temperature is reached, (2) a maximum temperature plateau for the time needed to carry out the appropriate ceramic characteristics and finally (3) the cooling phase.

Colors

The role of color is evidently aesthetic and can be intrinsic to the ceramic material or be produced by the addition of suitable dyes in form of oxides or specific salts. The color control was one of the great successes of the technological evolution of ceramic production. Ceramic coloring depends mainly on managing redox conditions of firing atmosphere, and therefore on the ability to make combustion takes place in the most appropriate conditions.

Surface coatings

Surface coatings are obtained by different techniques and compounds. The principal ones are slips, pigments, paints, washes, colorants and glazes. The need to waterproof a porous material like pottery has essentially led to two types of solutions: vitrification of the whole mixture, or the use of a glazed coating able to seal at least one of the two surfaces of the artifact. Typically, the coating is thinner than the ceramic body and it can completely or partially cover the object and it can be clayey or glazed. The former is porous and opaque, named engobe or slip, the latter is waterproof and shiny, and it is named glaze when transparent and enamel opaque. The slip or engobe is a fluid suspension of clay in water. It can be an intermediate step in pottery manufacturing as a surface coating. It is prepared by diluting in water, a clay material which is often the same paste used for the ceramic body itself. Then the coarse fraction is excluded, and the homogeneous suspension is applied to the vase which surface becomes homogeneous and smooth. Usually, kaolin and lean clays are used as plastic components. White slips result from the absence of metallic colorants while colored slips are commonly obtained using iron.

Another clay coating is the gloss. It is an opaque or more or less glossy thin layer. Raw material is the finest fraction of illitic clays, separated by sedimentation. Most representative examples are black and red figured pottery in Attic ceramic.

Vitreous coatings provide impermeability and shininess to the artifact. Glazes range from being completely transparent to being completely opaque. The former constituent of most ceramic glazes is silica (SiO_2). To produce glazes at lower temperature, fluxes or modifiers must be added. The glaze can be applied using aqueous suspensions which contain insoluble oxides in a suitable quantity. During firing, the oxides melt giving rise to the vitreous phase. This technique is called glazing and is used for leaded glazes containing lead oxides such as litharge (PbO) or minio (Pb_3O_4) finely divided. In the case of opaque glaze, it is necessary to add compounds capable of opacifying the mixture. In the past cassiterite (SnO_2) was the most used opacifier. Alkaline glazes are usually transparent and shiny, and their main fluxes are sodium and potassium. Lime-felspar glazes are the hardest, most durable and resistant when compared to other glazes illustrated.

The presence of some metals in their native state (e. g. gold, silver, copper) makes coatings very shiny. Metal nanoparticles express unique light-scattering characteristics leading to extraordinary coloring. Glaze coloring can be also be obtained by the technique of metallic luster that is produced inserting Ag, Au or Cu oxides in the glaze, later reduced during the cooling phase.

Degradation phenomena

Ceramics are stable materials thanks to their inorganic compounds that become inert after irreversible reactions occurring during firing phase. These mixtures are kinetically and thermodynamically stable, and slightly or no soluble. Consequently, the deterioration is mainly due to pottery fragility. Excluding breakings, deterioration is due to natural phenomena, very slow and barely stoppable. Therefore, conservation interventions tend to slow down all the processes responsible for the degradation itself, which causes are closely linked to the conservation of the object. Excluding thermal changes and wind erosion, water directly or indirectly is responsible for any other cause of degradation. It can penetrate the porous structure of the object and it can create an overpressure able to overcome the cohesion forces of the material with consequent breaking and weakening of the structure. Water is also linked to phenomena connected to the presence of salts. Salts migrate in porous media and can exceed solubility limits and precipitate as crystals. Precipitation generates the phenomenon of efflorescence, a powdery coating. On the contrary, subefflorescence forms under the material surface and results in significant damage, for example, the detachment of outer layers or coating.

Another event regard the characteristic of some salts to crystallize involving different amount of water molecules, that causes an increase in their molar volume. Also, leaching of alkali metals must be considered since leads to the formation of a colloidal silica layer and its consequent dissolution.

Generally, ceramic resists sufficiently to the aggression of acidic solutions (excluding HF) whereas the basic medium is more dangerous being able to cause the matrix dissolution.

Some general rules should be observed to best preserve pottery: (1) conservation should save rather than change an artifact, (2) each treatment should be reversible, (3) every chemical or physical treatment should be meticulously recorded, and (4) unstable specimens should be kept in an environment similar to that they come from.

Chemistry

Ceramic is mainly composed of silicon, aluminum, calcium and alkaline metals, which are the most common elements present on the Earth's crust, further supporting its abundant production.

Clay minerals are basically phyllosilicates, the non-clay ones can be other silicates, oxides or hydroxides, salts and occasionally organic compounds.

The fundamental structure of the phyllosilicates is based on interconnected six member rings of $(SiO_4)^{-4}$ tetrahedra extending in infinite sheets. Three out of the 4 oxygens from each tetrahedra are shared with other tetrahedra. This leads to a basic structural unit of $(Si_2O_5)^{-2}$.

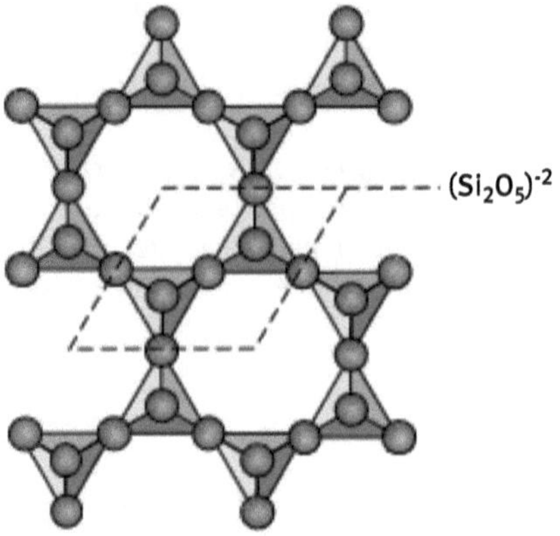

Figure 1.1: Phyllosillicates.

Analyses

- OM, SEM-EDS
- XRF
- XRPD, DTA, TGA, Mössbauer
- FTIR, Raman Spectroscopy
- ICP-MS, ICP-OES, INNA

References

Freestone I, Gaimster DR, editors. Pottery in the making: world ceramic traditions. British Museum Publications, 1997.

Noll W, Heimann RB. Ancient old world pottery. materials, technology, and decoration. Schweizerbart Science Publishers, 2016.

Rice PM. Pottery analysis – a sourcebook. Chicago, USA: University of Chicago Press, 1987.

Rye OS. Pottery technology. Principles and reconstruction. Manual of archeology, Taraxacum, Washington D.C., USA, 1981.

Tite MS. Ceramic production, provenance and use–A review. Archaeometry 50(2):216–31.

Glass

Raw materials and production

Naturally, glass is present in nature as a volcanic material (obsidian) and can be manufactured accidentally whenever a fusion of a siliceous material (such as silica-rich plants and sand) and an alkali (e. g. plant ashes) occurred.

When glass is manufactured for cultural heritage purposes, lead oxide (**PbO**) and specially silicon dioxide (**SiO_2**, generally named silica) are the most used as main component of glass. The necessary additional components to obtain a glass easily workable and stable are called **modifiers**. Mixing the powdered silica with fluxes (alkali metal salts), a mixture that liquefies at a temperature lower than that required to melt silica alone is obtained. Stabilizer like limestone is added to sand and flux to increase structural integrity and chemical resistance.

Most of glasses were SiO_2-Na_2O-CaO type from its first appearance until the entire Roman period, with different proportions of oxides (approximately: SiO_2 68–73 %, Na_2O 14–17 %, CaO 7–10 %). Different percentages depended on place and production period.

The silica-soda-lime type dominated the European glass production up to the eleventh century CE, when in northern Europe the soda was replaced by potash, more easily obtainable from wood ash.

The raw materials should be ground to a small size to increase the surface area to volume ratio and facilitate accretion, fluxing and melting.

The temperature needed to make the glass workable depends on its composition and vitrifying/flux ratio. Approximately, it requires 600–800 °C to start and 900–1200 °C to be terminated. Archaeological findings have shown that, in some production sites, glass formation took place in two distinct phases, carried on in different furnaces.

The first treatment, at a lower temperature, was used to react the vitrifying agent and the modifiers. At this point, the furnace temperature was not sufficient to cause the reaction product, the so-called "frit". The complete fusion took place in a second furnace, built to obtain significantly higher temperatures.

The millennial evolution of glass technology has always been based on attempts: every modification of ancient traditions has probably happened by chance or following the empirical experimentation of some talented glassmaker.

Only since the sixteenth century, thanks to the understanding of processes involved in the formation of glass and relations existing between structure and properties, the manufacturing becomes well reasonable and able to obtain materials with particular characteristics.

Forming

Glass can be shaped using different techniques like:

- Casting
  It is a very ancient technique: glass objects are cast by directing molten glass into a mold where it solidifies.
- Blowing
  Inflating molten glass into a parison with the aid of a blowpipe.
- Pressing
  Glassware formed by placing a blob of molten glass in a mold and pressing it to form the inside shape.
- Stretching
  In the past, it was produced by pouring the molten material onto a cast iron table and flattening it using a roller.

Recyling in antiquity

Glass recycling was a practice even preceding blowing invention. At the beginning, it was on a limited scale and carried on without re-melting fragments. Glassmakers re-used glass fragments of precious colors and decorated in gold leaf.

Following the discovery that glass could be fully recast, a targeted collection started. The most requested were natural color or colorless ones, in order to avoid the risk of obtaining an indistinct brownish compound during recast.

Decorations

Glass artworks can be decorated in different ways, the most common are engraving, painting and use of acids. Incisions can be made through different techniques (e. g. using abrasive rotating disks).

It is possible painting glass using enamels, which are then fused with the glass in a low-temperature furnace. The action of acids and sandblasting produce a frosted effect.

The golden glass is obtained by applying on the surface leaves, paint or gold powder, which are then embedded by firing.

Colors

The color of glass changes by adding especially metallic oxides to the batch.

Examples of common colorants include:

Iron – Green glass.
Copper – Light blue glass.
Manganese dioxide – Can decolorize colored glasses. In higher amounts, purple and even black color can be obtained.
Cobalt – Dark blue glass.
Gold – Deep red glass.

Opacifiers – Tin or antimony compounds (e. g. calcium antimonate, lead antimonate, lead pyroantimonate, tin oxide, lead stannate)

Degradation phenomena

Weathering

The main agent responsible for the degradation of glass is water, or rather, the aqueous solutions naturally occurring, which can cause changes in the composition and structure of the surface.

It usually involves the leaching of alkali from the glass, leaving behind siliceous weathering products that are often laminar.

There are two main processes of degradation: an acid attack, a term which improperly indicates the alteration caused by solutions with $pH \leq 7$ and a basic attack, due to the contact with solutions at $pH > 7$. In the acid attack, the alkaline ions and to a lesser extent the alkaline earth ions, are replaced by

H_3O^+ ions by forming silanol groups. Therefore, the surface is enriched with a thin layer of silica gel which can, in the case of high chemical resistance glasses, protect the glass. Conversely, it is possible that the chemical attack leads to the complete dissolution of glass when hydrofluoric acid, and in some cases phosphoric acid, are responsible for the attack.

The basic attack is the worst one, the hydroxyl ions OH^- enter the lattice of the glass, attacking the silicon-oxygen bond in a chain reaction that, at the end of the process, leads to the production of a soluble compound.

Usually, the modification of glass composition and structure causes changes in the surface appearance, generally iridescence, discoloration and a loss of transparency. Moreover, the contraction of the volume associated with ion exchange can lead to cracking and detachment.

Certainly, the chemical attack that occurs depends on composition of the glass, temperature, surface condition, contact time and obviously composition of the aqueous solution.

Chemistry

Glass is a non-crystalline solid. It can be described as constituted by a random network, where a certain structural order is present only at short range and not at medium-long one. This property can be easily verified using a technique able to detect crystal phases (X Ray Diffraction): no signal will be observed.

Its amorphous nature implies the presence of bonds whose energy covers a rather wide range. Consequently, the breakage of the bonds progressively occurs at various temperatures and the transition from solid to liquid is achieved through a progressive decrease in viscosity. During this transformation, the glass is easily workable, regaining its rigidity when it is returned to room temperature.

Formers make up the largest percentage of the mixture to be melted. In typical soda-lime-silica glass the *former* is SiO_2 in the form of sand.

Fluxes lower the temperature at which the formers will melt. Soda (sodium carbonate) and potash (potassium carbonate) are common fluxes. Potash glass is slightly denser than soda glass.

Stabilizers make the glass strong and water resistant. Calcium carbonate, often called calcined limestone, is a stabilizer. Without a stabilizer, water and humidity attack and dissolve glass.

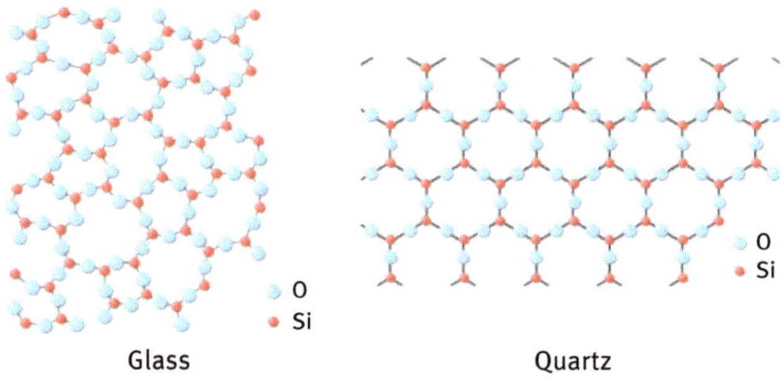

Figure 1.2: Glass Quartz.

The elementary crystalline cell of SiO_2 consists of a silicon atom placed at the center of a tetrahedron whose vertices are oxygen atoms. In the structure of solid glass, the tetrahedral structural units linked together by oxygen atoms are still present, but in a random disordered arrangement.

Properties

- Mechanically Strong
- Hard
- Elastic
- Chemical Corrosion-Resistant. Affected by few chemicals (HF). Resists most industrial and food acids.
- Thermal Shock-Resistant
- Heat-Absorbent
- Optical Properties: reflects, bends, transmits and absorbs light.
- Electrical Insulating

Analyses

- SEM-EDS
- XRD
- UV-Vis-NIR
- Raman Spectroscopy
- ICP-MS, ICP-OES, INNA

References

Degryse P, Scott RB, Brems D. The archaeometry of ancient glassmaking: reconstructing ancient technology and the trade of raw materials. Perspective (France) 2(2):224–38, 2014.

Frank S. Glass and archaeology. London, UK: Academic Press, 1982, ISBN 9780122656200.

Henderson J. Ancient glass: an interdisciplinary exploration. Cambridge, UK: Cambridge University Press, 2012.

Metal

Raw materials and production

Archaeological and radiochemical dating propose the beginning of metals use in the period 6000 to 7000 BCE. Metallic artworks have been found in soils of various compositions, caves, tombs, graves, etc. Dry environments appear to be optimum for metal preservation, but some metals have survived in shipwrecks for thousands of years.

Considering geological time scale, few metals are able to resist alterations induced by natural agents (e. g. oxidation) and this is the reason why only the less reactive ones have been found as native. A metal found in nature in its metallic form, either pure or in the form of an alloy is named **native metal**.

Native metals were the only source of access to metals for prehistoric men before mining. However, they are only found in small quantities so they cannot be used extensively. Usually, gold and copper were found in streams and rivers downstream of their ore deposits.

Generally, an **alloy** is an intentional mixture of two or more metals to obtain a material with properties different from those of the starting ones, nevertheless metal combinations also occur in nature.

Gold, silver, copper, lead, iron, tin and mercury are considered the seven metals of antiquity.

Gold

Easily worked, ductile, malleable, it does not tarnish or corrode. During antiquity it was almost always used as an ornament. It is able to survive in different soils, dry tombs and shipwrecks, at most suffering from encrustations or mild tarnish. Gold alloys with silver, platinum, copper, etc., also have been durable, although metal separation has occurred.

Cupellation process is likely both the oldest and most efficient method of separating gold from base metals. Precious metals do not oxidize or react chemically, unlike the base metals, therefore when they are heated at high temperatures, they remain apart while base metals react forming slags or other compounds.

Silver

Ductile and malleable but very soft if compared to copper or gold. It is rarely found in the native state as nuggets. It can be easily smelted from its chloride ores or produced from argentiferous galena (lead and silver were separated by smelting and then by cupellation).

Lead

Very malleable and dense, easily worked into sheets or cast into objects. It is simply extracted from its most abundant ore, galena.

Casting process occurs when a liquid material is poured into a mold, which contains a hollow cavity of the desired shape, and then allowed to solidify.

Copper

Found as native metal. It was first used as an ornament, then it was discovered that, when cold-worked and hammered, it became stronger and harder, and its purpose utilitarian.

Tin

Generally, not used in antiquity except as a constituent of its alloy with copper – **bronze**. Bronze is composed mostly of copper (80–95 %), tin (20–5 %), and often smaller amounts of other metals (e. g.

zinc and lead). It is corrosion resistant and harder, more easily cast and has a higher tensile strength than copper but it is less malleable. Therefore, it was used for many purposes: tools, ornamental objects, weapons and statues.

Iron

Most important metal of antiquity. It is one of the most abundant elements in the earth' s crust. Iron ores are frequent, wide-spread and easily available. It reacts with oxygen which implies that many early iron artifacts have likely disintegrated over time. It is the base metal of **steel**, an alloy with carbon, with increased malleability, toughness and ductility.

Mercury

Its primary use was in the purification of gold and silver via amalgamation.

Table 1.1: Main ores.

	MAIN ORES
Silver	1. Argentite 2. Chlorargyrite 3. Pyrargyrite 4. Galena
Copper	1. Cuprite 2. Malachite 3. Azurite 4. Chrysocolla 5. Tenorite
Lead	1. Galena 2. Cerussite 3. Anglesite
Tin	1. Cassiterite 2. Stannite
Iron	1. Magnetite 2. Hematite 3. Siderite
Mercury	1. Cinnabar

Metallurgy phases

These steps are characterized by many processes going hand by hand, considering no specific metal and complex discoveries.

1. Native metal used as found. Native metals were worked like stones.
2. Native metal phase. Cold working by hammering, tempering, cutting and grinding and likely involving copper, gold, silver, and meteoric iron.
3. Ore phase. Hot working by melting, casting, welding, etc. and alloys production. This stage would lead to the isolation and working of lead, silver, copper, antimony, tin, bronze, and brass.
4. Iron phase. Hot working by hammering, tempering, quenching and annealing and development of wrought iron, steel and cast iron.

Forming

If a metal can be melted, it is possible producing shapes by casting in a mold. Casting is mostly effective in case of complex shapes which could not be succeeded by forging. Actually, technology required to work metals is generally strongly subject to melting points. Indeed, metals of antiquity have low melting points (iron excluded) and usually alloys have lower melting points than that of their main component (e. g. bronzes and brasses melt at a lower temperature than copper). Alloying lead to changes in physical and mechanical properties which means that final use of an object influences pure metal or allow choice (e. g. refined jewelry).

Colors

Metals are crystalline materials, generally malleable, ductile and shiny. Their properties have allowed several applications, whose main ones in antiquity were decorative, military and utilitarian.

They (mercury excluded) are solids at room temperature with melting temperatures range very wide. Copper, silver and gold become molten in the region of 1000 °C, requesting specific tools and fuel. Conversely, tin and lead have low melting temperatures. Others (arsenic and zinc) boil at relatively low temperatures, forming vapors at high temperatures.

Most metals are good conductors of both heat and electricity. Atoms are bound together in a crystal lattice while outer electrons able to move rather freely.

Table 1.2: Melting metals.

Name (Symbol)	Melting	Boiling
Mercury (Hg)	−39 °C	357 °C
Tin (Sn)	232 °C	2602 °C
Lead (Pb)	328 °C	1749 °C
Arsenic (As)	Sublimes at 617 °C	
Zinc (Zn)	420 °C	907 °C
Antimony (Sb)	631 °C	1587 °C
Silver (Ag)	962 °C	2162 °C
Gold (Au)	1064 °C	2856 °C
Copper (Cu)	1085 °C	2927 °C
Iron (Fe)	1538 °C	2861 °C

Degradation phenomena

Corrosion of metals is mainly due to electrochemical reactions. Most metals corrode because they react with oxygen in the atmosphere, particularly under moist conditions, phenomenon named oxidation. Ferrous metals are particularly predisposed to oxidation, demanding constant care.

Conversely, some non-ferrous metals are resistant to corrosion thanks to the formation of strong oxides coating on their surfaces (e. g. lead) able to protect metal from additional oxidation.

Bronze disease is the corrosion process going on when chlorides come into contact with bronze or copper alloys. Usually, treatment consists of chlorides removal and a successive item isolation.

Patina is a thin coating that forms on the surface of metals like copper, bronze and similar. Usually, it provides a protective layer to materials preventing corrosion or weathering events and it is also aesthetically appealing and desirable. Time, weathering, accumulation of particulate matter, corrosion and use are all possible causes of patina development and the chemical compounds thus produced on the surface are oxides, carbonates, sulfides, sulfates. The appearance of patina can be very different, some of the most characteristic ones are: the almost powder green verdigris of copper, bronze and brass and unlike bronze disease; verdigris serves to protect metal.

Coatings have been used since ancient times. They can be functional or decorative. Gold leaf was one of the earliest techniques used by Egyptians. Pliny describes the use of white of eggs as adhesive for gold foil. Copper and bronze were treated with mercury to form surface amalgams. Tinning was used to protect iron and steel, but also non-metallic coatings were applied. The preservatives included liquid tar, bitumen, lead and gypsum.

Analyses

- Radiography
- Metallography
- SEM-EDS
- Mossbauer Spectrometry
- Isotopic Analysis
- XRF, XRD
- Microbeam techniques (EPMA, PIXE, SIMS)
- ICP-MS, ICP-OES, etc.

References

Craddock P. Early metal mining and production. Edinburgh: Edinburgh University Press, 1995.

Dillmann P, Beranger G, Piccardo P, Matthiessen H. Corrosion of metallic heritage artefacts. Investigation, conservation and prediction of long term behaviour. Woodhead Publishing, 2007.

Forbes RJ. Metallurgy in antiquity: a notebook for archaeologists and technologists. Leiden, The Netherlands: E. J. Brill, 1950.

Mainz VV. The metals of antiquity and their alloys in chemical technology in antiquity, Rasmussen, Seth C., ACS Symposium Series. Washington, DC: American Chemical Society, 2015.

Rehren T, Pernicka E. Coins, artefacts and isotopes-archaeometallurgy and archaeometry. Archaeometry. 2008;50:232–48.

Scott D. 1991. Metallography and microstructure of ancient and historic metals. The Getty Conservation Institute.

Bayley J, Crossley D, Ponting M. Metals and metalworking. A research framework for archaeometallurgy, Occasional Publication No 6, M. Ponting, J. Bayley Publisher: Historical Metallurgy Society, 2008.

Proteins

Occurrence

- animal glues (from mammal hide, cartilage and bones or parts of fish)
- egg (whole egg, egg yolk or egg white)
- casein and milk

Animal glues are colloidal dispersions of collagen – proteins showing a helix structure. They are obtained from collagen-based materials and have frequently been used as adhesives – to join wood panels or to fix canvas on wood – or as binder in ground layers (with gypsum, chalk or lead white) or in paint for certain pigments. In conservation animal glue is used for facings or *strappo*. The drawback of animal glue is its hygroscopic nature causing weakening of the ground layers or detachment of the paint.

Egg has been widely used as a paint binder, mixed with water (*tempera*), oil (*tempera grassa*) or natural resins. Whole egg and egg yolk contain both lipids and proteins and upon drying transform into an irreversible gel and film with excellent cohesive and adhesive properties. Egg white is essentially composed of an aqueous solution of proteins and salts and the substantial lack of lipids causes egg white films to be fragile and water sensitive. It has mainly been used as binder in illuminated manuscripts and inks, as well as in preparation layers for gilding.

Casein is the main phosphoprotein in mammal milk. Ammonium caseinate has been used as paint binder in mural and panel paintings, whereas calcium caseinate, obtained from casein and slaked lime, has been mainly employed as adhesive for wood panels and objects and sometimes as consolidant for plaster, or lining of canvas paintings. Casein tempera is characterized by rapid drying and formation of opaque films, but has scarcely been used due its fragile and cracked films. Milk, being composed of casein in emulsion with lipids, was more frequently used as paint binder, creating stable films, appearing similar to egg tempera.

Degradation phenomena

Swelling and contraction due to variation of temperature and humidity may cause loss of adhesion and cohesion of the paint film or adhesive. Biological processes such as the growth of fungi and other microorganisms are favored by humidity, whereas chemical degradation is mostly induced by light, humidity and interaction with other chemical species. These processes may alter the cohesive and optical properties of the binder. Proteins are relatively stable to oxidation, but are sensitive to water or aggressive solvents.

Chemistry

Proteins are macromolecules (≥5000 Da) composed of aminoacids (21 naturally occurring) linked by peptide bonds. Aminoacids are bifunctional organic compounds containing a carboxylic acid group (–COOH) and a basic amine group ($-NH_2$) on the same carbon atom and with a lateral chain (–R) which differentiates the various aminoacids. The number, type and sequence of the aminoacids are characteristic of each protein and determine structure, chemical and physical properties.

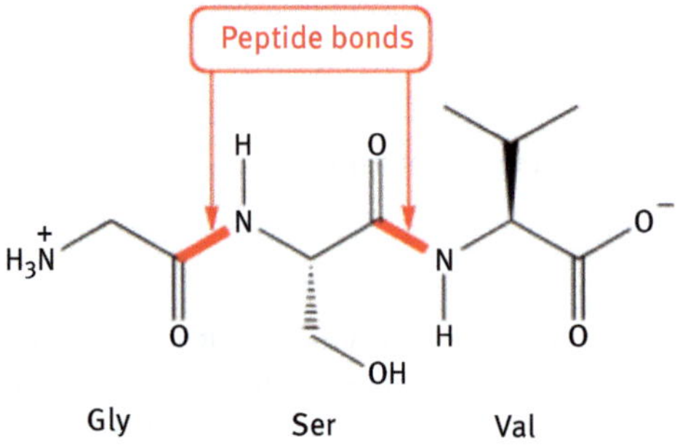

Figure 1.3: Peptide bonds.

The tridimensional structure of a protein is articulated on four levels: primary structure (specific sequence of aminoacids); secondary structure (conformation of the protein in certain regions: helix, globular, etc.); tertiary structure (total configuration: globular, fibrous, etc.); quaternary (association of two or more polypeptides linked by weak bonds). Hydrogen bonds, disulfur bridges, ionic bonds and hydrophobic interactions may cause folding of the proteins. Alteration of the conformation may lead to denaturation and loss of function. The denaturated protein will maintain its primary structure and will lose its solubility. This is positive when the protein is used as a binder, but presents a real drawback when used as an adhesive or consolidant (loss of reversibility).

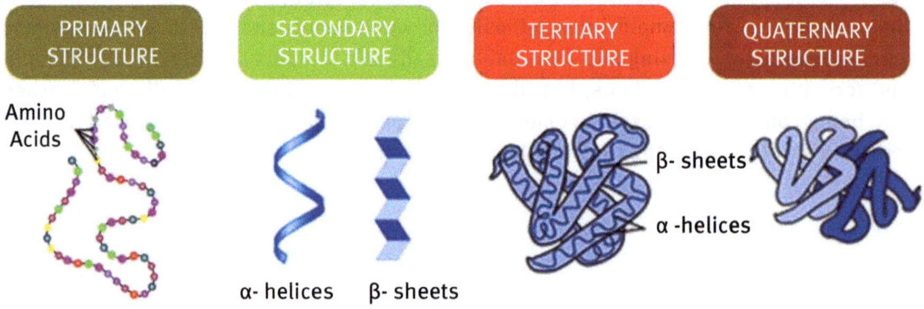

Figure 1.4: Protein structures.

Analyses

- Proteomics: MALDI-MS, LC-MS(MS)
- GC-MS
- Immunological methods
- FTIR spectroscopy

References

Dallongeville S, Garnier N, Rolando C, Tokarski C. Proteins in art, archaeology, and paleontology: from detection to identification. Chem Rev. 2016;116:2–79.

Mills JS, White R. Organic chemistry of museum objects, 2nd ed. 1994:84–94.

Carbohydrates

Occurrence

- vegetal gums
- honey
- starches and dextrin

Vegetal gums are obtained from the solidified exudates of certain plants or seeds. The gels that are formed by the gums in contact with water are apt to be used as adhesives and paint binders. They have been largely employed in watercolors, illuminated manuscripts, gilding mordants and inks. Vegetal gums are used alone or in emulsion with proteinaceous and lipid binders. Starch, honey or molasses (syrup) (plastifiers) and antifermentatives are often added. Historically, the most frequently used gums are gum Arabic, gum tragacanth and gums from different fruit trees. Gum Arabic is obtained from species of *Acacia*. The first examples of the use of gum Arabic date back to 5000 years ago in the Egyptian tombs. Gum tragacanth is an exudate of the *Astragalus gummifer*, which can be found in Greece, Iran and Minor Asia. It is in use since about 2000 years and represents an excellent colloid able to stabilize emulsions and dispersions. The gums from fruit trees include the exudates of different *Prunus* species (plum, cherry, peach, apricot). Their use is more limited and their films are very fragile.

Honey is produced by bees and for centuries it has been used as plasticizer for aqueous binders such as vegetal gums and albumin. It has the property of retaining a certain amount of water reducing the fragility of the paint films during drying.

Starch can be found in all the seeds and bulbs of plants. The main four providers of starch are: mais, potatoes, rice and wheat. Also legumes like beans are rich in starch. It is used as adhesive, for lining or in some wall painting preparations, and as a binder and support for organic pigments in Japanese prints. Thanks to its adhesive properties, starch is very important in paper production.

Dextrins are obtained from the acid treatment of starch and are rather soluble in water. They are sometimes used for watercolors in addition to vegetal gums, but their main use concerns the paper industry for reinforcement or coating.

Degradation phenomena

Due to the high amount of hydroxyl groups, carbohydrates are hydrophyllic and therefore very sensitive to variations of environmental humidity. Polysaccharides are subject to dehydration, reticulation, oxidation and hydrolysis leading to depolymerization. The solubility in water favors leaching and therefore it might be difficult to detect vegetal gums in ancient wall paintings, especially outdoors. Next to water, microorganisms represent a major cause of degradation.

Chemistry

Carbohydrates may be classified into:

- *Monosaccharides and uronic acids.* General formula: $C_nH_{2n}O_n$. Classified as aldoses if containing an aldehyde group, as ketoses if bearing a ketone group. They can be subdivided into trioses (glyceraldehyde and dihydroxyacetone), pentoses (arabinose, xylose, ribose, etc.), hexoses (glucose, fructose, mannose, galactose, rhamnose, fucose, etc.), glucuronic acid and galacturonic acid. The most abundant monosaccharide in nature is glucose, which exists as α-glucose and β-glucose.

- *Oligosaccharides*. Derive from the union of some disaccharide molecules, such as lactose, saccharose and maltose.
- *Polysaccharides*. Polymers of monosaccharides linked by glucoside bonds. They include cellulose, starches, mucilages and vegetal gums. Cellulose is formed from polycondensation of α-glucose, whereas by linking β-glucose molecules amylose and amylopectin (both contained in starches) are formed. These are thus homopolysaccharides (one type of monomer) with a linear (cellulose) or branched (starches) structure. In vegetal gums and mucilages the molecules are more complex with major molecular weight.

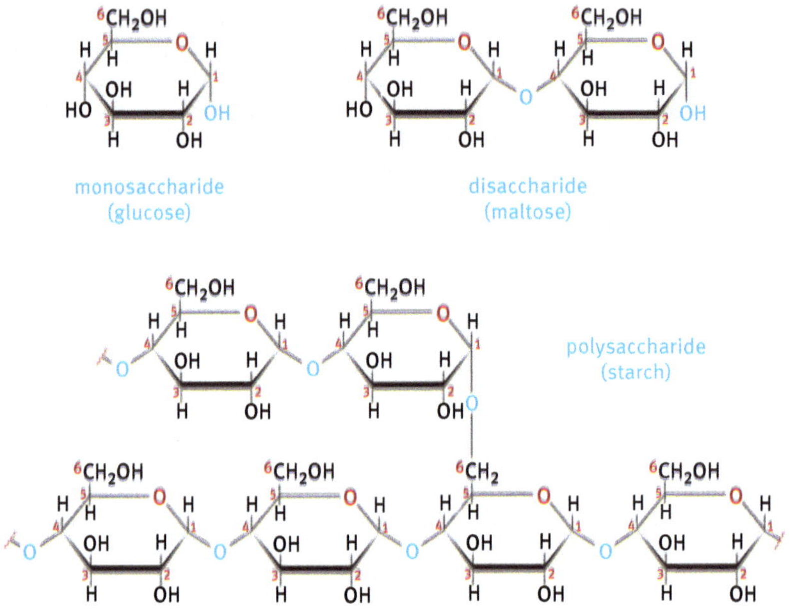

Figure 1.5: Carbohydrates.

Analyses

- GC-MS
- MALDI-MS
- FTIR spectroscopy

References

Mills JS, White R. Organic chemistry of museum objects, 2nd ed. 1994:69–83.

Granzotto C, Arslanoglu J, Rolando C, Tokarski C. Plant gum identification in historic artworks. Sci Rep. 2017;7:44538.

Lluveras-Tenorio A, Mazurek J, Restivo A, Colombini MP, Bonaduce I. The development of a new analytical model for the identification of saccharide binders in paint samples. PlosOne. 2012;7:e49383.

Lipids

Occurrence

- oils
- egg yolk
- fats

Siccative and semi-siccative oils are used as paint binders. They contain triglycerides and small amounts of sterols. The percentages of the various fatty acids – with double bonds differing in position and number – determine the properties of the oils. Linseed oil is the most widely used paint binder and is produced from *Linus usitatissimum*. Stand oil is made at temperatures above 250 °C in absence of air. This oil is drying more slowly, but forms a more stable and less yellowing film. Walnut oil is extracted from *Junglas regia* and is more siccative than poppyseed oil, due to its higher content of linolenic acid. Poppyseed oil is obtained from *Papaverum somniferum* and is less siccative but yellows less. Other oils used in paints are safflower, sunflower, rapeseed and castor oil.

Egg yolk is an emulsion of a colloidal aqueous solution of proteins and lipids, stabilized by emulsifiers of whom the most abundant is lecitin. The lipids are composed of triglycerides (65 %), phospholipids (30 %) and cholesterol (5 %). These lipids are not siccative, but act as plasticizers and confer hydrophobicity and irreversibility to the paint film.

Fats are composed of triglycerides with low amounts of polyunsaturated fatty acids.

Table 1.3: Fatty acid percentage composition of fresh vegetable oils and of animal lipids.

Oil/lipids	FA-C16	FA-C18	FA-C18:1	FA-C18:2	FA-C18:3	Other fatty acids
Linseed	6–8	3–6	14–24	14–19	48–60	
Walnut	3–7	0.5–3	9–30	57–76	2–16	
Poppyseed	8–12	2–3	12–17	55–65	3–8	
Sunflower	5–6	4–6	17–51	38–74		
Castor	1–2	1–2	3–6	4–7		83–89[1]
Rapeseed	2–6	1–3	20–30	17–22	6–10	13–16[2]; 20–40[3]
Hen's egg	25–27	9–12	38–44	13–15	0–1	
Lard	20–27	13–19	37–45	7–10	0–1	

FA-C16 Palmitic acid (hexadecanoic acid); FA-C18 Stearic acid (octadecanoic acid); FA-C18:1 Oleic acid (9-octadecenoic acid); FA-C18:2 Linoleic acid (9,12-octadecadienoic acid); FA-C18:3 Linolenic acid (9,12,15-octadecatrienoic acid); [1]Ricinoleic acid (12-hydroxy- (Z)9-octadecenoic acid); [2]Gondoic acid (11-eicosenoic acid); [3]Erucic acid (13-docosenoic acid).

Degradation phenomena

The drying of a (semi-)siccative oil film consists of a polymerization process continuing during years after application. Hardening and increase of refractive index lead to cracking and higher transparency. Yellowing is influenced by degree of purity, treatments, pigments, siccatives, humidity and light. Oil paint films become more sensitive to humidity with ageing due to oxidation and hydrolysis into mono- and diglycerides, and free fatty acids.

For egg yolk crosslinking phenomena are less important, but some auto-oxidation of unsaturated lipids may occur, leading to the loss of free triglycerides and phospholipids. Instead, degradation products are found.

Chemistry

Glycerolipids are composed of esters of glycerol and saturated and unsaturated fatty acids (triglycerides or triacylglycerols, TAGs).

Phospholipids contain two hydrophobic fatty acid "tails" and a hydrophilic "head" consisting of a phosphate group. The two components are joined together by a glycerol molecule. The phosphate groups can be modified with simple organic molecules such as choline, ethanolamine or serine.

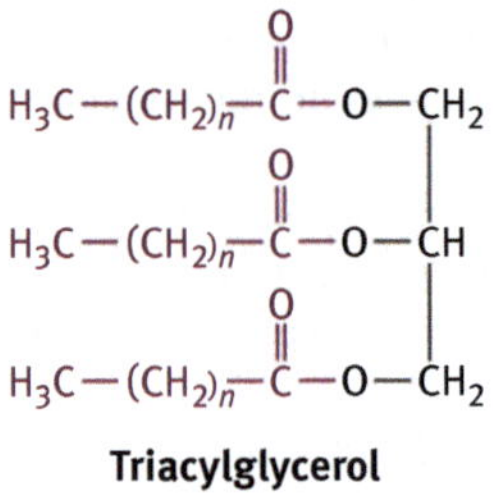

Figure 1.6: Triacylglycerol.

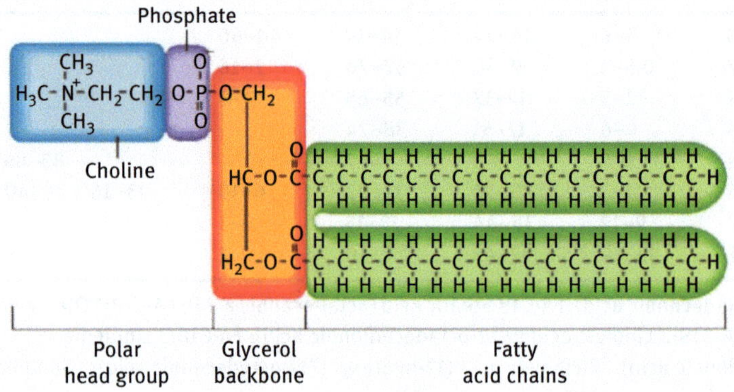

Figure 1.7: Phospholipids.

Unsaturated fatty acids undergo oxidation and cross-linking reactions when exposed to light and oxygen, leading to the formation of an oil network. These processes – polymerization (crosslinking reactions) and oxidation – are in competition. Oxidation results in the oxidative cleavage of the fatty acids hydrocarbon chains and the formation of hydroxyacids, low molecular weight aldehydes and ketons, and α,w-dicarboxylic acids with nonanedioic (azelaic) acid as the most abundant. Certain metals (pigments) play a primary role in catalyzing these reactions. During ageing hydrolysis of the ester bonds of the TAGs also occurs. Cations in zinc, copper and lead containing pigments may react with free fatty acids and other carboxylic acid groups to form metal soaps agglomerates or ionomeric structures.

Analyses

- (Py-)GC-MS
- LC-MS(MS)
- MALDI-MS
- FTIR spectroscopy

References

Mills JS, White R. Organic chemistry of museum objects, 2nd ed. 1994:31–48.
Colombini MP, Modugno F, editors. Organic mass spectrometry in art and archaeology. Wiley, 2009:6–9.

Waxes

Occurrence

- Animal waxes
- Vegetal waxes
- Mineral and artificial waxes

The main **animal wax** is beeswax. It is produced by the European *Apis mellifica* L. or asiatic species and has been used for writing (waxed wooden tablets), painting (*encausto*), fabrication of metal objects (*cera persa* technique), protective and sealing coatings.

Chinese wax is produced by the *Coccus ceriferus* insect, whereas spermaceti and lanolin are obtained from whales and sheep skin, respectively.

Carnauba wax is a **vegetal wax** and is gathered from a palm (*Copernica cerifera*) growing in Brazil. It is currently used in restoration treatments also mixed with beeswax or paraffin. Candelilla wax is from two *Euphorbia* species (Mexico) and is characterized by a high melting point.

Table 1.4: Melting points of various waxes.

Wax	Melting point
Animal	
Beeswax	66–71 °C
Chinese wax	80–83 °C
Spermaceti	42–50 °C
Lanolin	35–42 °C
Ambergris	60–80 °C
Vegetable	
Carnauba	82–86 °C
Candelilla	67–79 °C
Japan wax	50–60 °C
Mineral	
Ceresin	54–77 °C
Montan	76–92 °C
Paraffin	46–68 °C

Mineral and artificial waxes are montana wax, which is extracted from lignites, paraffin, from scists, lignites and petroleum, ceresine and microcristalline waxes. The latter is used for wax polishing and finishing coatings of various materials.

Degradation phenomena

One of the main characteristics of beeswax is its high stability with time. This is due to the fact that the main bonds are covalent and saturated. Prolonged heating might induce a decrease of acidity leading to conspicuous losses of hydrocarbons.

Chemistry

Natural waxes are composed of a mixture of esters from fatty acids and long-chain alcohols, fatty acids, free alcohols and long chain hydrocarbons in various proportions depending on the type of waxy materials.

Artificial waxes, also referred to as microcrystalline waxes, contain hydrocarbons deriving from refined petroleum.

Table 1.5: Composition of various waxes.

Components in wt%	Beeswax	Chinese wax	Spermaceti	Lanolin	Carnauba wax	Candelilla	Ceresine	Montan	Paraffin
Hydrocarbons	14[a]	x		x	1	50[g]	100	3	100[i]
Esters	67[b]	83[c]	92–99[d]	14–24	84[f]	28–29		53[h]	
Free acids and Alcohols	13	x	1–8	x	12	7–9		18–19	
Sterols				45–65[e]					
Terpenes				4–5		12–14		20–23	
Ketones								3–6	
Non-identified	6								

[a]C_{25}-C_{35} hydrocarbons (mainly uneven); [b]mainly esters of palmitic acid with C_{24}-C_{32} alcohols; [c]mainly esters of hexacosanoic acid with hexacosanol; [d]mainly esters of palmitic acid and myristic acid with hexacosanol; [e]lanosterol and cholesterol; [f]including esters of p-hydroxycinnamic acid (23 %) and of p-methoxycinnamic acid (7 %); [g]C_{29}-C_{33} hydrocarbons, mainly C_{31}; [h]esters of C_{22}-C_{32} acids; [i]linear C_{20}-C_{36}, isoalkanes and cyclic alkanes C_{18}-C_{36}.

Analyses

- FTIR spectroscopy
- (Py-)GC-MS
- DTMS

References

Mills JS, White R. Organic chemistry of museum objects, 2nd ed. 1994:49–55.

Colombini MP, Modugno F, editors. Organic mass spectrometry in art and archaeology. Wiley, 2009:10–12.

Regert M, Langlois J, Colinart S, Characterisation of wax works of art by chromatographic procedures. J Chrom A. 2005;1091:124–36.

Natural resins

Occurrence

Natural resins are polymeric materials produced by trees (vegetal resins) or insects (animal resins) used for their filmforming, adhesive and hydrophobic properties.

Vegetal resins are generally distinguished into diterpenoid resins and triterpenoid resins (see Chemistry).

The main **diterpenoid resins** are produced by the *Pinaceae* genera *Pinus*, *Larix*, *Picea* and *Abies*. *Pinus* resin (colophony) has largely been used in varnishes mixed with oil or other more precious resins. Venetian turpentine is obtained from *Larix decidua* and has been mentioned in ancient recipes for preparation of varnish and copper resinate pigment. Sandarac is from *Tetraclinis articulate,* a conifer growing in the south of Spain and North Africa, and has often been used in oil based varnishes.

The most commonly used **triterpenoid resins** are mastic and dammar resin. Mastic is obtained from *Pistacia lentiscus var. chia*, growing on the isle of Chios in Greece. However, the exudates of other species of *Pistacia* are also often defined as mastic resin. Since ancient times mastic has been used in mummification balsams, as adhesive and sealing agent. More recently, it has been employed as picture varnish in mixture with siccative oils. Dammar resins are exudates of *Dipterocarpaceae* trees (*Hopea* and *Shorea*) growing in South East Asia. The introduction of dammar in Europe dates back to the 1830s. It is used in solvent varnishes (frequently in oil of turpentine).

Amber (produced 45–10 million years ago) is a fossil resin only partially soluble in oil. Although mentioned in ancient recipes as a varnish ingredient, it is not apt for this use. *Copals* are classified as semifossil resins and are produced by *Araucariaceae*, *Caesalpinaceae* and tropical trees. The main types of copal are Congo, Zanzibar, Kauri, Sierra Leone and Manila copal. They have been occasionally used in picture varnishes.

Shellac is produced by an Indian scaled insect (*Laccifer lacca* Kerr, also known as Kerria lacca). It is commonly used as varnish for furniture, metal objects and scientific instruments. It can also be found as adhesive for ceramics.

Degradation phenomena

Natural resins are severely affected by ageing and their chemical composition is highly influenced by exposure to light. The degradation mechanisms are complex and mainly involve oxidation and polymerization reactions leading to the formation of oligomers and polymers and highly oxidized species. As a consequence varnishes often show yellowing (darkening), cracking and loss of solubility with an increase of acidity and polarity. Degradation of natural resin varnishes is one of the main restoration problems of paintings. Highly polar solvents are often required for removal.

If the pine resins are subject to thermal treatment pitches are formed typically containing aromatized molecules like retene.

Chemistry

Natural resins are composed of complex mixtures of terpenoids: mono-, sesqui-, di- and triterpenoids containing, respectively, 10, 15, 20 and 30 carbon atoms with hydroxyl, carboxyl and carbonyl groups. Mono- and sesquiterpenoids are contemporary present in the majority of resins, while di- and triterpenoids are never present together. The latter usually persist with time, although subject to oxidation and polymerization, and can be used for identification purposes.

Diterpenoid resins contain abietanes, pimaranes and labdanes, whereas triterpenoid resins are composed of dammaranes, ursanes and oleananes, next to a polymer fraction of polycadinene (dammar) or poly β-myrcene.

Shellac contains 70–80 % of resin, 4–8 % of colorant and 6–7 % of wax. The resin is a complex mixture of oligomers formed by esterification of polyhydroxycarboxylic acids.

Table 1.6: Botanical origin and chemical composition of terpenic resins.

Class	Family	Genus (type of resin)	Composition
Coniferales	*Pinaceae*	*Pinus* (pine resin, colophony)	Abietadienic acids, pimaradienic acids
		Abies (Strasbourg turpentine)	Abietadienic acids, pimaradienic acids, cis abienol
		Larix (Larch turpentine)	Abietadienic acids, pimaradienic acids, epimanool, larixol, larixyl acetate
	Cupressaceae	*Juniper, Cupressus, Tetraclinis articulata* (sandarac)	Pimaradienic acids (sandaracopimaric acid), communic acid, totarol
Guttiferales	*Dipterocarpaceae*	*Hopea* (dammar)	Dammaranes (hydroxydammarenone, dammaradienol), ursanes (ursonic acid, ursonaldehyde)
Terebinthales	*Anacardiaceae*	*Pistacia* (mastic)	Euphanes (masticadienonic acid, isomasticadienonic acid), oleananes (oleanonic acid, moronic acid), dammaranes
	Burseraceae	*Commiphora* (myrrh) *Boswellia* (olibanum or frankincense) *Canarium* (elemi)	α and β amyrin, euphanes, oleananes

Analyses

- (Py-)GC-MS
- FTIR spectroscopy

References

Mills JS, White R. Organic chemistry of museum objects, 2nd ed. 1994:95–128.

Colombini MP, Modugno F, editors. Organic mass spectrometry in art and archaeology. Wiley, 2009:215–35.

Synthetic polymers

Occurrence

From the beginning of the twentieth century synthetic polymers have been developed. These materials can be found in paint binders, coatings, adhesives, consolidants, and in modern art objects.

Synthetic paint binding media were first used for household paint (alkyd resins), but after the 1930s artists started employing these new materials. Acrylic paint came also into use with the advantage of fast drying and the possibility of using water emulsions avoiding the use of toxic solvents.

Acrylic resins, such as Paraloid B-72, are also frequently used as coatings, varnishes and consolidants. Polyvinylacetates are generally employed in emulsion as adhesives in restoration treatments. Epoxy resins are mainly used as adhesives in restoration of glass and stone objects. Ketone resins (Laropal K80, MS2A) represent an important class of synthetic resins used as varnishes.

In modern art objects a wealth of synthetic materials can be found ranging from polycarbonate, polypropylene to polystyrene.

Degradation phenomena

Synthetic resins generally show a higher resistance to oxidative change than natural resins. Some resins, however, may degrade more easily, like cellulose nitrate and cellulose acetate. Acrylic resins are sensitive to sunlight. Bond cleavage (methacrylates) and cross-linking (acrylates) are observed. Polyurethanes show yellowing and depolymerization. Epoxy resins are resistant to acids, alkali and solvents, but tend to yellow with ageing.

Chemistry

Synthetic resins can be classified in two main categories based on the polymerization mechanism: addition polymers and condensation polymers.

Addition polymers are based on ethylene (two carbon atoms with a double bond) monomers with different lateral groups and are also defined as vinyl polymers. Main vinyl polymers are polyethylene, polypropylene, polystyrene, polyvinyl acetate, polyvinyl alcohol, polyacrylates, polyvinyl chloride and polyethylene oxide.

Figure 1.8: Addition polymers.

Condensation polymers are formed by monomers with two functional groups. Linear polymers are generally obtained, but branched polymers can also be formed if more than two functional groups are present, or by cross-linking of linear polymers upon degradation. The main condensation polymers are polyesters (alkyd resins, polycarbonates), polyamides, epoxy resins, phenol-formaldehyde resins and cyclohexanone resins.

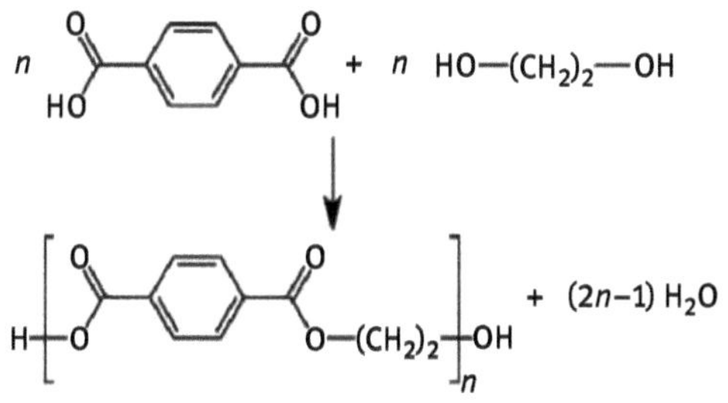

Figure 1.9: Condensation polymer.

Analyses

- FTIR spectroscopy
- Py-GC/MS

References

Mills JS, White R. Organic chemistry of museum objects, 2nd ed. 1994:129–40.

Colombini MP, Modugno F, editors. Organic mass spectrometry in art and archaeology. Wiley, 2009:26–8.

Learner T. Analysis of modern paints. Los Angeles: Getty Publications.

Dyestuffs

Occurrence

From ancient times animal and vegetal colored organic compounds are used for the dying of textiles and as pigments, as such or on an inorganic substrate as lakes. From the second half of the nineteenth century onwards synthetic dyes have gradually substituted the natural dyes.

Colorants can be classified according to the coloring mechanism:

- Direct dyes: taken up directly by cellulosic fibers (saffron, curcuma).
- Vat dyes: applied on wool or cotton as a solution of the reduced colorless (leuco) form. Oxidation by atmospheric oxygen or oxidizing agents yield the colored dyestuff (indigo).
- Acid dyes: applied on animal fibers in an acid dye bath (modern synthetic dyes for wool).
- Mordant dyes: dyes linked to the fiber via an organic compound (tannins) or dyes requiring a metallic hydroxide (alum) precipitated onto the fiber. The color depends on the metal ion (natural red and yellow dyes).

Natural dyestuffs are produced by plants and insects and can be red, yellow, blue and brown.

The chromophores exhibit flavonoid, anthraquinoid, indigoid, gallotannin, carotenoid, benzoquinone and antocyanine structures.

The synthetic organic dyestuffs have been produced since the 1850s with Perkin having patented the famous mauveine and the synthesis of alizarin in 1868. Synthetic colorants are classified according to their chemistry. The most important are triarylmethane dyes, diazo dyestuffs, anthraquinone vat dyes and the very stable phthalocyanines.

Degradation phenomena

Dyestuffs are generally extremely sensitive to degradation. Fading – caused by chemical oxidation – is related to the amount of light received, wavelength, level of humidity, type of substrate and the mordant. Ozone and nitrogen oxide have significant effects on various dyestuffs.

In general natural dyes are less stable than modern synthetic dyes.

Chemistry

Table 1.7: Origin and chemical composition of various natural dyestuffs.

Common name	**Botanic name**	**Main chromophores**					
Anthraquinoid dyes							
		Carminic acid	Kermesic acid	Flavokerm-esic acid	Dc II	Laccaic acid A	Laccaic acid B
Cochineal	*Kermes vermilio*		x	x			
	Porphyrophora polonica	x	x	x	x		
	Porphyrophora hameli	x	x	x	x		
	Dactylopius coccus Costa	x	x	x			
	Kerria lacca Kerr		x	x		x	x

(continued)

Table 1.7 (continued)

Common name	Botanic name	Main chromophores					
		Alizarin	Purpurin	Xanthopurpurin	Munjistin	Pseudopurpurin	Other compounds
Madder	*Rubia tinctorum*	x	x	x	x	x	
	Rubia peregrina		x			x	
	Rubia cordifolia	x	x	x	x	x	Morin
	Rubia akane Nakai		x			x	
Lady's bedstraw	*Galium verum* L.	x	x	x		x	
	Galium mollugo L.	x	x	x	x	x	Rubiadin
Relbunium	*Relbunium hypocarpium* L.		x	x	x	x	
Morinda	*Morinda citrifolia* L.	/x					morin
Indigoid dyes							
		Indigotin	indirubin	monobromoindigotin	dibromoindigotin	dibromoindirubin	Other compounds
woad	*Isatis tinctoria*	x					Quercetin, kaempferol
indigo	Various *Indigofera* species	x	x				Kaempferol
Indigo carmine	*Indigofera tinctoria*						indigocarmine
purple	*Muricidae and Rapaninae*	x		x	x	x	
orchil	*Rocella and Ochrolechia*						orcein
Flavonoid dyes							
		fisetin	morin	maclurin	kaempferol	quercetin	Other compounds
Young fustic	*Cotinus coggyria*	x					Fustin, sulfuretin
Old fustic	*Clorophora tinctoria* L.		x	x	x		
Quercitron bark	*Quercus velutina* L.					x	Quercitrin
Persian berries	*Rhamnus family*				x	x	Rhamnetin
		luteolin	apigenin	genistein	3-methyl-quercetin		
weld	*Reseda luteola* L.	x	x				
Dyer's broom	*Rhamnus frangula* (cortex)						Glucofrangulin A and B, franguline, emodin
Rhamnus bark	*Serratula tinctoria* L.	x			x		
		Barzilin	Brazilein	Hematoxylin	Hematein		

(continued)

Table 1.7 (continued)

Common name	**Botanic name**	**Main chromophores**					
Brazilwood	*Caesalpina* species	x	x	x	x		
Logwood	*Hematoxylum campechianum*			x	x		
Tannin-based dyes							
		Gallic acid	quercetin	Emodin	Quercitrin	tannins	Other compounds
Black alder	Alnus glutinosa	x	x	x	x		
Sumac	*Rhus genus*	x					Ellagic acid, kaempferol
Black walnut	*Juglans species*						juglon
Walnut galls	Galls from *Cynips* species, *Quercus infectoria* Oliv.					x	
Silver birch	*Betula* species					x	
Other dyes							
Henna	*Lawsonia inermis*	Lawson					
Sandalwood	*Pterocarpus santalinus*	Santalin A and B					
Safflower	*Carthamus tinctorius* L.	Carthamina, Safflower yellow A and B					
Annatto	*Bixa Orellana* L.	Bixin					
Barberry	*Berberis vulgaris* L.	Berberin					
Saffron	*Crocus sativus* L.	Crocin, crocetin					
Turmeric	*Curcuma longa*	Curcumin, demethoxycurcumin, bisdemethoxycurcumin					

Analyses

- HPLC-DAD-(MS)
- Raman spectroscopy
- Py-GC/MS

References

Mills JS, White R. Organic chemistry of museum objects, 2nd ed. 1994:141–59.

Colombini MP, Modugno F, editors. Organic mass spectrometry in art and archaeology. Wiley, 2009:22–5.

Pasquale Acquafredda

2 XRF technique

Abstract: The techniques of chemical investigation by X-ray fluorescence (XRF) are widespread since the 50s of the last century. Depending on the accuracy of the desired data and on the artifact characteristics, they can be used as partially destructive or as absolutely non-destructive and non-invasive techniques. The archeomaterials that can be analyzed are the most disparate: minerals, rocks, metals, building materials, pigments, and so on; practically almost everything that is solid, liquid or gelatinous can be analyzed by XRF. The theoretical physical principles and the main components of X-ray spectrometers, in energy dispersion (ED) and wavelength dispersion (WD), are described, also comparing the advantages and disadvantages of each analytical technique. In the last decades, the diffusion of the ED silicon drift detectors, together with the development of very accurate and high specialized software for quantitative analysis, has given a new impulse to the diffusion of the portable spectrometers offering new possibilities for *in situ* and very rapid archeomaterial characterizations. Case studies related to different artworks, like ceramics, necklaces, coins, obsidians and other lithic artifacts will also be presented: they show the important contribution that X-ray spectrometer technique gives to solve problems related to the characterization, restoration and to the source identification of the raw materials.

Keywords: ED-XRF, WD-XRF, elemental analysis, non-destructive analysis, archeomaterials, petroarchaeometry, Cultural Heritage

2.1 Introduction

X-ray fluorescence (XRF) is a spectrochemical technique which allows the determination of the elemental composition as well as the quantitative analysis of a wide range of inorganic materials. Results can be expressed as atom percent or weight percent. Often, by custom, especially in the analysis of minerals and rocks, the data are expressed as oxides, stoichiometrically binding the oxygen to the dosed cation. XRF spectrometric analysis can be carried out in wavelength dispersion (WD) or in energy dispersion (ED). In the case of WD analyses (WD-XRF), the technique is normally considered destructive (but rare exceptions), because the sample normally has to be ground and subsequently treated with organic binders or with other components in order to obtain a powdered pellet to submit to analysis. The indisputable advantage remains, compared to the cases of the analyses that require the solubilization of the sample, that the prepared specimen, can be reanalyzed, either

This article has previously been published in the journal *Physical Sciences Reviews*. Please cite as: Acquafredda, P. XRF technique *Physical Sciences Reviews* [Online] **2019**, 4. DOI: 10.1515/psr-2018-0171.

https://doi.org/10.1515/9783110457537-002

with another XRF spectrometer or with other analytical instrumentation, even after many years, provided that it has been well preserved. The amount of sample required for chemical analysis by WD-XRF varies, depending on the kind (qualitative/quantitative) of the searched information and depending on the method and the instrumentation used, from a few tens of milligrams (about 40 mg [1]) up to about 12 g. In some cases, as in provenance determination of obsidians studies, it is possible to use the WD-XRF method also in an absolutely non-destructive way [1].

In the case of ED analyses (ED-XRF), the specimen is normally analyzed as it is and therefore the technique is considered non-destructive. As a matter of fact, the ED counters can provide quantitative or semi-quantitative results even on samples with irregular surface because they suffer much less than the WD techniques from the sample geometry. A wide variety of materials from rocks, to minerals, industrial products, building materials, metals, pigments, etc., can be analyzed by XRF; practically almost everything that is a solid substance and in many cases also liquid or gelatinous substances.

Sampling and sample preparation will depend on the used analytical method non-invasive analysis (no sampling) is also possible by XRF and this opens up a very wide range of applications, especially in the field of Cultural Heritage, where the characterization of precious objects cannot justify their partial destruction (sampling) which would greatly reduce their value.

Moreover, in recent years, the development of portable XRF spectrometers [2–10], is more and more often allowing to analyze objects in their natural sites (restoration sites, museums, archeological excavations) as well as artifacts which cannot be moved for legal reasons or difficult to transport due to their dimensions. It is worthy to note that the analyses carried out with portable equipments should be validated on suitably chosen objects, comparing the analytical results with those obtained through instruments of qualified (certified) laboratories [11, 12].

2.2 Physical principles and theory

The XRF spectrochemical analysis technique uses, in the most common cases, a sample irradiation process by primary X-rays in order to subsequently analyze the secondary X-ray spectrum emitted by the sample itself.

X-rays are electromagnetic radiation with wavelength (λ) between 0.01 and 10 nm. Their energy (E) can be expressed in electron volts (eV) and is directly proportional to the frequency of vibration and inversely proportional to the wavelength, except for some constants:

$$E_{eV} = h \cdot c/\lambda \cdot e$$

Where h is the Plank constant, c and λ are respectively the speed of light in the vacuum and the frequency and wavelength of the X-rays, and e is the charge of the electron.

2.2.1 Generation of X-rays

In XRF spectrometers, X-rays are generated in a tube, under high vacuum conditions, called X-ray tube.

The operation of an X-ray tube can be schematized as follows:

1. A tungsten filament (Figure 2.1) is heated by the passage of direct current and therefore produces around it a dense cloud of electrons due to thermionic effect.
2. Part of these electrons are accelerated along a focusing path by means of a high potential difference (high voltage of the X-ray tube, expressed in kilovolt: kV) established between the filament and a metal target operating as an anode. The part of electrons not involved in the acceleration and focusing process is captured by a trap placed around the filament itself.
3. The electrons hit the anode and their kinetic energy is transformed into thermal energy, which is dissipated by a cooling system (generally water circulation), and partly in X radiations that escape from the tube through a window constituted from a thin sheet of material as much as possible transparent to X-rays. The thickness of the window must also take into account the heating effect of the material subjected to intense interaction with X-rays and of the backscattered electrons, which in large quantities are generated on the anode. The window can be positioned at the end of the tube (end-window X-ray tube) or sideways (side-window X-ray tube).

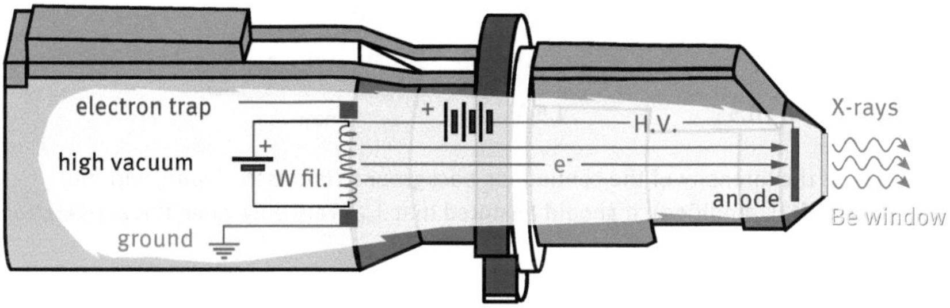

Figure 2.1: Schematic representation of the main parts constituting an end-window X-ray tube. H.V. = high voltage; W fil. = tungsten filament.

2.2.2 X-ray spectrum

The X-ray spectrum of an X-ray tube (primary source) or the one returned by an irradiated sample (secondary fluorescence X-rays) consists of a continuous radiation (*bremsstrahlung*) and characteristic radiations (Rayleigh lines).

The emission of a continuous radiation, or braking radiation, is a consequence of the deceleration of the incident electrons with those belonging to the atoms of which the X-ray tube anode is constituted.

The distribution of the intensity of the continuous radiation is a function of the energy possessed by the incident electrons and by the way this energy is transferred, for example, through a single interaction or through several successive interactions. In particular, if all the kinetic energy possessed by an electron is transferred in a single interaction, the corresponding emitted radiation has, within the continuous background, the maximum value of energy and the minimum value of wavelength. For increasing wavelength values, the intensity of the continuous background rapidly increases to a maximum value and subsequently decreases, with a hyperbolic trend (Figure 2.2).

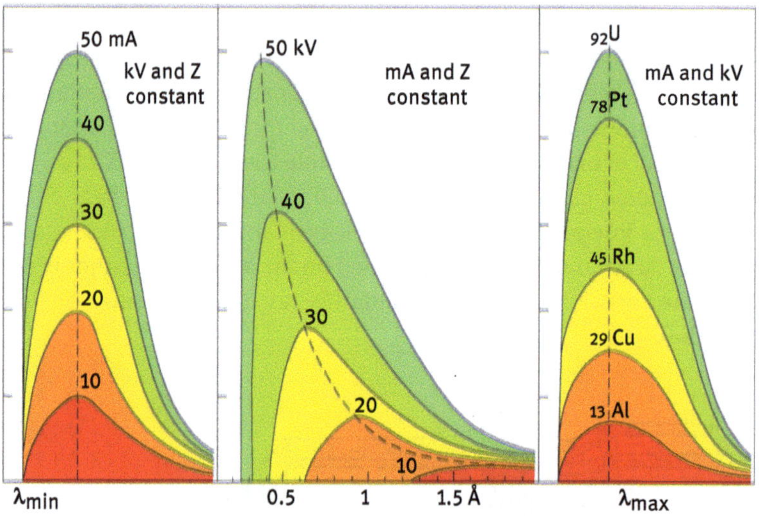

Figure 2.2: Variation of the intensity of the continuous background (*bremsstrahlung*) with the variations of V, A and Z ([13], modified); it should be noted that λ_{min} vary only when the acceleration voltage of the electrons varies.

The continuous radiation emitted by an X-ray tube contributes to the excitation of X fluorescence in the irradiated atoms.

The distribution of the intensity of the continuous background, in relation to the different wavelengths, depends on the current I of the filament (expressed in milliamps), the applied voltage V (expressed in kilovolt) and the atomic number Z of the element constituting the anode (Figure 2.2).

The characteristic radiations are generated as a consequence of the energy rearrangement process of the atom, when its excitation has led to the expulsion of one or more electrons from its internal shells. The wavelength of the characteristic radiations is independent of the cause that generated the hole in the electronic orbits, while it is strictly dependent: (1) on the type of electronic transition that leads to the energetic resettlement of the atom; (2) on the atom of the chemical element subjected to excitation.

The energy of the different radiations emitted is equal to the energy difference of the atom before and after the electronic transition. For a given chemical element, with reference to the lines of the K series, the energy difference for a re-arrangement from M to K is greater than the energy difference for a re-arrangement from L to K; consequently the energy of the radiation Kβ is greater than the energy of the radiation Kα, i. e. the wavelength of the radiation Kβ is lower than that of the radiation Kα.

In general terms, the wavelength of the radiation of a given chemical element progressively increases from $K\beta_1$ to $K\alpha_1$ to $L\gamma_1$ to $L\beta_1$ to $L\alpha_1$ to $M\gamma_1$ to $M\beta_1$ to $M\alpha_1$, and so on (Figure 2.3).

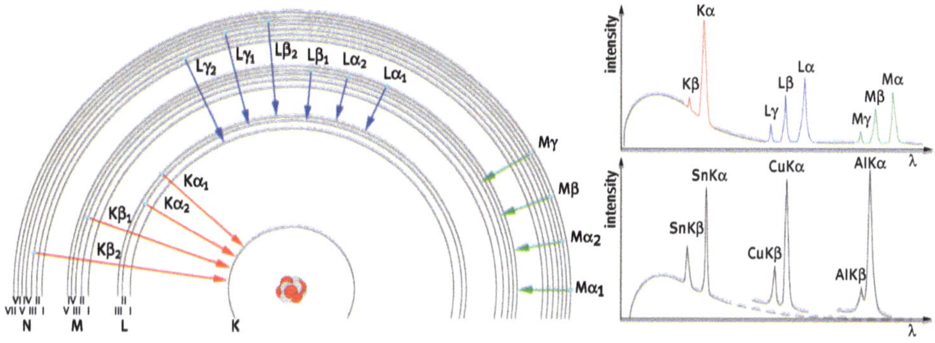

Figure 2.3: In the left part of the figure, simplified energy levels (orbitals or shells) of an atom between which the electrons move generating the characteristic X-ray lines. In the upper right part of the figure a schematic representation of the continuous background and the main characteristic radiations (lines) of a heavy element in function of λ. In the lower right part of the figure a schematic example of Kα/Kβ ratios variation for elements with different atomic number.

Concerning the distribution of the intensities of the different characteristic lines, it should be noted that it depends on the probability with which each electronic transition can occur and, therefore, the emission of the corresponding line. In particular, within the spectral K range, the transition from M to K is much less probable than the transition from L to K and hence the Kβ line is less intense than the Kα line. The $I_{K\alpha}/I_{K\beta}$ ratio increases as the atomic number decreases; approximately its value is 3 for Sn, 5 for Cu and 25 for Al (Figure 2.3): it follows that a conspicuous alteration of these intensity ratios is generally indicative of interferences between spectral lines of different chemical elements.

In quantitative XRF analytic procedures, the intensity of the characteristic lines emitted by the different chemical components is usually measured. The intensity of these characteristic lines depends, in addition to the reasons already described, and on the possible concentration of the chemical components, on the excitation efficiency of the incident radiation.

In particular, the characteristic lines of a given spectral series, for a given chemical element, are excited more effectively the higher is the intensity possessed by the incident primary radiation for wavelengths close to those re-emitted by the excited element, better if close to its absorption edges (Figure 2.4).

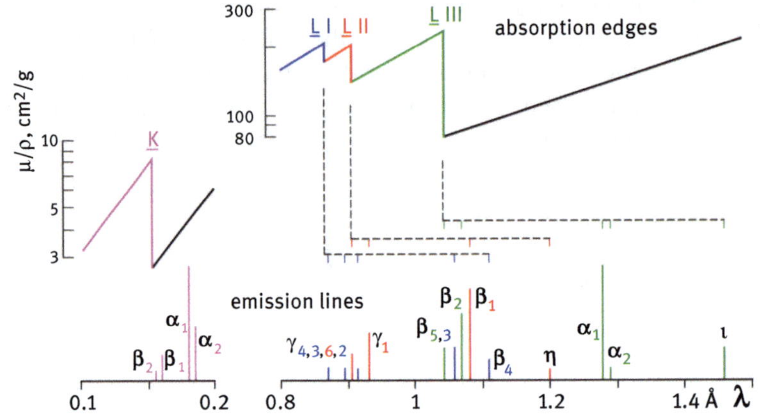

Figure 2.4: Some absorption edges and emission lines of gold. In the upper part of the figure are represented the K and L absorption edges with their relative mass absorption coefficient, μ/ρ, expressed as cm^2/g. In the lower part of the figure are represented in purple the X-ray emission lines of the electrons that falls to the K shell and in blue, in red and in green, respectively the X-ray emission lines of the electrons that falls to the L I, L II and L III shells ([13], modified).

The excitation of XRF radiation is caused both by the characteristic lines and by the continuous background of the X-ray tube anode with which the samples are irradiated. Thus, the choice of the anode of the X-ray tube, that is, in practice the spectrum of the incident radiation, is of particular importance for the most effective excitation of the elements under examination.

For example, with respect to the element W, the L lines constitute 25 % of the intensity of the whole spectrum whereas, relative to the element Cr, the K lines represent 75 % of the intensity of the whole spectrum.

To simplify, it can be said that X-ray tubes with anode of heavy elements favor the excitation of XRF of heavy elements while X-ray tubes with anode of light elements favor the excitation of XRF of light elements.

2.2.3 X-ray absorption

The law on X-ray absorption is expressed by the Lambert equation:

$$I = I_0 \cdot e^{-\mu t}$$

where I_0 is the intensity of a monochromatic radiation incident on a sheet of material with linear absorption coefficient μ and thickness t, while I is the intensity of transmitted radiation.

In terms of mass absorption coefficient μ/ρ, where ρ is the density of the material expressed in g/cm^3, the Lambert equation becomes:

$$I = I_0 \cdot e^{-(\mu/\rho)\cdot\rho t}$$

where ρt (g/cm^2) is the concentration of matter per unit area. The value of mass absorption coefficient μ/ρ is defined by the Bragg-Pierce law:

$$\mu/\rho = KZ^4\lambda^3$$

where λ is the wavelength expressed in centimeters, Z is the atomic number of the chemical element and K is a proportionality factor that varies according to the electronic orbits involved in the incident radiation.

In particular, considering the relationship between μ/ρ and λ, it is observed that for the same element (Z = constant) the mass absorption coefficient differs according to the discontinuity considered (K, L, M etc.), in particular it increases from the discontinuity K to those due to the most external shells (Figure 2.4). At the same energy considered (λ = constant) the mass absorption coefficient is higher for elements with a higher atomic number.

Since the mass absorption coefficient depends on Z, a matrix of heavy elements absorbs the X radiations (both the incident and the fluorescence emitted by the sample) much more than a matrix formed by light elements. Evidently, a greater absorption determines a lower intensity of the emitted fluorescence radiation and therefore the ratio between intensity of the characteristic radiation and concentration of the element to be determined is disturbed. For example, if a same quantity of Rb is present in a matrix of Fe_2O_3, or in a matrix of Al_2O_3, the intensity of the RbKα line is somewhat lower in the former than in the latter.

Furthermore, there are interference effects between elements, caused by the mutual position of the characteristic lines and the absorption edges. For example, if we suppose that in a sample there are two chemical elements A and B in such concentration that the ratio of the intensities of the Kα lines is R (i. e. $I_{AK\alpha}/I_{BK\alpha} = R$), the presence of a third chemical element C can disturb sensibly this ratio of intensity if its absorption edge is placed at an intermediate wavelength between that of the two Kα lines that are being analyzed. The selective absorption of the element C will determine a ratio $I'_{AK\alpha}/I'_{BK\alpha} = R' \neq R$. It is reported, as an example, the case of nickel, which is used as monochromator for the Kα radiation of the copper because its K absorption edge (1.49 Å) is in an intermediate position between the CuKα radiation (1.54 Å) and the CuKβ radiation (1.39 Å).

A more detailed discussion on the X-ray absorption can be found in specialistic books [13, 14]; a complete collection of all the elements X-ray emission lines, their absorption edges and the energies of the different atomic shells can be found in papers published by specialized journals [15–17].

2.2.4 Detection and measurement of X-rays

The X-ray detection consists in converting the electromagnetic radiation into a signal that can be measured and integrated in a finite time interval. There are numerous systems to achieve this purpose and each of them is based on the capacity of X-rays to interact with solid or gaseous matter.

One of the oldest system was based on the photographic method, that of radiographs, in which the photochemical action of X-rays determines the transformation of a silver salt into metallic silver. The total intensity of X-rays, determined by the quantity of metallic Ag produced, can be measured by the degree of blackening of the film.

The future is surely represented by a new generation of low-cost X-ray detectors that are highly efficient (up to 400 times the actual ones) and are based on perovskite wafers [18].

Today, for elemental chemical analyses, two types of detectors are basically used: energy dispersive (ED) or wavelength dispersive (WD).

2.2.4.1 Energy dispersive X-ray detection

The X radiation can be detected in energy dispersion with solid state counters (Figure 2.5). These counters are made up of a semiconductor: it is generally a silicon crystal, appropriately doped with various elements such as P, Al, B or Li, but other crystals, such as germanium or diamond, can also be used. Doping consists in the substitution of a certain number of atoms in the crystal lattice with atoms of the neighboring groups in the periodic table: with one more valence electron (n-doping)

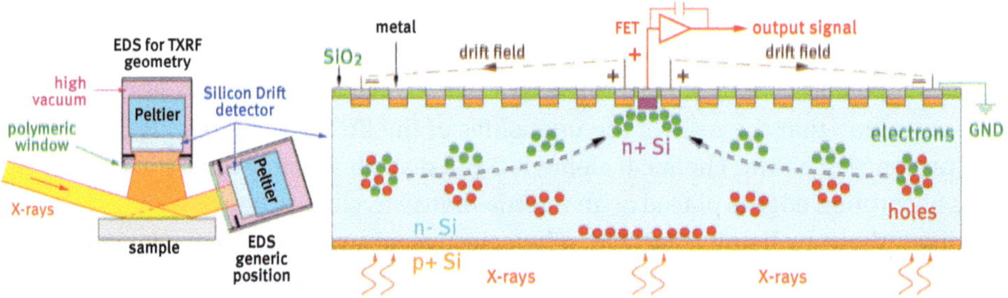

Figure 2.5: Schematic representation of an energy dispersive detector (ED) that can be mounted with any geometry with respect to incident X-ray beam (normally the incidence angle is of about 40 °); for low-angle incident X-rays (about 0.1 °) the signal collection can be performed just above the sample as in the case of Total reflection X-Ray Fluorescence (TXRF) spectrometers. On the right, the enlarged detail of an ED Silicon Drift detector (FET = field-effect transistor; GND = ground).

or one less (p-doping) than the starting material. When a X photon strikes one of these semiconductors it generates inside the material an "electron–hole" pair (e–h); for the silicon it is necessary an energy of 3.8 eV to generate an e–h couple while for germanium an energy of 2.9 eV is needed. The electrons move toward the positive pole leaving behind the holes: i. e. apparently, electrons move toward the positive pole and holes toward the negative pole.

Therefore, if an X photon with energy equal to 8,048 eV, corresponding to the $CuK\alpha_1$ radiation, hits a silicon semiconductor, it will generate 8,048 eV/3.8 eV = 2,118 e–h pairs. The possibility of counting e–h pairs will allow to know the energy, and therefore the wavelength, of the incident radiation.

In the last 30 years, ED Si(Li), germanium, Pentafet (i. e. with 5 silicon-lithium crystals coupled together), Silicon Drift detector (SDD) and finally the recent SDD Droplets (SDD^3) counters have been developed. Technological efforts aimed to increase the number of counts detectable, at the same time maintaining the detector resolution around 115–125 eV. In SDD or SDD^3 silicon counters (the latter is constituted by four or more SSD meters mounted together, with asymmetric collection of electrons) the back of the signal collection system consists of concentric circles with different potentials: the electrons are collected on a very small area from a large collecting front surface, and this allows to maintain a good resolution, about 125 eV, of the detector (Figure 2.5).

The big advantage of the ED counters is to analyze in real time the whole X-ray spectrum emitted by the sample. The drawbacks are essentially linked to three causes: (1) the detector must be refrigerated, with liquid nitrogen, as in the old Si (Li) or Ge meters, or with a Peltier system, as in the new SDD, to prevent damaging in the crystal semiconductor structure; (2) the resolution in energy (115–150 eV) often does not allow to efficiently separate some X-ray spectral lines; (3) it is not possible to irradiate the sample with particularly high power because, on increasing the number of detected X-ray photons, it also increases the detector dead time, which results in few counts coming out of the amplification chain: having less counts leads to a greater error in the measure.

The multi-channel character of the energy dispersion meters makes them particularly useful for qualitative analyses; moreover, the extremely simple construction geometry and the higher sensitivity with respect to WD counters make them irreplaceable in the assembly of portable spectrometers in which they are coupled to transportable X-ray sources, thus with low irradiation capacities, such as radionuclides or X-ray microtubes. The ED counters diffusion has increased considerably since their refrigeration was implemented through Peltier systems; the evolution of polymeric windows particularly transparent to X-rays has also allowed their use for the detection of light elements.

ED detectors find unique application in XRF spectrometry operated in Total Reflection Geometry (Total Reflection X-Ray Fluorescence, TXRF): in these spectrometers the incident X-ray beam angle is close to the critical reflection angle (about

0.1 °); under these conditions the X radiation to be read is that produced on the sample surface by the incident beam which does not penetrate the sample substrate (Figure 2.5). The advantage of this method is that it allows a sample surface analysis, it is not affected by matrix effects and therefore quantitative data can be obtained using only an internal standard. TXRF is particularly useful for the analysis of liquids as well as of solids treated as thin films, such as the pigments layers.

2.2.4.2 Wavelength dispersive X-ray detection

Wavelength dispersive spectrometers (WD-XRF) require that the fluorescence spectrum is spread into discrete wavelengths using a dispersion device before the detection step; the dispersion device is a crystal, organic or inorganic, which takes the name of diffracting crystal or diffractor.

The process of dispersion of X-ray radiation, which is in fact a diffraction, is regulated by the law of Bragg:

$$2d \sin \theta = n\lambda$$

where d is the lattice parameter of the crystal, θ is the value of the incidence and diffraction (reflection according to Bragg) angle, n is the order of the reflection and λ is the wavelength of the incident and the reflected radiation. Ultimately, the objective of the XRF analysis is to determine, given the d value of a diffracting crystal, the value of λ, characteristic for the various chemical elements, by measuring the θ value of the spread wavelengths.

The wavelength dispersive spectrometers are basically of two types: (i) flat and (ii) curved diffracting crystal spectrometers.

In the flat diffracting crystal spectrometers, the only expected movements are a rotation of a θ angle of the diffracting crystal and consequently a rotation of a 2θ angle of the detection system. There are no translations of both diffracting crystal and detector and therefore the construction geometry of these spectrometers is quite simple. During their rotation the sample, the diffracting crystal and the detector are not on the same focusing circle and then the equipment works in parafocalization geometry: the X-ray beam is focused on the diffracting crystal and on the detector, respectively, by a primary and a secondary collimator (Figure 2.6).

The X-rays produced on the sample are sensitively absorbed by the collimators but the benefits of a simpler geometry make these spectrometers of particular interest for the analysis of samples having centimetric dimensions.

The curved analyser crystal spectrometers, which work in perfect focusing geometry, have a more complex geometry in that both the diffracting crystal and the detector are rotated and translated into space, as well as being rotated by θ and 2θ, so that they are always on the same focusing circle. These spectrometers are normally used in microanalytical equipments, like Electron Probe MicroAnalysis (EPMA) or

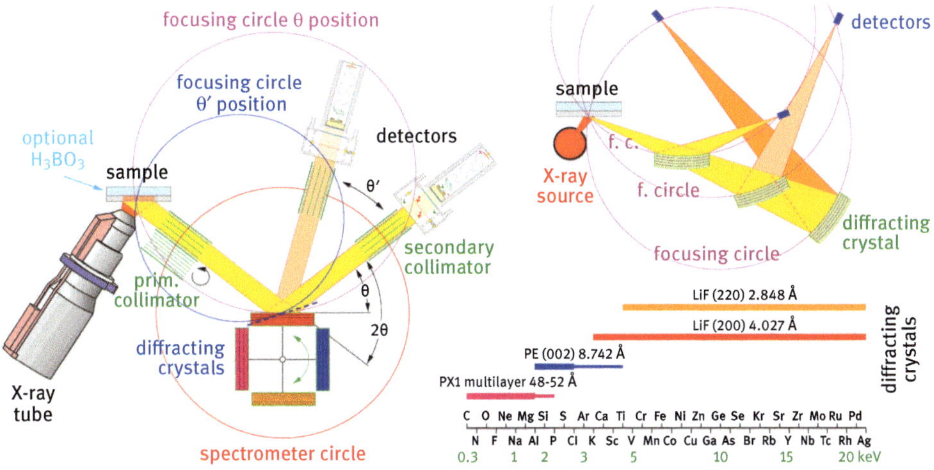

Figure 2.6: On the left part of the figure the scheme of a WD spectrometer with plane diffracting crystal that work in parafocalization geometry. On the right part of the figure the scheme of a curved diffracting crystal spectrometer: the simultaneous angular movement of diffracting crystal and detector, also associated with a translation of both objects, allows the spectrometer to work in perfect focusing. In the lower right part of the figure are reported the usual diffracting crystal used during the WD analysis of chemical elements.

scanning electron microscopes (SEM), where it is very essential to have no significant signal loss of the X-ray intensity along the sample-detector path (Figure 2.6).

Both the flat diffracting crystal and the curved diffracting crystal spectrometers can mount one or more X-ray detectors. Among the widely used detectors there are those that use the ionization process of a gas contained inside the detector itself; this counter consists of a chamber whose inner walls are held to a negative potential with respect to a central filament. The X-rays enter the counter through a window closed by a thin sheet of a material particularly transparent to the X-ray radiation, usually a polymeric sheet of Mylar ©, and cause the ionization of the gas atoms contained in it. The electrons and the positive ions produced in this way, moving under the influence of the electric field, are respectively directed toward the central filament (positive pole) and toward the walls (negative pole). The gain in the amplification of the X signal, consequent to the secondary ionizations, can reach values of 10^6 maintaining the proportionality between the amplitude of the output pulses and energy of the incident radiation.

One of the most used gas counter in the WD-XRF spectrometers is the flow gas detector: the gas consists of a mixture of argon (Ar = 90 %) and methane (CH_4 = 10 %) and flows continuously and with constant pressure inside of the chamber, coming from a tank (Figure 2.7).

Another widely used detector in XRF equipments is the scintillator (Figure 2.7) in which the X-rays interact with a solid-state counter consisting of sodium iodide

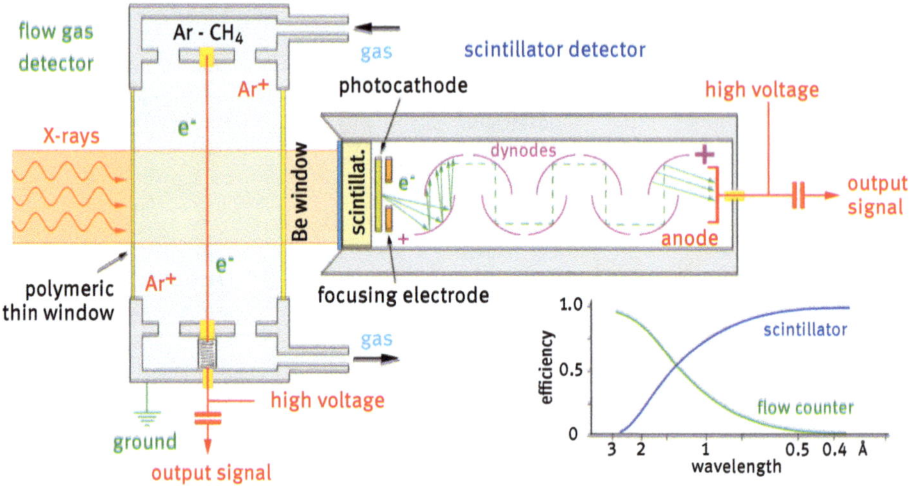

Figure 2.7: Scheme of a flow gas detector and of a scintillator detector; in the right bottom part of the figure is indicated the different efficiency of the two detectors according to the collected wavelength.

activated with about 1% of thallium. These type of crystals, hit by X-rays, emit a number of photons proportional to the energy of the X-ray absorbed. The phosphor rear face is welded to a photomultiplier tube consisting of a photocathode and a series of dynodes, with increasing positive potential. The photocathode, once hit by photons, emits a number of electrons proportional to the energy of the photons themselves; the electrons thus produced are attracted and multiplied by the series of dynodes, until reaching the anode where they discharge through the detection circuit. In this case too, as in the case of the gas flow meter, the gain, at the end of the amplification chain, is about 10^6.

XRF spectrometers are almost always equipped with two counters: the flow gas and the scintillator, in order to improve the instrumentation versatility. As a matter of fact, gas flow meters are particularly efficient in the range of high wavelength values (ionization of a gas), while scintillators are particularly efficient in the range of short wavelength values (ionization of solid matter). However, there is a wavelength range (between about 0.75 and 2 Å) in which the gas flow meter has already lost much of its efficiency while the scintillator has not yet reached its maximum efficiency (Figure 2.7). Unfortunately, in this wavelength range, the Kα radiations of many elements are commonly measured (Mn, Fe, Co, Ni, and so on), whose atomic numbers range from 25 to 40.

This lack of efficiency has been overcome equipping the flow gas meter with an exit window for the radiations that have not lost their energy in the ionization process; these radiations are collimated onto the scintillator (normally located behind the flow gas meter outlet window) and the instrumentation provides the sum of the pulses recorded by both meters.

It is convenient to exclude the gas flow meter at very short wavelengths and the scintillator at very high ones because their limited efficiency would lead to increase the background noise due to the electronic of amplification circuits.

In state-of-the-art spectrometers, gas-sealed meters, such as the Neon, Krypton and Xenon detectors, coupled with traditional gas flow and scintillation counters, fully cover the detection of the whole spectrum of X-ray emission of a sample.

It is worth pointing out that normally the X-ray tube, sample, diffracting crystal and counter (surely the flow gas) of a WD spectrometer are contained in a chamber in which a vacuum is made of about 2 Pa in order to prevent the X-rays of the light elements from being absorbed by the air.

2.3 Measurement modalities

The XRF chemical analyses are part of the so-called "indirect methods" according to which the concentration of a given chemical element is calculated starting from the measurement of an appropriate physical quantity, whose intensity is proportional to the concentration of the chemical element itself.

In XRF analyses, the intensity that is measured is that of a characteristic X-ray lines (Kα, Kβ, Lα ...) emitted by the chemical element contained in a sample irradiated by primary X-rays. This intensity depends, firstly, on the number of atoms of that chemical element interacting with X-rays. In other words, quantitative analysis in XRF is based on a simple equation like:

$$C = m \cdot I$$

where C is the concentration of the element of interest, I the X-ray intensity, m a proportionality coefficient.

Unfortunately, the proportionality coefficient is not always simple to determine because, except for the infinite dilution technique, it varies with the matrix of the investigated sample: approaches are needed, such as the mathematical matrix correction procedures, to obtain a reliable value of the element concentration from the X-ray intensity measurement.

2.3.1 Qualitative ED analyses

X-ray energy dispersive analysis is one of the most efficient techniques as to qualitative analysis, in that in a few seconds, about 50, the whole spectrum of emission in energy of the sample can be obtained (Figure 2.8). Element recognition is attempted using the appropriate tables of conversion from energy to chemical element: today, specific software are available for an automatic indexing of the peaks of the ED spectrum.

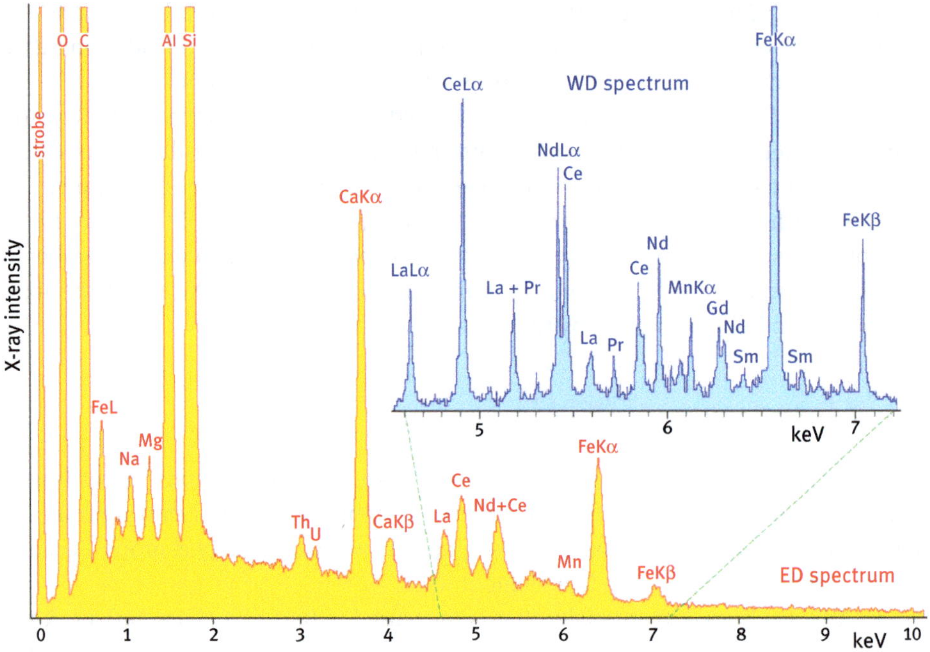

Figure 2.8: ED spectrum of a rare earth rich mineral (an allanitic epidote); the strobe is the reference peak of the system for zero energy. Note the strong overlap of the L lines of light rare earth elements (La, Ce, Nd): in cases like this, it is very difficult to state the presence of an element if it is in low concentration and there are strong overlaps of X-ray lines of contiguous elements. In the enlarged detail, in blue color, the WD spectrum of the same mineral (collected with a LiF 200 diffracting crystal in the 4.55–7.25 keV range): it is very clear the higher spectral resolution of a wavelength dispersive scansion.

It is important, in order not to make rough errors, to keep in mind the intensity ratios between the various lines as well as the occurrence that L lines of heavy elements can overlap completely or partially with K lines of lighter elements. As an example, the case of the xenotime (YPO_4) can be cited, for which the L lines of yttrium overlap the K lines of phosphorus giving spectrally a single peak. However, it must be underlined that strong asymmetries in the detected peaks, can suggest the overlap of several spectral lines.

2.3.2 Quantitative ED analyses

Quantitative ED XRF analyses consider the simultaneous acquisition of the whole sample emission spectrum in a well-defined time interval. Analysis is based either on the peak height or on the peak area measurement. This step is not so simple and implies, preliminarily, processing of the spectrum and background subtraction:

today, data treatment software are provided by instruments manufacturers, to convert X-ray intensities into elemental concentrations.

The calibration of the instrumentation can be performed by recording the spectra of pure compounds or standards with a known concentration; alternatively, "standardless" calibrations are provided by most of the manufacturers of the instrumentation.

Some new ED spectrometers can mount polycapillary lenses which exploit the phenomenon of total reflection for focusing X-rays on a very small area (down to about 10 μm); polycapillary lenses allow to obtain chemical analyses of very little areas of the sample, with a detectability that can reach the tens of parts per million, and permits to achieve high performance with low-powered X-ray tubes as the portable ones. Moreover, polycapillary lenses coupled with software controlled moving sample holder or spectrometer give the possibility of acquiring, through X-ray maps, the elemental distribution of the sample surface.

2.3.3 Qualitative WD analyses

Qualitative WD-XRF analyses can be performed using more than one diffracting crystal (Figure 2.6); the energy is generally accomplished, first using the smaller 2d crystals and then the larger 2d ones.

At the same two detectors (flow counter + scintillator) are generally employed in order to read X-rays both at low and high energy (Figure 2.6).

For each diffracting crystal, there are conversion tables from 2θ to λ which allow to index the various peaks relative to the elements contained in the sample. During the interpretation of a wavelength dispersive X-ray spectrum, however, analyst should keep in mind that, due to the Bragg's law ($2d \sin\theta = n\lambda$), at the same 2θ angle both a first order Kα line of a light element ($\lambda = x$ ed $n = 1$) and second order Kα line of a heavier element ($\lambda = x/2$ ed $n = 2$) can be detected.

2.3.4 Quantitative WD analyses

Quantitative WD-XRF analysis involves the extremely accurate measurement of the intensity of characteristic lines of the chemical elements that are present in the sample from which the relative elemental concentrations may be calculated.

With regard to each chemical element to be dosed, the following operations must be carried out:

1. select the highest intensity characteristic lines of the elements;
2. set the most appropriate instrumental conditions for the excitation and detection of the radiation under examination, that is: the anode of the X-ray tube, the type of diffracting crystal, the type of collimator and the type of detector;
3. experimentally measure the 2θ angle corresponding to the maximum intensity of the analytical line. It should be noted that the experimental value may be slightly

different from the theoretical value, as a function of the positioning accuracy of the collimators and the diffracting crystals, as well as the real 2d values of the diffracting crystal, especially for multilayer ones;
4. set the detector on correct 2θ angle and select a proper measurement time, both on the peak and on the background to collect a number of impulses sufficient to minimize counting errors. The optimal fixed time can be selected by calculating the relative standard deviation (σ), which is given by:

$$\sigma\% = \frac{100}{\sqrt{\text{time}}} \frac{1}{\sqrt{\text{peak}} - \sqrt{\text{background}}}$$

σ will be the smaller, the longer is the time interval of counting and/or the higher is the net intensity of the peak. Moreover, during an X-ray acquisition, the statistic error ε depends from the total counts C collected:

$$\varepsilon = C \pm \sqrt{C}$$

The intensity of the continuous background should be subtracted in order to get the net peak intensity, however, even under the same instrumental conditions, it varies with the variation of the wavelength under examination, and so with the variation of the 2θ angle, and with the variation of the sample matrix. In particular, if the value of the *bremsstrahlung* is practically constant in a small interval around the peak, it is sufficient to measure only one point of the background in the immediate vicinity of the peak itself.

On the other hand, if the *bremsstrahlung*, even if expressed by a linear function, shows a marked intensity variation near the peak, it will be necessary to measure its values at two or more different points, around the peak position, and then use the mean, arithmetic or weighted as appropriate, as a reference value of the 2θ background angle.

In some cases, when the background cannot be expressed through a linear function, the representative equation of the whole curve of the *bremsstrahlung* must be experimentally determined, so that its intensity can be calculated for each value of the 2θ angle involved in the analytical measurements. This is, for example, the case of the background for some elements such as Rb, Sr, Y, Zr, Nb (Figure 2.9). The curve that best fits background is an exponential function of the type:

$$I_B = e^{-a\lambda + b}$$

where I_B is the intensity of the background, λ the wavelength of the background, **a** and **b** two constants that depend on the instrumental conditions and the composition of the matrix [19]. The two constants **a** and **b** must be determined experimentally by sampling more backgrounds around a determined λ and in any case the operation

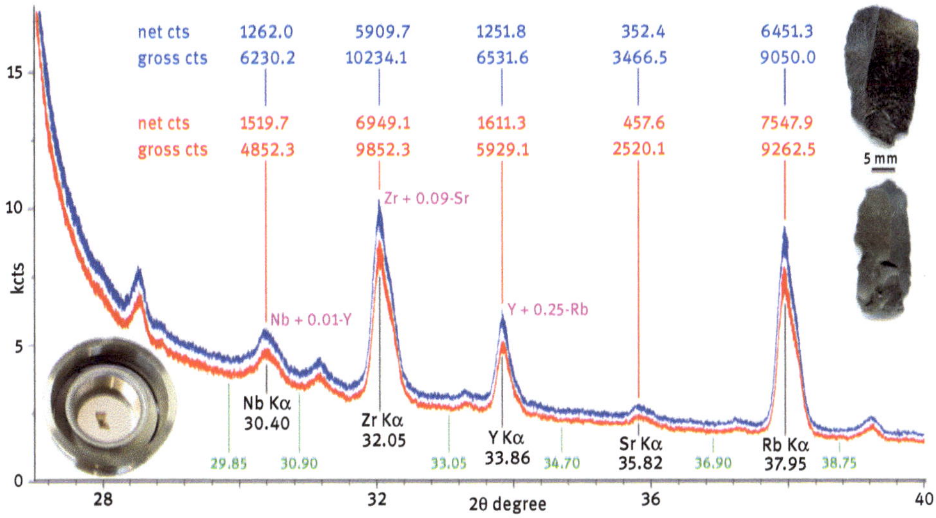

Figure 2.9: Plot of two WD-XRF scan between 27 and 40 2*θ* degrees (about the frequency range from 0,067 to 0.097 nm) of the X-ray emitted by two different obsidian samples, one bigger (X-ray scan in blue) and one smaller (X-ray scan in red); analytical conditions were: Rhodium X-ray tube set at 60 kV and 66 mA, LiF 220 diffracting crystal and scintillator detector. In black are reported the 2*θ* degree of the Kα trace element emission and in green the 6 angular positions for background sampling. The gross counts per second measured on the peak position of the trace elements and their relevant net counts (background and interference free) are also reported for the bigger obsidian (in blue) and for the smaller one (in red). In pink is indicated the contribution of the interfering line to the measured peak ($YK\beta_{1,3}$ on $NbK\alpha_{1,2}$, $SrK\beta_{1,3}$ on $ZrK\alpha_{1,2}$ and $RbK\beta_{1,3}$ on $YK\alpha_{1,2}$), as calculated for the specific WD spectrometer spectra obtained by Pasquale Acquafredda (Dipartimento di Scienze della Terra e Geoambientali, University of Bari, Italy).

must be performed sample by sample: interpolating the *bremsstrahlung* data, the intensity of the background can be calculated exactly at the peak position of each emitting element.

Once the net peak intensities of the chemical elements constituting the sample have been measured with extreme precision after background subtraction, various analytical procedures for the determination of the concentrations can be used; among these procedures, those most widely used are: Standard Addition, Linear Regression, Infinite Dilution and Mathematical Matrix Correction.

2.3.4.1 Standard addition procedure

The standard addition procedure consists in adding to the sample a known quantity of the element to be determined; the sample "as it is" and those with the various additions are submitted to the XRF analysis. In this way, you will have three or more equations, depending on the number of additions; for two additions:

$C_x = m \cdot I$ for the original sample,
$(C_x + C_{n1}) = m \cdot I'$ for the sample with the first addition,
$(C_x + C_{n2}) = m \cdot I''$ for the sample with the second addition.

C_x, C_{n1} and C_{n2} are respectively the unknown and known concentration of the element to be measured; I, I' and I'' are the collected intensities; m is the coefficient of proportionality (angular coefficient of the straight line) that is supposed to be the same in the analyzed specimens.

The algebraic resolution of the equations or the graphical one (the unknown concentration is read as negative value on the abscissas; Figure 2.10) allow to obtain the unknown concentration of the element to be determined.

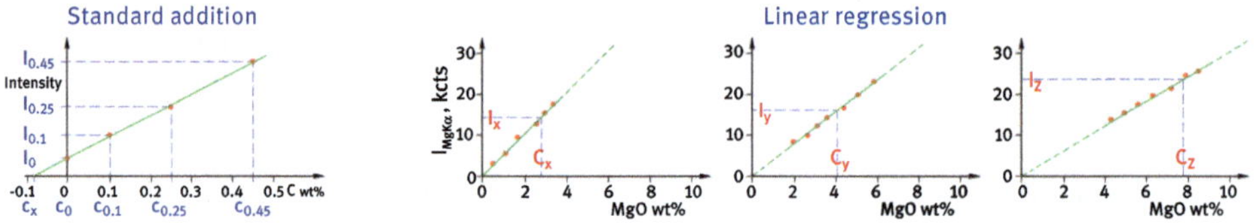

Figure 2.10: On the left part of the figure the graphical resolution to calculate the concentration of an element using the standard addition procedure; on the right part of the figure an example of three linear regression straight line (the slope of the green straight line varies to indicate a significant variation of the μ/ρ of the matrix), obtained by measuring the same element, present at different concentrations and in different matrices, on certified standards.

This procedure is valid only when the proportionality coefficient m can be considered actually the same in the original sample and in the one to which the addition was made: in practice, it can be used for measuring trace elements (contained in concentrations lower than 0, 1 %), or at most for minor elements (contained in concentrations between 0.1 and 1 %), where the addition does not substantially modify the composition of the matrix.

A handicap of the procedure lies in the necessity of add the element to be measured to all the samples to be analyzed compromising in part their reuse as original samples, apart from the risk of introducing errors (and contamination) in this preliminary manipulation phase.

2.3.4.2 Linear regression procedure

The linear regression procedure involves the use of numerous certified standards. The XRF intensities of the X-ray spectral lines of the certified elements are measured; is necessary to group samples with a matrix as similar as possible and at the same time containing a given element in concentration as varied as possible.

The XRF intensities of each element are plotted against its concentration in each standard and the regression straight line is plotted for the considered standard population: for each element, more regression straight line will be obtained depending on the family of matrices in which the element itself is contained (Figure 2.10).

To know the concentration of an element in an unknown sample it will be sufficient to read the intensity of its characteristic X-ray line and, using one of the experimentally obtained regression straight lines, calculate the corresponding concentration: take care to use only the linear interpolation (continuous green line in Figure 2.10) and not the extrapolation parts (broken green line in Figure 2.10) of the straight lines.

It is evident that the concentration value obtained will depend on the group of standards with which the regression straight line has been calculated: the more these will have a matrix similar to that of the sample to be analyzed, the more accurate the determinations will be.

This procedure, to be reliable requires many certified standards and, above all, implies a fairly knowledge of the chemical composition, i. e. of the matrix, of the sample to be analyzed.

2.3.4.3 Infinite dilution procedure

The infinite dilution procedure consists in diluting the sample in a very light matrix so that the constituent atoms do not interfere significantly with each other when they emit X-rays. Lithium metaborate ($LiBO_2$) or tetraborate ($Li_2B_4O_7$) are usually used as dispersing agent; both these compounds also have the advantage of lowering the melting point of the powder sample (between 1,000 and 1,200 °C), when glassy disc (glass bead) of the sample have to be prepared (Figure 2.11(g) and Figure 2.11(h)). Experimental data indicate that an infinite degree of dilution is reached for a sample/diluent ratio of about 1:18 [20]; in these dilution conditions the matrix of a standard with known concentration can be considered identical to that of an unknown sample. This analytical method, if on the one hand eliminates both the problems related to the matrix and those related to the granulometry and mineralogy of the samples, on the other hand it is not recommended to measure the trace elements, contained in concentrations of parts per million, due to the considerable dilution of the sample. It is instead perfectly efficient for the determination of major elements (concentrations > 1 wt. %) and minor elements (concentrations between 0.1 and 1 wt. %).

2.3.4.4 Mathematical matrix correction procedures

The mathematical matrix correction procedures involve calculating the concentration of an element taking into account all the absorption and/or reinforcement effects that contiguous elements present in the sample can produce. One of the mathematical formula that make it possible to transform the XRF intensities of the major elements into concentrations is:

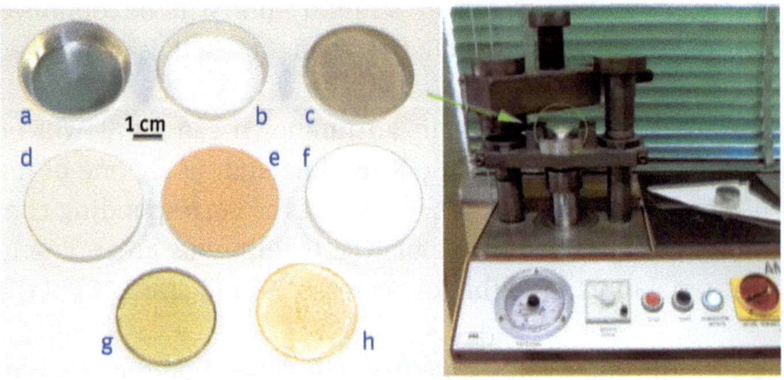

Figure 2.11: Preparation of a powdered pellet samples using an aluminum holder (a): the bottom of the aluminum cup can be filled with boric acid (b), in order to be sure to have reached the infinite thickness (the sample has infinite thickness when the X-rays do not reach the bottom of the specimen); the dust filled aluminum holded (c) is then compacted with an hydraulic press (right image), to obtain discs with a perfectly flat surface. The binder of the powder may be polyvinyl alcohol (d), elvacite (e) or beeswax (f). The different color of the three powdered pellets does not depend on the type of binder but indicates a different composition of the samples. In some cases, it may be useful to melt the sample, thus obtaining a homogeneous glass disc (glass bead) (g) avoiding air bubbles formation (h).

$$C_i = I_i \sum_{J=1}^{N} K_{i,J} C_J$$

where C_i are the concentrations of the sample elements, N the number of the elements, I_i the intensity of the measured characteristic X-ray line, $K_{i,J}$ the coefficients that take into account the matrix effects, and C_J the apparent concentrations before the matrix correction. The values of the $K_{i,J}$ coefficients were calculated experimentally measuring the X-ray intensities on numerous international certified standards [21]. In a very simple way, it can be said that the concentration of each of the major elements (C_i) depends on its X-ray line intensity (I_i) taking into account the matrix in which it is contained ($\sum K_{i,J} C_J$).

For more details on how to apply this analytical method, both to the major elements and to the trace elements, refer to the specific paper in which the procedure is described [19, 21].

2.4 Sample preparation

The necessity to carry out absolutely non-destructive analyses, as in the field of Cultural Heritage, must favor the use of ED spectrometers because measurements are less affected by errors induced by an irregular geometry of the sample surface [1, 22, 23], though with a less reproducible and less accurate data.

When sampling is allowed, WD spectrometers can be used which allow high accuracy and precision in the analysis; this is generally the case of investigations on ceramic fragments or lithic artifacts where is possible to pick-up a part of the sample without compromising the aesthetics of the sample itself (Figure 2.12).

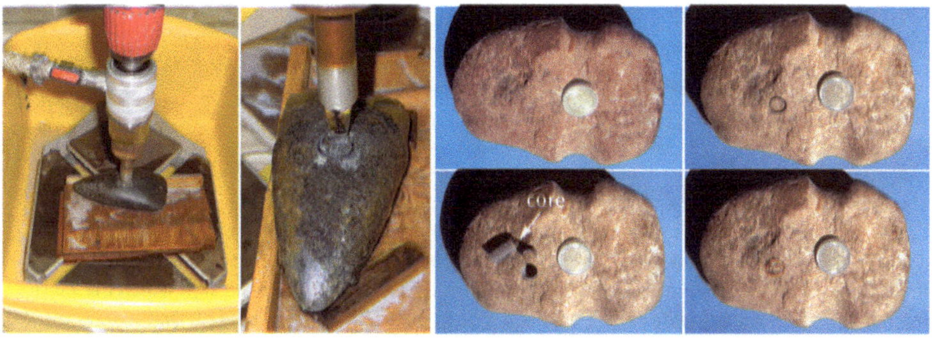

Figure 2.12: Coring of an archeological sample using a microdriller with the edge enriched in diamond powder; the driller is cooled with water and mounted on a column drilling machine. The hole was performed on the less important surface of the artifact, the one not to be placed in view during its musealization. The hole generated by coring can be filled with inert and non-hygroscopic material and closed with a little portion of the cored sample itself. A paint stroke and the inventory number completely mask the pick-up point of the specimen.

When WD-XRF analysis is performed on powdered samples, it must be ensured that the specimen is very well grinded, because the sample granulometry influences the distribution of full and empty spaces and so the emission efficiency: on decreasing the grain size, the empty spaces decrease and therefore the number of atoms which interact with the incident radiation increases. Better results can be obtained by compacting the sample with a hydraulic press in order to reduce, as much as possible, the presence of empty spaces (Figure 2.11).

Furthermore, for the preparation of powdered samples it is also necessary to consider the mineralogical nature of the specimen: if sheet silicates (micas, clay minerals, and so on) are present, whose lamellar structure favors their iso-orientation, the chemical elements present in their structure generally show a concentration higher than the actual one; as a consequence, the concentration of the elements present in the other mineralogical components, is masked by the shielding effect consequent to the iso-orientation of the phyllosilicates.

In other cases, in which absolutely non-destructive analysis are needed (consider that LA-ICP MS produces little crater pits in the sample and EPMA requires that at least a very little part of the specimen is flatted), like those for obsidian provenance investigation, analyses can be performed without any manipulation using a plexiglass sample holder or using a very thin sheet of Mylar (Figure 2.13) which suspends the artifact on the X-ray beam pathway [1].

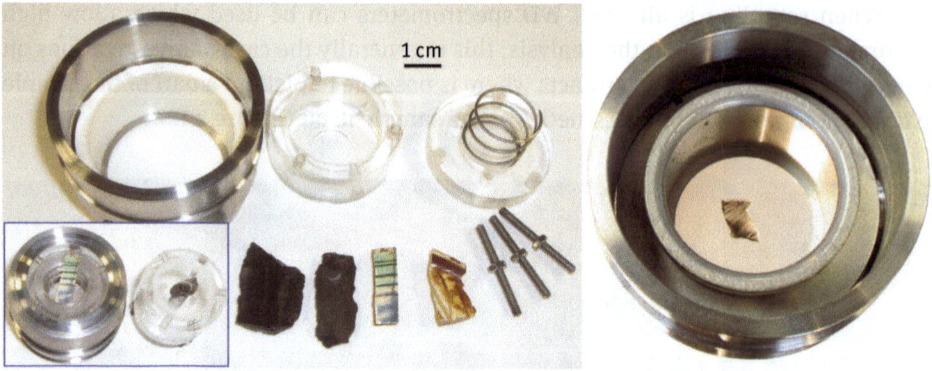

Figure 2.13: WD-XRF spectrometer sample holder equipped with plexiglas capsules in which the samples (two obsidians and two peaces of ceramic) can be fixed (see the inset in the left lower part of the figure); the plexiglas boxes were built with different diameter holes through which the more regular surface of the artifacts can be exposed to the spectrometer X-ray source. On the right the WD-XRF sample holder closed at the bottom with a very thin Mylar© polyester film on which a little obsidian specimen is placed, as suspended in the air.

Suppletive informations about measurements, calibration modalities and sample preparations regarding X-ray spectrometers can be found in other specific books [14, 24].

2.5 Case studies

The use of XRF stectrometers in the field of Cultural Heritage varies from the analysis of rocks to that of ceramics, from binders (cement, mortar, etc.), to pigments, from natural and artificial glass, to metals and alloys, to minerals, to products of alteration, and so on.

ED and WD X-ray spectrometers can be either individual and autonomous apparatus (both bench and portable instruments) or can be coupled with other instruments like Scanning and Transmission Electron Microscopes (SEM and TEM) or EPMA.

Until the end of the last century, the WD spectrometers were used in very reliable laboratories for bulk chemical analyses or for high quality microanalyses if mounted on SEM or EPMA and the ED spectrometers were essentially dedicated to qualitative analyses.

In the last few years, in addition to the cumbersome and expensive laboratory instruments, which normally involve complex calibrations, the use of economic instrumentation, easily transportable, able to give good results in extremely short times, is becoming increasingly widespread. This the reason why the archeometric use of ED X-ray spectrometers [2–10], that is portable X-ray Fluorescence (pXRF) [10], Field-portable X-ray Fluorescence (FpXRF) and Hand Held portable X-ray Fluorescence

(HHpXRF) [7], and especially the apparatus equipped with the new SDDs, has offered new potential for very rapid and cheap characterization of archeomaterials.

Very often portable X-ray spectrometers operate in the air, so they allow the intermediate and heavy elements to be detected; light elements up to magnesium or even up to fluorine, can be measured if the instruments operate under helium flow or are equipped with a small vacuum chamber.

Particularly useful have been the employment of the ED detectors for the diffusion of the portable spectrometers offering new possibilities for *in situ* and very rapid archeomaterial characterizations. Moreover these ED portable equipments are furnished with a software that allows to get easily and quickly elemental composition of the specimen, often with a standardless routine analysis though an accurate check of the results is necessary using very good (and, unfortunately, very expensive) reference standards.

As to portable ED XRF instruments, however, in agreement with Framm [11] a general warning should be given, they are particularly useful in the training of students [4] and are very precious for an initial survey and sorting artifacts, but they cannot replace and compete in performances with bench instruments [12, 25–27]. It is also true, on the other hand, that accurate quantitative analysis on many kinds of samples, including lithic archeomaterials [23, 28, 29] and pigments [30], have been performed with ED XRF detectors (coupled with SEM), in many cases with results comparable with other analytical techniques such as LA-ICP MS [31].

In the following, some examples of archeometric investigations carried out on completely different materials with ED or WD spectrometers are reported: the choice of the detector is mainly conditioned by the possibility to perform minimally invasive analyses and by the sample isotropy.

2.5.1 Ceramics

WD-XRF chemical analyses of powdered ceramic fragments, supported also by other investigation techniques, like optical microscopy and powder X-ray diffraction, give very useful information about the sedimentary rocks source, the manufacturing techniques and the eventual purification process of the raw materials [32, 33].

Quantitative image analysis of ceramic thin sections through petrographic investigation gives information concerning the additives (temper) and natural inclusions (silt and sand) in the ancient ceramic production. In addition to traditional optical microscopic identification of mineral inclusions, the acquisition of ED X-ray maps of ceramic thin sections allows to evaluate the quantity of quartz, feldspars and calcite in the artifacts; the technique, in combination with algorithms for detecting regions of interest and their segmentation, provides also the size distribution of the minerals [31]. Recent research on two Late Acacus (Mesolithic) potsherds (8,900–4,200 BP) demonstrated that the acquisition of ED X-ray maps on ceramic thin sections improves the accuracy of modal optical microscope analysis [34].

2.5.2 Necklaces

Here is reported the case concerning the chemical ED-XRF analyses carried out on a necklace to ascertain its authenticity (Figure 2.14). In this specific case, the analysis of the beads composition (artificial sodium glass), together with the analysis of the chain metal (24 carats pure gold), initially suggested that the necklace could be a jewel of Roman age. The ED chemical analyses, performed directly on the natural surface of the artifact, were repeated at least three times and the mean and standard deviation were calculated (Figure 2.14). Finally the optical stereomicroscope analysis, carried out by a goldsmith, highlighted the presence of cutting signs typical of modern tools, definitively excluding the assignment of the necklace to Roman age.

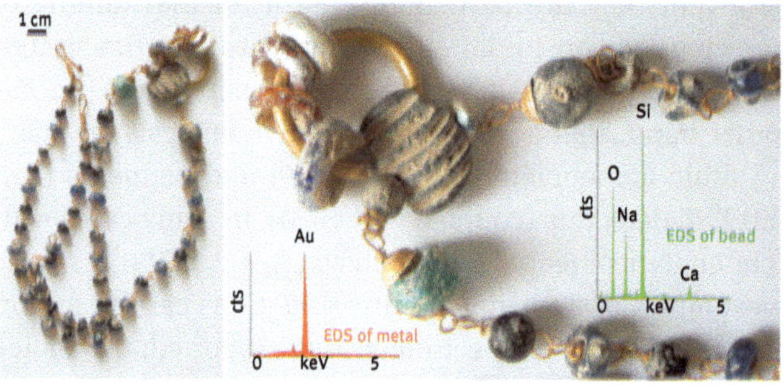

EDS analyses of one bead

	1	2	3	mean	st.dev.
SiO_2	70.55	70.50	70.65	70.65	0.08
Al_2O_3	1.76	1.81	1.78	1.78	0.03
FeO	0.30	0.24	0.43	0.32	0.10
MgO	0.37	0.34	0.37	0.36	0.01
CaO	7.76	7.84	7.77	7.79	0.05
Na_2O	18.05	17.99	17.72	17.92	0.17
K_2O	0.48	0.47	0.47	0.48	0.00
PbO	0.73	0.82	0.81	0.78	0.05

Figure 2.14: Necklace with glass beads and gold chain; in green the EDS spectrum of one bead and in orange that of the gold, which turns out to be pure; on the right the ED analysis (values in wt%) of one of the glass bead, repeated three times to calculate mean and standard deviation values.

2.5.3 Coins

The study of the X-ray maps on coins, is of great value not only to the historians and archeologists but also to numismatists, economists, and conservators because their chemical composition may offer valuable information regarding their manufacturing technology, age, minting places and authenticity.

It is necessary to analyze the artifacts using non-destructive techniques like ED spectrometry that allows to obtain X-ray maps of the elemental distribution. The case study here reported concerns the characterization of a "sestino" coin, dated back to a period between the second half of the fifteenth and sixteenth centuries, of the fortified settlement of Rocca Montis Dragonis located in Campania region (Italy). Analysis performed with an ED silicon drift detector (Figure 2.15), highlighted the presence of copper as the main element. The presence, on the surface, of calcium and silicium and of oxidation compounds is certainly linked to the interaction of the sample with the soil [35].

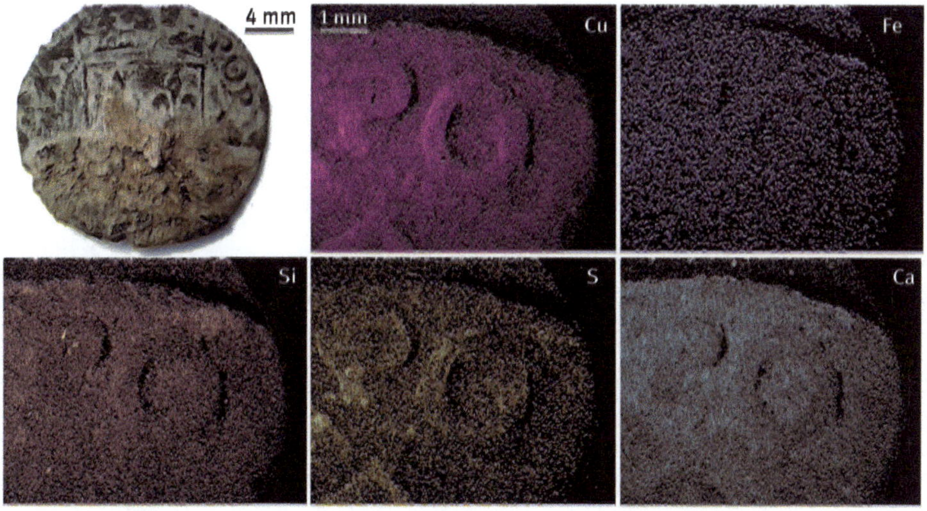

Figure 2.15: Macrophotograph of the "sestino" coin of the fortified settlement of Rocca Montis Dragonis (Campania, Italy); the relevant five EDS maps (CuKα, FeKα, SiKα, SKα and CaKα X-ray lines) refer to a particular of the coin, the letters PO [35].

2.5.4 Obsidians artifacts

In ancient times, obsidian rocks were essentially used for manufacturing cutting and ornamental tools. An accurate knowledge of obsidian artifacts sourcing area can provide important information on cultural, social and economic relationships between peoples as well as on trade routes in prehistoric times.

The obsidian artifacts provenance can be determined with minimally destructive or non-destructive techniques such as XRF energy dispersive spectrometers [2, 7], using peak intensity ratios of various elements [1, 36, 37], EPMA [38], scanning electron microscopy coupled with energy dispersive spectrometer microanalysis [23, 28], and laser ablation inductively coupled plasma-mass spectrometry [39]; in the last 30 years, our research group developed absolutely non-destructive analytical methods to solve sourcing recognition of obsidian artifacts.

Obsidians are rocks essentially constituted by glass with a few quantity of microphenocrysts. Non-destructive chemical analysis on the specimen can be made either on the whole sample or on the glass portion or microphenocrysts. Due to the non-regular surface of the obsidians, the use of the ED detectors is preferred [1, 22].

The very similar composition of obsidians makes difficult the distinction of the source areas on the basis of major elements WD-XRF determination on the whole rock (Figure 2.16); on the contrary, very useful indications can be obtained if trace elements are determined [28].

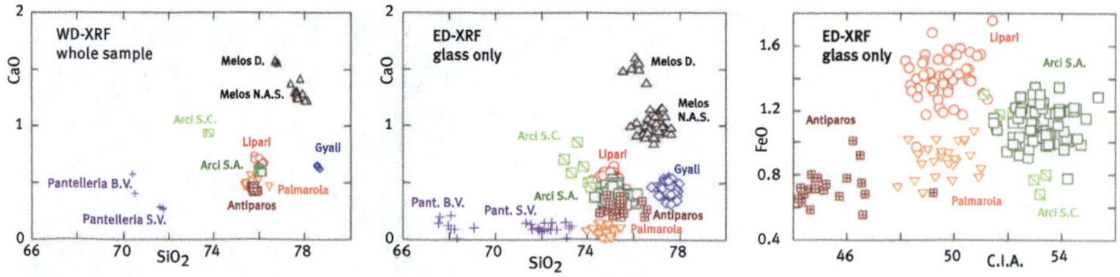

Figure 2.16: WD-XRF and ED-XRF analyses (wt.%) of obsidians from different geological obsidian sources of the Mediterranean basin. Sub-sources outcrops: Monte Arci, S.C. = Perdas Urias; S.A. = Conca Cannas, Canale Perdera and Riu Solacera. Pantelleria, B.V. = Bagno di Venere, S.V. = Saltò la Vecchia and Balata dei Turchi. Melos, D. = Demenegaki; N.A.S. = Nihia, Adamas and Sarakiniko. C.I.A. = Chemical Index of alteration [$Al_2O_3/(Al_2O_3 + CaO + Na_2O + K_2O)$ in mol.%]. ([28, 40], modified).

Alternatively, the composition of the glass can be determined with ED spectrometers coupled with scanning electron microscope [28] or equipped with polycapillary lens that focus the X-ray beam on a very small area (about 10 µm) of the sample. In some cases it is possible to identify not only the obsidian source areas but also their sub-sources by analyzing the microphenocrysts present in the glass [29]. To this purpose, the acquisition of X-ray maps of the surface of the sample can help, in a few minutes, to highlight presence and type of phenocrysts (Figure 2.17).

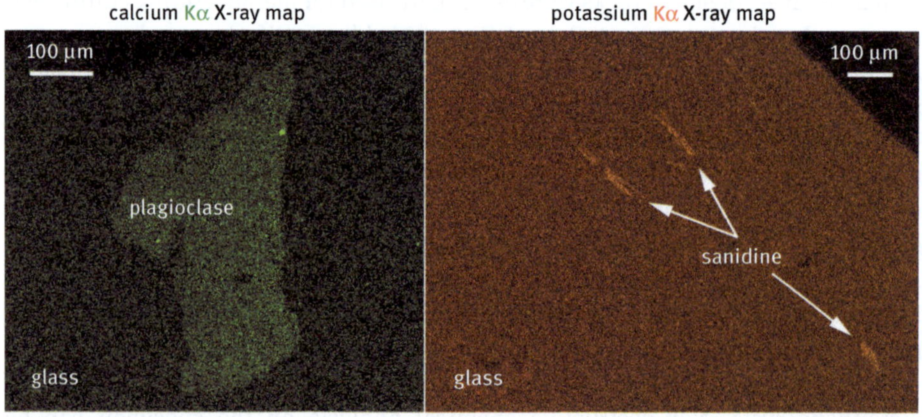

Figure 2.17: Feldspar maps present in the glass of Antiparos obsidians; the X-ray maps allow to recognize the presence and the type of microphenocrysts present in the obsidian glass ([40], modified).

Lastly, the obsidian can be sourced by analyzing the whole sample (glass and microphenocrysts) by XRF in an absolutely non-destructive way, also using the peak intensity ratios of some trace elements [1, 36, 37]. This XRF analytical procedure is particularly useful when the surface of the artifact is slightly altered or covered by very thin carbonate film formed during burial [1, 29]: the obsidian is placed in a sample holder closed at the bottom by a very thin Mylar polyester film (Figure 2.13). The Kα lines of trace elements, such as Rb, Sr, Y, Zr and Nb (Figure 2.9) can be detected, which are practically not absorbed by the Mylar or the incrustant carbonate films thanks to their high frequency [1]. When the peak intensity ratios of some trace elements is used as sourcing technique (Figure 2.18) a special attention must be paid to a very accurate measurement of the net intensities of trace elements: they must be perfectly polished from the contribution of the background and from interelemental interferences [1].

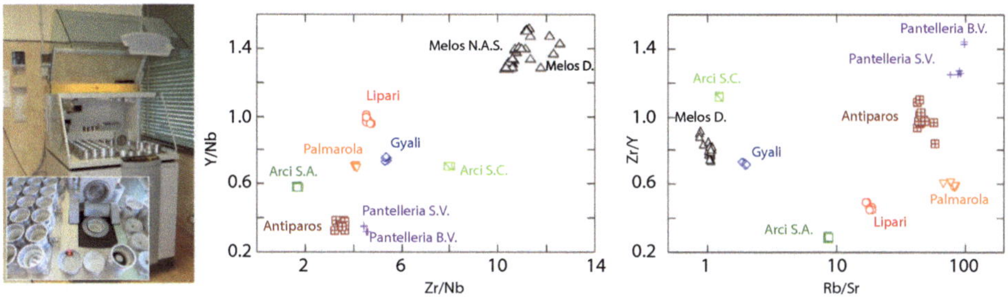

Figure 2.18: WD-XRF spectrometer (in the inset the sample holders) of the Dipartimento di Scienze della Terra e Geoambientali (University of Bari, Italy) used to measure the X-ray intensity ratios of some contiguous trace elements (the two plots of the figure) of geological samples from various obsidian sources and sub-sources of the Mediterranean basin; same symbols as in Figure 2.16 ([1, 40], modified).

2.5.5 Other lithic artifacts

The data here presented concern artifacts acquired during an archological research campaign held in the Prehistoric site of Pertosa Auletta caves (Salerno, Southern Italy) in 2013. Among the numerous archeological finds, four querns realized with volcanic rocks were chosen: a minimally invasive technique was used to pick-up a representative quantity of the artifacts, coring them with a microdriller with the edge enriched in diamond powder (Figure 2.12).

A representative fragment of the drilled specimen was finely grinded and the powder was treated with elvacite (2,5 g of powder mixed with 1 cc of a 20 % elvacite solution in pure acetone) before to be placed in an aluminum cup with the bottom filled with boric acid. The powdered pellets, obtained compacting the aluminum cups with a hydraulic press, were analyzed by a WD-XRF stectrometer [41]. The chemical analyses, major and trace elements (Figure 2.19) indicate two different provenance for the four volcanic querns: two of them are basanites and related to the volcanic activity of the Monte Vulture (Potenza, Southern Italy) and the other

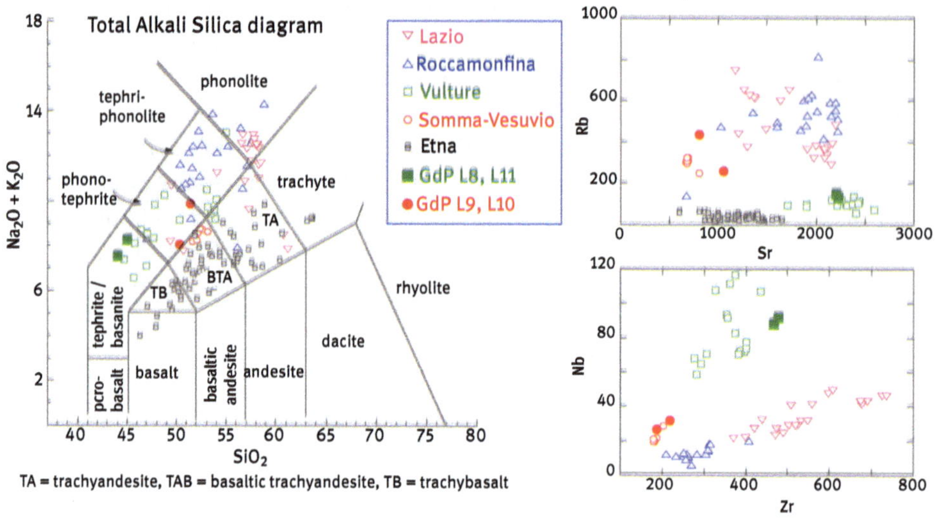

Figure 2.19: On the left the plot of major elements (wt%) in the total alkali silica diagram and on the right two plots of trace elements (ppm) determined on volcanic artifacts (four querns) found in the Prehistoric site of Pertosa Auletta caves (Salerno, Southern Italy). The four artifacts were classified as basanites from Vulture, Potenza, Southern Italy (samples GdP L8 and GdP L11) and as phonotephrites from the Somma-Vesuvius, Napoli, Southern Italy (samples GdP L9 and GdP L10); reference data of volcanoes of Central and Southern Italy are also reported ([41], modified).

two samples are phonotephrites and related to the volcanic activity of the Somma-Vesuvius (Napoli, Southern Italy).

2.6 Conclusions

XRF technique, as energy dispersive or as wavelength dispersive spectrometry, is a very versatile analytical tool, able to provide chemical information, both qualitative and quantitative, on very different types of materials. It finds large application in the characterization of works of art: in the chapter, strength and weakness points are highlighted.

ED XRF analysis requires no sample preparation, is non-destructive and can be easily performed also with very low-cost and portable equipments, particularly useful for on-site analyses.

WD XRF technique can imply the destruction of part of the sample, however granting highly reproducible and reliable analyses of major and trace elements; when used to characterize the natural surface of an artifact, important archeometric issues can be addressed, such as provenance attributions, just relying on peak intensity ratios measurements of some trace elements.

Case studies are described which stress pros and cons of the technique in the investigations of artifacts and materials of interest in Cultural Heritage.

Acknowledgements: Thanks are due to customer service support of Panalytical, particularly to Domenico Vigilante, and to Mauro Pallara that during last 30 years take care of the XRF spectrometers of the Dipartimento di Scienze della Terra e Geoambientali (University of Bari, Italy).

References

[1] Acquafredda P, Muntoni IM, Pallara M. Reassessment of WD-XRF method for obsidian provenance shareable databases. Quat Int. 2018;468:169–78.

[2] Phillips SC, Speakman RJ. Initial source evaluation of archaeological obsidian from the Kuril Islands of the Russian Far East using portable XRF. J Archaeol Sci. 2009;36:1256–63.

[3] Nazaroff AJ, Prufer KM, Drake BJ. Assessing the applicability of portable X-ray fluorescence spectrometry for obsidian provenance research in the Maya lowlands. J Archaeol Sci. 2010;37:885–95.

[4] Shackley. MS. Portable X-ray fluorescence spectrometry (pXRF): the good, the bad, and the ugly. Archaeol Southwest Mag. 2012;26. https://www.archaeologysouthwest.org/pdf/pXRF_essay_shackley.pdf.

[5] Forster N, Grave P. Non-destructive PXRF analysis of museum-curated obsidian from the Near East. J Archaeol Sci. 2012;39:728–36.

[6] Millhauser JK, Rodríguez-Alegría E, Glascock MD. Testing the accuracy of portable X-ray fluorescence to study Aztec and Colonial obsidian supply at Xaltocan. Mex J Archaeol Sci. 2011;38:3141–52.

[7] Frahm E, Doonan RC, Kilikoglou V. Handheld portable X-ray fluorescence of Aegean obsidians. Archaeometry. 2014;56:228–60.

[8] Frahm E, Schmidt B, Gasparyan B, Yeritsyan B, Karapetian S, Meliksetian K, et al. Ten seconds in the field: rapid Armenian obsidian sourcing with portable XRF to inform excavations and surveys. J Archaeol Sci. 2014;41:333–48.

[9] Conrey RM, Goodman-Elgar M, Bettencourt N, Seyfarth A, Van Hoose A, Wolff JA. Calibration of a portable X-ray fluorescence spectrometer in the analysis of archaeological samples using influence coefficients. Geochemistry: Exploration, Environment, Analysis. http://dx.doi.org/10.1144/geochem2013-198, 2014.

[10] Milić M. PXRF characterisation of obsidian from central Anatolia, The Aegean and Central Europe. J Archaeol Sci. 2014;41:285–96.

[11] Frahm E. Validity of "off-the-shelf" handheld portable XRF for sourcing Near Eastern obsidian chip debris. J Archaeol Sci. 2013;40:1080–92.

[12] Acquafredda P, Laviano R. Fluorescenza di raggi X: caratterizzazione chimica di materiali mediante spettrometria di raggi X. In: Mercurio M, Langella A, Di Maggio RM, Cappelletti P, editors. Analisi Mineralogiche in ambito forense. Canterano, RM: Aracne editrice, 2019:83–131. ISBN 978–88–255-2235-8.

[13] Bertin EP. Principles and practice of X-ray spectrometric analysis. New York: Plenum Press, 1975.

[14] Beckhoff B, Langhoff N, Kanngiesser B, Wedell R, Wolff H. Handbook of practical X-ray fluorescence analysis. Berlin: Springer, 2006.

[15] Bearden JA. X-ray wavelengths. Rev Modem Phys. 1967;39:78–124.

[16] Bearden JA, Burr F. Reevaluation of X ray atomic energy levels. Rev Modem Phys. 1967;39:125–42.

[17] Hubbell JH, Seltzer SM. Tables of X-ray mass attenuation coefficients and mass energy-absorption coefficients. National Institute of Standards and Technology, Gaithersburg (USA), 1995. ark:/13960/t3mw7z463.

[18] Shrestha S, Fischer R, Matt GJ, Feldner P, Michel T, Osvet A, et al. High-performance direct conversion X-ray detectors based on sintered hybrid lead triiodide perovskite wafers. Nature Photonics. 2017 6 19;11:436–40. DOI: 10.1038/nphoton.2017.94

[19] Leoni L, Saitta M. X-ray fluorescence analysis of 29 trace elements in rock and mineral standards. Rendiconti della Società Italiana di Mineralogia e Petrologia. 1976;32:497–510.

[20] Fabbi BP. A refined fusion x-ray fluorescence technique, and determination of major and minor elements in silicate standards. Am Mineral. 1972;57:237–45.

[21] Leoni L, Menichetti M, Saitta M. Ricalibrazione di una metodologia in fluorescenza-X per l'analisi di minerali e rocce su campioni di polvere. Atti Società Toscana di Scienze Naturali, Memorie. 2004;109:13–20.

[22] Ruste J. X-ray spectrometry. In: Maurice F, Meny L, Tixier R, editors. Microanalysis and scanning electron microscopy. Summer school, 11–16 Sep 1978. St-Martin d'Hères, France: Les Editions de Physique, Orsay, 1979:215–67.

[23] Acquafredda P, Paglionico A. SEM-EDS microanalyses of microphenocrysts of Mediterranean obsidians: a preliminary approach to source discrimination. Eur J Mineral. 2004;16:419–29.

[24] Potts PJ. A handbook of silicate rock analysis. New York: Chapman and Hall, 1987.

[25] Chataigner C, Gratuze B. New data on the exploitation of obsidian in the southern Caucasus (Armenia, Georgia) and eastern Turkey, Part 1: source characterization. Archaeometry. 2014;56:25–47.

[26] Chataigner C, Gratuze B. New data on the exploitation of obsidian in the southern Caucasus (Armenia, Georgia) and eastern Turkey, Part 2: obsidian procurement from the Upper Palaeolithic to the late Bronze age. Archaeometry. 2014;56:48–69.

[27] Frahm E. Characterizing obsidian sources with portable XRF: accuracy, reproducibility, and field relationships in a case study from Armenia. J Archaeol Sci. 2014;49:105–25.

[28] Acquafredda P, Andriani T, Lorenzoni S, Zanettin E. Chemical characterization of obsidians from different Mediterranean sources by non-destructive SEM-EDS analytical method. J Archaeol Sci. 1999;26:315–25.

[29] Acquafredda P, Muntoni IM. Obsidian from Pulo di Molfetta (Bari, Southern Italy): provenance from Lipari and first recognition of a Neolithic sample from Monte Arci (Sardinia). J Archaeol Sci. 2008;955:947–55.

[30] Caggiani MC, Ditaranto N, Guascito MR, Acquafredda P, Laviano R, Giannossa LC, et al. Combined analysis of enamelled and gilded glassware from Frederick II Castle at Melfi (Italy) to identify technology and raw materials. X-ray Spectrom. 2014;44:191–200.

[31] Giannossa LC, Caggiani MC, Laviano R, Acquafredda P, Rotili M, Mangone A. Synergic analytical strategy to follow the technological evolution of campanian medieval glazed pottery. Archaeol Anthropological Sci. 2017;9:1137–51.

[32] Eramo G, Laviano R, Muntoni IM, Volpe G. Late roman cooking pottery from Tavoliere area (Southern Italy): raw materials and technological aspects. J Cult Heritage. 2004;5:157–65.

[33] Dell'Aquila C, Laviano R, Vurro F. Chemical and mineralogical investigations of majolicas (16th–nineteenth centuries) from Laterza, southern Italy. In: Geomaterials in cultural heritage Maggetti M, Messiga B, editors. Geological Society, vol. 257. London: Special Publication, 2006:151–62. ISBN 1-86239-195-5.

[34] Aprile A, Castellano G, Eramo G. Classification of mineral inclusions in ancient ceramics: comparing different modal analysis strategies. Archaeological and Anthropological Sciences, 2018. DOI: 10.1007/s12520-018-0690-y

[35] Gargiulo B, Sogliani F, Scrano L, Acquafredda P. Application of image analysis in archeology: the case of the copper coins recovered on Montis Dragonis fortress. 41st International Symposium on Archaeometry (ISA) 15–21 May 2016, abstracts. Kalamata, Greece, 2016:302.

[36] Nelson DE, D'auria JM, Bennett RB. Characterization of Pacific Northwest coast obsidian by X-ray fluorescence analysis. Archaeometry. 1975;17:85–97.

[37] De Francesco AM, Crisci GM, Bocci M. Non-destructive analytic method using XRF for determination of provenance of archaeological obsidians from the Mediterranean area: a comparison with traditional XRF methods. Archaeometry. 2008;50:337–50.

[38] Tykot RH. Chemical fingerprinting and source tracing of obsidian: the central Mediterranean trade in black gold. Acc Chem Res. 2002;35:618–27.

[39] Gratuze B, Blet-Lemarquand M, Barrandon JN. Mass spectrometry with laser sampling: a new tool to characterize archaeological materials. J Radioanal Nucl Chem. 2001;26:869–81.

[40] Acquafredda P, Micheletti F, Muntoni IM, Pallara M, Tykot RH. Petroarchaeometric data on Antiparos obsidians (Greece) for provenance study by SEM-EDS and XRF. Open Archaeol. 2019;5:18–30.

[41] Acquafredda P, Larocca F, Pallara M. Manufatti litici dalle ricerche 2013. Aspetti tipologici e petroarcheometrici. In: Larocca F, editor. Tra pietra e acqua. Archeologia delle grotte di Pertosa-Auletta, studi e ricerche 2004-2016. Bari, Italy: Pubblicità & Stampa srl – Modugno, 2017:86–105. ISBN 978-88-908337-6-2.

Josefina Pérez-Arantegui and Francisco Laborda

3 Inorganic mass spectrometry

Abstract: Inorganic mass spectrometry has been used as a well-known analytical technique to determine elemental/isotopic composition of very diverse materials, based on the different mass-to-charge ratios of the ions produced in a specific source. In this case, two mass spectrometric techniques are explained and their analytical properties discussed: inductively coupled plasma mass spectrometry (ICP-MS) and thermal ionisation mass spectrometry (TIMS), since they are the most used in art and archaeological material studies. Both techniques combine advantageous analytical properties, like low detection limits, low interferences and high precision. The use of laser ablation as sample introduction system in ICP-MS allows to avoid sample preparation and to perform good spatial-resolution analysis. The development of new instruments, improving the mass separation and the detection of the ions, specially multicollection detectors, results in high-precision isotopic analysis. A summary of the important applications of these mass spectrometric techniques to the analysis of art and archaeological materials is also highlighted.

Keywords: mass spectrometry, ICP-MS, TIMS, laser ablation, trace elements, isotopic analysis

3.1 Introduction

At present, inorganic mass spectrometry (MS) [1] represents a group of well-established and well-known analytical techniques. The term can correspond to the definition by Koppenaal as "the mass spectrometric measurement of atomic (as opposed to molecular or polyatomic) ions, for the primary purpose of elemental and/or isotopic composition determinations" [2]. Then, the term of atomic MS is also used to refer these techniques.

Inorganic MS is based on the different physical properties of the atom nuclei. A chemical element can have a fixed number of protons (atomic number: Z), but a variable number of neutrons, and then different isotopes (several mass numbers, A, protons + neutrons). For instance, iron (Fe, Z = 26) has four isotopes, ^{54}Fe (A = 54), ^{56}Fe (A = 56), ^{57}Fe (A = 57) and ^{58}Fe (A = 58). Only 20 elements are mono-isotopic. Moreover, the natural abundance of each isotope for a chemical element is known, therefore, each chemical element is characterised by its isotopic pattern (e. g. ^{54}Fe (5.845%), ^{56}Fe (91.754%), ^{57}Fe (2.119%) and ^{58}Fe (0.282%)). If we obtain ions

This article has previously been published in the journal *Physical Sciences Reviews*. Please cite as: Pérez-Arantegui, J., Laborda, F. Inorganic Mass Spectrometry. *Physical Sciences Reviews* [Online] **2019**, 4. DOI: 10.1515/psr-2018-0003.

https://doi.org/10.1515/9783110457537-003

from these isotopes, we will be able to separate them by means of electric and magnetic fields, and their different measured masses (or mass-to-charge ratios, m/z, z = charge of ion) will allow to identify the chemical element.

All the analytical techniques included as inorganic MS share a similar working setup (see Figure 3.1), where the main difference is usually the ion source and, thus, the way to obtain atomic ions to be transported to the mass separation system (sometimes also called mass analyser). In the ion source, the sample is vaporised, atomised and ionised. Several systems are mostly used as ion sources for inorganic analysis, namely inductively coupled plasma (ICP) (or also laser ablation-ICP, LA-ICP), glow discharge (GD), secondary ion (SI) source, and thermal ionisation (TI) source. The gas-phase ions are then separated in the mass separator based on their mass-to-charge ratio (m/z). This separation is achieved through the use of electromagnetic fields because heavier ions with greater momentum will be deflected less than lighter ions (assuming equivalent ion charges).

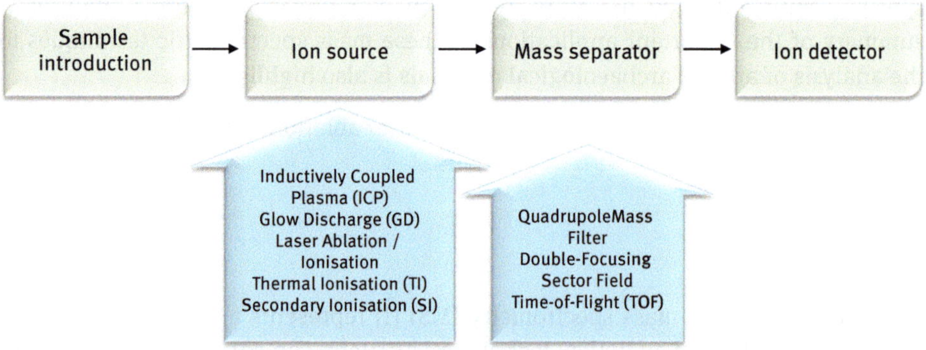

Figure 3.1: General working setup in inorganic mass spectrometry.

In the last decades, especially since the 1990s, the development of reliable instrumentation has represented the improvement in its analytical achievements and an outstanding growth in the use of inorganic MS, mainly because of the sensitivity of these spectrometric techniques and the possibility of determining isotopes. However, not all the analytical methods included in inorganic MS have been exploited to characterise cultural heritage materials. For instance, secondary ion mass spectrometry (SIMS) was first applied to studies in art and archaeology later in the 1980s, because it is a surface analysis method with very good spatial resolution, however it is not a very common technique in cultural heritage [3, 4]. In SIMS the chemical composition of solid surfaces and thin films is analysed by sputtering the specimen with a focused primary ion beam (0.1–20 keV) and collecting and detecting ejected secondary ions. SIMS is a "nano-destructive" technique for the simultaneous identification of inorganic and organic compounds, with particular interest in polyatomic ions. It has been applied to the analysis of metals and alloys, glass and glassy materials, human residues and pigments (in fact, in the characterisation of paintings, one of the major advantages of SIMS is its capability to link both inorganic and organic analysis [4–6]). Moreover, thanks to its excellent lateral resolution (micrometer), SIMS has become a reference technique for chemical imaging by MS [7].

This is why only some of these inorganic mass spectrometric techniques will be emphasised here. The chapter has been organised to provide the basic principles and instrumentation of the two most popular inorganic MS techniques in this field: inductively coupled plasma mass spectrometry (ICP-MS) and TIMS, with some examples of their analytical possibilities (like isotopic analysis) and applications in art and archaeology, and including a summary of sample preparation and introduction methods. Organic mass spectrometric techniques, like ESI-MS, MALDI-MS or chromatography coupled to MS, will be presented in other chapters of this textbook.

3.2 Inductively coupled plasma mass spectrometry

3.2.1 Principles and instrumentation

ICP-MS provides information for most elements in the periodic table (only noble gases, H, N, O, F and C are excluded), with very low detection limits. On a routine basis, ICP-MS is considered one of the most versatile atomic spectrometric techniques, providing rapid multielement analysis, from major to ultra-trace components, with detection limits down to the part-per-trillion level, for a large range of samples [1, 8, 9].

The ICP-MS spectrometer combines two analytical facilities to produce an instrument with powerful potential in the field of multielement trace analysis. The argon (Ar) ICP is used as a source of ions. Atoms are ionised in the plasma and then separated and detected in a mass spectrometer. The extraction of ions from the plasma source, which is at atmospheric pressure and 5,000–10,000 K, into the mass spectrometer involves an interface consisting of a series of orifices and chambers held at progressively lower pressures (down to 10^{-6} torr) and ambient temperature. The technique, therefore, combines the freedom from matrix interferences characteristic of the ICP with the very favourable signal-to-background ratios obtainable by mass spectrometry. As a mass spectrometry technique, ICP-MS is also capable of determining individual isotope ratios. There is, therefore, considerable potential in applying isotope dilution procedures for the quantitative determination of the selected elements.

A basic diagram of a typical ICP-MS instrument is represented in Figure 3.2. Although samples can be analysed in any aggregation state, analytes are commonly introduced in solutions or suspensions using a nebulisation system. It consists of a nebuliser and a spray chamber, which produces an aerosol of small droplets. Once the droplets, containing the analyte, pass into the plasma, solvent evaporates,

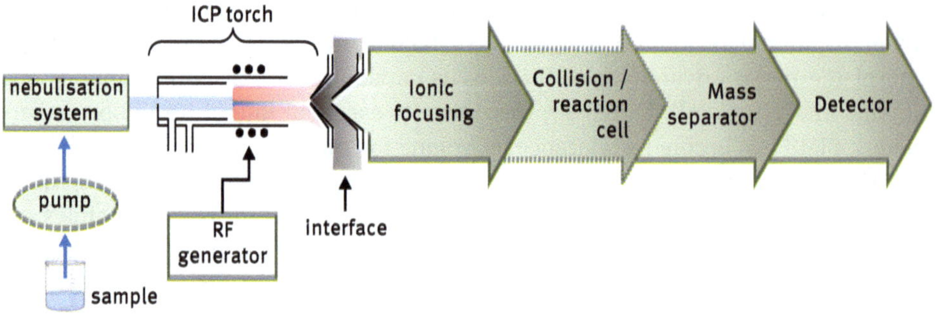

Figure 3.2: Basic diagram of a typical ICP-MS instrument (collision/reaction cell can be an optional facility).

forming solid particles, which in turn are vaporised and their elements atomised and consequently ionised.

The ICP ion source consists of the argon plasma itself, comprising torch, induction or radio-frequency load coil and radio-frequency generator. The plasma torch is configured horizontally, so the plasma is directed along the horizontal axis at mass spectrometer interface. As an ion source, samples are very efficiently ionised in an ICP. Moreover, the ion spectrum is buffered by the first ionisation energy of argon (15.76 eV). Therefore, singly-charged ions of almost all the elements are effectively excited by an argon plasma. Few elements have second ionisation energies, which fall below 15.76 eV, and consequently interferences in the mass spectrum due to doubly-charged ions will normally be negligible. The third ionisation energy is completely out of the range of excitation by an argon plasma and accordingly, triple-charged ions will be absent from mass spectra.

Analyte ions are confined in the central channel of the plasma, being sampled along with the plasma gases through the apertures of the sampling and skimmer cones (interface unit). The design of this interface is critical to instrument performance. Mass spectrometers require a very high vacuum for successful operation (usually better than about 10^{-6} Torr). The sampling interface can, therefore, be considered to be a controlled leak, permitting plasma gases to enter the mass spectrometer for analysis. The design of this interface must take into account the interactions that occur when plasma gases at a pressure of one atmosphere and a temperature of 6,000–7,500 K expand through an orifice machined in a cool metal plate into a vacuum beyond. Now, most of the instruments operate in a continuum sampling mode, with a configuration of sampler cone, expansion chamber and skimmer cone. The plasma gas generated in the ICP source, collimated by sampling the skimmer cone, then passes through a number of plates that focus the gas-phase species in a form suitable for transmission through the mass filter.

Positive ions are focalised and separated from neutral species by an ion optic system, consisting of one or more electrostatically controlled lens components. Next, ions are introduced into the mass separator, where they are separated according to their mass-to-charge ratio (m/z). Quadrupoles, double-focusing sector field or time-of-flight (TOF) separators are used in commercial ICP-MS instruments. Quadrupole instruments are available with one or more additional multipoles (quadrupoles and/or octopoles), placed between the ion optics and the mass separator, which work in combination with a gas (He, O_2, H_2, CH_4, NH_3) as collision or reaction cells, reducing or eliminating the occurrence of spectral interferences [10]. Finally, ion detection is performed using electron multipliers or Faraday cups [11]. Electron multipliers working in pulse counting mode are used for ultimate sensitivity. The output count pulses due to detected ions are stored in a multichannel analyser. Sometimes, when the simultaneous determination of accurate isotope ratios measurement is required, a multiple ion collector (MC) device is used as detection system; usually, a system of several Faraday cups and/or electron multipliers is applied.

The mass separator and detector system and the associated data-collection electronics permit rapid scanning of selected mass ranges between 0 and 300 amu (atomic mass unit). The entire mass range may be scanned by the detector in less than one second, working in a conventional instrument (or in some tens of microseconds with a TOF mass separator). Several sweeps are accumulated in the different channels and they give rise to a typical channel-by-channel mass spectrum. Standard measurements in ICP-MS provide a mass spectrum whereby the intensity of the ion peaks at the corresponding m/z is plotted.

3.2.2 Some analytical features of ICP-MS

In an ICP-MS spectrum, the occurrence of the ion peaks along with the knowledge of the isotopic pattern of the elements allow to get qualitative information about the elemental composition, whereas elemental quantitative information is obtained from the ion peak intensity of the isotope/s selected for each element. Quantification involves the calibration with dissolved element standards or the use of isotope dilution.

The argon plasma source in ICP-MS provides mass spectra with two main features: (i) its high temperature facilitates the volatilisation and dissociation of the sample into its constituent atoms, and then a large proportion of singly-charged atomic ions is produced; (ii) the resulting mass spectra are very simple (compared with those observed in some organic mass spectrometric techniques), although polyatomic ions formed from the plasma gas and/or the sample matrix can be present, particularly below 80 m/z. Both characteristics results in low detection limits and low-interference analytical methods. The spectral interferences, where they occur, can be predicted with confidence. There are, in fact, only 211 mass lines in total contributed by all the elements in the periodic table. Furthermore, all elements, except indium, have at least one line free of overlap if background peaks and double-charged interferences are ignored.

Background interferences can arise from five sources: (1) the argon plasma gas (Ar^+, Ar_2^+), (2) the solvent in which the analysed species were dissolved (H_2O^+, H_3O^+, OH^+, ArH^+), (3) air entrained in the plasma and impurity gases in the argon itself (N^+, O_2^+, NO_2^+), (4) material eroded off the aperture/skimmer assembly (copper, nickel and molybdenum), or (5) reagent contamination. Collision and reaction cells (with a variety of reaction gases) have been used in ICP-MS to promote reactive and non-reactive collisions, with resulting benefits in interference reduction, isobar separation, and thermalisation/focusing of ions in ICP-MS [10].

These characteristics, together with the uniform sensitivity towards elements with first ionisation energies below 10 eV, give this technique a potentially low detection limit for a wide range of elements. Solution detection limits are typically on the order of 1 $\mu g \cdot L^{-1}$ for light elements, and some $ng \cdot L^{-1}$ for heavy elements. In general, typical precision is on the order of 1–5% for most elements.

Another analytical advantage of ICP-MS in quantitative determinations is that response curve calibrations are linear over several orders of magnitude, so that the same sample solution may be used for measuring a range of analyte concentrations. Therefore, a wide range of elements can be determined by ICP-MS in a sample with no element pre-concentration or chemical separation.

3.2.3 Sample introduction in ICP-MS methods

In ICP-MS analysis, samples must first be taken up into solution and are then aspirated into the argon plasma, generally in a conventional manner by means of a pump. Therefore, the sample is previously dissolved by fusion methods, acid attack, or digested with microwave assistance, according to the sample matrix requirements. Sample amount uses to be between 100 mg to 1–2 g, depending on the analytes of interest and the sample type.

Another way of sample introduction in ICP-MS methods is by means of laser ablation (LA-ICP-MS), especially interesting in art and archaeological studies [12–14]. The laser ablation device, which can remove a tiny solid material from a spot much less than 1 mm in diameter, is able to generate a gas-phase sample that is directly send to the ICP source. LA-ICP-MS is becoming the method of choice for solid-sample analysis due to the advantage of direct solid sampling and its capability of providing micro-scale information. While the small amount of sample makes this technique minimally destructive, samples must nevertheless fit inside the ablation chamber.

However, although this method saves all the preparation time involved in dissolving a sample, results are not highly precise, and detection limits are "only" in the parts-per-million (ppm) range, even if the precision could be improved for homogeneous materials, like glass. Therefore instrumental developments have been focused on improving the laser ablation regime, on increasing plasma stability, and especially on developing multiple ion detection systems. With the introduction of the multiple ion collector (MC) instruments, LA-ICP-MS has become an established method for high-precision isotopic analysis with a performance comparable to MC-TIMS, and the possibility of additional multi-element analysis at trace and ultra-trace level.

LA-ICP-MS uses the ablation of solid material by means of a focused laser beam in an inert gas (helium, He, or argon, Ar) atmosphere under normal pressure in a closed chamber (ablation cell), and transferring ablated material into the ICP ion source of the mass spectrometer. Different commercial laser ablation systems have been coupled to the ICP source: mostly solid state Nd-YAG laser (working at 266 nm or 213 nm), excimer laser (argon-fluorine operating at 193 nm and the fluorine at 157 nm), or solid state 193-nm laser. Also since about 2000, femtosecond lasers have been used as an alternative to ultraviolet nanosecond ones for LA-ICP-MS, because their ultrashort pulse duration provides insufficient time to heat and induce melting at the ablation site.

Other instrumental sections that also play an important role in the final analytical performance of the LA-ICP-MS are the ablation cell design and the way that the aerosol is transferred to the plasma torch (the transfer line). Furthermore, two main issues related to the reliability of the LA-ICP-MS quantitative results are the representativeness of the ions formed in the plasma (regarding the elemental composition of the sample) and the calibration strategy, because solid calibration standards (especially with a chemical composition similar to heritage materials) are seldom available.

Finally, LA-ICP-MS combines the highest-sensitivity detector available for inorganic analysis with a direct solid introduction system capable of high spatial resolution. It offers good spatial resolution, laterally and in-depth (typically tens of microns in both cases). Today it is possible to work with a laser spot of ~ 20 microns. Improving lateral resolution at the nanometre range would be possible by the application of the near-field effect in laser ablation.

Nevertheless, there still are two limitations for a broad use of LA-ICP-MS in ancient material analysis: the ablation-cell dimensions (even if current commercial cell are up to 15 × 15 cm) and the possibility to become portable instruments, although different new alternative approaches are going to be developed [15].

3.3 Thermal ionisation mass spectrometry

3.3.1 Principles and instrumentation

Another inorganic MS technique also applied in cultural heritage studies is thermal ionisation mass spectrometry (TIMS). It is one of the oldest mass spectrometric techniques, but less usual than ICP-MS. The thermal surface ionisation source is a quite simple ion source, especially useful for elements with relatively low ionisation energy (in general lower than 7 eV), although for elements with higher ionisation energy some special methods have been also developed to improve their ionisation [1, 16].

In TIMS, a small volume of sample solution is loaded in a thin metal (e. g. rhenium, Re) filament and evaporated to dryness, then, by increasing the temperature of the metal surface, the deposited atoms are evaporated and ionised. A basic arrangement of TIMS is shown in Figure 3.3. One or two filaments are used to evaporate and to ionise the sample. TIMS operates under high vacuum conditions, and the ions are then extracted by an ion optical system to the single-focusing mass spectrometer (magnetic sector mass separator), where ions are separated according to the mass-to-charge ratio (m/z). TIMS instruments can also work in multi-collection mode.

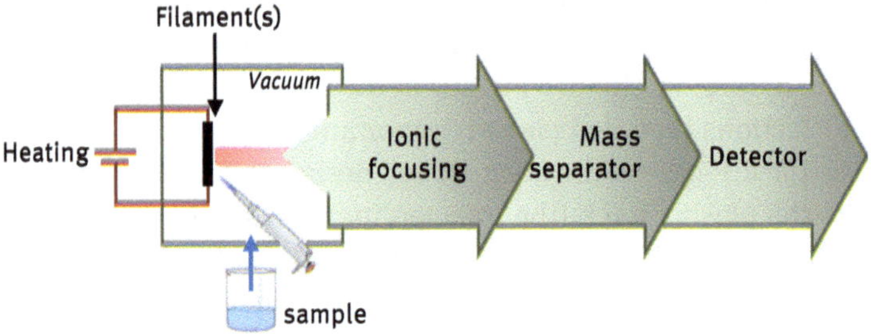

Figure 3.3: Basic arrangement of a TIMS instrument.

TIMS ion sources exploit the Langmuir effect to obtain ions: when an atom is absorbed on a metal surface, it will be ionised if the work function for electrons leaving the metal surface is larger than the first ionisation energy of the atom.

A main feature of TIMS, compared with ICP-MS, is the possibility of working with positively and also negatively charged ions in the mass spectrometer. Elements or species with a relatively-high electron affinity (>2 eV) can be identified by negative TIMS (NTIMS). Furthermore, in TIMS we are able to produce relatively stable polyatomic ions, also used in isotope ratio measurements in both forms, positively and negatively charged, and decreasing the occurrence of spectral interferences from impurities. One of the main analytical advantages of TIMS is the small amount of sample required to do the analysis: only some microlitres of sample solution containing some analyte nanograms o micrograms are enough. However, TIMS is especially recognised by the high precision of its isotopic ratio measurements, better than 0.01% working in a conventional instrument. The major TIMS analytical limitation is the required previous separation of the investigated elements following sequential extraction methods, and then the longer analysis times of the samples.

Like in other mass spectrometric techniques, in TIMS, qualitative information of the isotope composition is obtained from the ion peak positions in the mass spectrum, whereas elemental quantitative information is obtained from the ion peak intensity of the isotope/s selected for each element. TIMS analysis has actually become a routine method, especially in topics related to geochemistry, geochronology, the nuclear industry and environmental research.

3.3.2 Sample preparation in TIMS

Like in the case of routine ICP-MS analysis, samples to be analysed by TIMS should be in solution. Similar methods can be applied to dissolve solid samples and to obtain a proper solution, depending on the chemical nature of the sample. For the purpose of the isotope ratio measurements, some milligrams of sample are used in TIMS methods.

Due to spectral interferences and other mass bias in TIMS, most applications require complete dissolution of the sample followed by liquid chromatography to isolate elements of interest. For instance, in lead (Pb) isotope ratio measurements this element is extracted from the sample and purified using standard anion exchange procedures, e.g. ion exchange chemistry in HBr and HCl media can be used to purify the Pb to a level suitable for TIMS analysis. In the determination of copper (Cu) isotopes ($^{63}Cu/^{65}Cu$ ratio), a combined technique can be also used to purify the sample, e. g. its solution in HCl media can pass through an ion exchange column to remove major matrix components, and further purification can be effected by cathodic electrodeposition onto a platinum electrode in HNO_3 media.

Finally, direct loading onto the metal filament is the simplest form of sample introduction in TIMS. In this case, a small volume of sample (between 0.5 and 2 microlitres) is directly deposited onto a degassed filament with a micropipette and then dried. Other sample loading methods can be by electrodeposition, ion-exchange resin bead loading, or micro-porous ion emitter (PIE). For some analytes, filament additives (H_3PO_4, silica gel, H_3BO_3, graphite,. . .), also referred to as "ionisation enhancers", are also used because it is believed that this addition modifies the dominant mechanism of ionisation from the filament surface and it leads to greater levels of ionisation.

3.4 Isotope-ratio determinations by MS

Though the abundance of stable isotopes of a chemical element is generally known, in nature this is not constant since several sources of possible variations have been observed (isotope fractionation effects, isotope exchange reactions, geological and cosmogenic processes, weathering effects, etc.). Moreover, the radiogenic decay of unstable isotopes results in changes of natural isotope abundances. Therefore, the knowledge of isotope abundance data can be a key information to interpret some processes or origin of solid materials, and required in different research fields. In fact, variation of an isotopic ratio is usually presented as δ value of an isotope, expressed as *per mil* (‰), and calculated as the proportion between the isotopic ratio of two isotopes in the sample and in the standard.

Although accurate isotope analysis was firstly performed for the study of light elements, such as H, C, O, Li, B or S, isotope variations have also been described for other heavier elements, like Ca, Fe, Cu, Sr, Nd, Hf, Os, Pb, or U. Elements with one or more radiogenic isotopes (like Sr, Nd, Hf, and Pb) have been determined by TIMS for decades. H, C and O isotopes are usually determined by gas source mass spectrometric techniques. Nowadays, multiple ion collector (MC) mass spectrometers are the most precise and accurate instruments (such as SIMS, TIMS and ICP-MS) for isotopes measurements of all the other heavier elements [17].

In isotope-ratio determination, the use of the right reference materials and the availability of certified standards is crucial to obtain accurate analytical results. Nevertheless, in the last decades, the development of the MS instruments has experimented important improvements in sensitivity, detection limits, precision and accuracy (more efficient and powerful ionisation techniques, improved ion separation systems, and more sensitive ion detectors).

TIMS is a valuable analytical technique in routine isotopic analysis for high-precision isotope-ratio determinations, especially for elements with ionisation potentials lower than 7 eV (e. g. alkali, earth-alkali and rare-earth elements). The precision with MC-TIMS achieved up to 0.001%, making this technique ideal for fine isotope-ratio studies in age dating.

But TIMS is being replaced by MC-ICP-MS for isotope measurements of many elements, from lithium to uranium, because of the advantages of ICP-MS: more simple, less time-consuming sample preparation, higher sensitivity, similar precision using MC detectors, no restriction for atomic ionisations (even chemical elements with high first ionisation potential), possible multi-elemental determination in the same sample solution, etc. One of the main application of isotope-ratio measurements by MC-ICP-MS is geochronology, for instance, following the radioactive decay of ^{87}Rb, ^{238}U or ^{235}U, ^{232}Th, or ^{187}Re, and then the study of geological materials related to palaeontology and archaeology.

ICP-MS instruments offer not only the possibility of different mass separators and ion detectors (quadrupole mass filter, double-focusing sector field, multiple ion collector), but also diverse sample introduction systems, such as laser ablation. LA-ICP-MS allows direct determination of precise and accurate isotope ratios in solid samples [1, 13, 18]. The most important source of uncertainty in LA-ICP-MS is the appearance of different isobaric interferences of studied isotopes with singly or doubly charged atomic or molecular ions, however, it can be solved by the application of double-focusing sector field ICP-MS at the required mass resolution, or by the application of ICP-MS with collision/reaction cells. Another limitation in isotope-ratio determination by LA-ICP-MS is mass discrimination effects, but internal corrections can be applied. A third limiting problem is fractionation effects in laser ablation. Elemental fractionation does not play an important role for the determination of intra-element isotope ratios, but elemental fractionation has to be considered and minimised or avoided when inter-element isotope ratios (e. g. $^{206}Pb/^{238}U$, or $^{187}Re/^{188}Os$) are measured. As a result, ICP-MS is today the most frequently used analytical technique for isotopic analysis [19].

3.5 Applications of inorganic MS in art and archaeology

Because of the large range of topics, it is quite difficult to summarise all the potential inorganic MS applications in cultural heritage research, but it can be helpful to note some significant ones.

The first analytical approach using MS in art and archaeology is the possibility to know the multielemental composition of different materials, from major to ultra-trace components, as qualitative data as well as quantitative results. For this purpose, nowadays ICP-MS is the most useful and most utilised technique because of its analytical features. Ancient materials, such as obsidian, ceramics, metals, stones, glasses, animal and human remains. . ., have been analysed in order to investigate specially their origin. Provenance studies assume that the chemical composition (elements, isotopes or isotopic ratios) of the ancient material is similar and indicative of the chemical composition of the raw material from which it is made.

One of the earliest ICP-MS analysis of archaeological objects was that performed by Longerich et al. in 1987 [20], where 12 trace elements were determined in native silver artefacts using sample weights as small as 0.5 mg. Since the 1990s, rapid multi-element techniques, such as ICP-MS, became more commonly available and its capabilities were exploited for a diverse range of “fingerprinting” applications, paying special attention to minor and trace-element data.

The use of laser ablation for sample introduction in ICP-MS (LA-ICP-MS) has widely contributed to the investigation of diverse materials such as obsidian, glass, pottery, ceramics, human remains, written heritage, metal objects, pigments and other miscellaneous stone materials [14, 15]. LA-ICP-MS added the advantages of direct analysis of the solid sample (without preparation) and the small amount of ablated material (minimally destructive) to the other analytical abilities of ICP-MS. However, this technique has a significant limitation in dealing with inherent sample heterogeneity of ancient materials at the volume used for analysis. Therefore, for provenance investigations, LA-ICP-MS measurements have been focussed on more homogeneous materials, such as obsidian, porcelain, glazes and glass [21, 22]. Also, since the beginning, researchers have exploited LA-ICP-MS capability of high spatial resolution as direct solid introduction system, and then it has been used for spatially resolved trace analysis of archaeological materials, such as metals, usually heterogeneous [23, 24]. The presence or absence of some minor or trace elements is also the signature of some materials and then LA-ICP-MS analysis can be used to verify the source or to detect art forgeries [25].

No doubt, the most important application of inorganic MS in art and archaeology is the opportunity of isotopic analysis. As noted before, one of the main application of isotope-ratio measurements is geochronology, and then the dating of geological materials related to palaeontology and archaeology. Isotope ratios have been mainly used to trace the provenance of some ancient materials (specially metals and glass [26]). Also the analysis of the isotopic compositional variation of radiogenic elements, like $^{206}Pb/^{204}Pb$ and $^{87}Sr/^{86}Sr$, in teeth, bones and soils is an important topic for the study of the origin and migration of populations. The high precision of TIMS results contributed to the wide use of this method for isotope ratio determination (specially for the study of Pb isotope ratios from different ancient ore sources [27, 28]). However, since the beginning of the ICP-MS for isotopic analysis, this technique became also a competitor (e.g. to measure lead isotope composition in archaeological skeletal remains, soils and cosmetic pigments [29]), and especially since the introduction of the high-precision multi-collector detector (MC-ICP-MS) [30]. The two classical elements to determine their isotope ratio in art and archaeology have been lead and strontium, but others like neodymium, bore, tin and antimony have been also prospected [15]. The application of natural variations of the isotopic composition of copper to archaeological materials has been explored too [31].

The ability for high spatial resolution analysis of LA-ICP-MS has been exploited to map, not only trace elements, but also some isotopes in ancient samples where the distribution of these isotopes can be of interest, like U and Th in teeth or bones [32]. Combining the LA-ICP-MS lateral-resolution property with its analytical sensitivity for some isotopes, using minute samples, this technique has been even applied to investigate the potential of Pb isotope abundance ratios of white lead pigments in authentication and origin assignment studies of painting materials [33].

3.6 Conclusions

Inorganic MS has demonstrated to be an important analytical tool to determine the elemental composition of ancient materials, and specially to develop isotopic analysis. The use of ICP-MS and TIMS methods, and in particular the introduction of LA-ICP-MS and MC-ICP-MS instruments, combine the analytical advantages of low detection limits, low interferences and high precision, as well as the possibility of a good spatial resolution analysis. Inorganic MS methods have been applied to the analysis of metals, glass, glazes, ceramics, obsidian, soils, bones, teeth, and pigments, for the study of art and archaeological materials. The investigation on provenance, mineralogical origin, chronology, diet, or authenticity are among the most usual aims of these studies.

References

[1] Becker JS. Inorganic mass spectrometry: principles and applications. Chichester: Wiley, 2007.

[2] Koppenaal DW. Atomic mass spectrometry. Anal Chem. 1988;60:113R–31R.

[3] Spoto G. Secondary ion mass spectrometry in art and archaeology. Thermochim Acta. 2000;365:157–66.

[4] Dowsett M, Adriaens A. The role of SIMS in cultural heritage studies. Nucl Instruments Methods Phys Res B. 2004;226:38–52.

[5] Calvano CD, van der Werf ID, Palmisano F, Sabbatini L. Revealing the composition of organic materials in polychrome works of art: the role of mass spectrometry-based techniques. Anal Bioanal Chem. 2016;408:6957–81.
[6] Adriaens A. Non-destructive analysis and testing of museum objects: an overview of 5 years of research. Spectrochim Acta Part B. 2005;60:1503–16.
[7] Spoto G, Grasso G. Spatially resolved mass spectrometry in the study of art and archaeological objects. Trends Anal Chem. 2011;30:856–63.
[8] Thomas R. Practical guide to ICP-MS: a tutorial for beginners. Boca Raton: CRC Press, 2013.
[9] Montaser A. Inductively coupled plasma mass spectrometry. Chichester: Wiley, 1998.
[10] Koppenaal DW, Eiden GC, Barinaga CJ. Collision and reaction cells in atomic mass spectrometry: development, status, and applications. J Anal At Spectrom. 2004;19:561–70.
[11] Koppenaal DW, Barinaga CJ, Bonner Denton M, Sperline RP, Hieftje GM, Schilling GD, et al. MS detectors. Anal Chem. 2005;77:418A–27A.
[12] Sylvester PJ, Jackson SE. A brief history of laser ablation inductively coupled plasma mass spectrometry (LA–ICP–MS). Elements. 2016;12:307–10.
[13] Pickhardt C, Dietze H-J, Becker JS. Laser ablation inductively coupled plasma mass spectrometry for direct isotope ratio measurements on solid samples. Int J Mass Spectrom. 2005;242:273–80.
[14] Giussani B, Monticelli D, Rampazzi L. Role of laser ablation-inductively coupled plasma-mass spectrometry in cultural heritage research: a review. Anal Chim Acta. 2009;635:6–21.
[15] Degryse P, Vanhaecke F. Status and prospects for quasi-non-destructive analysis of ancient artefacts via LA-ICP-MS. Elements. 2016;12:341–6.
[16] Aggarwal SK. Thermal ionisation mass spectrometry (TIMS) in nuclear science and technology – a review. Anal Methods. 2016;8:942–57.
[17] Wieser ME, Schwieters JB. The development of multiple collector mass spectrometry for isotope ratio measurements. Int J Mass Spectrom. 2005;242:97–115.
[18] Woodhead JD, Horstwood MSA, Cottle JM. Advances in isotope ratio determination by LA-ICP-MS. Elements. 2016;12:317–22.
[19] Vanhaecke F, Degryse P, editors. Isotopic analysis: fundamentals and applications using ICP-MS. Weinheim: Wiley-VCH, Verlag GmbH & Co, 2012.
[20] Longerich HP, Fryer BJ, Strong DF. Trace analysis of natural alloys by inductively coupled plasma-mass spectrometry (ICP-MS): application to archeological native silver artifacts. Spectrochim Acta Part B. 1987;42:101–9.
[21] Neff H. Analysis of mesoamerican plumbate pottery surfaces by laser ablation-inductively coupled plasma-mass spectrometry (LA-ICP-MS). J Archaeol Sci. 2003;30:21–35.
[22] Pérez-Arantegui J, Resano M, García-Ruiz E, Vanhaecke F, Roldán C, Ferrero J, et al. Characterization of cobalt pigments found in traditional Valencian ceramics by means of laser ablation-inductively coupled plasma mass spectrometry and portable X-ray fluorescence spectrometry. Talanta. 2008;74:1271–80.
[23] Devos W, Senn-Luder M, Moor C, Salter C. Laser ablation inductively coupled plasma mass spectrometry (LA-ICP-MS) for spatially resolved trace analysis of early-medieval archaeological iron finds. Fresenius J Anal Chem. 2000;366:873–80.
[24] Wanner B, Moor C, Richner P, Bronnimann R, Magyar B. Laser ablation inductively coupled plasma mass spectrometry (LA-ICP-MS) for spatially resolved trace element determination of solids using an autofocus system. Spectrochim Acta Part B. 1999;54:289–98.
[25] Smith K, Horton K, Watling RJ, Scoullar N. Detecting art forgeries using LA-ICP-MS incorporating the in situ application of laser-based collection technology. Talanta. 2005;67:402–13.
[26] Degryse P, Lobo L, Shortland A, Vanhaecke F, Blomme A, Painter J, et al. Isotopic investigation into the raw materials of Late Bronze Age glass making. J Archaeol Sci. 2015;62:153–60.

[27] Stos-Gale ZA, Gale NH, Annetts N. Lead isotope analyses of ores from the Aegean. Archaeometry. 1996;38:381–90.

[28] Stos-Gale ZA, Gale NH, Houghton J, Speakman R. Lead isotope analyses of ores from the Western Mediterranean. Archaeometry. 1995;37:407–15.

[29] Ghazi AM. Lead in archaeological samples: an isotopic study by ICP-MS. Appl Geochem. 1994;9:627–36.

[30] Baker J, Stos S, Waight T. Lead isotope analysis of archaeological metals by multiple-collector inductively coupled plasma mass spectrometry. Archaeometry. 2006;48:45–56.

[31] Gale NH, Woodhead AP, Stos-Gale ZA, Walder A, Bowen I. Natural variations detected in the isotopic composition of copper: possible applications to archaeology and geochemistry. Int J Mass Spectrom. 1999;184:1–9.

[32] Duval M, Aubert M, Hellstrom J, Grün R. High resolution LA-ICP-MS mapping of U and Th isotopes in an early Pleistocene equid tooth from Fuente Nueva-3 (Orce, Andalusia, Spain). Quat Geochronol. 2011;6:458–67.

[33] Fortunato G, Ritter A, Fabian D. Old Masters' lead white pigments: investigations of paintings from the sixteenth to the seventeenth century using high precision lead isotope abundance ratios. Analyst. 2005;130:898–906.

Demetrios Anglos

4 Laser-induced breakdown spectroscopy in heritage science

Abstract: Laser-induced breakdown spectroscopy (LIBS) is a versatile analytical technique that can be used to probe the elemental composition of materials in diverse types of heritage samples, objects or monuments. The main physical principles underlying LIBS are presented along with analytical figures of merit and technical details concerning instrumentation. In practice, LIBS analysis does not require any sample preparation and the technique is nearly non-invasive, offering close to microscopic spatial resolution and the possibility for depth profile analysis. These features are, at present, available in a number of compact or transportable instruments that offer versatility and enable the use of LIBS for the analysis of a broad variety of objects/samples at diverse locations and this can be highly valuable at several stages of archaeological investigations and conservation campaigns. Representative examples are presented indicating how LIBS has been used to obtain compositional information for materials in the context of archaeological science, art history and conservation.

Keywords: LIBS, laser, elemental analysis, cultural heritage

4.1 Introduction

Elemental analysis methods have long been used as tools for the characterization and study of a diverse variety of heritage objects, monuments and works of art [1–3]. The identity and composition, qualitative and/or quantitative, of different types of materials such as pigments, metal alloys, stone, glass, pottery, soil or even biomaterials can be determined by known elemental analysis techniques including, x-ray fluorescence (XRF) spectrometry [4, 5], particle-induced x-ray emission (PIXE) [6], electron microscopy coupled to X-ray microanalysis [7], atomic absorption/emission spectrometry or inductively-coupled plasma mass spectrometry (ICP-MS) [8, 9]. However, with the exception of XRF [10, 11], and, in part, PIXE [12], the techniques mentioned are applied by use of laboratory instrumentation while in addition they require sampling and sample preparation, which can often involve elaborate chemical procedures.

In response to the need for an analytical tool that offers versatility in application, ease of data interpretation, speed of analysis and mobility, several researchers have

This article has previously been published in the journal *Physical Sciences Reviews*. Please cite as: Anglos, D. Laser-Induced Breakdown Spectroscopy in Heritage Science. *Physical Sciences Reviews* [Online] **2019**, 4. DOI: 10.1515/psr-2018-0005.

https://doi.org/10.1515/9783110457537-004

investigated the use of Laser-induced breakdown spectroscopy (LIBS) in the context of heritage science. LIBS is an atomic emission spectrochemical method that exploits the light given off by a micro-plasma, produced upon focusing of a laser pulse on the surface of a solid sample/object [13, 14]. Monitoring the characteristic spectral emission lines from atomic species in the plasma plume enables one to obtain information about the elemental composition of the material investigated. It is considered as an optical micro-probe technique and has been used in a wide variety of analytical applications for the qualitative, semi-quantitative and quantitative analysis of materials. It offers certain features which are quite important in relevance to requirements arising in heritage materials analysis [15–17]. As a method, LIBS is a straightforward, versatile and simple one and thus can be used even by non-specialized users. It is fast and provides results in real time, which is important when decisions for further action need to be made, for instance how to proceed with an excavation or how to deal with object conservation. More importantly, LIBS does not require sampling, so analysis can be performed directly on the object or monument and because of the tight beam focusing it ensures good spatial resolution with spot sizes nearly invisible to the naked eye. Additionally, all these features can be combined in compact portable instruments that enable field deployment of LIBS for the analysis of a broad variety of objects/samples at diverse locations [18–20].

4.2 Physical principles and analysis

The main principle underlying LIBS relies on the interaction of laser radiation, confined in space and time, with a solid surface which leads to formation of plasma [13]. Spatial confinement of the light energy is achieved by optical focusing of the beam down to a typical spot of diameter on the order of 0.1 mm while temporal confinement is ensured by use of a pulsed laser source, operating typically in the few nanoseconds range. The result of such a spatiotemporal confinement is a rapid deposition of a small amount of energy (typically 1–10 mJ) on the sample surface and this produces a power flux on the order of 1×10^9 Wcm^{-2} or greater, capable of initiating the formation of a micro-plasma that sustains itself for a few microseconds in ambient atmosphere. The process is collectively termed laser ablation and can be viewed as a sequence of steps, overlapping in time, that involve laser light absorption (energy deposition) within a minute volume of material, less than 0.1 mm^3, material breakdown and generation, just above the irradiated spot, of a plasma plume which subsequently expands and decays in a few microseconds depending on the amount of energy in the laser pulse. The ablated mass is in the range of a few ng (1 ng = 10^{-9} g).

The plasma consists of an ensemble of electrons, neutral and ionized atoms and small molecules and as such conveys compositional information that reflects the stoichiometry of the solid target ablated by the laser pulse. The plasma gas immediately upon its generation is quite hot due to the highly energetic collisions among electrons, ions and neutrals (temperature in the range of 15,000 K) but as it evolves in time, it cools and emits radiation in the UV, Visible and near IR arising from the relaxation of the excited neutrals and ions in the plume. As a result, optical recording of the wavelength-resolved plasma emission, preferably in a time-resolved manner, gives rise to characteristic atomic spectra that provide the analytical information in a LIBS measurement.

As seen in Figure 4.1, the spectrum collected from a red paint composed of red lead (Pb_3O_4), a pigment used extensively in painting, is dominated by distinct emission lines assigned to specific transitions of excited lead atoms, shown in a Grotrian diagram (Figure 4.2) with the corresponding energy state information listed in Table 4.1. On the basis of such characteristic spectral lines one obtains reliable information on the qualitative composition of the material analysed. The interpretation of the data is facilitated by use of spectral libraries, nowadays available on a PC or accessible over the internet [21]. It is noted that in most cases a single laser pulse is sufficient to produce a clean spectrum which can lead to quick determination of the qualitative composition of a sample. With this in mind and considering that the ablation depth per pulse is approximately on the order of 1 µm or less, it is then straightforward to realize that multiple pulses, namely successive LIBS measurements on the same spot, yield a depth profile analysis of the sample, very useful when complex stratigraphies are investigated.

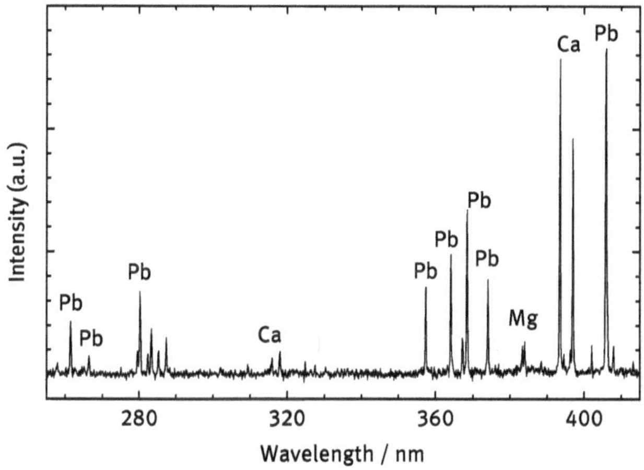

Figure 4.1: Typical emission spectrum obtained in LIBS analysis of red paint on stone sculpture. Characteristic lead (Pb) emission lines are labelled. Calcium (Ca) and magnesium (Mg) lines originate from the stone background.

The peak intensity or more properly the integrated emission of a spectral line is proportional to the number density of the corresponding elemental species in the plume and this, in turn, with the concentration of specific elements in the ablated

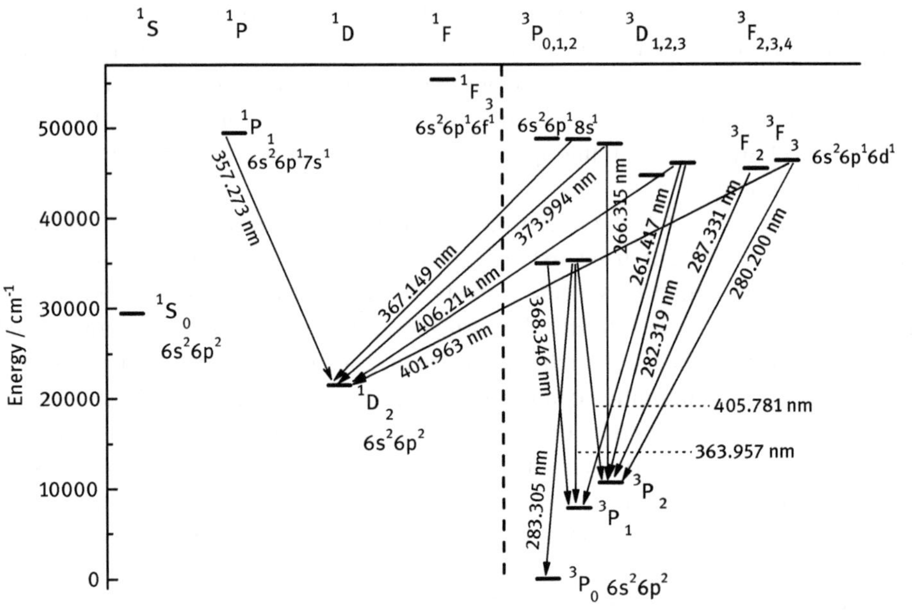

Figure 4.2: Energy diagram (Grotrian) showing some of the transitions observed experimentally (see Figure 4.1) from lead and indicating electron configuration and terms for the states involved in the transitions.

Table 4.1: Atomic emission spectral lines for neutral lead (Pb I)[a].

λ/nm	E_i/cm^{-1} [b] Configuration (Term)[c]	E_k/cm^{-1} [b] Configuration (Term)[c]
261.417	46,068.436 $6s^2\ 6p^1\ 6d^1$ ($^3D_2$)	7819.263 $6s^2\ 6p^2$ ($^3P_1$)
266.315	48,188.630 $6s^2\ 6p^1\ 7s^1$ ($^3P_2$)	10,650.327 $6s^2\ 6p^2$ ($^3P_2$)
280.200	46,328.667 $6s^2\ 6p^1\ 6d^1$ ($^3F_3$)	10,650.327 $6s^2\ 6p^2$ ($^3P_2$)
282.319	46,068.436 $6s^2\ 6p^1\ 6d^1$ ($^3D_2$)	10,650.327 $6s^2\ 6p^2$ ($^3P_2$)
283.305	35,287,224 $6s^2\ 6p^1\ 7s^1$ ($^3P_1$)	0 $6s^2\ 6p^2$ ($^3P_0$)
287.331	45,443.171 $6s^2\ 6p^1\ 6d^1$ ($^3F_2$)	10,650.327 $6s^2\ 6p^2$ ($^3P_2$)
357.273	49,439.616 $6s^2\ 6p^1\ 7s^1$ ($^1P_1$)	21,457.798 $6s^2\ 6p^2$ ($^1D_2$)
363.957	35,287.224 $6s^2\ 6p^1\ 7s^1$ ($^3P_1$)	7819.263 $6s^2\ 6p^2$ ($^3P_1$)
367.149	48,686,934 $6s^2\ 7p^1\ 8s^1$ ($^3P_1$)	21,457.798 $6s^2\ 6p^2$ ($^1D_2$)

(continued)

Table 4.1 (continued)

λ/nm	E_i/cm^{-1} [b] Configuration (Term)[c]	E_k/cm^{-1} [b] Configuration (Term)[c]
368.346	34,959.908 $6s^2\,6p^1\,7s^1\,(^3P_0)$	7819.263 $6s^2\,6p^2\,(^3P_1)$
373.994	48,188.630 $6s^2\,6p^1\,7s^1\,(^3P_2)$	21,457.798 $6s^2\,6p^2\,(^1D_2)$
401.963	46,328.667 $6s^2\,6p^1\,6d^1\,(^3F_3)$	21,457.798 $6s^2\,6p^2\,(^1D_2)$
405.781	35,287.224 $6s^2\,6p^1\,7s^1\,(^3P_1)$	10,650.327 $6s^2\,6p^2\,(^3P_2)$
406.214	46,068.436 $6s^2\,6p^1\,6d^1\,(^3D_2)$	21,457.798 $6s^2\,6p^2\,(^1D_2)$

a. The data correspond to the spectral lines shown in the LIBS spectrum in Figure 4.1 and the transitions indicated in the Grotrian diagram (Figure 4.2)
b. E_i, E_k: Energy (in units of cm^{-1}) of the upper and lower quantum states involved in the transitions observed at the corresponding λ.
c. Electron configuration and spectroscopic term of each quantum state (the dominant term in the notation corresponding to the L-S coupling scheme has been used even though Pb excited states can be described following the jj coupling scheme).

material. As a result, the strength of individual emission lines coming from different elements in a sample can be used to obtain quantitative compositional data. Assuming the plasma has reached a state of local thermodynamic equilibrium, the integrated intensity of a spectral emission line from a single species in the plasma, is given by eq. (4.1), as a function of the number density of the species, relevant spectroscopic parameters and the plasma electron temperature:

$$I_{\omega}^{ik} = N \frac{h\omega}{8\pi^2} \frac{g_i A_{ik} e^{-(E_i/k_B T_e)}}{Z(T_e)} \tag{4.1}$$

I_{ω}^{ik}: intensity of emission line at frequency $\boldsymbol{\omega}$ corresponding to a transition from the upper quantum state i to the lower k; N: species number density in the point of observation within the plasma; h: Plank's constant; g_i: statistical weight of state i; A_{ik}: transition probability for spontaneous emission from energy level i to k; E_i: energy of quantum state i with respect to the ground state of the emitting species; k_B: Boltzmann's constant; T_e: plasma electron temperature; $Z(T_e)$: partition function for the quantum states of the emitting species.

Commonly, quantitative analysis is performed by analysing samples against calibration standards or reference samples of known analyte composition, following the calibration curve method (Figure 4.3). It is important to note that matrix matched samples provide more reliable quantitative results. It is understood, however, that because of the different types of samples encountered, their varying compositions and material heterogeneity, it is not always feasible to obtain proper standards for

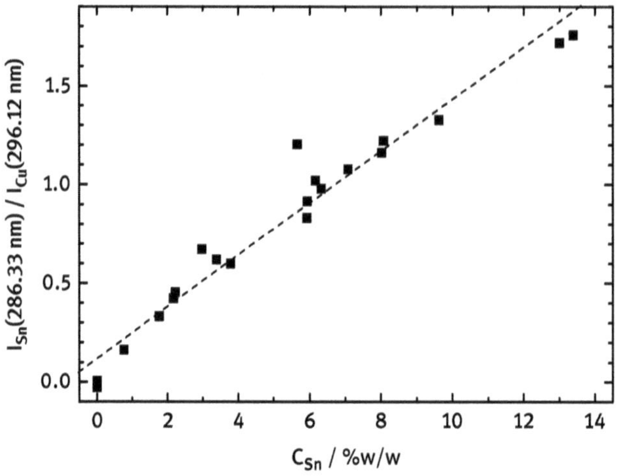

Figure 4.3: Calibration curve for analysis of archaeological Cu-Sn alloys representing the spectral intensity ratio of two atomic emission lines for Sn (286.332 nm) and Cu (296.116 nm) respectively over the concentration of Sn (mass of Sn in 100 g of alloy). The dashed line indicates the result of a least-square linear fit of the data.

comparison and in such cases one should be aware of this when reporting quantitative analysis data. In certain cases, reliable calibration-free LIBS methods have been proposed and used [22].

Concerning sensitivity, in the context of the analysis of complex matrices such as those encountered in heritage investigations, LIBS performs very well with major (> 5% w/w) and minor elements (0.1–5% w/w) even with single-pulse spectra, as long as analytical lines corresponding to the latter are not overwhelmed by those of the former. This might be the case, for example, with iron rich materials given the multitude of spectral lines observed in iron emission. Analysis of trace elements, in the range of 1–100 ppm, is more demanding and requires the use of strong analytical lines and averaging several tens of single-pulse spectra. Considering the ablation taking place during analysis, such an approach (spectral averaging) would work with compositionally homogeneous materials but would be not appropriate for heterogeneous substrates or layered structures. Likewise, it would not be accepted for the analysis of sensitive or precious objects.

4.3 Instrumentation

The instrumentation for conducting LIBS analysis is quite straightforward and either in research-level systems, prototypes or commercial ones, it comprises: (a) a pulsed laser source, (b) a spectrometer and (c) optics necessary to deliver the laser beam to the sample and collect the plume emission into the spectrometer. The schematic in

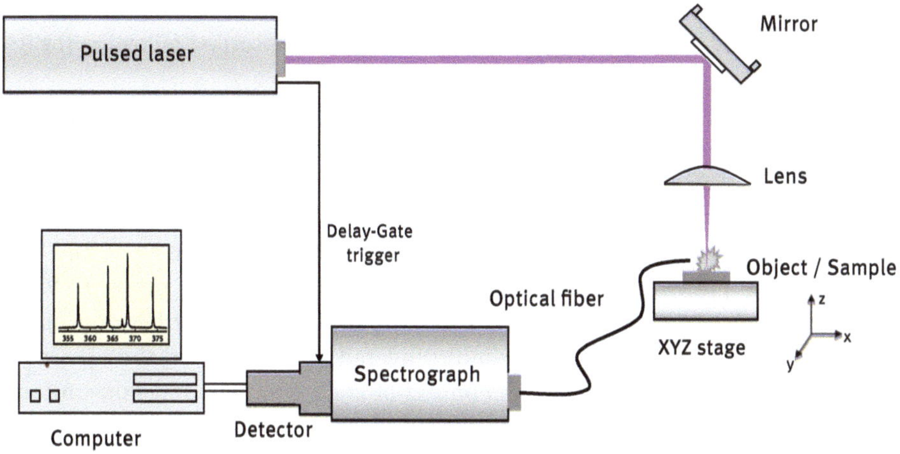

Figure 4.4: Schematic diagram of a typical LIBS experimental arrangement illustrating the main instrumentation components and optics.

Figure 4.4 sketches out a basic experimental layout for a LIBS instrument and shows graphically how a measurement is performed.

Typically, a Q-switched Nd:YAG laser emitting pulses, a few nanoseconds in duration, is used at its fundamental (1064 nm) or its harmonics (532, 355, 266 nm). Other nano- as well as pico- and femto-second pulsed lasers have been used, but mostly in research studies. The laser pulse is focussed by use of a convergent lens with focal length, typically between 50 and 200 mm, which is a normal working distance that permits one to perform analysis not only on samples but also on whole objects or even in the proximity of a monument surface if a mobile instrument is used. Focussing under these conditions concentrates the pulse energy down to a spot of diameter in the range of 50–150 μm, depending on the lens focal length and the quality of the laser beam profile. With typical values of the laser pulse energy on the order of 1–25 mJ, and appropriate choice of laser spot size on the sample surface, the energy density irradiating the sample surface falls in the range of 1–50 Jcm^{-2} (or equivalent to a power flux of 0.1–5 × 10^9 Wcm^{-2}) sufficient for producing a plasma useful for elemental analysis.

The plasma plume emission is collected through a lens system or directly by use of an optical fibre into a spectrograph, where light is analysed and recorded on a CCD (charge-coupled device) detector producing a typical LIBS spectrum. In general, a single laser pulse is adequate for collecting a spectrum with high S/N (see Figure 4.1). Gating the detector permits recording of the plasma emission within a time window, which is delayed with respect to the laser pulse, and this allows for rejection of the strong continuum background present at the early phases of laser plasma formation. Most spectrometers available today provide this gating functionality. Intensified-CCD detectors offer simultaneously with gating the option to amplify the emission signal. An additional approach that leads to significantly improved signal intensity is called double-pulse LIBS and employs for plasma formation a pair of laser pulses, separated from one another by approximately 1 μs, with the emission recorded following the 2nd pulse [23, 24].

In general, a few practical issues need to be taken into account during the application of LIBS in heritage samples and objects. Prior to irradiation, the area for analysis must be carefully defined, preferably through optical magnification, considering the surface heterogeneity of heritage objects. This can be critical for interpreting the analytical results in particular when very fine features are to be probed or when superficial contamination is present. Many common cases, for instance, corroded metals, stone and pottery bearing burial deposits and paintings are effectively multi-layered structures and the LIBS data obtained need to be explained in the context of such complex and non-uniform stratigraphies.

4.4 Mobile LIBS

An obvious requirement when performing studies of works of art and historical/archaeological objects/samples is the capability of the analytical instrument to "travel" [25]. This permits analysis to be conducted on site, in the conservation laboratory, the museum or even close to or at an excavation site overcoming practical and regulatory constraints that prevent transportation of art and archaeological objects. This requirement has generated an interest for the use of portable instrumentation and indeed examples gaining from the use of mobile LIBS equipment have been increasingly presented in recent years. Some are listed in Table 4.2 along with relevant literature references where further details can be found concerning mobile LIBS technology and its application in heritage studies. Expanding on the concept of

Table 4.2: Mobile or transportable LIBS systems used in heritage analytical applications.

Instrument (Laboratory/Company)	Specs (Laser/Detector/Weight)	Ref.
LMNT-I (IESL-FORTH)	Nd:YAG (1064 nm)/ICCD/45 Kg	[26]
LMNT-II (IESL-FORTH)	Nd:YAG (1064 nm)/CCD/10 Kg	[27]
Portable LIBS (LRMH)	Nd:YAG (1064 nm)/CCD/20 kg	[28]
(Smart)Modi (Applied Laser Spectroscopy Laboratory, Istituto per i Processi Chimico Fisici, CNR/Marwan Technology)	Nd:YAG (1064 nm) (Single- or Double-pulse)/CCD/50 kg	[29]
Portable LIPS (Istituto di Fisica Applicata "Nello Carrara"- CNR)	Nd:YAG (1064 nm)/ICCD/14 kg	[19]
Portable LIBS (University of Malaga)	Nd:YAG (1064 nm)/CCD/5 kg (main unit, excluding laser power supply)	[18]
Stand-off LIBS (University of Malaga)	Nd:YAG (1064 nm)/ICCD/250 kg	[30]
Nano-LIBS, portable, handheld instrument (B&W Tek, Newark, DE, USA)	Diode-pumped Nd:YAG (1064 nm, 500 ps, 1–5 kHz)/CCD/1.8 kg	[31]

mobile LIBS instruments, researchers have introduced systems capable of remote or even under-water operation, highly promising for further developments and applications. Within the context of remote sensing applications open-path remote LIBS is capable of providing rapid elemental analysis on remote targets [30, 32, 33]. This approach enables scanning of large surfaces, such as for example, statues, monuments, building facades from as far as 50–100 meters without the need for special scaffolding. The stand-off LIBS systems are considerably larger than the compact portable ones, however mobility between different locations is managed by use of a small transportation van, while on location mobility is based on wheels. Furthermore, remote LIBS instrumentation has been demonstrated to be capable of performing analysis even underwater enabling examination of archaeological remains or shipwrecks and their assessment [34].

Next, a portable LIBS instrument is described in more detail in order to illustrate some of the design and operational principles of mobile LIBS technology and its application for conducting analyses in the context of cultural heritage campaigns in museums and monuments. The system, LMNT II+, is shown in Figure 4.5 and has been developed and constructed at IESL-FORTH. It consists of the following basic sub-units:

a. an optical probe head (1) housing a compact laser source (a nanosecond pulsed Nd:YAG laser operating at 1064 nm) and appropriate optics for laser beam focusing and for collecting the plume emission into an optical fiber; an integrated mini-camera enables object viewing and proper positioning of the head with respect to the spot to be analysed.
b. a dual spectrometer – detector (covering the spectral range: 200–460 nm and 415–660 nm) (2) fibre-coupled to the probe head and
c. a power supply and laser trigger unit (3).

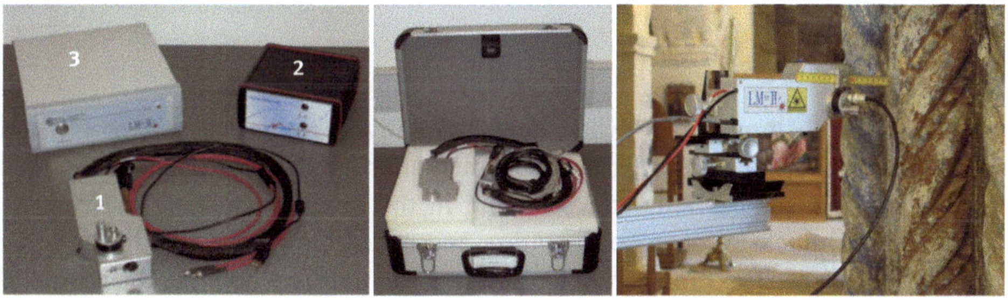

Figure 4.5: Main subunits of LMNT-II+ (left) and instrument in carrying case (right).

The instrument fits in a compact case (dimensions of $46 \times 33 \times 17\ cm^3$) and weighs less than 9 kg. It is fully operated and controlled through a laptop computer via a custom-made software. A micrometre translation stage is used for accurate pointing and analysis while, in certain cases, a tripod or appropriate scaffolding is used.

In brief, the laser is a compact passively Q-switched Nd:YAG, emitting at 1064 nm (10 mJ/pulse, 10 ns). It is housed within the probe head and delivers single pulses upon activation of a manual trigger switch located on the power supply unit. Laser pulses go through a convergent lens (f = + 70 mm) at the front side of the probe head and focused on the surface of the object investigated. The system is equipped with a miniature CCD camera that enables visualization of the object during analysis and aiming of the laser beam. In a typical analysis the probe head position is adjusted by means of a XYZ translation stage, in a way which brings the object/sample at the focal plane of the lens (Z-axis adjustment). Then, a spot is selected for analysis (XY adjustment) based on a cross-hair-like feature pre-aligned with the laser focal spot and embedded on the image of the object viewed through the CCD camera. This way the laser is only fired once for performing the analysis. A circular area corresponding to a diameter of about 0.2 mm is probed by the laser beam. A dual fiber optic compact spectrograph is used to record the emission spectra across a wavelength range extending from 200 to 660 nm, with resolution of about 0.2–0.3 nm. All spectra are collected with a time-delay of 1.3 μs with respect to the laser pulse. A single pulse is adequate for obtaining a spectrum with high signal-to-noise ratio (S/N). If necessary, additional pulses (to a maximum of 5) are delivered at each point, and separate spectra are recorded, particularly when depth profiling information is sought for. The instrument is supplied with a user friendly interface and can be operated by non-technical personnel following basic training. The software includes a reference library and provides an option for spectra interpretation. It also provides the option of overlaying the acquired spectra with reference ones. The reference library can be continuously updated with relevant data. It is also noted that the viewing camera permits photographic documentation of the probed spots during all stages of analysis (before, after the 1st, 2nd, 3rd, etc. pulse and after the end of analysis). Documentation images are stored along with the spectral data.

4.5 Applications of LIBS

Several research groups worldwide have been investigating the use of LIBS in heritage science as a diagnostic and documentation tool and at this stage a good number of published reports can be found in the literature, mainly in scientific journals, and much fewer in conservation and archaeology journals. It has to be noted that despite the promising results the potential user communities are still hesitant of using LIBS as a standard or routine technique or even unaware of its capabilities. A number of reasons can be listed, such as the lack of commercially available, user friendly LIBS instruments and standard procedures relevant to heritage studies that can be employed by non-experts. Moreover, potential users classify LIBS as an invasive technique (despite the fact it is actually micro-invasive) and would prefer to avoid using it for the analysis of CH objects/samples even though at the same time a sampling procedure might be followed.

To provide a brief overview concerning applications of LIBS in heritage science, a few case studies are described in more detail next while more examples from indicative applications along with representative literature references are listed in Table 4.3.

Table 4.3: Examples of LIBS applications in heritage studies.

Applications	Reference
Analysis of pigments in easel paintings, icons, wall paintings, manuscripts	[15, 23, 27, 28, 35, 36]
Archaeological/historical metal and metal alloy objects and sculpture (e. g. various objects, sculpture, jewellery, coins, daguerreotypes)	[19, 37–40]
Corrosion of metal objects	[41]
Glass objects	[42, 43]
Stone, Marble sculpture	[31, 44–47]
Ceramics and pottery	[48–51]
Geological materials	[52, 53]
Biominerals and Fossils	[54–56]
Paper and parchment	[57]

4.5.1 Pigments and colorants

Object and building painting has been continuously used since ancient times in various contexts: artistic, religious, architectural, decorative and many more. The study of pigments provides diverse information as regards provenance of objects and materials, technology used, current state of preservation and potentially dating. Most pigments found in paintings, from antiquity to relatively modern times, are inorganic compounds that have been either naturally available as coloured minerals or synthetic ones. As a result, pigment identification can be quite straightforward if through LIBS analysis the metals found in the colorant are related to known pigment materials.

The following example refers to a rather more complex situation related to the study of traces of paint on stone sculpture [35] and demonstrates the use of a mobile instrument on site. Optical examination of the stone surface showed scarce remains of a yellow paint and the LIBS probe was aimed at an area showing a pale yellow coloration. A single-laser pulse spectrum was obtained (Figure 4.6) dominated by emission lines corresponding to calcium (Ca), originating from the stone substrate. However, a rather strong emission line at 405.78 nm and a few additional ones of lower intensity, prove the presence of lead (Pb) which signifies the use of a lead-based pigment. Among these is lead chromate ($PbCrO_4$) or chrome yellow and indeed its presence is verified by several weak but distinct lines due to chromium (Cr). A more thorough examination of other lines in the range of 250–350 nm, reveals further the presence of Ag, Au and Cu. The co-existence of these three metals is evidence for

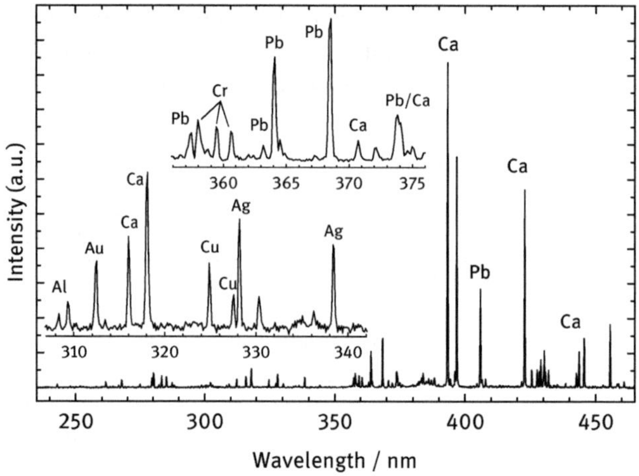

Figure 4.6: Emission spectrum obtained in LIBS analysis of traces of yellow paint on stone sculpture.

the presence of a gold alloy likely used in decorative features no longer visible but still detected by LIBS.

4.5.2 Stone sculpture and monuments

In the case of stone and other lithotype materials, elemental analysis can be useful in determining the provenance of stone or characterizing encrustations over the stone surface and relating those to specific burial environments in the case of excavated objects or to pollution processes in sculpture, buildings and monuments exposed to urban environment. The possibility to carry out LIBS analysis in situ based on portable equipment is quite attractive.

In a detailed study [47] LIBS has been employed in the analysis of different types of white marble (from Naxos (Greece), Proconnesos (Turkey) and Carrara (Italy)) examining both the bulk stone and the surface encrustation. Quantitative analysis data was obtained for the main constituents (Ca, C, Al and Si) as well as for minor elements (Fe, Mg, Mn, Ti, Ba, Cu) present at the ppm level based on the calibration curve method and the use of reference samples composed of $CaCO_3$ matrices doped with certified soils.

An important question in the context of archaeological or historical stone monument conservation and preservation relates to understanding the factors influencing the formation of weathering crusts on the monument. For these reasons, the determination of the crust thickness and the distribution of certain pollutants as a function of depth from the surface can be quite illuminating. In a recent study investigating a number of limestone masonry blocks from the entrance gate of the historic Castello Svevo (Bari, Italy), double-pulse LIBS was employed in depth profiling mode and revealed the stratigraphy of the pollution crust on the basis of the distinctly higher concentration of iron (Fe) and manganese (Mn) in the crust compared to the unweathered stone [44]. This compositional step was used to define the interface between the crust and the underlying limestone and was found to be complementary to the stratigraphic behaviour of calcium (Ca) and strontium (Sr) that showed an increase on the transition from the crust to the stone.

4.5.3 Metal objects

LIBS is quite an efficient method for the analysis of metals and metal alloys exploiting the fact that metal ablation and plasma formation is quite favorable process given the strong coupling of laser radiation with metal surfaces. This makes it suitable for the characterization of archaeological and historical metallic artifacts such as sculpture, tools, weapons, home utensils, jewellery and coins. The main metallic materials used in antiquity include copper and bronze - copper-tin alloys (starting in the Bronze Age) - and later iron. Other metals used include lead and zinc while precious metals such as silver or gold alloys are more often associated with jewellery and for decorating different objects.

Simple, qualitative LIBS analysis can be a helpful tool for quickly classifying or discriminating among like objects, for instance, various types of bronze materials or different types of coins. More importantly, quantitative analysis of metal alloys can reveal valuable information about metallurgical technology, period of manufacture and possibly provenance of the raw materials on the basis of the major, minor and trace elements detected. For example, by means of quantitative LIBS analysis researchers managed to obtain a chronological classification of a number of bronze objects from the Iberian peninsula, dating from the Bronze to the Iron Age, on the basis of the elemental content of the alloys and particularly the concentration of arsenic (As), the presence of which is a key indicator of the Early Bronze Age [37]. Cluster analysis enabled reliable sorting of the objects across the different periods. In a similar study [38] archaeological bronze artefacts from a burial site in Southern Tuscany, Italy, dated in the period of 2500–2000 B.C.E. were examined using micro-LIBS analysis and by use of the calibration-free LIBS method concentrations of the elements in the alloy were obtained. These results, with the help of principal component analysis (PCA) led to a typological grouping of the objects and in certain cases indicated a classification different from that suggested by the traditional typology offering archaeologists a deeper view into the origin and use of the objects.

Concerning quantitative analysis of archeological metal alloys, a critical factor that needs to be taken into account is the presence of superficial corrosion, often of significant thickness. In such a case of a thick corrosion layer present, based on LIBS analysis some conclusions can still be drawn about the elemental composition of the metal object, however an accurate quantitative result is rather difficult to get unless the corrosion layer is removed either by successive laser pulses or by other mechanical means. On the other hand, LIBS contributes to the characterization of superficial contamination and corrosion layers on metal artefacts providing information about object burial history or informing about certain types of corrosion that might require specific conservation treatment. A thin corrosion layer is usually etched away following irradiation with a small number of pulses allowing the laser beam to probe the bulk metal [19, 41].

4.5.4 Pottery

Ceramic objects are the most common among archaeological and historical findings. Analysis of the fabric (main body of the ceramic) provides evidence about the clay composition revealing technological and potentially provenance information. Surface analysis may yield artistic and technological information concerning the slip (original paint layer) or the glaze in the case of glazed pottery. LIBS has been used in several studies of pottery such as, for example, the characterization of a collection of ancient Roman pottery sherds, *Hispanic Terra Sigillata*, that originated from different production sites and dated back to the first to fifth century C.E [49]. On the basis of the elemental content of sherds and with the aid of linear correlation analysis of LIBS spectra, it was possible to classify the sherds according to their production site. Furthermore, depth profiling studies were carried out in order to distinguish between the slip and the fabric of the ceramic sherds and elements such as calcium and iron were used as markers for the transition from the one layer to the other. In a similar study [51], selected pottery sherds coming from three different excavations in Eastern Turkey and dated from the Early to Middle Iron Age were examined as regards to their composition by using LIBS. Comparison of results from LIBS analysis of the coloured slip (paint layer) and on the fabric (bulk ceramic), revealed a strong correlation of the red colour with the iron (Fe) content of the slip layer. Furthermore, test studies, using PCA on LIBS spectra, showed that the analytical information contained in the raw spectral data (no quantitative analysis done) can be used for differentiating among sherds coming from different sites known to have used different clay sources and pottery making technology.

4.5.5 Archaeological and historical glass

Vitreous materials (faience, glass, glaze) are made of quartz (or quartz sand), lime, alkalies (soda or potash) and metal oxides. Chemical analysis of ancient glass is a way to reveal the composition, selection of raw materials, as well as the coloring agents. Furthermore, the analysis of weathered glass helps to study the decay processes and understand mechanisms.

In a recent study [58] LIBS results obtained on a set of a historical Late Roman glasses from a recently unearthed graveyard located close to Madrid (Spain) contributed to the identification of major glass components used to classify the objects into main historical glass groups as well as minor elements that were related to chromophores, decolouring agents or degradation products. Quantitative results were obtained on the basis of the calibration curve method applied with certified glass standards and local standards and were comparable to those obtained using XRF spectrometry. The LIBS methodology was proposed as part of a protocol that can be applied for the in situ characterization of historical glasses.

4.5.6 Geological materials

Various types of geological materials have been used in early times to make simple tools, utensils, seals, decorative objects or sculpture. Their elemental composition, and particularly that of minor and trace elements, has been related to provenance of the object. In this respect, such analysis can be of importance in uncovering, for example, the sources of raw materials used to make certain objects by comparison of compositional patterns between the object in question and the original raw materials. Quick screening can be performed on the basis of qualitative analysis, if differences in elemental content are significant [52] and this can be followed by informed selection of certain objects or samples for further analysis exploring their trace elemental profile that can be related to provenance.

In a different context, LIBS was used to study rock alterations on samples collected from a cave in the South West of France currently used as an experimental facility for observing the growth of calcitic formations on the rock that overcoat possible existing rock art [53]. A portable LIBS system was tested for its ability to discriminate among two main types of rock alterations and this was found to be the case on the basis of the different contents of the samples in Si, Mg and Sr Clearly portable LIBS did not provide a full analysis of the rock samples but it showed that it can be used in campaigns in caves for quick analysis and screening of different types of rocks and alterations based on the minor elements present in them.

4.5.7 Palaeoclimate studies

Recent reports have shown that LIBS can provide an efficient route to fast measurements of palaeotemperature proxies on geomaterials such as speleothems or sediments and biominerals, for example, marine mollusc shells or fossilized animal or tree remains. The simplicity and speed of analysis offered by LIBS compared to other techniques (for example, LA-ICP-MS or SIMS) enables data collection from extensive areas on the specimens investigated and thus yields proxy maps across surfaces as wide as a few cm^2 and at a resolution as low as a few micrometers. This provides statistically important information hence more reliable paleoclimatic and environmental data. In this context, J. O. Cáceres and co-workers [59] collected mega-pixel multi-elemental (Ca, C, Si, Fe, Mg, Al, Mn, Sr) maps by scanning a large fraction of a speleothem and a branch coral sample by use of an auto-focusing micro-LIBS system providing pixel resolution in the range of 10–20 μm.

Likewise, Hausmann et al [60] working with diverse types of mollusc shells (both calcitic and aragonitic) have produced detailed maps reflecting variations of Mg over Ca across sections of the shells (Figure 4.7). These maps enable one to assess quickly whether the observed variations of the ratio Mg/Ca reflect an environmental signal or result from other processes that might distort the expected seasonal variability. Recent preliminary studies have shown that in cases in which the environmental

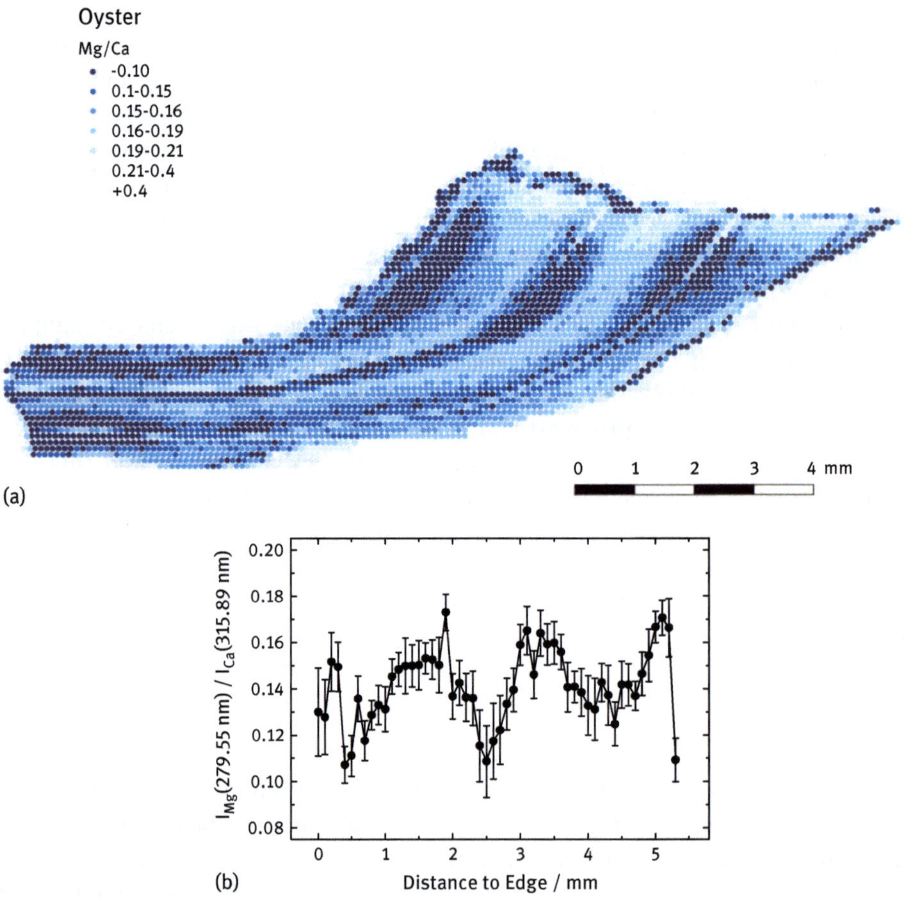

Figure 4.7: (a) LIBS mapping on oyster cross section indicating variations of the Mg/Ca line intensity ratio. (b) Line scan along the top-right part of the image following the shell growth line.

variability impact on the Mg/Ca is regular LIBS maps fit well with temperature data obtained from independent analysis of $^{18}O/^{16}O$ isotopes.

4.5.8 Biomaterials

Analysis of biomaterials is an area of potential interest to archaeological research focused on human remains, aiming to identify certain diseases, nutritional habits or deficiencies or the presence of potentially toxic elements or animal remains and fossils. The characterization of calcified tissues employing LIBS is an example of such a quantitative analysis of traces of aluminium, strontium and lead in human bones or teeth [54]. It was demonstrated in that study that the above trace elements can be quantitatively detected and, in fact, surface-mapped in calcified tissue samples. These results indicate strongly the potential advantages of LIBS in the analysis of bio-archaeological samples. In a more recent test study on fragments of human bones and teeth, LIBS data processed through Neural Networks (NN) led to quite accurate discrimination of the fragments on the basis of the individual they originated from [61]. This finding, even though the study was not performed on ancient samples, shows that the proposed methodology could be quite appropriate in supporting archaeological and historical excavation campaigns concerning sorting of bone or teeth fragments.

In a different type of investigation, samples of parchment (collagen) from different sources were analyzed by LIBS and on the basis of the concentration of a number of elements (Ca, Na, K, Mg, Fe, Cu, and Mn), it became possible to achieve a clear discrimination between modern and historical samples. Furthermore, correlations were found between the animal type and the ratio of Mg over Cu emission line intensity [57].

4.6 Conclusions — outlook

Laser-induced breakdown spectroscopy is a versatile analytical technique that can be used to probe very different types of heritage materials and provide valuable information on their elemental composition. In practice, LIBS analysis requires minimal or no sample preparation and the technique is nearly non-invasive, offering close to microscopic spatial resolution and the possibility for depth profile analysis. These features can be available in compact, mobile instruments that offer versatility and enable the use of LIBS for the analysis of a broad variety of objects/samples at diverse locations and this can be highly valuable at several stages of archaeological research and conservation campaigns. Early analysis during excavation could provide feedback to the archaeologist and possibly guide an on-going excavation campaign. At the conservation laboratory LIBS can be a handy tool for classifying objects, identifying surface alterations and/or previous interventions helping the conservator to decide on optimum intervention methodologies and even monitor their progress. Detailed studies of the elemental composition of materials can finally help archaeologists, historians and conservators to place objects within a certain context or chronological period and answer questions about raw materials, manufacturing and processing. At this point a close collaboration between the heritage (archaeologists, historians, conservators) and analytical spectroscopy communities is necessary for fully exploiting the potential of LIBS. To this end, on-going advances in laser, spectrometer and detector technologies are expected to enhance the capabilities and lower the cost of LIBS systems and make such instruments attractive to small or medium scale laboratories enabling conservators and archaeologists to make routine use of LIBS and assess its practical value for addressing analytical challenges in archaeological science and conservation.

Acknowledgements: The author gratefully acknowledges scientific interactions throughout these years with students, co-workers and colleagues at IESL-FORTH and the University of Crete as well as in the broader LIBS and heritage science communities. Financial support in relevance to Heritage Science activities has been provided through a number of projects: (a) EC, H2020-INFRAIA-2014-2015, IPERIOn CH project, GA 654028, (b) EC, H-2020, Marie Skłodowska-Curie Individual Fellowship, ACCELERATE project, GA 703625, (c) NSRF 2014-2020, POLITEIA-II (MIS-5002478) and HELLAS CH (MIS 5002735) projects, funded by the General Secretariat for Research and Technology, Greece and the European Regional Development Fund/ European Commission (Operational Programme: Competitiveness, Entrepreneurship and Innovation, National Strategic Reference Framework, NSRF 2014-2020).

References

[1] Ciliberto E, Spoto G, Modern analytical methods in art and archaeology, chemical analysis, a series of monographs on analytical chemistry and its applications, vol. 155. Winefordner JD, editors. New York: Wiley, 2000.

[2] Spoto G, Torrisi A, Contino A. Probing archaeological and artistic solid materials by spatially resolved analytical techniques. Chem Soc Rev. 2000;29:429–39. DOI: 10.1039/a903358k.

[3] Stuart BH. Analytical techniques in materials conservation. New York West Sussex, UK: Wiley, 2007.

[4] Mantler M, Schreiner M. X-ray fluorescence spectrometry in art and archaeology. X-Ray Spectrom. 2000;29. DOI: 10.1002/(SICI)1097-4539(200001/02)29:1<3::AID-XRS398>3.0.CO;2-O.

[5] Dik J, Janssens K, Van Der Snickt G, van der Loeff L, Rickers K, Cotte M. Visualization of a lost painting by Vincent van Gogh using synchrotron radiation based X-ray fluorescence elemental mapping. Anal Chem. 2008;80:6436–42. DOI: 10.1021/ac800965g.

[6] Dran J-C, Salomon J, Calligaro T, Walter P. Ion beam analysis of art works: 14 years of use in the Louvre. Nucl Instr Meth Phys Res B. 2004;219–220:7–15. DOI: 10.1016/j.nimb.2004.01.019.

[7] Schreiner M, Melcher M, Uhlir K. Scanning electron microscopy and energy dispersive analysis: applications in the field of cultural heritage. Anal Bioanal Chem. 2007;387:737–47. DOI: 10.1007/s00216-006-0718-5.

[8] Gratuze B, Blet-Lemarquand M, Barrandon JN. Mass spectrometry with laser sampling: A new tool to characterize archaeological materials. J Radioanal Nucl Chem. 2001;247. DOI: 10.1023/A:1010623703423.

[9] Giussani B, Monticelli D, Rampazzi L. Role of laser ablation –inductively coupled plasma –mass spectrometry in cultural heritage research: A review. Anal Chim Acta. 2009;635:6–21. DOI: 10.1016/j.aca.2008.12.040.

[10] Kantarelou V, Karydas AG, Sokaras D, Mahfouz L, Qurdab A, Al-Saadi M, et al. In situ scanning micro-XRF analyses of gilded bronze figurines at the national museum of Damascus. J Anal At Spectrom. 2015;30:1787–98. DOI: 10.1039/c5ja00079c.

[11] Hocquet FP, Calvo Del Castillo H, Cervera Xicotencatl A, Bourgeois C, Oger C, Marchal A, et al. Elemental 2D imaging of paintings with a mobile EDXRF system. Anal Bioanal Chem. 2011;399:3109–16.

[12] Romano FP, Pappalardo L, Masini N, Pappalardo G, Rizzo F. The compositional and mineralogical analysis of fired pigments in Nasca pottery from Cahuachi (Peru) by the combined use of the portable PIXE-alpha and portable XRD techniques. Microchem J. 2011;99. DOI: 10.1016/j.microc.2011.06.020.

[13] Cremers DA, Radziemski LJ. Handbook of laser-induced breakdown spectroscopy. New York, USA: Wiley, 2006.

[14] Musazzi S, Perini U, editors. Laser-induced breakdown spectroscopy. Theory and applications Springer Series in Optical Sciences Vol. 182. Berlin Heidelberg: Springer-Verlag, 2014.

[15] Anglos D. Laser-induced breakdown spectroscopy in art and archaeology. Appl Spectrosc. 2001;55:186A-205A. DOI: 10.1366/0003702011952398.

[16] Giakoumaki A, Melessanaki K, Anglos D. Laser-induced breakdown spectroscopy (LIBS) in archaeological science-applications and prospects. Anal Bioanal Chem. 2007;387:749–60. DOI: 10.1007/s00216-006-0908-1.

[17] Spizzichino V, Fantoni R. Laser induced breakdown spectroscopy in archeometry: a review of its, application and future perspectives. Spectrochim Acta B. 2014;99:201–9. DOI: 10.1016/j.sab.2014.07.003.

[18] Fortes FJ, Cuñat J, Cabalín LM, Laserna JJ. In situ analytical assessment and chemical imaging of historical buildings using a man-portable laser system. Appl Spectrosc. 2007;61:558–64. DOI: 10.1366/000370207780807722.

[19] Agresti J, Mencaglia AA, Siano S. Development and application of a portable LIPS system for characterising copper alloy artefacts. Anal Bioanal Chem. 2009;395:2255–62. DOI: 10.1007/s00216-009-3053-9.

[20] Rakovsky J, Cermak P, Musset O, Veis P. A review of the development of portable laser induced breakdown spectroscopy and its applications. Spectrochim Acta B. 2014;101:269–87. DOI: 10.1016/j.sab.2014.09.015.

[21] https://www.nist.gov/pml/atomic-spectra-database Accessed: 18 Oct 2018.

[22] Ciucci A, Corsi M, Palleschi V, Rastelli S, Salvetti A, Tognoni E. New procedure for quantitative elemental analysis by laser -induced plasma spectroscopy. Appl Spectrosc. 1999;53:960–4. DOI: 10.1366/0003702991947612.

[23] Scaffidi J, Michael Angel S, Cremers DA. Emission enhancement mechanisms in dual-pulse LIBS. Anal Chem. 2006;78:24–32. DOI: 10.1021/ac069342z.

[24] Ferretti M, Cristoforetti G, Legnaioli S, Palleschi V, Salvetti A, Tognoni E, et al. In situ study of the Porticello Bronzes by portable X-ray fluorescence and laser -induced breakdown spectroscopy. Spectrochim Acta B. 2007;62:1512–18. DOI: 10.1016/j.sab.2007.09.004.

[25] Anglos D, Detalle V. Cultural heritage applications of LIBS, chapter 20. In: Musazzi S, Perini U, editors. Laser-induced breakdown spectroscopy, theory and applications, series in optical sciences. vol. 182. Berlin Heidelberg: Springer-Verlag; 2014:531–554.

[26] Melessanaki K, Mastrogiannidou A, Chlouveraki S, Ferrence SC, Betancourt PP, Anglos D. Analysis of archaeological objects with LMNTI, a new transportable LIBS instrument. In: K. Dickmann, C. Fotakis, J. F. Asmus, editors. Proceedings, 5th International Conference Lasers in the Conservation of Artworks in Lasers in the Conservation of Artworks, LACONA V Proceedings, vol. 100. Osnabrueck, Germany: Springer Proceedings in Physics, Sept. 15-18, 2003: 443–51.
[27] Westlake P, Siozos P, Philippidis A, Apostolaki C, Derham B, Terlixi A, et al. Studying pigments on painted plaster in Minoan, Roman and early Byzantine Crete. A multi-analytical technique approach. Anal Bioanal Chem. 2012;402:1413–32. DOI: 10.1007/s00216-011-5281-z.
[28] Duchene S, Detalle V, Bruder R, Sirven JB. Chemometrics and laser induced breakdown spectroscopy (LIBS) analyses for identification of wall paintings pigments. Curr Anal Chem. 2010;6:60–5. DOI: 10.2174/157341110790069600.
[29] Bertolini A, Carelli G, Francesconi F, Francesconi M, Marchesini L, Marsili P, et al. Modi: a new mobile instrument for in situ double-pulse LIBS analysis. Anal Bioanal Chem. 2006;385:240–7. DOI: 10.1007/s00216-006-0413-6.
[30] Gaona I, Lucena P, Moros J, Fortes FJ, Guirado S, Serrano J, et al. Evaluating the use of standoff LIBS in architectural heritage : surveying the Cathedral of Malaga. J Anal At Spectrom. 2013;28:810–20. DOI: 10.1039/c3ja50069a.
[31] Senesi GS, Manzini D, De Pascale O. Application of a laser -induced breakdown spectroscopy handheld instrument to the diagnostic analysis of stone monuments. Appl Geochem. 2018;96:87–91. DOI: 10.1016/j.apgeochem.2018.06.008.
[32] Gronlund R, Lundqvist M, Svanberg S. Remote imaging laser -induced breakdown spectroscopy and remote cultural heritage ablative cleaning. Opt Lett. 2005;30:2882–4. DOI: 10.1364/OL.30.002882.
[33] Tzortzakis S, Gray D, Anglos D. Ultraviolet laser filaments for remote laser-induced breakdown spectroscopy (LIBS) analysis: applications in cultural heritage monitoring. Opt Lett. 2006;31:1139–41. DOI: 10.1364/OL.31.001139.
[34] López-Claros M, Fortes FJ, Laserna JJ. Subsea spectral identification of shipwreck objects using laser -induced breakdown spectroscopy and linear discriminant analysis. J Cult Heritage. 2018;29:75–81. DOI: 10.1016/j.culher.2016.12.015.
[35] Papliaka ZE, Philippidis A, Siozos P, Vakondiou M, Melessanaki K, Anglos D. A multi-technique approach, based on mobile/portable laser instruments, for the in situ pigment characterization of stone sculptures on the island of Crete dating from Venetian and Ottoman period. Heritage Sci. 2016;4:15. DOI: 10.1186/s40494-016-0085-2.
[36] Bicchieri M, Nardone M, Russo PA, Sodo A, Corsi M, Cristoforetti G, et al. Characterization of azurite and lazurite based by laser induced breakdown spectroscopy and micro-Raman spectroscopy. Spectrochim Acta B. 2001;56:915–22. DOI: 10.1016/S0584-8547(01)00228-2.
[37] Fortes FJ, Cortes M, Simon MD, Cabalin LM, Laserna JJ. Chronocultural sorting of archaeological bronze objects using laser -induced breakdown spectrometry. Anal Chim Acta. 2005;554:136–43. DOI: 10.1016/j.aca.2005.08.081.
[38] Corsi M, Cristoforetti G, Giuffrida M, Hidalgo M, Legnaioli S, Masotti L, et al. Archaeometric analysis of ancient copper artefacts by laser -induced breakdown spectroscopy technique. Microchim Acta. 2005;152:105–11. DOI: 10.1007/s00604-005-0388-6.
[39] Melessanaki K, Mateo M, Ferrence SC, Betancourt PP, Anglos D. The application of LIBS for the analysis of archaeological ceramic and metal artifacts. Appl Surf Sci. 2002;197–198:156–63. DOI: 10.1016/S0169-4332(02)00459-2.
[40] Anglos D, Melessanaki K, Zafiropulos V, Gresalfi MJ, Miller JC. Laser-induced breakdown spectroscopy for the analysis of 150-year-old daguerreotypes. Appl Spectrosc. 2002;56:423–32. DOI: 10.1366/0003702021955079.

[41] Colao F, Fantoni R, Lazic V, Caneve L, Giardini A, Spizzichino V, et al. LIBS as a diagnostic tool during the laser cleaning of copper based alloys: experimental results. J Anal Atom Spectrom. 2004;19:502–4. DOI: 10.1039/b315488b.

[42] Müller K, Stege H. Evaluation of the analytical potential of laser -induced breakdown spectrometry (LIBS) for the analysis of historical glasses. Archaeometry. 2003;45:421–33. DOI: 10.1111/1475-4754.00119.

[43] Carmona N, Oujja M, Rebollar E, Romich H, Castillejo M. Analysis of corroded glasses by laser induced breakdown spectroscopy. Spectrochim Acta B. 2005;60:1155–62. DOI: 10.1016/j.sab.2005.05.016.

[44] Senesi GS, Nicolodelli G, Milorim DMBP, De Pascale O. Depth profile investigations of surface modifications of limestone artifacts by laser -induced breakdown spectroscopy. Environ Earth Sci. 2017;76:565. DOI: 10.1007/s12665-017-6910-4.

[45] Anzano JM, Villoria MA, Gornushkin IB, Smith BW, Winefordner JD. Laser-induced plasma spectroscopy for characterization of archaeological material. Can J Anal Sci Spectrosc. 2002;47:134–40.

[46] Maravelaki-Kalaitzaki PV, Anglos D, Kylikoglou V, Zafiropulos V. Compositional characterization of encrustation on marble with laser induced breakdown spectroscopy. Spectrochim Acta B. 2001;56:887. DOI: 10.1016/S0584-8547(01)00226-9.

[47] Lazic V, Fantoni R, Colao F, Santagata A, Morona A, Spizzichino V. Quantitative laser induced breakdown spectroscopy analysis of ancient marbles and corrections for the variability of plasma parameters and of ablation rate. J Anal Atom Spectrom. 2004;19:429–36. DOI: 10.1039/b315606k.

[48] Colao F, Fantoni R, Lazic V, Spizzichino V. Laser-induced breakdown spectroscopy for semi-quantitative and quantitative analyses of artworks - application on multi-layered ceramics and copper based alloys. Spectrochim Acta B. 2002;57:1219–34. DOI: 10.1016/S0584-8547(02)00054-X.

[49] Lopez AJ, Nicolas G, Mateo MP, Ramil A, Pinon V, Yanez A. LIPS and linear correlation analysis applied to the classification of Roman pottery Terra Sigillata. Appl Phys A. 2006;83:695–8. DOI: 10.1007/s00339-006-3556-6.

[50] Anzano J, Gutierrez J, Villoria M. Direct determination of aluminum in archaeological clays by laser -induced breakdown spectroscopy. Anal Lett. 2005;38:1957–65. DOI: 10.1080/00032710500232810.

[51] Erdem A, Çilingiroğlu A, Giakoumaki A, Castanys M, Kartsonaki E, Fotakis C, et al. Characterization of iron age pottery from eastern Turkey by laser -induced breakdown spectroscopy (LIBS). J Arch Sci. 2008;35:2486–94. DOI: 10.1016/j.jas.2008.03.019.

[52] Harmon RS, DeLucia FC, McManus CE, McMillan NJ, Jenkins TF, Walsh ME, et al. Laser-induced breakdown spectroscopy - an emerging chemical sensor technology for real-time field-portable, geochemical, mineralogical, and environmental applications. Appl Geochem. 2006;21:730–47. DOI: 10.1016/j.apgeochem.2006.02.003.

[53] Bassel L, Motto-Ros V, Trichard F, Pelascini F, Ammari F, Chapoulie R, et al. Laser-induced breakdown spectroscopy for elemental characterization of calcitic alterations on cave walls. Environ Sci Pollut Res. 2017;24:2197–204. DOI: 10.1007/s11356-016-7468-5.

[54] Samek O, Beddows DCS, Telle HH, Kaiser J, Liska M, Caseres JO, et al. Quantitative laser -induced breakdown spectroscopy analysis of calcified tissue samples. Spectrochim Acta B. 2001;56:865–75. DOI: 10.1016/S0584-8547(01)00198-7.

[55] Rusak DA, Marsico RM, Taroli BL. Using laser-induced breakdown spectroscopy to assess preservation quality of archaeological bones by measurement of Calcium-to-Fluorine ratios. Appl Spectrosc. 2011;65:1193. DOI: 10.1366/11-06364.

[56] Suliyanti MM, Sardy S, Kusnowo A, Pardede M, Hedwig R, Kurniawan KH, et al. Preliminary analysis of C and H in a “Sangiran” fossil using laser -induced plasma at reduced pressure. J Appl Phys. 2005;98:093307. DOI: 10.1063/1.2121930.
[57] Dolgin B, Chen Y, Bulatov V, Schechter I. Use of LIBS for rapid characterization of parchment. Anal Bioanal Chem. 2006;386:1535–41. DOI: 10.1007/s00216-006-0676-y.
[58] Oujja M, Sanz M, Agua F, Conde JF, García-Heras M, Dávila A, et al. Multianalytical characterization of late Roman glasses including nanosecond and femtosecond laser induced breakdown spectroscopy. J Anal At Spectrom. 2015;30:1590–9. DOI: 10.1039/c5ja00150a.
[59] Cáceres JO, Pelascini F, Motto-Ros V, Moncayo S, Trichard F, Panczer G, et al. Megapixel multielemental imaging by laser-induced breakdown spectroscopy, a technology with considerable potential for paleoclimate studies. Sci Rep. 2017;7:5080. DOI: 10.1038/s41598-017-05437-3.
[60] Hausmann N, Siozos P, Lemonis A, Colonese AC, Robson HK, Anglos D. Elemental mapping of Mg/Ca intensity ratios in marine mollusc shells using laser -induced breakdown spectroscopy. J Anal At Spectrom. 2017;32:1467–72. DOI: 10.1039/c7ja00131b.
[61] Moncayo S, Manzoor S, Ugidos T, Navarro-Villoslada F, Caceres JO. Discrimination of human bodies from bones and teeth remains by laser induced breakdown spectroscopy and neural networks. Spectrochim Acta B. 2014;101:21–5. DOI: 10.1016/j.sab.2014.07.008.

Marcello Picollo, Maurizio Aceto and Tatiana Vitorino

5 UV-Vis spectroscopy

Abstract: UV-Vis reflectance spectroscopy has been widely used as a non-invasive method for the study of cultural heritage materials for several decades. In particular, FORS, introduced in the 1980s, allows to acquire hundreds of reflectance spectra *in situ* in a short time, contributing to the identification of artist's materials. More recently, microspectrofluorimetry has also been proposed as a powerful non-invasive method for the identification of dyes and lake pigments that provides high sensitivity and selectivity. In this chapter, the concepts behind these spectroscopic methodologies will be discussed, as well as the instrumentation and measurement modes used. Case studies related with different cultural heritage materials (paintings and manuscripts, textiles, carpets and tapestries, glass, metals, and minerals), which show the usefulness of UV-Vis reflectance spectroscopy and microspectrofluorimetry applied to the study of artworks, will also be presented.

Keywords: non-invasive, UV-Vis spectroscopy, reflectance, FORS, microspectrofluorimetry, paintings and manuscripts, textiles, carpets and tapestries, glass, metals, minerals

5.1 Introduction

In order to provide the appropriate tools for curators, conservators, and conservation scientists to understand and preserve the cultural heritage, it is important to analyse, identify, and characterise the materials used in the production and restoration of artworks, which are often complex systems [1, 2]. Presently, the study of the materials constituting artworks can be performed using invasive and/or non-invasive approaches including spectroscopic methods which can be classified according to the region of the electromagnetic spectrum used or produced in the measurement [2].

In its broadest sense, ultraviolet-visible spectroscopy is concerned with interactions between electromagnetic radiation in the ultraviolet-visible region and matter. The ultraviolet (UV) region covers approximately the 10–380 nm range of the electromagnetic spectrum. It is commonly divided in three main sub-regions, which are: UVA in the 320–380 nm; UVB in the 280–320 nm; and UVC in the 100–280 nm. In addition, the 10–200 nm range is also named as vacuum ultraviolet (VUV), even though it is only explored if measurements are done in vacuum. The visible (Vis) region comprehends the 380–750 nm spectral range.

This article has previously been published in the journal *Physical Sciences Reviews*. Please cite as: Picollo, M. UV-Visible Spectroscopy. *Physical Sciences Reviews* [Online] **2019**, 4. DOI: 10.1515/psr-2018-0008.

https://doi.org/10.1515/9783110457537-005

UV-Vis spectroscopy is related to excitation of the outermost electrons of the atoms, which are involved in the formation of molecules, and is, therefore, often referred to as "electronic spectroscopy" [3]. Measurements in the UV-Vis region are usually performed in transmittance, reflectance and photoluminescence (fluorescence and phosphorescence) modes. Transmittance and reflectance measurements need to be recorded against a reference material while photoluminescence acquisitions can be considered absolute measurements.

In the art conservation field this spectroscopic technique, unlike the most traditional applications of UV-Vis spectroscopy [3], is mainly applied *in situ* by using portable devices and employing non-invasive methodologies. The studied artworks have, excluding a few cases such as glass items and stained glass windows, an opaque appearance that force them to be measured in reflectance mode. Spectra acquired in reflectance mode on bulk samples or masstone pigments and dyes are usually more difficult to interpret than the corresponding transmittance spectra due to several factors affecting reflectance measurements, such as the surface roughness and texture, particles dimension and distribution, packing density, compositional homogeneity and the thickness of the paint layers.

Spectroscopic measurements in the Vis were introduced to the cultural heritage field by Rawlins at the National Gallery in London in the 1930-1940s [4]. At the same time, Vis reflectance spectroscopy was applied to characterise paints and pigments for the first time during the 1930s at the Department of Conservation and Technical Research of the Fogg Art Museum, in Cambridge, Massachusetts [5]. After the World War II, in Rome at the Istituto Centrale del Restauro (now Istituto Superiore per la Conservazione ed il Restauro) Manlio Santini developed a precise and reliable non-invasive methodology for the acquisition of reflectance spectra in the Vis on the painting Maestà by Duccio di Boninsegna to monitor the cleaning procedure during its conservation treatments [6]. In the same years, Vis reflectance spectroscopic methodologies were strongly improved by researchers working mainly in paint formulation laboratories and paint industries [7]. In the years that followed this methodology was further developed at the Conservation Laboratories of the National Gallery in London [8]. Lastly, since the early 1980s, when fibre optical devices became available and starting from the pioneering work of Mauro Bacci at the Istituto di Ricerca sulle Onde elettromagnetiche of the Consiglio Nazionale delle Ricerche (which became Istituto di Fisica Applicata "Nello Carrara" in [9]), the non-invasive technique of fibre optic reflectance spectroscopy (FORS) has been applied and improved in this field including the use of portable instruments and expanding the operating range from the UV up to the near-infrared (NIR, 750–1100 nm) and short-wavelength infrared (SWIR, 1100–2500 nm) regions [10–16].

After the work of E. René de la Rie in [17], fluorescence spectroscopy has also become a powerful analytical technique in the field of cultural heritage [17, 18]. More recently, microspectrofluorimetry, which combines the advantages of fluorescence measurements with those of a confocal microscope, was introduced into the art conservation field particularly for the study of organic colourants and binders [19–24]. Maria João Melo and co-workers have been pioneers of the development of microspectrofluorimetry, combined with data from other analytical techniques, to characterise dyes and lake pigments, and of the creation of a reference database to overcome the absence of a molecular fluorescence fingerprint [25–29]. Microspectrofluorimetry provides high sensitivity and selectivity, combined with good spatial and spectral resolution and fast data acquisition [20]. Moreover, for movable objects that can be transported into the laboratory, it can be used *in situ* without any contact with the artwork or (micro) sample under study [20].

In this chapter, the use of UV-Vis reflectance spectroscopy and microspectrofluorimetry in the study of cultural heritage will be discussed. The principles and theory behind these spectroscopic methodologies will be presented, followed by a section on the instrumentation and measurement modes used. The chapter will end with the presentation of some relevant case studies related with different cultural heritage materials (paintings and manuscripts, textiles, carpets and tapestries, glass, metals, and minerals), which show the usefulness of UV-Vis reflectance spectroscopy and microspectrofluorimetry as *in situ* methods applied to the study of artworks.

5.2 Principles and theory

Ultraviolet and visible radiation that hit a surface can interact with matter in different ways: it can be transmitted, transmitted in a diffuse way, reflected (in a specular and diffuse way, as discussed below), absorbed, absorbed and emitted as photoluminescence (fluorescence and phosphorescence), or laterally diffused at wavelengths different from those of the incident monochrome radiation (diffusion or Raman effect) [30]. If the frequency of incident radiation corresponds to the difference of energy of a transition between two energy levels that specific frequency is absorbed by the material causing an excitation of resonance which generates a variation in the distribution of the electronic density (electronic spectroscopy). This effect can be represented by the Bohr model:

$$\Delta E = E_2 - E_1 = h\nu \rightarrow h\nu = hc\bar{\nu} = hc/\lambda$$

where E_1 and E_2 are the initial and final energies, respectively; h = Planck's constant (6.62×10^{-34} Js); c = light velocity in vacuum (2.99×10^8 m/s); ν = frequency; λ = wavelength; $\bar{\upsilon}$ = wavenumber. This phenomenon causes, for example in the case of electronic spectroscopy, a variation in the distribution of the electronic density in the outermost orbitals of an individual atom or, in molecular orbitals, an electronic transition from the occupied levels to the unoccupied levels. The Bohr's relation therefore connects the atomic or molecular electronic energy levels with the frequency of the incident radiation.

In general, when matter absorbs part of the incident radiation and is excited to higher energy levels, this state is not stable and the species returns to its initial equilibrium. This ground state can be reached through "deactivation" processes that can take place in several ways: without emission of radiation (for example, internal conversion of energy in order to obtain a thermal equilibrium, i. e. dissipation as heat), with spontaneous emission of radiation with a lower energy than that of the absorbed radiation (fluorescence, phosphorescence), with photochemical processes, etc. Depending on the molecular environment and the "deactivation" process, the excitation states can exist for 10^{-13}-10^{-3} sec. Lambert and Beer developed a relation that correlates the intensity of an absorption with the concentration and thickness of the absorbing species (Beer-Lambert Law or Beer's Law, which is strictly observed only for absorbing materials in diluted homogeneous medium):

$$A = \varepsilon bc = -\log T = \log(1/T)$$

where A = absorbance; **ε** = molar absorptivity or molar absorption coefficient; b = thickness of the sample; c = concentration; T = transmittance.

When dealing with measurements on opaque materials, different parameters play an important role in the definition of the spectral features [3, 31]. The reflected radiation from a generic surface, or a paint layer, that can be considered as a multicomponent polycrystalline assemblage, is generally called "total reflectance" (R_T), which is the sum of the specular (or surface, R_S), diffuse superficial (R_{DS}), retrodiffused (R_{RD}), and diffuse (or volume, R_V) reflectance (Figure 5.1). The specular component is the portion of the reflected radiation that has not penetrated into the paint layer, whereas the diffuse component is that portion of radiation which has penetrated into the paint layer and interacts with one or more pigment's particles [32]. According to Vincent and Hunt, R_V depends on several parameters, such as

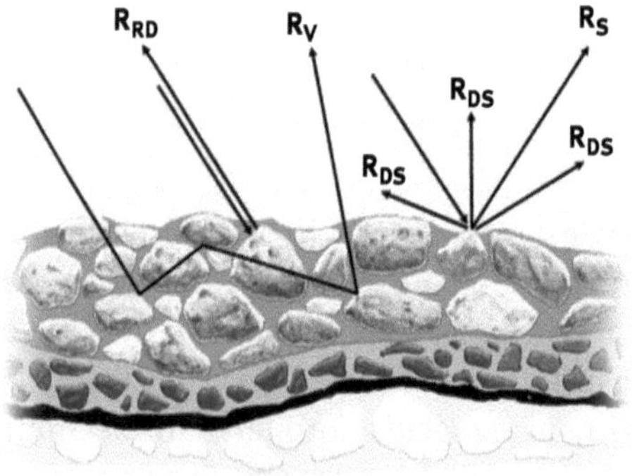

Figure 5.1: Total reflection modes (courtesy of Miguel Ángel Herrero Cortell).
No credits it's our own figure.

wavelength, particle dimensions, packing density, real refractive index n and absorption coefficient k, scattering coefficient s, and optics geometry [33].

The volume reflectance carries with it the major part of spectral information that is used to identify the materials. If one is interested in quantitative information, such as in the measurement mode in transmittance, the reflectance spectra can be expressed as apparent absorbance:

$$A' = \log(1/R)$$

or as Kubelka-Munk (KM) function [34], $f(R_\infty)$, from a homogeneous material with infinite thickness (R_∞). The KM infinitive reflectance function is given by:

$$f(R_\infty) = (1 - R_\infty)^2/2R_\infty = k/s$$

where the parameters k and s are the absorption and scattering coefficients, respectively. This equation assumes that the incident radiation is monochromatic, the scattering processes show no wavelength dependence, and the two parameters k and s are constant in a homogenous medium. Both formulae try to represent the relationship between the intensity of absorbance and the concentration of the compound in a linear way. A' is usually more favourable in the NIR-SWIR region and for highly absorbing samples (low dilution), while KM is applied when condition of infinite dilution in a non-absorbing matrix (high dilution) is present. Even though the KM phenomenological theory [34] is widely used for multiple scattering calculations in the paint industry, it implies certain assumptions and hence has several limitations. The most significant assumption is the isotropic scattering within the sample. Each particle that composes the paint layers, such as pigments, dyes, fillers, etc., has to be far enough apart from one another in order to scatter or absorb the incident radiation independently from all the others, which means that they have to obey to the "no dependent scattering" rule. Moreover, the paint layer is assumed to have plane, parallel surfaces, sufficient in extent to ensure that edge effects can be ignored in the measured reflectance spectra. The incident radiation and its reflected component have to be diffuse, and no generation of radiation, such as fluorescence and related processes, within the paint layer is expected. All these assumptions make the application of KM function to reflectance measurements acquired on paintings, and artworks in general, very problematic as the paint layers constituting these objects barely fit any KM theory requirements.

5.2.1 Electronic transitions

In the UV-Vis spectral range transitions between electronic energy levels can be observed, which determine the absorption bands in the UV-Vis region. An electron is excited when the frequency of the incident electromagnetic radiation is the same as the difference of energy between two electronic states. This difference of energy depends on the electronic structure of the molecule and of its "environment". For a transition to happen after absorption of radiation it is necessary to have a dislocation of charge and some rules, such as Laporte Selection Rule, Spin Multiplicity Selection Rule, and Coupling Interaction with Neighbour Cations, have to be respected [32].

The electronic transitions can be related to transitions between molecular orbitals and they are often classified according to the nature of the molecular orbitals involved. Based on the symmetry of the orbitals involved and the electromagnetic field, it is also possible to assess the intensity of a transition. One of the most important rules in this case is related with the spin multiplicity of the orbitals involved in the transition. Transitions between states that have the same multiplicity of spin are allowed, whereas transitions between levels with different spins are forbidden. The consequence of this rule is that the allowed transitions produce spectral characteristics which are more intense, while the forbidden transitions are absent or create weaker absorption bands.

The absorption bands of pigments are generally related to Ligand-Field (LF), Charge Transfer (CT), and Valence-Conduction band transitions (VC), also known as Band theory, while for organic materials delocalised Molecular Orbital (MO) band transitions are considered [30, 32, 35]. In many cases, the spectral features enable the identification of the nature of the compounds analysed and, in particular, to identify pigments, dyes, alteration products, and some other organic compounds.

LF involves inorganic compounds containing metal ions with unpaired electrons in *d* or *f* orbitals, such as the *3d* transition metals: titanium (Ti), vanadium (V), chromium (Cr), manganese (Mn), iron (Fe), cobalt (Co), nickel (Ni), and copper (Cu) [30, 32]. It explains absorption of the UV and Vis radiation by the transition-metal compounds or impurities of many minerals, gems, and pigments, such as azurite, ruby, and chromium sesquioxide (chrome green or chromium oxide green). The outermost electrons of the transition metals located in orbitals *3d* are responsible for the transitions between the energy levels. Several features cause the degeneracy of the *d* or *f* orbitals of the transition-metal ions, such as the anions or dipolar groups (defined as ligands) which surround the transition metals, the symmetry and geometrical arrangement of the ligands as well as their number and type. These features determine the number of permitted transitions and the line shape of the registered spectra, which can lead to different absorption spectra (and therefore different transmittance or reflectance) for the same transition metal (for example as can be observed for Cu^{2+} in azurite and in malachite). However, not all possible transitions between levels have the same intensity. As previously stated, the selection rules determine which transitions, and their intensities, are permitted and measured. The LF transitions are usually not so intense since they are only partially permitted by the selection rules [32].

CT generally produces intense absorptions since these transitions are fully allowed by the selection rules [30]. Here, processes in which the absorbed energy determines the migration of one electron between close ions or between an ion and ligand are found. Although there is a real transfer of an electron, it remains located in its new position, which means that this electron is linked to one specific ion. In the crystalline structure, an electron mainly located in an orbital of a ligand can be excited and transferred to an orbital mainly located in the metal ion or vice-versa (CT between ligand-metal, LMCT, or metal-ligand, MLCT, respectively), as in chrome yellow pigment (lead chromate, $PbCrO_4$), where the absorption of radiation causes the transfer of one electron from one oxygen of the chromate ion to the chrome ion. These transfers can also happen between ions adjacent to the same metal in different valence states, and in this case the process is essentially a photochemical oxidation-reduction reaction. This is often the case of ion pairs such as Fe^{2+} and Fe^{3+}, Mn^{2+} and Mn^{3+}, or Ti^{3+} an Ti^{4+}, when they are placed next to each other, as for example in Prussian blue pigment, $Fe_4[Fe(CN)_6]_3 \cdot nH_2O$, where the charge transfer occurs between two non-equivalent Fe ions ($Fe_A^{2+} + Fe_B^{3+} \rightarrow Fe_A^{3+} + Fe_B^{2+}$).

VC transitions are characteristic of metals and some inorganic pigments that show properties of semiconductors [30]. In some crystalline lattices, the discrete energy levels of the external electrons of the ions constituting the lattice, are extended to energy bands that exist close to them. There are therefore two energy bands in which the electrons can exist: one at lower energy called "valence band" and one at a higher energy called "conduction band". The electrons in the conduction band have higher energy than those in the valence band and, in practice, are not linked to any ion, but are free to move in the crystalline lattice, from which comes the definition "free electrons" or "conduction electrons". Between the valence band and the conduction band there is an energy gap, called "band gap", in which the electrons cannot be found. In the semiconductor materials, such as vermilion, HgS, red lead, Pb_3O_4, and cadmium yellows, CdS, (Cd,Zn)S, the width of the forbidden band is in between that of the metals and the dielectric compounds, and the passage of electrons from the valence band to the conduction band is seen as an intense absorption in the UV-Vis range. This type of transition produces characteristic absorption bands with an "S" shape and the transition energies can be approximately defined by their inflection points, i.e. the point of the curve at which the sign of the curvature changes, which is determined by looking at the first derivative of their reflectance spectra [16].

Finally, delocalised molecular orbital (MO) theory explains the electronic transition in organic molecules [30]. The electronic orbitals in covalent or weakly ionic bonding are capable of overlapping between themselves and form molecular orbitals called bonding molecular orbitals σ or π and anti-bonding molecular orbitals σ^* or π^*. In this representation, the bonding orbitals represent the maximum positive overlapping between symmetric wave functions and are more stable than the individual atomic orbitals of the single atom. The corresponding anti-bonding molecular orbitals, on the other hand, are less stable than the individual atomic orbitals of the single atom. Absorption of radiation of appropriate energy can promote one of the σ or π electrons to an anti-bonding orbital σ^* or π^* (transition $\sigma \rightarrow \sigma^*$ or $\pi \rightarrow \pi^*$), respectively.

The vast majority of dyes, natural and synthetic, can be viewed as containing an extended conjugated chromophore system to which are attached electron donor and electron acceptor groups (auxochromes). A chromophore is the atom or group of atoms of a molecule in which the electronic transition responsible for a given spectral band is approximately localized; the term referred originally to the groups in the molecule that are responsible for its colour (IUPAC definition,http://goldbook.iupac.org). Chromophores include C = C bonds as well as the azo group (-N = N-), thio group (C = S), and nitro group (-N = O). In turn, an auxochrome is an atom or group of atoms which, when introduced into a chromophore, causes a bathochromic shift (red shift) and/or a hyperchromic effect in a given band of the chromophore, usually in that of lowest frequency (IUPAC definition). Auxochromes have the capability to donate to, or accept from, one or more electrons, through σ- or π-bonds. Those belonging to the first group are electron donor groups while those in the second group are electron acceptor groups. The choice of σ- or π-bonds depends on the electronegativity, the presence of multiple bonds or the existence of lone pair of electrons. The substituents $-NH_2$, X (F, Cl, Br, I), $-CH_3$, Ar, $-NR_2$ and -OR, where R represents an organic group, are electron donors through π-bonds, while $-NO_2$, -R(C = O), -C ≡ N, $-CO_2H$ are electron acceptors. To have colour generation, the energies involved in the MO transitions have to be moved from the ultraviolet into the visible, which can be done by increasing the size of the conjugated system and/or with the addition of auxochromes. For example, a red shift induced by -OH groups is observed from the colourless anthraquinone to the coloured alizarin (1,2-dihydroxy-9,10-anthraquinone) and purpurin (1,2,4-trihydroxy-9,10-anthraquinone). Generally speaking, the absorptions occurring in organic coloured molecules are based on π bonding in the extended conjugated systems and the colours observed are modulated by the energy gap between the ground and excited electronic states involved in the transition.

5.2.2 Microspectrofluorimetry

When a species is excited through the absorption of a photon, it causes electronic transitions and brings the absorbing species into an electronic excited state. In the case of organic molecules, the emission of photons accompanying de-excitation is called photoluminescence, which can be divided into two phenomena depending on the nature of the excited state and the lifetime of the phenomena: fluorescence (excited singlet states) and phosphorescence (triplet excited states) [36]. The phenomenon of fluorescence as an emission of radiation following absorption of radiation was demonstrated by George Gabriel Stokes in the middle of the nineteenth century [37]. Stokes stated that the emitted radiation is always of longer wavelength (lower energy) than the exciting one (Stoke's law).

Fluorescence is measured by exciting a species at an absorption wavelength, also called the excitation wavelength, and measuring the emission at a longer wavelength called the emission wavelength. The characteristics of fluorescence of a given species are affected by any excited-state process involving interactions of the excited species with its close environment, which can provide information not only on the emitting species but also on the respective microenvironment. For this reason, and for being highly sensitive, fluorescence spectroscopy has been widely used in many scientific and technological fields, such as physics, chemistry, materials science, biology, and medicine.

Spectrofluorimeters hyphened to confocal microscopes give the possibility to acquire both fluorescence excitation and emission spectra. In the excitation spectrum, that resembles an absorption spectrum, fluorescence emission is collected at a fixed emission wavelength, which usually corresponds to the maximum fluorescence, while the excitation wavelength is scanned. The fluorescence emission spectrum is obtained by keeping a fixed excitation wavelength (commonly corresponding to the maximum of the excitation/absorption spectrum) while the emission wavelengths are scanned. Emission is generally highly sensitive and selective since detection of different species can be achieved using the appropriate excitation and/or emission wavelengths. The high sensitivity is important for detecting materials which may be present in low concentrations due to, for example, degradation processes. Moreover, the confocal microscope helps to avoid emission from the support or other layers possibly present in the sample. This is achieved with the use of a pinhole aperture at a focus point that coincides with the focal plane of the objective lens. The light from the focal plane of the objective lens will pass through the pinhole aperture, while light which does not come directly from the focal plane (out-of-focus) gets blocked by the pinhole [38].

5.3 Instrumentation, accessories and measurement modalities

The basic optical configuration for a spectrometer working in the UV-Vis region includes one or more polychromatic light sources, one or two wavelength selectors for splitting the radiation into its different components (wavelengths), and one or more detectors for measuring the radiation that interacts with the sample [3, 39, 40]. Depending on the type of instrumentation, an internal or external sample holder can also be present. Reflectance spectra are usually measured using an integrating sphere linked to the spectrometer for working in total and/or diffuse reflectance modes. Most of the compact-portable devices are also equipped with fibre optic bundles to send radiation to the measured spot and to collect its reflected/transmitted component. Spectrometers are split in dispersive and non-dispersive. The latter are based on interferometers, such as the Michelson interferometer, while the former need one or two monochomators for spectrally dispersing the incident radiation. However, in the UV-Vis regions the interferometric-based systems have not been very widespread for routine analysis. Dispersive spectrometers are divided in spectrophotometers and spectroanalysers, the sequence of the optical path of the former being defined as a) light source, b) monochromator, c) sample, and d) detector, while that of spectroanalysers is characterised for having the sample placed right after the light source, before the monochromator. Traditional spectrometers are also grouped in "single beam" and "double beam" devices. The latter group of instruments have two light paths, through the sample and the reference, which are automatically and continuously interchanged during the acquisition of the spectra. The single beam spectrometers make the acquisition of the sample and reference at different times, which means that the devices require well stabilised light sources and electronics to keep the measurement conditions steady during the acquisition of the spectra. Starting from the 1990s a new generation of spectroanalysers also became available. They include multiplex detectors, made of linear photodiode or charge-coupled devices (CCDs) arrays, which allow to measure a complete spectrum in fractions of second up to few minutes, depending on the device, experimental set-up, and the analysed sample. In the last years, Texas Instruments has produced a new set of low cost micro electromechanical system (MEMS) digital micromirror devices (DMD) working in the UV-Vis and NIR-SWIR ranges. The core of these devices, the DMD, consists of an array of thousands/millions of tiny micromirrors, which pick the individual wavelengths by selectively turning columns of mirrors on or off, in order to reflect only the desired wavelengths to the single detector [41].

The most common detectors and light sources are reported in Table 5.1.

Table 5.1: Most common detectors and light sources used in UV-Vis spectroscopy.

Detectors	Sensitivity	Radiation source	Emission range
Photomultiplier tube	< 190–850 nm	Tungsten-Halogen filament lamps With the addition of iodine, a higher filament temperature is obtained	320–3200 nm
Silicon photodiodes	200–1100 nm	Deuterium lamp	185–400 nm
Silicon charge-coupled devices (CCD)	200–1100 nm	Xenon arc lamp	270–1000 nm

As previously stated, reflectance measurements are "relative measurements" which means that these spectra need to be referenced against a reflectance standard, such as barium sulphate (barite) plates or fluoropolymer (PTFE) targets, which have high diffuse reflectance and are virtually free of spectral features over the UV-Vis and NIR-SWIR ranges [32, 42]. These standards have to be stable with time, temperature, etc., show a homogenous and smooth surface, present no fluorescence and be opaque (no transparency).

Fluorescence measurements are made with the aid of fluorimeters or spectrofluorimeters, the latter comprising an excitation and emission monochromator and recording both excitation and emission spectra [3]. Spectrofluorimeters hyphened to confocal microscopes offer the possibility to acquire fluorescence spectra *in situ*, from samples placed directly on a programmable x-y-z microscope stage [22]. A continuous xenon lamp is used, which is directed into a double-grating excitation monochromator that selects the desired wavelength of the excitation beam. The beam passes through a dichroic mirror and all-reflective optic mirrors which direct the incident light to the microscope. Within the microscope, a pinhole-turret containing holes of various diameters controls the spatial resolution collimating the illumination, which then illuminates the sample on the programmable microscope stage. Fluorescence from the sample is directed back up into the microscope, passing through the optic mirrors and dichroic mirror to a spectrometer or spectrograph with one or two detectors (a photomultiplier tube and a CCD array). The dichroic filters which transmit certain wavelengths and reflect others enable the separation of emission and excitation light. Fluorescence signals are expressed in "counts per second" (cps). Optimization of the signal is made through mirror alignment in the optic pathway of the microscope using strong emitters references such as rhodamine.

5.4 Case studies

Most of the applications of UV-Vis diffuse reflectance spectroscopy in the field of cultural heritage are carried out using fibre optic systems (FORS). Other applications are carried out using handheld systems working in contact with the object or, to a lesser extent, inside laboratories using systems equipped with integrating spheres.

5.4.1 Paintings, manuscripts

A major part of FORS applications are related to paintings and illuminated manuscripts, for a couple of reasons. Firstly, these artworks are generally flat, which is a favourable situation for analysis; secondly, the identification of coloured materials is usually straightforward if the artist's technique, the geographic area and the period in which the object was produced are known. Moreover, FORS can contribute to the identification of organic colourants such as dyes and lake pigments in a non-invasive way, which is difficult to do with other analytical techniques. Therefore, there are several works in which FORS has been applied in the identification of colourants on historical paintings, such as easel paintings [11, 16] and mural paintings [15, 43–45], and on illuminated manuscripts [46–49]. While most of these FORS studies on colourants aim at a qualitative identification, some evaluate the possibility of obtaining quantitative data on the colourants present in a paint [50, 51], using specific mathematic algorithms.

In the analysis of paintings and illuminated manuscripts, the spectral range used is the most important feature. Vis range instruments are able to identify some of the "coloured" colourants, but will not provide information on certain groups of materials, such as the white and black pigments. The extension into the UV region (200–380 nm) adds spectral features useful for discrimination among white pigments [16, 52]. On the other side, extension into the NIR region adds spectral features useful for identification of specific materials, such as irongall ink, which can be distinguished from carbon pigments according to its typical rising above 700 nm [53]. Finally, the possibility of extending the spectral range up to the SWIR range strongly widens the diagnostic power of the technique, adding information provided by vibrational band overtones and combinations which allows identifying pigments containing hydroxyls, water molecules, and carbonates [14].

A different situation is the analysis of contemporary art. The number of colour materials used by artists is perhaps hundreds of times larger than those used from the past up to the beginning of the twentieth century. Moreover, many synthetic compounds show chromatic features similar or identical to historic artist's materials. One example is shown in Figure 5.2 in which the inflection points of vermillion and other four modern colourants (alizarin crimson, irgazin scarlet, permanent red, cadmium red) are all located at ca. 600 nm. Despite these premises, some studies recently proposed the use of FORS in the characterisation of contemporary painted artworks [54, 55], at least as a complementary technique. In these cases, the existence of spectral databases is particularly useful [56].

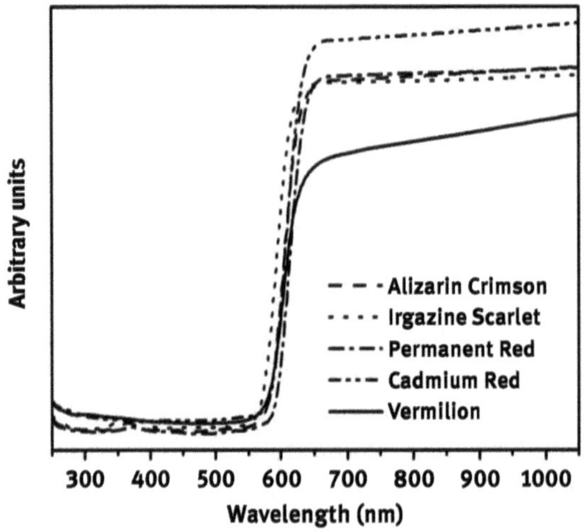

Figure 5.2: FORS spectra of vermillion and modern red colourants.
No Credits.

Of great interest in the field of conservation of paintings is the possibility of monitoring colour changes by means of periodic reflectance measurements [8, 52, 57]. Data can be collected at regular intervals and possible degradation trends can be easily highlighted. Some recent works have also evaluated the potential of FORS in the characterisation of polymers used in paintings and manuscripts, with particular concern to binders [58, 59] and plastics [55, 60].

Complementary to reflectance spectroscopy, microspectrofluorimetry has been particularly applied to the identification of dyes and lake pigments in illuminated manuscripts and paintings, including analysis in paint cross-sections and *in situ* [19, 24]. It has been used to study brazilwood [27], see Figure 5.3, lac dye [28], cochineal [29], alizarin, purpurin and eosin [22], and orcein [25]. A recent work has applied microspectrofluorimetry together with a chemometric approach to discriminate between the different chromophores in a reference database [26].

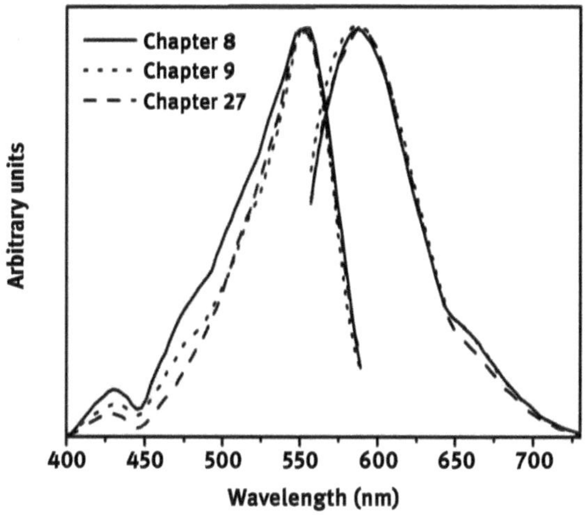

Figure 5.3: Excitation (λem 610 nm) and emission (λex 530 nm) spectra of Brazilwood lake pigments produced from different medieval recipes in *Livro de como se fazem as cores* [27].
No Credits.

5.4.2 Textiles, carpets, tapestries

The main application of FORS to textile materials is related with the identification of the dyes used to give colour to fibres. From the analytical point of view, there is similarity between the identification of dyes on a painting, illuminated manuscript or textile since the chromophores involved are mostly the same. The analysis of textiles is however more difficult for at least two reasons: firstly, the surface of a textile artwork is usually more irregular than that of a painting or manuscript, which is an unfavourable situation when using a surface technique such as reflectance spectroscopy; secondly, in the art of textiles the most used dyes, the so-called mordant dyes (e.g. madder, kermes, cochineal, weld, logwood) were applied on fibres with different mordants (copper, tin, iron and aluminium salts), a feature that can introduce variability in the spectral characteristics useful for identification. An example of the great variability in the identification of cochineal is reported in a study by Morales and Berrie [61], which includes both absorbance and fluorescence spectroscopies.

Due to the great performances of FORS as a survey method, in the last years several works have been published concerning the analysis of carpets [62, 63], clothes [64] or various garments [65–67]. The fast response of FORS can address further investigation such as sampling and analysis with more powerful techniques (HPLC-MS, SERS, etc.). Ancient textiles have also been successfully studied with microspectrofluorimetry [19].

5.4.3 Glass

The use of FORS in the characterisation of coloured vitreous materials cannot yield information on their global composition. However, it can provide information on the nature of the chromophore system in a simple way, including the oxidation state of the species responsible for light absorption, an information that would need more powerful and expensive analytical techniques such as Mössbauer, Auger or X-ray photoelectron spectroscopies. The characterisation of the oxidation state of the chromophores can also yield information on the firing conditions used in glassmaking, in particular with concern to Fe^{2+}/Fe^{3+} ratios. Most of the chromophore systems in glasses can be identified due to their absorption positions and spectral shapes. Table 5.2 presents a selection of these data.

Several studies on glass artworks exploited the identification of chromophores by means of absorption spectrophotometry, using either the transmission or reflectance mode. The pioneering works by Bamford [71] and by Schreus and Brill [73] set the basis for the measurement of the absorption properties of a glass system in transmission mode. Subsequent works exploited the same method [12, 16, 75], using either sunlight or specific light sources, while other studies exploited the reflectance mode [76–78]. Both transmittance and reflectance, in fact, display the same electronic transitions [79]. The lack of scattering centres renders the reflectance mode less advantageous for transparent glasses, while it is more advantageous (or even mandatory) when studying translucent or opaque glass systems such as enamels [80].

5.4.4 Metals

Due to the intrinsic features of FORS, which is a technique mostly devoted to the characterisation of coloured artworks, its contribution in the study of metals is poor.

Table 5.2: Absorption maxima of chromophore systems in glass.

Ions	Absorption maxima	References
Ag(0)	420–500 nm depending on particle size	[68]
Au(0)	525–550 nm depending on particle size	[68]
Co^{2+}	three bands in the 525–650 nm range	[12]
Cr^{3+}	430–460 nm, bands in the 630–680 nm range	[69]
Cr^{6+}	365 nm	[69]
Cu(0)	band centred at 450 nm, peak at 570 nm	[70]
Cu^{+}	no bands	[71]
Cu^{2+}	700–900 nm	[72]
Fe^{2+}	450 nm (sh), band in the 900–1200 nm range	[72]
Fe^{3+}	a series of bands in the 375–450 nm range	[72]
Fe^{3+}-S_2^{-}	405 nm	[73]
Mn^{2+}	430 nm, very weak	[71]
Mn^{3+}	470–520 nm	[69]
Mn^{4+}	450 nm	[69]
Ni^{2+}	three bands in the 430–640 nm range	[69]
UO_2^{2+}	416 nm, bands in the 410–475 nm range	[74]
V^{3+}	425, 645 nm	[71]
V^{4+}	1100 nm	[71]
V^{5+}	350 nm	[71]

Most metals used in artworks can be identified only when they are pure and unaltered [14]. In this context, a preliminary study proposed the possibility of monitoring the conservation state of metal swords [81].

5.4.5 Minerals

With concern to the identification of mineral phases, FORS has already demonstrated its potential as a preliminary technique. Several papers have shown its usefulness in the characterisation of iron oxides and hydroxides [82] and some databases are available online (https://speclab.cr.usgs.gov/spectral-lib.html). Yet, from the mineralogical point of view, the most relevant information provided by FORS in the field of cultural heritage is the identification of coloured gemstones. As in the case of glass, FORS provides the identification of the chromophore system which, in many cases, leads to the identification of the gemstone itself. It cannot be used, though, for the identification of uncoloured gemstones such as diamond and quartz. Another point in common with the analysis of glass artworks is the fact that transparent gemstones usually yield poor spectral responses, due to the lack of scattering centres, while much better spectra can be obtained when analysing opaque gemstones. Usually only the reflectance mode can be exploited in the analysis of gemstones, particularly when complete jewellery artworks are studied in which gemstones cannot be taken away from their settings. Table 5.3 lists the absorption features that can be useful for identification of the most precious gemstones by means of FORS.

Table 5.3: List of spectral features for the identification of gemstones.

Gemstone	Spectral features	Chromophore system	References
amethyst	540 nm (absorption)	Fe^{3+} (colour centre)	[83]
emerald	430/600–615 nm (absorption)	Cr^{3+} (ligand field)	[84]
garnet	500/525/575/695 (absorption)	Fe^{2+}, Fe^{3+}, Mn^{2+}, Mn^{3+}, Cr^{3+}	[85]
ruby	413/564 (absorption), 693/694 nm (fluorescence)	Cr^{3+} (ligand field)	[86]
sapphire	385/450/578 nm (absorption)	Fe^{2+}, Fe^{3+}, Ti^{4+} (ligand field, intervalence charge transfer)	[87]
turquoise	425/625–915 nm (absorption)	Fe^{2+}, Fe^{3+}, Cu^{2+} (ligand field)	[88]

5.4.6 General comments

Independently from the type of material analysed, it is important to point out that the characterisation given by reflectance spectroscopy is based on the position of the minimum/minima of reflectance, which become maximum/maxima values when the spectrum is turned into apparent absorbance or Kubelka-Munk coordinates. The reflectance maximum/maxima, in fact, only provide information concerning the colour of the area analysed, not the chromophore system. The reflectance spectra of cold colours (blue, green, turquoise, etc.) usually have one or more minima and the identification of the spectral features is relatively simple. When the spectrum has no minima, generally the characterisation is based on the inflection point, which is a typical situation for warm colours (red, orange, yellow, brown, etc.). In these cases, an accurate identification of the spectral features is based on the first derivative of the spectrum, where the inflection point becomes a maximum.

Given the features of reflectance systems, with concern mainly to portable FORS systems which can provide several spectra in short times, it can be useful to manage the spectral responses using pattern recognition chemometric methods. Mathematical techniques such as Principal Components Analysis (PCA), Cluster Analysis (CA) or Discriminant Analysis (DA) can be used for the purpose, allowing to group objects (reflectance spectra in this case) which are more similar and therefore providing a semi-automatic way to process large amounts of spectral information. This is particularly important in cases when reflectance spectroscopy is used as a survey method to address further, more powerful diagnostic techniques or the subsequent withdrawal of samples. The most critical point of the pattern recognition process is the normalisation of spectral data, in order to extract the best of the information contained in it. The most suitable methods are *range scaling* (scaling all spectra between 0 and 1), *mean centering* and *autoscaling* (mean centering and normalisation to the standard deviation) [89]. Some examples of the application of pattern recognition methods to spectral information obtained with FORS are reported in the literature with concern to paintings [90, 91] and textiles [9, 64].

5.5 Conclusions (advantages, drawbacks)

For the technical features that it presents, UV-Vis reflectance spectroscopy has been particularly suitable for diagnostic studies on a wide range of artwork typologies. Portable systems, in particular when coupled with fibres, can be used for easily acquiring hundreds of spectra in short time; this amount of information can then be used to address further measurements by means of more powerful techniques. Particularly relevant is the possibility of identifying colourants on paintings and textiles in a non-invasive way, a feature which is unique among spectroscopic techniques. The major drawbacks of reflectance spectroscopy can be resumed as follows: it is hard to manage spectral responses from mixtures of substances (e.g. mixtures of colourants); the spectral responses can be poor in artworks or samples with irregular surfaces or that are partially transparent; it is generally not possible to discriminate colourants having similar absorption features. In such cases, the complementary use of other analytical techniques such as Raman and infrared spectroscopy or mass spectrometry is mandatory. On the other hand, microspectrofluorimetry is also presented as a powerful technique for the non-invasive identification of colourants, particularly if supported by a reference database and by the complementarity of other analytical methods.

Acknowledgements: The authors wish to thank all the colleagues who have collaborated and contributed to the application of UV-Vis spectroscopy and microspectrofluorimetry in the field of cultural heritage. The authors particularly wish to acknowledge Mauro Bacci for his pioneering work on developing FORS applications in this field.

References

[1] Taft WS, Mayer JW. The science of paintings. New York: Springer-Verlag, 2000.
[2] Pinna D, Galeotti M, Mazzeo R, editors. Practical handbook on diagnosis of paintings on movable support, European project ARTECH. Firenze: Centro Di, 2009.
[3] Gauglitz G. Ultraviolet and visible spectroscopy. In: Gunzler H, Williams A, editors. Handbook of analytical techniques. Vol. 1. Weinheim: Wiley-VCH, 2001.
[4] Rawlins FI. Studies in the colorimetry of paintings: a note in conclusion. Tech Stud Field Fine Arts. 1942;10:230–31.
[5] Barnes NF. A spectrophotometric study of artists' pigments. Tech Stud Field Fine Arts. 1939;7:120–30.
[6] Santini M. Applicazione del metodo spettrofotometrico nella determinazione colorimetrica sui dipinti. Boll dell'Istituto Centrale del Restauro. 1959;23/24:95–129.

[7] Duncan DR. The identification and estimation of pigments in pigmented compositions by reflectance spectrophotometry. J Oil Colours Chem Assoc. 1962;45:300–24.

[8] Bullock L. Reflectance spectrophotometry for measurement of colour change. Natl Gallery Tech Bull. 1978;2:49–55.

[9] Cazenobe I, Bacci M, Picollo M, Radicati B, Bacci G, Conti S, et al. Non-destructive spectroscopic investigations of dyed textiles: an application to yellow dyed wool samples. In: Preprints of the 13th triennial ICOM meeting. Rio de Janeiro, ICOM Committee for Conservation, and Roy Vontobel Eds. London : James & James. 2002: 238–44.

[10] Dupuis G, Elias M, Simonot L. Pigment identification by fiber-optics diffuse reflectance spectroscopy. Appl Spectrosc. 2002;56:1329–36.

[11] Leona M, Winter J. Fibre optics reflectance spectroscopy: a unique tool for the investigation of Japanese paintings. Stud Conserv. 2001;46:153–62.

[12] Bacci M, Picollo M. Non-destructive detection of Co(II) in paintings and glasses. Stud Conserv. 1996;41:136–44.

[13] Delaney JK, Ricciardi P, Deming Glinsman L, Facini M, Thoury M, Palmer MR, et al. Use of imaging spectroscopy, fiber optic reflectance spectroscopy, and X-ray fluorescence to map and identify pigments in illuminated manuscripts. Stud Conserv. 2014;59:91–101.

[14] Aceto M, Agostino A, Fenoglio G, Idone A, Gulmini M, Picollo M, et al. Characterisation of colourants on illuminated manuscripts by portable fibre optic UV-visible-NIR reflectance spectrophotometry. Anal Meth. 2014;6:1488–500.

[15] Bacci M, Baldini F, Carlà R, Linari R. A color analysis of the Brancacci Chapel Frescoes. Appl Spectrosc. 1991;45:26–31.

[16] Bacci M, Picollo M, Trumpy G, Tsukada M, Kunzelman D. Non-invasive identification of white pigments on twentieth century oil paintings by using fiber optic reflectance spectroscopy. J Am Inst Conser. 2007;46:27–37.

[17] de la Rie ER. Fluorescence of Paint and Varnish layers (Part I). Stud Conserv. 1982;27:1–7.

[18] Romani A, Clementi C, Miliani C, Favaro G. Fluorescence spectroscopy: A powerful technique for the noninvasive characterization of artwork. Acc Chem Res. 2010;43:837–46.

[19] Claro A, Melo MJ, Seixas de Melo JS, van Den Berg KJ, Burnstock A, Montague M, et al. Identification of red colorants in van Gogh Paintings and Ancient andean textiles by microspectrofluorimetry. J Cult Heritage. 2010;11:27–34.

[20] Melo MJ, Claro A. Bright light: microspectrofluorimetry for the characterization of Lake Pigments and Dyes in works of art. Acc Chem Res. 2010;43:857–66.

[21] Mounier A, Lazare S, Le Bourdon G, Aupetit C, Servant L, Daniel F. LEDμSF: a new portable device for fragile artworks analyses. application on medieval pigments. Microchem J. 2016;126:480–87.

[22] Claro A, Melo MJ, Schäfer S, Seixas de Melo JS, Pina F, van Den Berg KJ, et al. The use of microspectrofluorimetry for the characterization of Lake Pigments. Talanta. 2008; 74:922–29.

[23] Matteini P, Camaiti M, Agati G, Baldo MA, Muto S, Matteini M. Discrimination of Painting binders subjected to photo-ageing by using microspectrofluorometry coupled with deconvolution analysis. J Cult Heritage. 2009;10:198–205.

[24] Melo MJ, Otero V, Vitorino T, Araújo R, Muralha VSF, Lemos A, et al. A Spectroscopic study of Brazilwood Paints in medieval books of hours. Appl Spectrosc. 2014;68:434–44.

[25] Melo MJ, Nabais P, Guimarães M, Araújo R, Castro R, Oliveira MC, et al. Organic Dyes in illuminated manuscripts: a unique cultural and historic record. Philos Trans R Soc A Math Phys Eng Sci. 2016;374:1–20.

[26] Nabais P, Melo MJ, Lopes JA, Vitorino T, Neves A, Castro R. Microspectrofluorimetry and chemometrics for the identification of medieval lake pigments. Heritage Sci. 2018;6:1–11.

[27] Vitorino T, Melo MJ, Carlyle L, Otero V. New insights into Brazilwood lake pigments manufacture through the use of historically accurate reconstructions. Stud Conserv. 2016;61:255–73.
[28] Castro R, Pozzi F, Leona M, Melo MJ. Combining SERS and microspectrofluorimetry with historically accurate reconstructions for the characterization of Lac Dye Paints in medieval manuscript illuminations. J Raman Spectro. 2014;45:1172–79.
[29] Vitorino T, Otero V, Leslie C, Melo MJ, Parola AJ, Picollo M, 2017, Nineteenth-century Cochineal Lake Pigments from Winsor & Newton: insight into their methodology through reconstructions. In: Bridgland J, editors. ICOM-CC 18th Triennial Conference Preprints. Copenhagen. Paris: International Council of Museums, 4-8 Sep 2017. 1–9.
[30] Nassau K. The physics and chemistry of color, the fifteen causes of color. New York: Wiley & Sons; 1983.
[31] Mitton PB. Opacity, hiding power, and tinting strength. In: Patton TC, editor. Pigment handbook volume III characterization and physical relationships. New York: Wiley & Sons, 1973: 289–329.
[32] Burns RG. Mineralogical applications of crystal field theory. 2nd ed. Cambridge Topics in Mineral Physics and Chemistry Vol. 5. Cambridge: Cambridge University Press, 1993.
[33] Vincent RK, Hunt GR. Infrared reflectance from mat surfaces. Appl Opt. 1968;7:53–59.
[34] Kubelka P, Munk F. Ein Beitrag Zur Optik Der Farbanstriche. Z Tech Phys. 1931;12:593–601.
[35] Hollas JM. Modern spectroscopy. 2nd ed. New York: John Wiley & Sons, 1992.
[36] Lakowicz JR. Principles of fluorescence spectroscopy. Berlin/Heidelberg: Springer Science+Business Media, LLC; 2006.
[37] Valeur B. Molecular fluorescence: principles and applications. Weinheim (Germany): Wiley-VCH Verlag GmbH, 2001.
[38] Semwogerere D, Weeks ER. Confocal microscopy. In: Wnek GE, Bowlin GL, editors. Encyclopedia of biomaterials and biomedical engineering. 2nd ed. Vol. I. Boca Raton, Florida, USA: Taylor & Francis, 2008: 705–14.
[39] Workman J. Ultraviolet, visible, and near-infrared spectrometry. In: Workman J, Springsteen A, editors. Applied spectroscopy. A compact reference for practitioners. San Diego: Academic Press, 1998: 29–48.
[40] Springsteen A. Reflectance Spectroscopy: an overview of classification and techniques. In: Workman J, Springsteen A, editor(s). Applied spectroscopy. A compact reference for practitioners. San Diego: Academic Press, 1998:193–224.
[41] Feather GA, Monk DW, 1995, The digital micromirror device for projection display. In: Proceedings IEEE International Conference on Wafer Scale Integration (ICWSI). San Francisco, California, USA: Tewksbury and Chapman (Eds.), 18-20 Jan 1995. 43–51.
[42] Springsteen A. Standards for reflectance measurements. In: Workman J, Springsteen A, editor(s). Applied spectroscopy. A compact reference for practitioners. San Diego: Academic Press, 1998:247–67.
[43] Appolonia L, Vaudan D, Chatel V, Aceto M, Mirti P. Combined use of FORS, XRF and Raman spectroscopy in the study of mural paintings in the Aosta Valley (Italy). Anal Bioanal Chem. 2009;395:2005–13.
[44] Bacci M, Baldini F, Carlà R, Linari R, Picollo M, Radicati B. A color analysis of the Brancacci Chapel Frescoes. part II. Appl Spectrosc. 1993;47:399–402.
[45] Bacci M, Picollo M, Radicati B, Casini A, Lotti F, Stefani L. Non-destructive investigation of wall painting pigments by means of fibre-optic reflectance spectroscopy. Sci Technol Cult Heritage. 1998;7:73–81.
[46] Aceto M, Agostino A, Fenoglio G, Baraldi P, Zannini P, Hofmann C, et al. First analytical evidences of precious colourants on mediterranean illuminated manuscripts. Spectrochim Acta A. 2012;95:235–45.

[47] Delaney JK, Ricciardi P, Deming Glinsman L, Facini M, Thoury M, Palmer MR, et al. Use of imaging spectroscopy, fiber optic reflectance spectroscopy, and X-ray fluorescence to map and identify pigments in illuminated manuscripts. Stud Conserv. 2013;59:91–101.
[48] Aceto M, Agostino A, Fenoglio G, Idone A, Crivello F, Griesser M, et al. Analytical investigations on the coronation gospels manuscript. Spectrochim Acta A. 2017;171:213–21.
[49] Cucci C, Bracci S, Casini A, Innocenti S, Picollo M, Stefani L, et al. The illuminated manuscript corale 43 and its attribution to Beato Angelico: non-invasive analysis by FORS, XRF and hyperspectral imaging techniques. Microchem J. 2018;138:45–57.
[50] Dupuis G, Menu M. Quantitative characterisation of pigment mixtures used in art by fibre-optics diffuse-reflectance spectroscopy. Appl Phys A. 2006;83:469–74.
[51] Fernandez Rodrıguez JM, Fernandez Fernandez JA. Application of the second derivative of the Kubelka–munk function to the semiquantitative analysis of Roman Paintings. Color Res Appl. 2005;30:448–56.
[52] Bacci M, Casini A, Cucci C, Picollo M, Radicati B, Vervat M. Non-invasive spectroscopic measurements on the Il ritratto della figliastra by Giovanni Fattori: identification of pigments and colourimetric analysis. J Cult Heritage. 2003;4:329–36.
[53] Aceto M, Calà E. Analytical evidences of the use of iron-gall ink as a pigment on miniature paintings. Spectrochim Acta A. 2017;187:1–8.
[54] Cucci C, Bartolozzi G, De Vita M, Marchiafava V, Picollo M, Casadio F. The colors of Keith Haring: a spectroscopic study on the materials of the mural painting Tuttomondo and on reference contemporary outdoor paints. Appl Spectrosc. 2016;70:186–96.
[55] Cucci C, Bigazzi L, Picollo M. Fibre optic reflectance spectroscopy as a non-invasive tool for investigating plastics degradation in contemporary art collections: A methodological study on an expanded polystyrene artwork. J Cult Heritage. 2013;14:290–96.
[56] Montagner C, Bacci M, Bracci S, Freeman R, Picollo M. Library of UV–Vis–NIR reflectance spectra of modern organic dyes from historic pattern-card coloured papers. Spectrochim Acta A. 2011;79:1669–80.
[57] Saunders D. The measurement of colour change in paintings. Eur Spectrosc News. 1986;67:10–18.
[58] Dooley KA, Lomax S, Zeibel JG, Miliani C, Ricciardi P, Hoenigswald A, et al. Mapping of egg yolk and animal skin glue paint binders in early renaissance paintings using near infrared reflectance imaging spectroscopy. Analyst. 2013;138:4838–48.
[59] Pallipurath A, Skelton J, Ricciardi P, Bucklow S, Elliott S. Multivariate analysis of combined Raman and fibre-optic reflectance spectra for the identification of binder materials in simulated medieval paints. J Raman Spectro. 2013;44:866–74.
[60] Cucci C, Bacci M, Picollo M, Olmi R. Non invasive techniques for identification and characterisation of polymers in contemporary artworks. In: Berchtold T, editor. Future talks 009. The conservation of modern materials in applied arts and design. Munich: Die Neue Sammlung, 2011: 178–84.
[61] Morales KM, Berrie BH. A note on characterization of the Cochineal Dyestuff on wool using microspectrophotometry. E Preserv Sci. 2015;12:8–14.
[62] Pozzi F, Poldi G, Bruni S, De Luca E, Guglielmi V. Multi-technique characterization of dyes in ancient Kaitag textiles from Caucasus. Archaeol Anthropological Sci. 2012;4:185–97.
[63] Angelini LG, Tozzi S, Bracci S, Quercioli F, Radicati B, Picollo M, 2010, Characterization of traditional dyes of the Mediterranean area by non-invasive UV-Vis-NIR reflectance spectroscopy. In: Conservation and the Eastern Mediterranean Proceedings of the IIC Istanbul Congress. Istanbul (Turkey),Sep 20-24 2010. 184–89.
[64] Gulmini M, Idone A, Davit P, Moi M, Carrillo M, Ricci C, et al. The Coptic textiles of the Museo Egizio di Torino: a focus on dyes through a multi-technique approach. Archaeol Anthropological Sci. 2016;9:485–97.

[65] Maynez-Rojas MA, Casanova-González E, Ruvalcaba-Sil JL. Identification of natural red and purple dyes on textiles by fiber-optics reflectance spectroscopy. Spectrochim Acta A. 2017;178:239–50.

[66] de Ferri L, Tripodi R, Martignon A, Ferrari ES, Lagrutta-Diaz AC, Vallotto D, et al. Non-invasive study of natural dyes on historical textiles from the collection of Michelangelo Guggenheim. Spectrochim Acta A. 2018;204:548–67.

[67] Gulmini M, Idone A, Diana E, Gastaldi D, Vaudan D, Aceto M. Identification of dyestuffs in historical textiles: strong and weak points of a non-invasive approach. Dyes Pigm. 2013;98:136–45.

[68] Quinten M. The color of finely dispersed nanoparticles. Appl Phys B. 2001;73:317–26.

[69] Terczynska-Madej A, Cholewa-Kowalska K, Laczka M. The effect of silicate network modifiers on colour and electron spectra of transition metal ions. Opt Mater. 2010;32:1456–62.

[70] Schreiber HD, Stone MA, Swink AM. Novel red–blue dichroic glass containing copper nanocrystals. J Non-Cryst Solids. 2006;352:534–38.

[71] Bamford CR. Colour generation and control in glass. Amsterdam: Elsevier, 1977.

[72] Meulebroeck W, Baert K, Ceglia A, Cosyns P, Wouters H, Nys K, et al. The potential of UV-VIS-NIR absorption spectroscopy in glass studies. In: Meulebroeck W, Nys K, Vanclooster D, Thienpont H, editors. Proceedings of SPIE Vol. 8422, integrated approaches to the study of historical glass - IAS12. Vol. 8422. Belgium: SPIE Brussels 2012: 842208–1-11.

[73] Schreus JW, Brill RH. Iron and sulfur related colors in ancient glasses. Archaeometry. 1984;26:199–209.

[74] Reiche I, Röhrs S, Salomon J, Kanngießer B, Höhn Y, Malzer W, et al. Development of a nondestructive method for underglaze painted tiles - demonstrated by the analysis of Persian objects from the nineteenth century. Anal Bioanal Chem. 2009;393:1025–41.

[75] Meulebroeck W, Cosyns P, Baert K, Wouters H, Cagno S, Janssens K, et al. Optical spectroscopy as a rapid and low-cost tool for the first-line analysis of glass artefacts: a step-by-step plan for Roman green glass. J Archaeol Sci. 2011;38:2387–98.

[76] Arletti R, Conte S, Vandini M, Fiori C, Bracci S, Bacci M, et al. Florence baptistery: chemical and mineralogical investigation of glass mosaic tesserae. J Archaeol Sci. 2011;38:79–88.

[77] Galli A, Poldi G, Martini M, Sibilia E, Montanari C, Panzeri L. Study of blue colour in ancient mosaic tesserae by means of thermoluminescence and reflectance measurements. Appl Phys A. 2006;83:675–79.

[78] Mirti P, Davit P, Gulmini M. Colourants and opacifiers in seventh and eighth century glass investigated by spectroscopic techniques. Anal Bioanal Chem. 2002;372:221–29.

[79] Möncke D, Papageorgiou M, Winterstein-Beckmann A, Zacharias N. Roman glasses coloured by dissolved transition metal ions: redox reactions, optical spectroscopy and ligand field theory. J Archaeol Sci. 2014;46:23–36.

[80] Agostino A, Aceto M, Fenoglio G, Operti L. Analisi non invasive sugli smalti limosini medievali di Palazzo Madama. In: Castronovo S, editor. Smalti di Limoges del XIII secolo. Savigliano, Italy: L'Artistica Editrice, 2014: 211–39.

[81] Grazzi F, Iannaccone R, Barzagli E, Civita F, Salvemini F, Picollo M, et al. 2011. UV-VIS-NIR specular reflectance spectroscopy: a non invasive and portable technique for the characterization of the conservation status of ancient swords. Application to the Japanese blade collection of the Stibbert museum. In: Proc. ART11, 10th International Conference on non-destructive investigations and microanalysis for the diagnostics and conservation of cultural and environmental heritage. Florence: CD-ROM. 13-15 Apr 2011.

[82] Torrent J, Barròn V. Diffuse reflectance spectroscopy of iron oxides. In: Hubbard AT, editor. Encyclopedia of surface and colloid science. New York: Marcel Dekker, 2002: 1438–46.

[83] Rossman GR. Colored varieties of the silica minerals. Rev Mineral. 1994;29:433–68.

[84] Wood DL, Nassau K. The characterization of beryl and emerald by visible and infrared absorption spectroscopy. Am Mineral. 1968;53:777–800.
[85] Izawa MRM, Cloutis EA, Rhind T, Mertzman SA, Poitras J, Applin DM, et al. Spectral reflectance (0.35–2.5 μm) properties of garnets: implications for remote sensing detection and characterization. Icarus. 2018;300:392–410.
[86] Liu Y, Lu T, Mu T, Chen H, Ke J. Color measurement of a ruby. Color Res Appl. 2013;38:328–33.
[87] Bristow JK, Parker SC, Catlow CRA, Woodley SM, Walsh A. Microscopic origin of the optical processes in blue sapphire. Chem Commun. 2013;49:5259–61.
[88] Qiu JT, Qi H, Duan JL. Reflectance spectroscopy characteristics of turquoise. Minerals. 2017;7:1–10.
[89] Aceto M, Agostino A, Fenoglio G, Gulmini M, Bianco V, Pellizzi E. Non invasive analysis of miniature paintings: proposal for an analytical protocol. Spectrochim Acta A. 2012;91:352–59.
[90] Aceto M, Agostino A, Fenoglio G, Picollo M. Non-invasive differentiation between natural and synthetic ultramarine blue pigments by means of 250–900 nm FORS analysis. Anal Meth. 2013;5:4184–89.
[91] Orlando A, Picollo M, Radicati B, Baronti S, Casini A. Principal component analysis of near-infrared and visible spectra: an application to a XIIth century Italian work of art. Appl Spectrosc. 1995;49:459–65.

Francesca Rosi, Laura Cartechini, Diego Sali and Costanza Miliani

6 Recent trends in the application of Fourier Transform Infrared (FT-IR) spectroscopy in Heritage Science: from micro- to non-invasive FT-IR

Abstract: The relevance of FT-IR spectroscopy in heritage science has experienced a constant grow in the last two decades owing to analytical peculiarities that make it an extremely useful tool to answer the questions posed by the study and conservation of art-historical and archaeological materials. High versatility, sensitivity and molecular specificity are, in fact, all requirements that FT-IR spectroscopy fulfils allowing for the investigation of the chemical properties of heritage materials spanning from the micro- to the macro-scale and offering a variety of approaches to minimize sample manipulation and maximize extracted information. Molecular identification and localisation at high lateral resolution of organic and inorganic components in micro-samples was, over recently, the mostly exploited use of FT-IR in heritage science; however, benefiting from technological progress and advances in optical materials and components achieved in the last decade, it now stands out also for non-invasive surface analysis of artworks by fully portable instrumentation.

Keywords: external reflection, ATR, hyperspectral imaging, micro-FT-IR, portable FT-IR

6.1 Introduction

Analytical chemistry plays a key role for addressing and solving complex and challenging issues related to knowledge and preservation of our world heritage treasures. Art-historical and archaeological questions, the evaluation of the state of conservation, the establishment of the best conservation practice to limit alterations (the so-called preventive conservation), as well as the monitoring of restoration interventions are all aspects commonly faced and successfully solved by analytical chemistry [1]. Above all, Fourier Transform Infrared spectroscopy (FT-IR) highly contributes to heritage science providing valuable information to tackle many of these issues. Versatility, high specificity and sensitivity are the main advantages of this technique for studying cultural heritage (CH) materials. Since the first paper published in 1966 [2], the exploitation of this technique for investigating paintings and ancient artefacts has

This article has previously been published in the journal *Physical Sciences Reviews*. Please cite as: Miliani, C., Rosi, F., Cartechini, L., Sali, D., Recent trends in the application of Fourier Transform Infra Red (FT-IR) spectroscopy in Heritage Science: from micro- to non-invasive FT-IR *Physical Sciences Reviews* [Online] **2019**, 4. DOI: 10.1515/psr-2018-0006.

https://doi.org/10.1515/9783110457537-006

grown exponentially favoured by the advent of FT-IR microscopy and the introduction of portable non-invasive spectrometers and hyperspectral camera.

The present chapter aims at giving an overview of both micro and non-invasive FT-IR spectroscopies which meet the specific requirements posed by the study of heritage materials characterized by an inherent chemical heterogeneity at both the micro- and macro-scale and by a preciousness that asks for minimally/non-invasive approaches. These two types of FT-IR analysis are typically based on the reflection acquisition modes, allowing to interrogate infrared optically thick samples. Thus, for convenience of the reader, the main principles of FT-IR reflection spectroscopy are summarized in the next first paragraph while the illustration of the general theory for FT-IR absorption spectroscopy is left to the wide available literature ([3, 4] and references therein).

6.2 Reflection measurements in FT-IR spectroscopy

FT-IR spectroscopy in reflection configuration owes its wide diffusion in many research contexts and diagnostic applications to its versatility in approaching the analysis of highly absorbing samples requiring minimal preparation with respect to transmission measurements.

Typically, reflection acquisition modes are divided in Attenuated Total Reflection (ATR), external reflection and transflection. In ATR, an Internal Reflection Element (IRE) with a high refractive index (n_{IRE}) is put in contact with an optically opaque sample with a lower refractive index (n_{sample}). Above a certain angle (the critical angle) the IR light totally reflects inside the IRE generating the so-called *evanescent wave*. It extends beyond the surface of the crystal into the sample held in contact with the IRE; the sample absorbs the *evanescent wave* at specific energy regions, then the attenuated IR beam exits from the IRE and reaches the detector [5]. The electric field of the *evanescent wave* exponentially decays with the distance from the surface of the sample and the penetration depth varies with the wavelength (λ) according to the following relationship:

$$d = \frac{\lambda}{2\pi n_{IRE}\sqrt{(\sin^2\theta - (n_{sample}/n_{IRE})^2}}$$

(where θ is the incident angle at the internal surface of the IRE) [6]. Thus ATR spectra show apparent higher absorption intensity at longer wavelengths (lower wavenumbers) that can be corrected and generally software packages for data management are equipped with dedicated algorithms [5]. The resulting spectral profiles resemble those acquired in transmission mode thus facilitating spectra interpretation by referencing to the available spectral databases.

Penetration depth depends also on the relative difference of the two refractive indexes. The sharp rise of the refractive index across an absorption band may cause the criterion of internal reflection to be lost resulting in an unpredictable distortion of the peak shape and shift of the band position [7, 8]. This type of distortions mainly affects strong IR absorbing materials (more intense IR bands) and generally cannot be corrected.

External reflection is a non-contact acquisition mode in which light propagates from a medium (i.e. air with n = 1) toward a sample (with n > 1). In these conditions two types of reflection take place: surface reflection and volume reflection [9].

Surface reflection is a specular reflection, being light reflected at the same angle of the incident beam; it characterizes optically flat surfaces in which the crystal dimensions are much larger than the wavelength of the incident radiation. Surface reflection, ruled by the Fresnel's law at a normal incidence depends on the real part (n, representing the light dispersion) and the imaginary part (ik, with k representing the absorption index) of the complex refractive index ñ (ñ = n + ik) as it follows:

$$R = \frac{(n-1)^2 + k^2}{(n+1)^2 + k^2}$$

where n has a typical dispersion with the wavenumbers and assumes a derivative-like profile in correspondence of the absorption bands; k is proportional to the absorption coefficient of the Lambert Beer's law thus it changes along the wavenumber axis similarly to an IR absorption profile [4].

When $k < 1$ (most of the organic compounds, including polymeric systems) R follows mainly the behaviour of n, namely inflection points replace the IR absorption bands. When $k \gg 1$ (inorganic compounds containing oxyanions, i.e. carbonates, sulphates, phosphates), R tends to unity. The strong absorption hampers radiation to penetrate the sample in that energy range due to the strong change of n; as a result, light is mainly reflected (at these wavenumbers the sample behaves as a metallic surface) giving rise to a maximum of reflection at the maximum absorption (inverted band); this phenomenon is better known as *reststrahlen* effect [10].

In volume reflection, IR penetrates the sample undergoing refraction, reflection, and scattering prior to be reflected out from the surface in all the directions (in fact it is also called diffuse reflection). Volume reflection gives rise to spectra similar to those recorded in transmission mode, but with variation in the band relative intensities and with band broadening mainly linked to the degree of light penetration.

This decreases with stronger scattering and absorption according to the relationship

$$d = \frac{1}{\sqrt{k + (k + 2s)}}$$

where s is the scattering coefficient and k the absorption coefficient. It explains why volume reflection is highly sensitive to weak absorbers [11, 12].

In external reflection mode, generally, CH materials show spectral distortions arising from the variable contributions of both volume and surface reflection, depending from the roughness and absorption properties of the sample surface. Common correction algorithms (Kubelka-Munk and Kramers Kronig [4, 12–15] are not able to properly correct these distortions for enabling identification via standard library search.

Transflection is a particular case of specular reflection not governed by the Fresnel's law. It generally occurs when a sample, with a thickness (d) of the order or slightly larger than the IR wavelengths $d > 0.5\ \lambda$, is deposited on an infrared reflecting substrate (most often a high reflecting metallic surface) [4]. First, the infrared light passes through the sample, then it is reflected from the substrate surface and is transmitted through the sample again before being detected. The resulting profile is equivalent to that of a sample of thickness 2d in transmission mode but with some contribution from the beam directly reflected at the sample air/ interface and connected distortions.

6.3 Micro-FT-IR spectroscopy: resolving the chemical complexity of CH samples at the micro-scale

The study of precious and irreplaceable objects demands for highly sensitive techniques capable of working at high lateral resolution in order to minimize the amount of the sample and to solve at the micro-scale the complex compositional profile that usually characterizes heritage materials. A typical example of these complex systems is represented by polychrome objects (i.e. paintings, works on paper, polychrome sculptures, etc.). A sample collected from a painted artwork usually shows a micro-scale multi-layered structure made of heterogeneous mixtures of organic and inorganic substances and the products of their possible deterioration. The identification and spatial location of the various components within the multi-layered structure still represent a challenging issue. To this purpose micro-FT-IR spectroscopy is a well-established method for in deep molecular investigation of the art objects micro-structure, offering a variety of approaches and solutions able to address specific analytical issues. Over the last two decades, the significance of micro-FT-IR spectroscopy in the field of heritage science strongly increased thanks to the tangible improvement in terms of automations and spectral sensitivity [16–20].

6.3.1 Instrumental set up: point by point mapping, line mapping and imaging micro-FT-IR

Different instrumental set ups are available allowing for point by point mapping, line mapping and imaging experiments to be performed. These are realized through the use of different optics and detectors which strongly impact the final characteristics of the micro-FT-IR measurement in terms of: measurement time, achievable lateral resolution and costs. Point by point mapping (often called also "raster scan", Figure 6.1(a)) is the simplest and less expensive acquisition set up being based on a FT-IR microscope equipped with a single element detector, generally a photoconductive Mercury Cadmium Telluride-MCT, and an automated sample stage, software controlled. Line scanning and imaging set ups (Figure 6.1(b) and Figure 6.1(c), respectively), on the other hand, use multiple element detectors (usually MCT) arranged in an array, with a linear or a rectangular/square format (Linear Array-LA and Focal Plane Array-FPA, respectively). The principle of working with LA and FPA detectors is based on projecting the image of a large IR beam (a few mm^2) on to the sample and each detector element records the spectrum of the specific projected pixel. This technology allows faster spectra collection according to the dimension of the multiple-element detector (which is roughly proportional to the total cost). Acquisition time decreases roughly of a factor corresponding to the number of pixels covered with the standard mapping procedure using a single MCT element detector (in the case of a square FPA with a 64 × 64 array it is possible to get 4.096 spectra in one second). LA and FPA detectors have a typical cut-off at 720 and 900 cm^{-1}, respectively, and the single pixel has a typical dimension of about 40 μm. The ordinary single element MCT detector has a typical dimension of about 250 μm (or

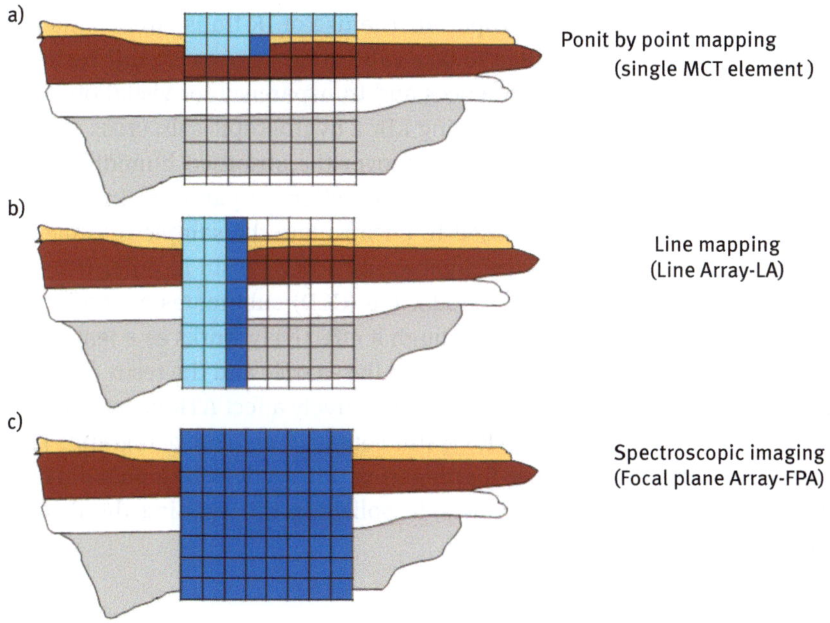

Figure 6.1: Sketch of the possible instrumental set ups for micro-chemical imaging: a) point by point mapping with single MCT element, b) line mapping with Line Array and c) spectroscopic imaging with Focal Plane Array.

smaller, down to 100 μm in particular cases), it is able to provide a broader spectral range down to 420 cm^{-1} [4, 20] and it has typically a better signal to noise ratio if compared to the single pixel of the FPA or LA detector.

6.3.2 Sample preparation

In micro-FT-IR the selection of the most appropriate acquisition mode (namely transmission or reflection), is done depending on the nature, amount, typology of the sample (powder, fragment) and the questions to be solved. While sample bulk analysis can be easily carried out with minimal sample preparation, in the case of stratigraphic studies the sample has to be adequately prepared as to expose the layering sequence. Cross- and thin- sections (for working in reflection and transmission mode, respectively) are then commonly prepared by immersing the multilayered sample in an embedding material which hardens in a short time. Its main function is to prevent any sample loss during cutting and polishing operations for exposing the whole stratigraphy. As main drawback of the procedure, it may happen that during embedding with an organic polymer (acrylate or polyester resins are the most common ones [21]) the resin penetrates into the stratigraphy producing spectral signals that may hinder the infrared bands of the sample components. This problem has been widely faced off by scientists and several alternative methods of preparing cross-sections have been proposed. The use of IR-transparent salts [18, 22–28], in particular KBr, demonstrated to be a valid alternative to organic resins in terms of elimination of spectral interferences and transparency for visual observation of the embedded sample. However, being KBr a hygroscopic salt, cross-section preservation over time is not guaranteed, moreover the absorbed humidity negatively affects spectral quality, shortly after cross-section preparation. In 2008 de Fonjaudran and co-workers [29] proposed to encapsulate the sample into melted cyclododecane before immersion in the organic embedding resin in order to avoid its penetration inside the sample. Cyclododecane (CCD) sublimates at room temperature once the sample is sectioned. Although it effectively works as a temporary barrier, after sublimation it leaves a gap between the sample and the resin. This gap reduces the sample stability and further it may negatively affect ATR measurements because of the not optimal sample-ATR crystal contact (see next paragraph) [26]. This method was recently adapted also for preparing thin sections by embedding the sample in CDD and performing a double-side polishing till reaching the desired section thickness [30].

6.3.3 Micro-ATR-FT-IR: high resolution chemical imaging at the micro-scale

Among the different acquisition modes, ATR is the most commonly employed capable to reach the highest lateral resolution achievable in FT-IR experiments.

The lateral resolution (the minimum distance at which two points can be resolved) in far field commercial FT-IR microscopes is mainly dominated by the diffraction limit (defined as $r = 0.61\lambda/NA$, with NA-the objective numerical aperture= $n\sin\alpha$, n the refraction index of the medium between point source and lens and α the half aperture angle of light). In ATR mode the lateral resolution is improved by the increased NA due to the high refractive index IRE [31–33] and reference therein]. As an example, using a standard FPA detector (single pixel size 40 μm) and a Germanium IRE, with 20× objective magnification and 1:1 optics, ATR spectra have a pixel resolution of 0.5 μm [20, 31]. Although the highest lateral resolution is generally achieved by using FPA detectors, published data [20, 34] demonstrated that it is possible to spatially resolve layers 5 μm thick also with a mapping device with single element MCT detector using state of the art IR microscopes with fully automated ATR operation.

In ATR measurements it is essential to ensure good optical contact between the surface of the sample and the crystal for achieving good spectral quality [34–36]. The necessary pressure for a good contact may cause an unwanted displacement of the sample thus measuring an area different from the originally selected one [26].

An important advantage is that ATR spectra can be converted to conventional absorption ones, critical condition for materials identification by cross-validation of data with reference databases. However CH materials often have high absorbing properties and in these conditions (as explained in Section 6.2) the internal reflection criterion may be lost resulting in spectral distortions that cannot be corrected.

As an example Rosi et al. [37] reported on the ATR spectral distortion of the strongest absorption band of the sulphate anion (ν_3) in $CaSO_4$ samples containing the anhydrous, emi and di-hydrate forms. The antisymmetric stretching of the sulphate anion was distorted and slightly shifted to lower wavenumbers with respect to transmission mode making impossible to distinguish the three $CaSO_4$ species in ATR mode.

Valuable applications of micro-ATR-FT-IR imaging (including FPA, LA and mapping modes) to CH materials are several: mural paintings [34], polychrome sculptures [18, 24], paintings [20, 38], photographic prints [39], for which it enables to unveil the distribution of organic and inorganic components within the different layers. Studies on degradation of painting materials also benefitted of the high lateral resolution and the high chemical specificity of micro-ATR-FT-IR imaging [16, 40]. As an example in Figure 6.2 the distribution of metal-carboxylates by micro-ATR-FT-IR imaging on a cross-section from a white paint (composed of a mixture of ZnO and TiO_2 in oil) of *Alchemy* by Jackson Pollock [16] is shown. Each chemical image was obtained by integrating the infrared bands diagnostic for the different molecular species; the values were, then, plotted for each spatial position. Chemical imaging enabled to find a widespread interaction of Zn with carboxylate groups having a characteristic infrared broad band at about 1590 cm^{-1} [41]. Furthermore newly formed zinc stearate aggregates ($\nu_{as}COO^-$ at about 1540 cm^{-1}) were identified and localized around particles of aluminum stearate (used as jellifying agent in the oil paint formulation, $\nu_{as}COO^-$ at 1588 cm^{-1}). The

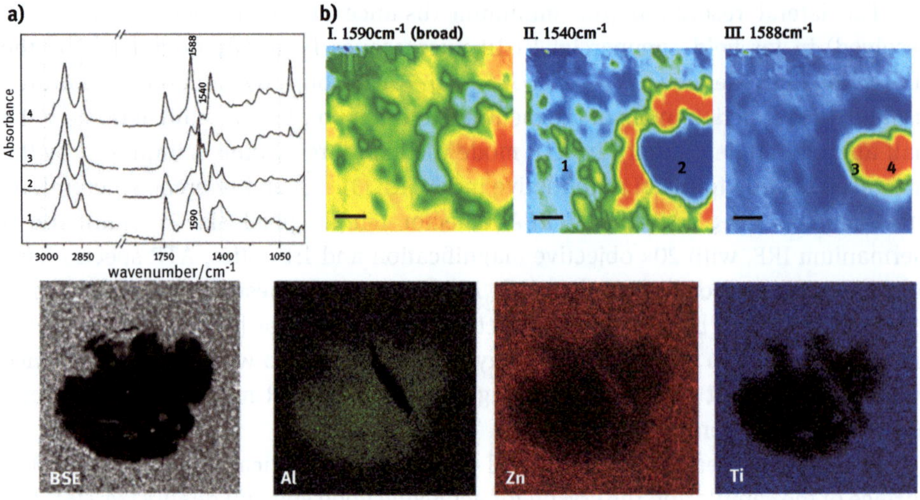

Figure 6.2: (a) Micro-ATR–FT-IR spectra extracted from single pixels of the FPA detector corresponding to different locations in a cross section of a white paint micro-sample from the painting *Alchemy* by J. Pollock. Numeric labels refer to their spatial localization within the corresponding chemical images reported in (b). (b) Chemical images resulting from the integration of (I) the broad band at 1590 cm^{-1}, (II) the band at 1540 cm^{-1} of zinc stearate, and (III) the band at 1588 cm^{-1} of aluminum stearate. (c) Back scattered electron-SEM and elemental images (Al, Zn, and Ti) in the same area of the cross section. In the micro-ATR–FT-IR images, the scale bar corresponds to 10 μm. Reprinted with permission from F. Gabrieli et al. Anal. Chem. 2017, 89, 2, 1283-1289. Copyright (2016) American Chemical Society [16].

provided information greatly contributed to understand the complex heterogeneous microstructure of *Alchemy* as a result of the pigment-oil interaction.

6.3.4 External reflection micro-FT-IR: contactless micro-chemical imaging

External reflection micro-FT-IR works in non-contact mode using a Cassegrain objective. After the pioneering work of Heeren [42] micro-FT-IR spectroscopy in reflection mode experienced a rather limited use in heritage science with respect to the ATR mode in spite of its potentialities in determining the chemical identification and molecular spatial distribution of painting stratigraphies as highlighted in research papers by Boon and co-workers [21, 43–46] and in few other publications [47, 48]. Owing to the fact that the microscope optical layout of the Cassegrain objective favours the collection of specular reflected radiation over the diffuse light, it is generally assumed that the reflection FT-IR profiles can be treated by the Kramer Kronig (KK) correction. Nevertheless, as already explained in Section 6.2, this correction may not work properly for heritage materials and lead to misleading conclusions.

In 2011 Rosi et al. [49] demonstrated the potentialities of working with external reflection micro-FT-IR spectroscopy on cross-sections through mapping experiments without applying any type of correction to the data. In order to facilitate spectral interpretation, the wide spectral range including the near infrared portion (until 6000 cm^{-1}) was explored enabling to unveil the nature of the different binders in a seventeenth century Italian wooden polychrome sculpture.

For the image re-construction, multivariate image processing (Principal Component Analysis-PCA) was successfully employed in order to treat the distorted reflection infrared profiles displaying complex convolutions of derivative-like and *reststrahlen* bands (Figure 6.3(a) [49]). The use of multivariate methods for data analysis in FT-IR imaging experiments (independently from the acquisition mode) is a very profitable approach to extract chemical and spatial information from large spectral datasets. In fact, due to the compositional complexity and heterogeneity of CH materials, the simple integration of the band (univariate method): (i) may be insufficient for representing the complete chemical composition, (ii) may not be able to manage spectral overlapping affecting spectral identification and spatial localization of the specific compounds, and (iii) may not be able to manage spectral distortions originated by the specific acquisition mode. A valid alternative to the univariate approach is the application of chemometrics [50, 51]. Above all, PCA is able to extract the relevant information from a large spectral dataset producing reliable chemical distributions of the materials in a complex stratigraphy. The chemical map is spatially re-constructed plotting the score values along the selected principal component (PC) thus obtaining the distribution of all the chemical species contributing to this PC (Figure 6.3(a-i) [49]).

Recently, Capobianco et al. [17] successfully used Hierarchical Cluster Analysis (HCA) -validated with a supervised Principal Component based k-Nearest Neighbor Analysis (PCA-kNN)- for the re-construction of chemical images obtained in FT-IR imaging experiments in reflection mode. The non-contact acquisition mode and the use of an FPA detector enabled to acquire a very large image ($1024 \times 640\ \mu m^2$) at very high lateral resolution (diffraction limited) thus obtaining a large overview of the chemical composition of the investigated sample (Figure 6.3(m)). The possibility of exploring large areas of a sample cross section is, in fact, the other main advantage of the external reflection acquisition mode with respect to the ATR, which contrarily suffers from limited field of view.

6.3.5 Enhanced micro-FT-IR techniques for CH materials

The solution of specific analytical issues demanding for improved spectral quality, high lateral resolution and reduced acquisition time for micro-FT-IR studies in heritage science as well as in other disciplines fostered the development of more sophisticated methods. A possible approach is based on the use of high brilliance synchrotron radiation (SR) sources in substitution of conventional thermal sources. The high flux density of SR provided in a highly focused beam allows for probing smaller areas at higher signal to noise ratio with respect to conventional thermal sources. Frequently,

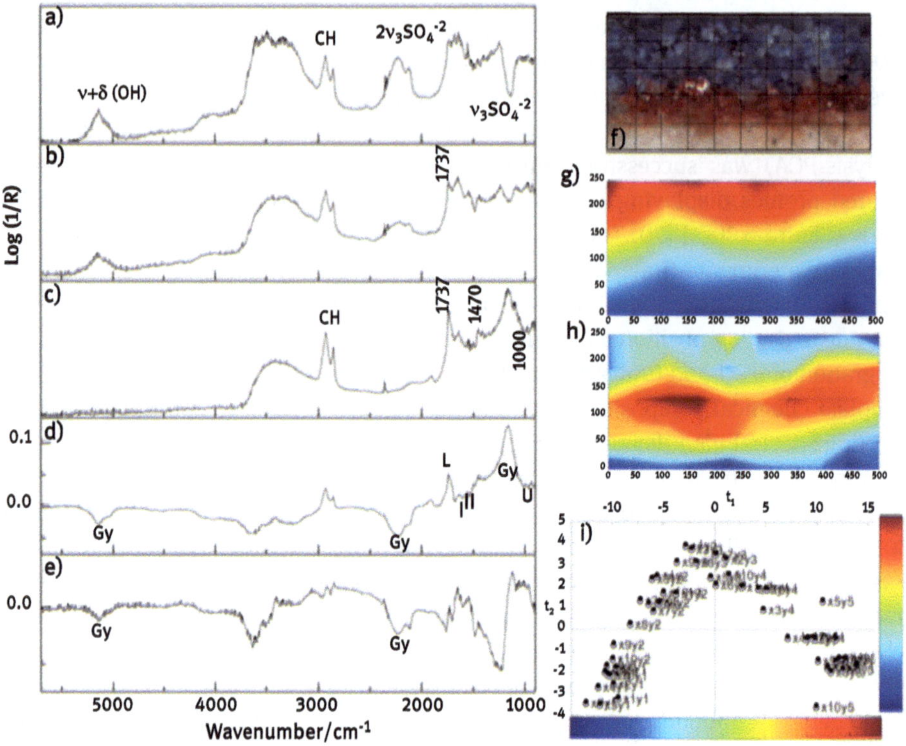

Figure 6.3: a-c) Reflection mode micro-FT-IR spectra from a cross section of a paint model made of a) a gypsum ground layer, b) a cinnabar-protein layer and c) an ultramarine-linseed oil layer; d) and e) loading profiles along the first-p_1 and second-p_2 principal components showing the vibrational modes/chemical components responsible of the differences highlighted by the PCA (labels: Gy=gypsum, L=lipid, U=ultramarine, I=amide I, II=amide II); f) OM image with the grid indicating the mapped area (single point measurement of dimension 50x50 μm^2); g) and h) chemical maps obtained by plotting the score values along each components, t_1-score map and t_2-score map, respectively, resulting from the PCA; i) t_1 vs. t_2 score plot with color bars referred to the score values along each components. The t_1-score map defines the whole material distribution: the second layer (ultramarine–linseed oil) is in red, the ground layer is in blue, and the first layer (cinnabar–animal glue) is in yellow–cyan colors. In order to better visualize the glue-containing layer, the second PC can be considered. l) Cross section of the patina sample from the Arch of Septimius Severus. The inset shows the analyzed area (1024 × 640 μm^2) by micro-FT-IR mapping in external reflection mode. m) Chemical image reconstructed by mapping the average spectra resulting from Hierarchical Cluster Analysis-HCA. The average spectra were used as reference to calculate the correlation coefficient of each spectrum of the original map respect to a reference spectrum. Every correlation map is assigned to a different color in the *Winner takes it all*-WTA image. Adapted with permission from Springer Nature Analytical and Bioanalytical Chemistry, Rosi F et al. 399:3133-45 (2011) [49]. Adapted with permission from Elsevier, Microchemical Journal, G. Capobianco et al. 132:69-76 (2017) [17].

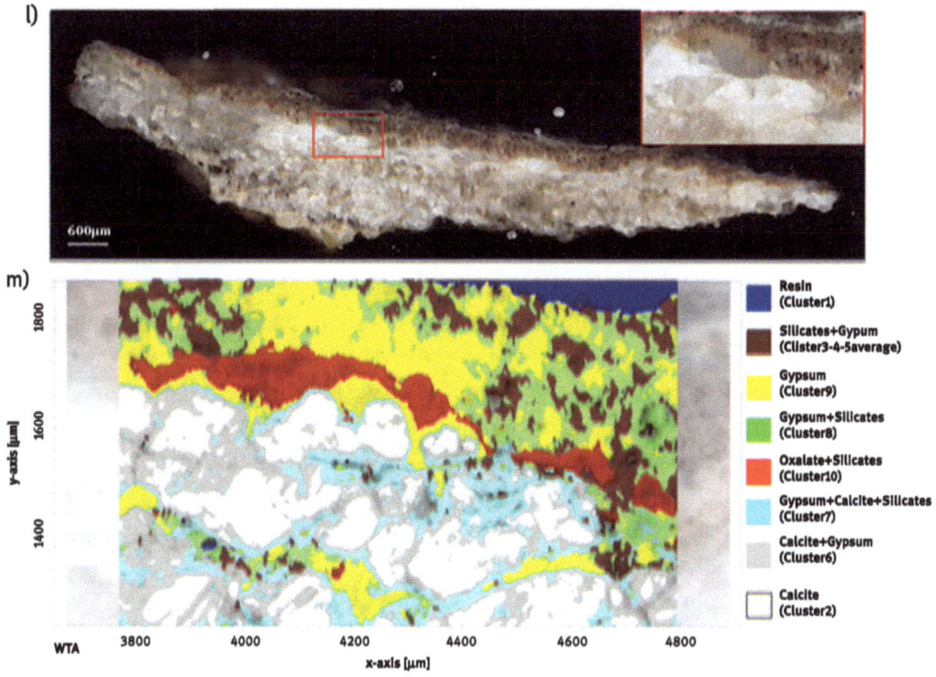

Figure 6.3: (continued)

the FT-IR experiments carried out with SR are performed in transmission mode being often preferred over reflection acquisition mode for the improved spectral quality and for benefitting of the coupling with SR X-rays based techniques [52–57]. On the other hand, working in transmission mode requires that the sample has to be reduced in thickness by squeezing or thinning it (down to few μm –generally below 4-5 μm– depending on the IR absorption properties of the sample) in order to avoid the complete absorption of the infrared radiation. If the aim is to study a painting stratigraphy, thin-section preparation is much more challenging than that of cross-section [29, 53, 54].

The most promising development is the use of SR combined with an FPA detector to perform micro-SR-FT-IR imaging in transmission mode with improved lateral resolution, signal to noise ratio and acquisition time [58–60]. At the IRENI beamline of the Synchrotron Radiation Center (Wisconsin, USA) the coupling of the FPA-fitted microscope with SR was possible by using an array of 12 SR beams to create a homogeneous beam over the large field of view of the FPA [58, 61]. This promising configuration, primarily employed for the study of biological samples, has been recently successfully applied for investigating both model paintings and historical samples [62].

More recent efforts are directed towards the development of nanoscale methods able to provide lateral resolution down to 50–100 nm. For example by coupling a pulsed IR laser and an Atomic Force Microscope (AFM) it was possible to investigate and map the collagen denaturation of altered parchments directly on individual collagen fibres (~1–5 μ m) or fibrils (~100 nm) [63].

When the spectroscopic analysis focuses on the identification of trace components (or low infrared absorbing materials) minimizing the required amount of sample, other enhanced FT-IR methods with improved sensitivity can be applied for studying CH materials.

Similarly to surface-enhanced Raman scattering (SERS), the infrared adsorption of a sample on metallic nanoparticles results enhanced by a combination of electromagnetic and chemical mechanisms. The whole process is known as surface-enhanced infrared reflection absorption-SEIRA and it has been already applied with positive outcomes for the analysis of dyes ([19] and references therein). Dyes identification represents a well-known challenging task in heritage science; dyes possess high tinting strength and are generally present in low concentration in the supporting matrix (paper, textile, painting). This latter partially or totally prevails over the IR bands of the dyes making conventional infrared spectroscopy not suitable for their identification. Differently, SEIRA, having a higher sensitivity, can be performed on sample micro-extracts eliminating the spectral interference of the strong IR bands of the supporting matrix [64].

In order to further reduce the amount of the sample for the analysis, Prati et al. [65] proposed also a new enhanced FT-IR method called metal underlayer ATR spectroscopy (MU-ATR). With this method few ngs of dyed fibre micro-extracts are spotted on a gold coated glass slide and analysed in ATR. Recently, MU-ATR has been combined with complementary SERS for identifying dye mixtures deposited on thin-layer chromatography (TLC) substrate [66].

Rosi et al. [67] proposed a new FT-IR enhanced method based on micro-transflection carried out through a diamond-coated metallic stick commercially available as an accessory for diffuse reflectance infrared Fourier transform (DRIFT) measurements-used here both as micro-sampling tool for painting surface materials and as a reflective substrate for micro-transflection measurements. The approach demonstrated to be very useful for the characterization of varnish coatings on painting surfaces before cleaning treatments. A very localized minimal sampling is carried out through a gentle scratch of the varnish surface by the diamond-coated stick (Figure 6.4). The small amount of varnish spread on the metal stick is then analysed in reflection mode through a FT-IR microscope: the incident infrared beam is transmitted through the sample, reflected by the metallic stick and transmitted again through the sample, approaching a transflection-like regime [4]. The micro-transflection approach has been positively combined with non-invasive reflection FT-IR spectroscopy in monitoring the effect of solvent cleaning on a sixteenth century painting by Vittore Carpaccio [67]. Non-invasive FT-IR permitted a wide survey of the painting surface by a representative number of measurements (before as well as after the cleaning), the minimally invasive micro-transflection approach helped in the molecular identification supporting the spectral interpretation of the reflection FT-IR data.

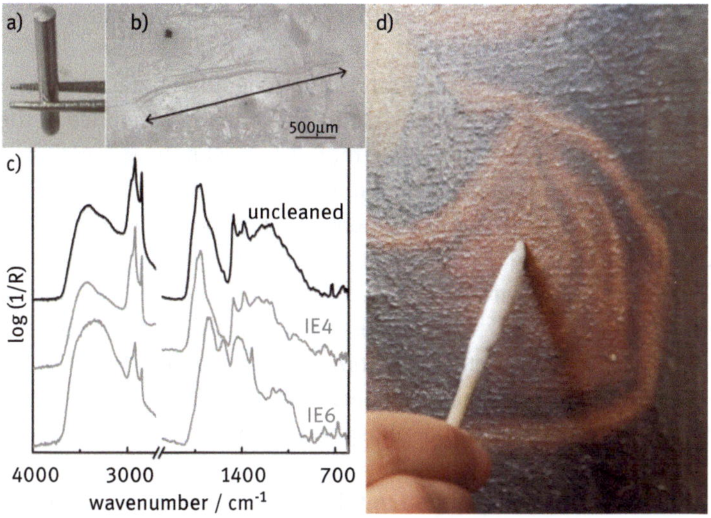

Figure 6.4: a) A diamond-coated metallic stick is employed as micro-sampling tool and as a reflective substrate for micro-transflection measurements of surface painting layers (i.e. varnishes). b) By a gentle scratch of the surface materials the diamond-coated stick allows very localized and minimal sampling. c) The small amount of varnish spread on the metal stick is then analyzed in reflection mode through an FT-IR microscope in a transflection-like regime. d) The method allows for identifying the nature of surface varnishes in order to support restorers for selecting the most appropriate cleaning method and for monitoring the treatment in combination with non invasive FT-IR reflection spectroscopy. Labels IE4 and IE6 indicate the selected cleaning solution (IE4=40:60 isooctane:ethanol and IE6=60:40 isooctane: ethanol). Adapted with permission from Springer Nature, Analytical and Bioanalytical Chemistry, Rosi et al. 409:3187-97 (2017) [67].

6.4 Portable and non-invasive IR spectroscopy: probing the molecular composition of CH objects at the macro-scale

6.4.1 Single point FT-IR spectroscopy

Since the last fifteen years the need of performing non-invasive analytical investigations in the full respect of the object integrity, working *in situ* without any risk and cost for artwork transportation, has boosted the development of portable non-invasive spectroscopic instruments working with different types of electromagnetic radiation, from X-ray to microwaves [68, 69]. In this contest, non-invasive FT-IR spectroscopy in external reflection mode proved to be a very powerful method to examine by point analysis the surface of works of art mainly thanks to the good performances of the available mobile instruments, which are today comparable to those of standard bench-top equipment. The main challenge arises from spectra interpretation which is complicated by the distortions owing to the concurrent collection of both surface and volume components of the reflected light. In what it follows an overview of strategies for maximising the diagnostic information provided by reflection FT-IR spectra of CH objects is given.

Reflection FT-IR profiles of inorganic compounds are generally characterized by *reststrahlen* signals originated by the surface reflected light in correspondence of the high k values of the fundamental bands (see paragraph 2 and references mentioned there). As an example of spectral distortions produced by the *reststrahlen* effect, reflection spectra of different carbonate-based pigments are reported in Figure 6.5(a). Typically, these distortions hamper the identification of inorganic compounds by using their fundamental bands (the antisymmetric CO_3^{-2} stretching in the example of Figure 6.5(a)). Nevertheless, Ricci et al. [70] demonstrated that the intensity of *reststrahlen* bands at specific ranges of concentration approaches a Lambert-Beer's behaviour allowing for a semi quantitative determination of a variety of alteration compounds (namely, sulphate, oxalates and phosphates) on the surface of stone sculptures (Figure 6.5(b-d)).

As valid alternative to the use of the fundamental bands, it has been proved that the combination and overtone bands can be profitably exploited in reflection FT-IR for the identification of a wide variety of inorganic species of interest in the field of cultural heritage [9, 37, 68, 71–74]. These forbidden bands have very small values of k index and are typically neglected in transmission mode spectra. Thanks to the enhancement of the low absorbing bands due to diffuse reflection, combination and overtone signals are well visible in reflection mode and can be exploited for molecular recognition bringing the same information of the fundamental bands. Figure 6.6 shows an example of the specificity and sensitivity of these bands for the identification of carbonate-pigments.

Furthermore the enhancement of low absorbing bands in reflection mode proved to be very effective for the identification of minor components used as markers for specific type of pigments. As an example, natural ultramarine (also known as *lapis lazuli*) can be distinguished from its synthetic form thanks to the detection of CO_2 traces (band at 2340 cm^{-1}) retained in the mineral owing to its geological formation [75] (Figure 6.7(a)). Guinet green, a variation of viridian, can be identified by the detection of two small bands at about 1290 and 1250 cm^{-1} assigned to residues of an insoluble chromium borate specifically related to its synthesis procedure [76] (Figure 6.7(b)). The distinction between ivory-black and lamp-black is typically achieved in transmission FT-IR spectroscopy by the identification of hydroxyapatite ($Ca_5(OH)(PO_4)_3$) which is the main component beside carbon (IR inactive) obtained by calcination of bones. While the fundamental vibrational modes of the PO_4^{-3} have a poor diagnostic reliability in reflection mode, the signal at 2013 cm^{-1} tentatively assigned to cyanamide apatite, a side product formed during the calcination [77], proved to be a robust spectral marker for non-invasive identification of ivory-black [78] (Figure 6.7(c)).

The high specificity and sensitivity of infrared spectroscopy for the recognition of organic compounds represent an added value for non-invasive studies of CH

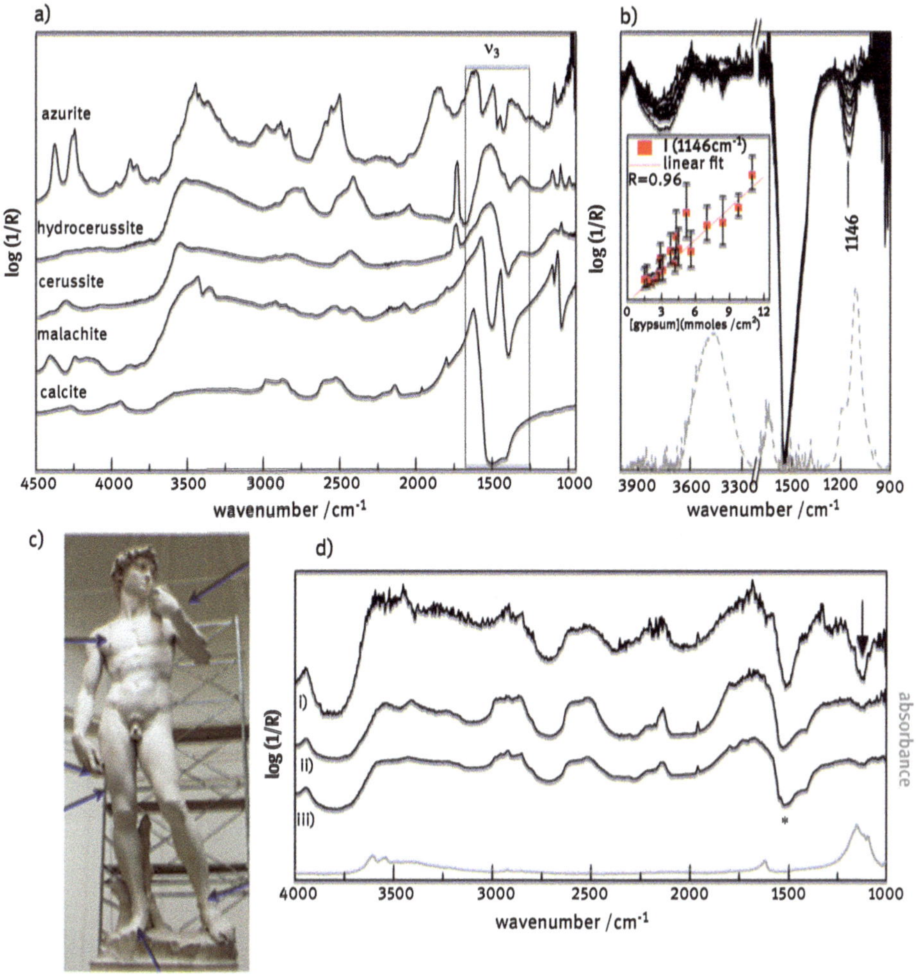

Figure 6.5: a) Reflection mode FT-IR spectra of carbonate-based pigments (azurite, hydrocerrussite, cerussite, malachite and calcite) in the mid-IR range. The strong $\nu_3(CO_3^{-2})$ mode (highlighted by a rectangle) appearing as *reststrahlen* band distorts a wide spectral range in the fingerprint region. In the high wavenumber side, combination and overtone bands are peculiar for each of the considered pigments. b) Reflection mode FT-IR spectra acquired on marble samples treated with gypsum ($CaSO_4 \cdot 2H_2O$) solutions at different concentration (black line); the dash gray line is the transmission-mode spectrum of a gypsum standard. In the inset the linear behavior (R=0.96) of the intensity of the $\nu_3(SO_4^{-2})$ *reststrahlen* band (at 1146 cm^{-1}) versus salt concentration is shown. c) Image of the Michelangelo's *David* sculpture; arrows indicate the sulphates distribution with example spectra shown in d): i) residual black crust on *forefinger*, ii) grey incrustation on the *right chest* and iii) not macroscopically altered region on the *left shoulder*. In grey the absorbance spectrum of a gypsum standard is reported. The star indicates the *reststrahlen* band of $\nu_3(CO_3^{-2})$ of the carbonate substrate and the arrow indicates the *reststrahlen* band of the sulphate anion. Adapted with permission from Elsevier Talanta, Ricci C et al. 2006;69(5):1221-26 [70].

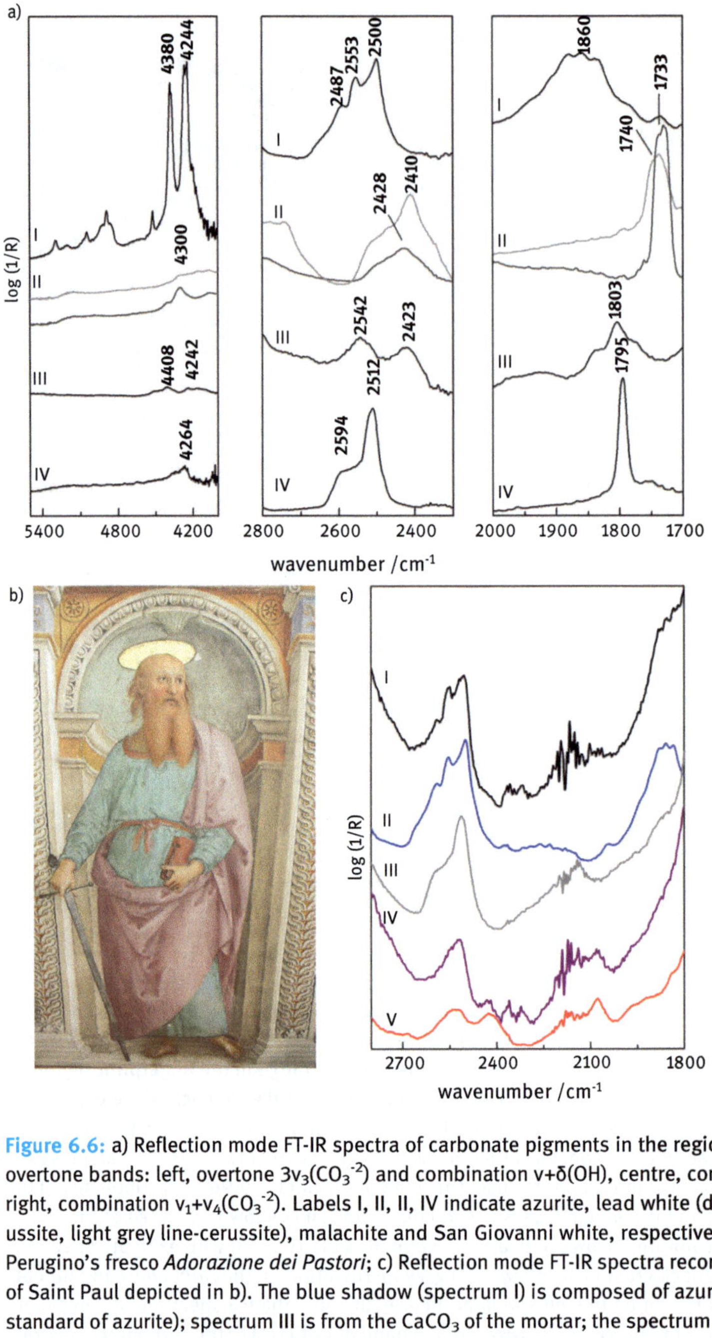

Figure 6.6: a) Reflection mode FT-IR spectra of carbonate pigments in the region of combination and overtone bands: left, overtone $3\nu_3(CO_3^{-2})$ and combination $\nu+\delta(OH)$, centre, combination $\nu_1+\nu_3(CO_3^{-2})$, right, combination $\nu_1+\nu_4(CO_3^{-2})$. Labels I, II, II, IV indicate azurite, lead white (dark grey line-hydrocerussite, light grey line-cerussite), malachite and San Giovanni white, respectively. b) A detail of Perugino's fresco *Adorazione dei Pastori*; c) Reflection mode FT-IR spectra recorded from the green robe of Saint Paul depicted in b). The blue shadow (spectrum I) is composed of azurite (spectrum II is from a standard of azurite); spectrum III is from the $CaCO_3$ of the mortar; the spectrum of the green shadow (IV) contains bands assigned to malachite (spectrum V is from a standard of malachite). Adapted with permission from Springer Nature, Applied Physics A: Materials Science & Processing. Miliani C et al. 106 (2):295-307 (2012) [9]. Adapted with permission from Springer Nature, Analytical and Bioanalytical Chemistry D. Buti et al. 405(8):2699-2711 (2013) [14].

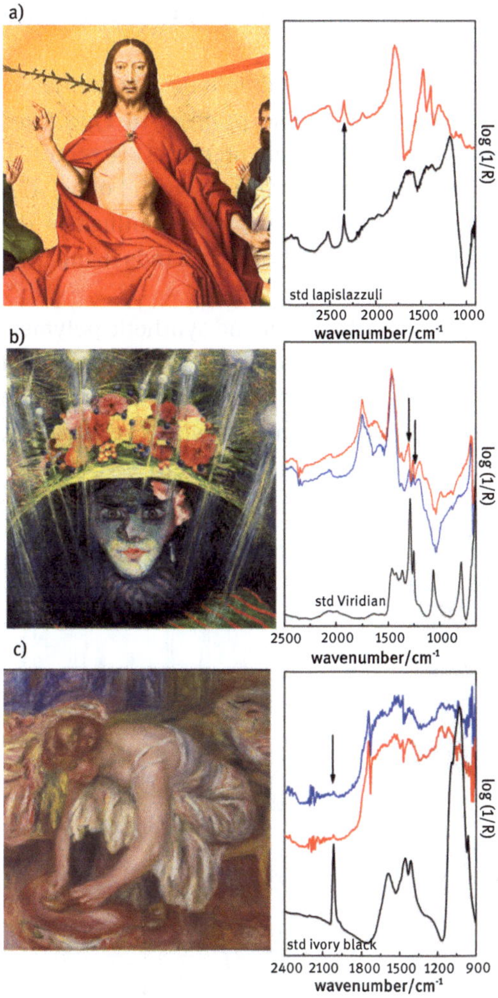

Figure 6.7: Identification of inorganic pigments by non invasive reflection-mode FT-IR spectroscopy: a) spectrum (red line) of the blue gem of the Christ's mantel in Memling's masterpiece *Last Judgment* [79]. The comparison with a reference spectrum (black line) indicates the use of *lapis lazuli* for the blue ornament. The arrow indicates the weak but peculiar band associated to the vibrational mode of CO_2 encapsulated in the sodalite cage of the natural pigment. b) spectra (red and blue lines) of the green paints in the futurist painting *Modern Idol* by Boccioni. The presence of a sharp and weak doublet at about 1290 and 1260 cm^{-1} suggests the use of viridian prepared with boric acid. The reference spectrum of viridian in transmission mode is also shown (black line). c) spectra (blue and red lines) of the dark red and grey paints in the impressionist painting *A woman at her toilette* by Renoir. The weak band at about 2010 cm^{-1} is visible also in the presence of strong IR-absorbing materials (in this case an hydrocarbon-based organic material is detected). The spectrum in black is from a standard of ivory black. Adapted with permission from Springer Nature, Applied Physics A: Materials Science & Processing, Miliani C et al. 89:849-56 (2007) [78].

materials since other non-invasive molecular spectroscopic methods such as Raman and UV-Vis spectroscopy show some shortcomings.

Reflection spectra of organic compounds are typically dominated by the refractive index change with frequency, showing derivative-like profiles. In Figure 6.8 some examples of non-invasive FTIR identification of organic pigments are reported. Two types of naphtol reds, PR3 and PR4, were detected in *Alchemy* by J. Pollock [73] and in *Bianco 1952* by A. Burri [71], respectively. Indigo dye was identified in the watercolour *Still Life* by P. Cézanne thanks to several specific sharp bands rising from the broad IR signals of the paper matrix.

The possibility of recognising different classes of natural and synthetic polymers non-invasively is a unique feature of reflection FT-IR spectroscopy. As a matter of fact derivative-like spectra maintain almost the same level of specificity of transmission spectra allowing for distinction among protein, lipid, wax, alkyds, vinyl, acrylics, etc.

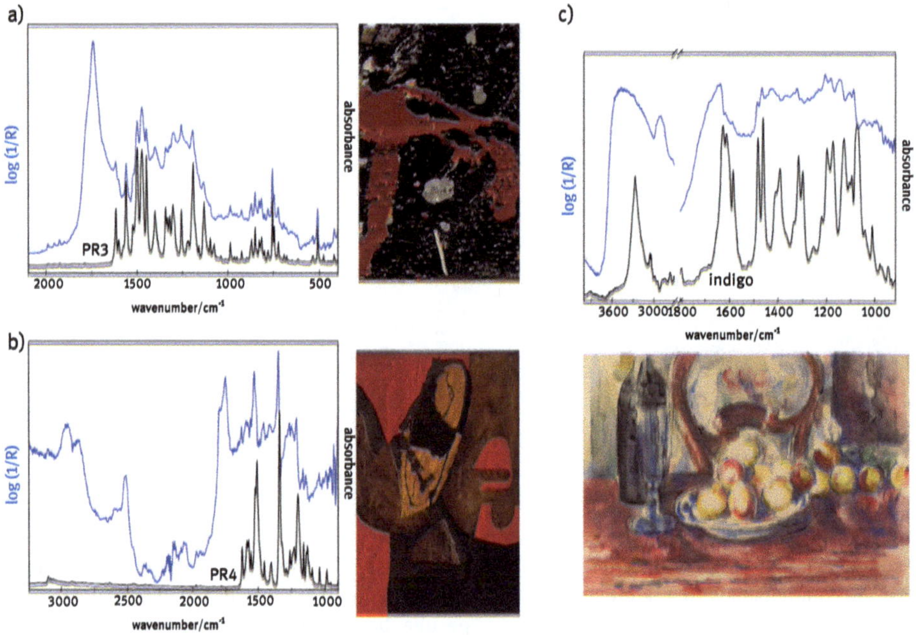

Figure 6.8: Identification of organic pigments by non invasive reflection mode FT-IR spectroscopy in modern paintings: a) spectrum of a dripping red paint of the painting *Alchemy* by Pollock compared with the transmission mode spectrum from a standard of naphtol pigment PR3; b) spectrum of the red paint of the painting *Rosso 50* by Burri compared with the transmission mode spectrum from a standard of naphtol pigment PR4; c) spectrum of a dark blue area of the watercolor *Still Life* by Cézanne compared with the transmission mode spectrum of a standard of indigo. In all the three cases the diagnostic bands of the organic pigments are clearly observed as sharp bands above the IR absorption of the other painting components. Adapted from Rosi F et al. Herit Sci. 2016;4:18 [73]; Adapted with permission from Springer Nature, Applied Physics A: Materials Science & Processing, Rosi F et al. 100:613-24 (2010) [71].

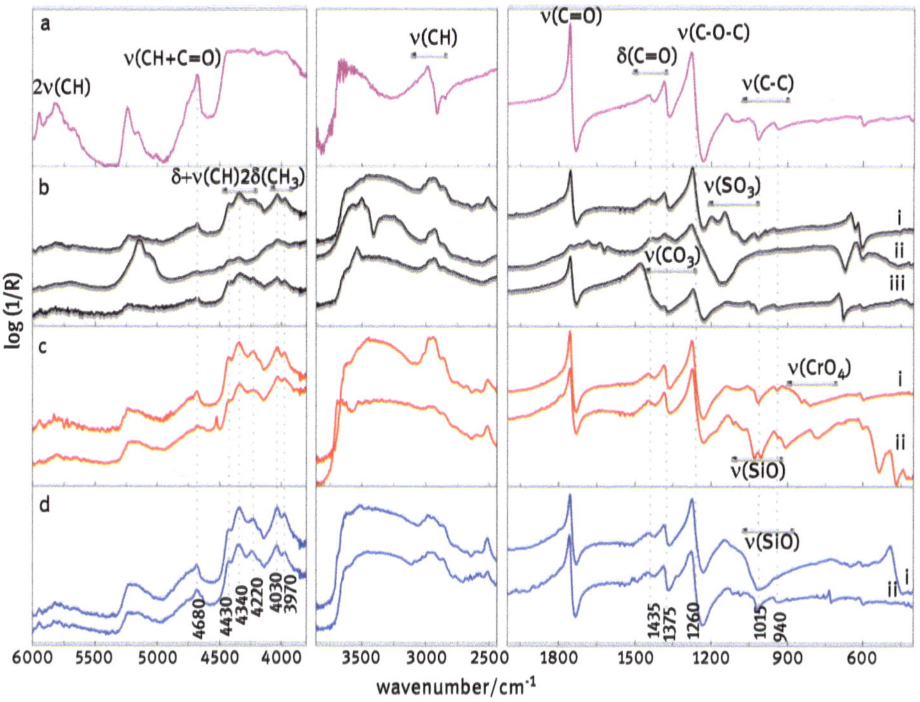

Figure 6.9: Reflection mode FT-IR spectra of paint reconstructions made of polyvinyl acetate resin alone a) and with different pigments: b) white pigments, i= lithopone, ii= gypsum, iii= lead white; c) yellow pigments, i = lead chromate, ii= ochre; d) blue pigments i = ultramarine, ii= phthalocyanine. Diagnostic bands are evidenced by dot lines. Reprinted with permission from Elsevier, Microchemical, Journal Rosi F et al. 124:898-908 (2016) [81].

[71–73, 80–87]. As an example, in Figure 6.9, the FT-IR reflection profiles of polyvinyl acetate (PVA) are reported showing the specific spectral marks that can be used for its identification also in mixture with strongly absorbing IR pigments.

As is in the case of transmission spectroscopy, the diagnostic capability of reflection FT-IR is strongly hampered for complex mixtures of organic polymers or when the target compound is in lower amount with respect to a strongly absorbing inorganic counterpart. In order to improve the sensitivity, the near infrared range can be also explored, using the undistorted combination and overtone bands for profitable recognition of different classes of organic and inorganic compounds [81].

In addition to limitations related to band overlapping, the interpretation of reflection spectra collected from complex matrices is further complicated by discontinuous spectral effects related to the optical and chemical properties of the target surface. A good example of the effect of surface morphology is given by the analysis of *Bianco 1952* by A. Burri in Figure 6.10. Here, the spectra of the same paint, a mixture of PVA and zinc white recorded on two areas which differ in the surface

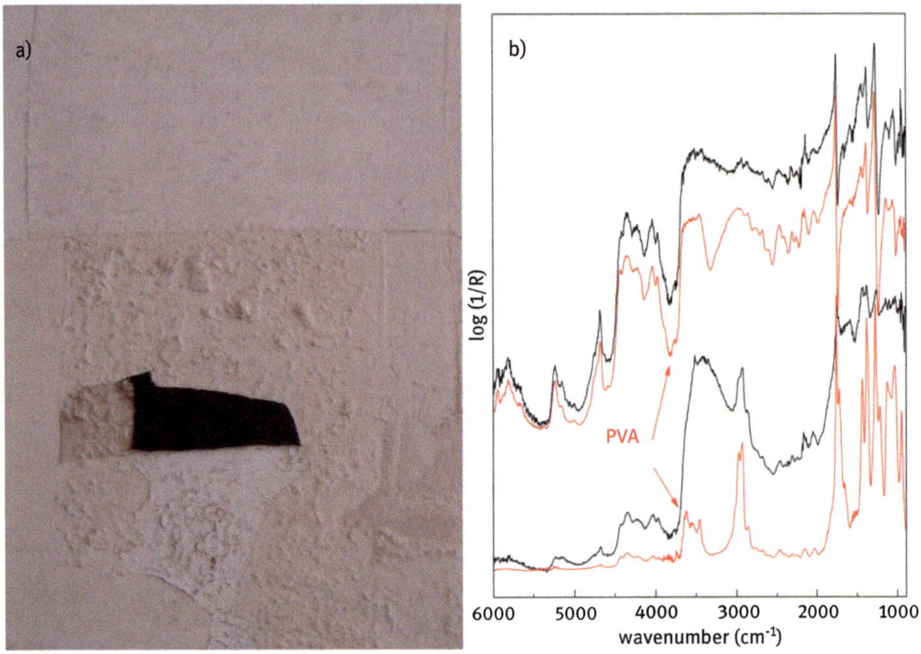

Figure 6.10: a) Detail of the painting *Bianco 52* by Burri; it is a monochrome painting with different surface morphologies and degrees of white brightness. b) in black, two reflection mode FT-IR spectra recorded on a smooth (upper spectrum) and rough (lower spectrum) white areas compared. The two profiles, although being quite different, both indicate the use of PVA manifested by: -derivative-like features in the smooth area (since the higher contribution of the surface reflection) and - positive broad bands in the rough surface (since the higher contribution of the volume reflection). In red the reflection mode FT-IR spectra of PVA paint standards with a flat (top) and a rough surface (bottom) are shown for comparison. Adapted with permission from Springer Nature, Applied Physics A: Materials Science & Processing, Rosi F et al. 100:613-24 (2010) [71]

morphology, are compared. The spectrum acquired on the smooth area is characterized by derivative like bands due to the higher contribution of the surface reflection of the optically flat surface. Diversely, the spectrum acquired on the rough area shows enhanced combination bands and positive broad peaks as a consequence of a higher contribution of the volume reflection.

Similarly to what discussed for applications of micro-FT-IR reflection spectroscopy, also in the non-invasive approach chemometric methods may help to deal with distorted spectral profiles as well as to manage large datasets acquired *in situ* [71, 72, 87].

Non-invasive reflection FT-IR is highly informative not only for characterization of original and overlaid materials but also for identification and localisation of surface alteration products. As demonstrated in a recent paper [41] it enables to distinguish the metal ion involved in the formation of oxalates (also distinguishing hydrated forms) and carboxylates as well as the type of the fatty acid involved in the carboxylate formation.

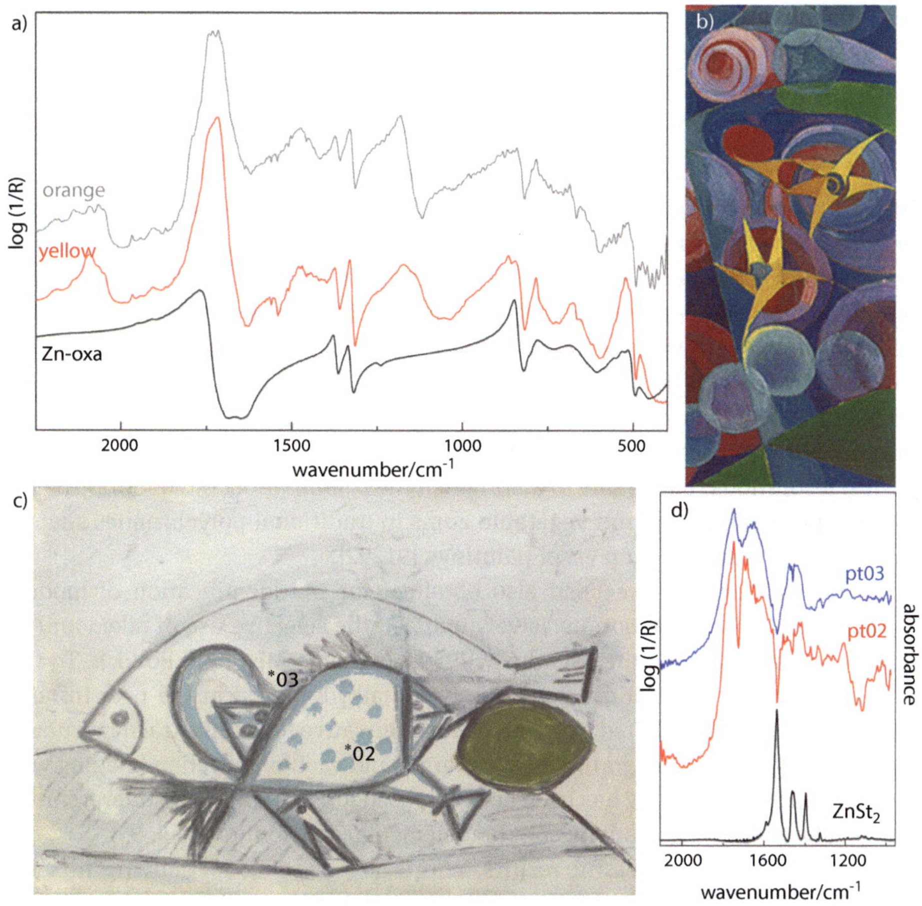

Figure 6.11: a) Reflection mode FT-IR spectra recorded on yellow and orange floral decorations (shown in b) of the futurist painting *Flora* by Dottori. In both the profiles the strong bands of Zn-oxalates are visible (see the comparison with the Zn-oxalate reference spectrum in black); b) *Still Life with Three Fish, Moray Eel and Lime on White Ground* (1946) by Picasso. The two investigated points (pt02 and pt03) are composed of zinc oxide. The reflection mode spectrum recorded on point 02 is further characterized by a sharp signal at 1540 cm^{-1} assigned to zinc stearate. Adapted with permission from [41] In: Metal Soaps in Art – Conservation & Research (Eds. F. Casadio, K. Keune, P. Noble, A. van Loon, E. Hendriks, S. Centeno, G. Osmond) Springer Nature book, 2018, in press.

As an example, shown in Figure 6.11(a,b), zinc oxalates have been clearly detected in the Futurist painting *Flora* by Dottori in correspondence of the yellow and orange Cd-based paints. In another painting, *Still Life with Three Fish, Moray Eel and Lime on White Ground*, dated 1946 by Picasso (Figure 6.11(c,d)), the white paint composed of ZnO showed in some areas the FT-IR bands of zinc carboxylates. Interestingly, ZnO white areas gave different spectral profiles (Figure 6.11(d)) characterized by sharp (blue line) or broad (red line) shapes of the zinc carboxylate band that suggested the use of different paint tube formulations within the painting [41].

6.4.2 Hyperspectral imaging infrared spectroscopy

The positive outcomes from the application of non-invasive reflection FT-IR in the heritage science field as well as the availability of new and more efficient cameras are fostering the development of methods for non-invasive molecular hyperspectral imaging at the macro-scale. To date, the range more easily accessible by hyperspectral cameras is the near infrared-NIR (10,000–4000 cm^{-1}, 1000–2500 nm). This range is of interest for both electronic transitions of some pigments and vibrational combination and overtone bands of some oxyanion compounds and organic binders [88–90]. The state of the art of NIR reflectance imaging systems can work at spectral resolutions that enable to distinguish the combination and overtone bands of lipids, proteins, waxes and vegetable gums in traditional polychromies such as illuminated manuscripts and easel paintings [91, 92].

Positive outcomes have been also obtained for the identification of modern binders; Dooley et al. [93] non-invasively mapped the selective use of alkyd and oil binding media in the painting *Number 1, 1950 - Lavender Mist* by J. Pollock. To this purpose, the weak spectral differences between alkyd and oil in the near infrared range [81] have been exploited by applying first derivative to the spectral data cube. A multi-modal approach integrating near infrared (and visible) reflection imaging with X-ray fluorescence mapping has been also proposed for supporting the interpretation of molecular maps with spatially coherent elemental distributions [94, 95].

Driven by these successful results, efforts are ongoing for moving infrared hyperspectral imaging toward the mid infrared region which would strongly improve its diagnostic capabilities. To achieve this goal, difficulties arise from the technical constrains in translating the robust micro-imaging technology of bench top systems to portable hyperspectral devices working at the macro scale. In addition, with respect to cameras working at shorter wavelengths, imaging systems exploring the mid-IR range are more expensive, since transmitting/optical materials and detection systems -generally made with highly sensitive semiconductors of MCT type- are very costly. Legrand et al. [96] circumvented these technical constrains mounting a conventional portable single-point FT-IR spectrometer (working range 8000–400 cm^{-1}) on a scanning system. The resulting chemical maps obtained for a nineteenth Century icon and a twentieth Century Antillean painting were extremely informative but the acquisition time was too long (i.e. 36 h for a 5.5 × 8 cm^2 area with a lateral resolution of 1 mm) for allowing an exploitation of the method beyond this proof of concept.

The first experiment of hyperspectral imaging of a painting in the mid-IR range has been published in 2013 [97]. In this paper, a camera (HI90, working in the range 1300–

860 cm^{-1} and developed for remote sensing of hazardous compounds [98]) was adapted for reflection measurements of paintings by the use of an external infrared radiation source. Technically, the imager combines a plane mirror interferometer with an FPA detector (256 × 256 pixels) obtaining high throughput of radiation and high lateral resolution at the same time. The camera can work 1 m far from the painting with a lateral resolution of about 1 mm, a field of view of each pixel of 0.69 mrad, measuring a total area of about 9 × 9 cm^2 in 80 s. Preliminary studies were carried out on dedicated paint models [99] comparing the single-pixel spectra (at 4 cm^{-1} of spectral resolution) with those recorded on the same sample by a single-point portable reflection FT-IR spectrometer. This feasibility study proved the potentials of the camera in terms of spectral and lateral resolution and sensitivity. The imager has been then successfully exploited for investigating *Sestante 10 (1982)* by Alberto Burri, unveiling the distribution of acrylic and vinyl resins as binders (Figure 6.12) and of silicate-based pigments along with barium and calcium sulphates as filler/additives of the paint formulations.

After this pioneering work, other interesting technical solutions have been reported in the literature but with limited operative spectral ranges. Gabrieli et al.

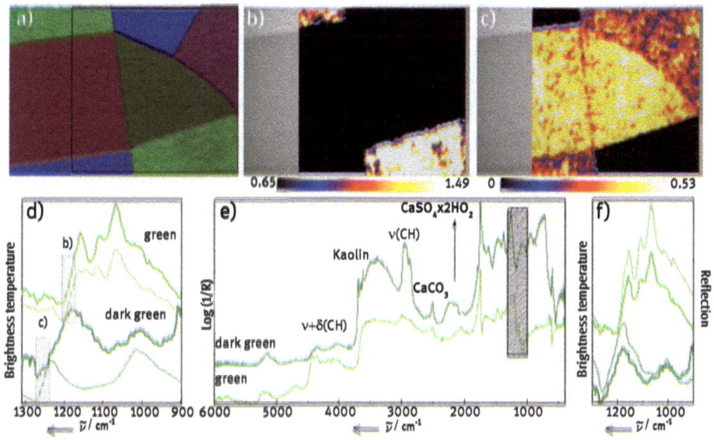

Figure 6.12: a) Color image of a portion of the painting *Sestante 10* by Burri. The black rectangle highlights the area investigated with the HI90 system (9x9 cm^2). b,c) Brightness temperature difference images (measured in Kelvin) in the range 1155–1159 to 1200–1203 cm^{-1} (b) and 1209–1215 to 1243–1249 cm^{-1} (c). d) Comparison of the spectral profiles representative of the two areas highlighted in (b) and (c) with reference spectra collected with the HI90 imager from laboratory paint models composed of lithopone white and acrylic resin (thin green line) and a silica-based pigment with a vinyl binder (thin dark-green line). e) Infrared reflection spectra acquired with the Alpha-R system with spectral assignment in the whole spectral range accessible with the single-point portable spectrometer for the green and dark-green sectors (the rectangle highlights the portion of the range accessible with the HI90 instrument). f) Comparison of the spectral profiles obtained with the HI90 instrument (thick lines) and the Alpha-R system (thin lines) in the green and dark-green sectors. Reprinted with permission from John Wiley and Sons, Angewandte Chemie International Edition, Rosi et al. 52:5258-61 (2013) [97].

[100] proposed an innovative set-up to record passive emissive spectra in the restricted range of 1240–760 cm^{-1} reducing the environmental thermal radiation along the light path between the painting and the detector by cooling. Recently, Daveri et al. [101] used a commercial hyperspectral camera equipped with an FPA detector and an external IR source to explore the range 3700–1800 cm^{-1}. The use of a 2D-scanner enabled to investigate large surfaces, however the slight accessible spectral window was suitable only for the identification of a limited class of pigments.

Laser-based technology has been also tested for exploring the mid infrared range [102, 103]. The spatial scanning is achieved by deflecting the laser beam across the surface by two galvanometer-mounted gold mirrors directing and collecting the reflected light to a single-element photovoltaic IR detector. The spectral tuning is achieved by the Optical Parametric Oscillator (OPO) technology which converts the radiation from a 1064 nm laser into a broadly tunable radiation in the mid-IR portion of the spectrum (3700–1820 cm^{-1}). The application of this highly innovative technology is however still limited since the long acquisition time which is strictly connected to the scanning in wavelength for covering the considered range with an appreciable spectral resolution.

All these examples clearly show the analytical potentialities of hyperspectral macroscale chemical imaging of surfaces in the mid-IR range. To date the main limitation is linked to the narrow spectral windows that can be explored and that limits the diagnostic capabilities. In order to fully exploit these potentialities, much effort still needs to be directed to enlarge the operative spectral range maintaining good spectral resolution and to shorten acquisition times. Also reduction of instrumental costs is crucial for a profitable diffusion of the technique in the CH field.

6.5 Conclusion

The research work and experience accumulated over years in the CH field allowed FT-IR spectroscopy to evolve from the initial focus on identification of original materials to the study of degradation processes, characterization of conservation products, monitoring of cleaning treatments.

Furthermore, the needs of non-invasiveness and of minimal impact on micro-samples, specific of the heritage science field, pushed efforts towards improvement of the analytical power of FT-IR spectroscopy in terms of full exploitation of the reflection acquisition mode in different set-ups, enlargement of the spectral window, boost of lateral resolution and exploration of advanced statistical treatments for large spectral datasets. This, undoubtedly, fostered the advancement of FT-IR spectroscopy also for transversal applications in contiguous scientific fields (biology, medicine, catalysis, interface phenomena, etc.).

The next striking improvement for FT-IR spectroscopy will be the achievement of infrared hyperspectral imaging at the macroscale in the mid-infrared range. This was, so far, restrained by the restricted operative spectral range, too long acquisition time, and prohibitive costs exhibited by the few devices today available. Nevertheless the analytical potentialities of macroscale chemical imaging by infrared spectroscopy are so appealing that great efforts are currently exerted by researchers to reach this goal in the near future.

Acknowledgements: The authors gratefully acknowledge the financial support of the project IPERION CH, Integrated Platform for the European Research Infrastructure on Cultural Heritage, H2020 RIA GA n. 654028, www.iperionch.eu.

References

[1] Domenéch-Carbó MT, Osete-Cortina L. Another beauty of analytical chemistry: chemical analysis of inorganic pigments of art and archaeological objects. Chem Texts. 2016;2:14.
[2] Casadio F, Toniolo L. The analysis of polychrome works of art: 40 years of infrared spectroscopic investigations. J Cult Herit. 2001;2:71–8.
[3] Derrick MR, Stulik DC, Landry JM. Infrared spectroscopy in conservation science. Los Angeles: The Getty Conservation Institute, 1999.
[4] Griffiths PR, de Haseth JA. Fourier transform infrared spectrometry. New Jersey: John Wiley & Sons, Inc. Publication, 2007.
[5] Milosevic M. Internal reflection and ATR spectroscopy. New Jersey: John Wiley & Sons, Inc. Publication, 2012.
[6] Averett LA, Griffiths PR. Effective path length in attenuated total reflection spectroscopy. Anal Chem. 2008;80:3045–9.
[7] Van Nimmen E, De Clerck K, Verschuren J, Gellynck K, Gheysens T, Mertens J, et al. FT-IR spectroscopy of spider and silkworm silks, part I : different sampling techniques. Vib Spectrosc. 2008;46:63–8.
[8] Miljković M, Bird B, Diem M. Line shape distortion effects in infrared spectroscopy. Analyst. 2012;137:3954–64.
[9] Miliani C, Rosi F, Daveri A, Brunetti BG. Reflection infrared spectroscopy for the non-invasive in situ study of artists' pigments. Appl Phys A. 2012;106:295–307.
[10] Korte EH, Röseler A. Infrared reststrahlen revisited: commonly disregarded optical details related to $n < 1$. Anal Bioanal Chem. 2005;382:1987–92.
[11] Fox M. Optical properties of solids. London: Oxford University Press, 2010.
[12] Milosevic M, Berets SL. A review of FT-IR diffuse reflection sampling consideration. Appl Spectrosc Rev. 2002;37:347–64.
[13] Hopfe V, Korte EH, Klobes P, Grälert W. Optical data of rough-surfaced ceramics: infrared specular and diffuse reflectance versus spectra simulation. Appl Spectr. 1993;47:423–9.
[14] Buti D, Rosi F, Brunetti BG, Miliani C. In-situ identification of copper-based green pigments on paintings and manuscripts by reflection FTIR. Anal Bioanal Chem. 2013;415:2699–711.
[15] Monico L, Rosi F, Miliani C, Daveri A, Brunetti BG. Non-invasive identification of metal-oxalate complexes on polychrome artwork surfaces by reflection mid-infrared spectroscopy. Spectrochim Acta A. 2013;116:270–80.
[16] Gabrieli F, Rosi F, Vichi A, Cartechini L, Pensabene Buemi L, Kazarian SG, et al. Revealing the nature and distribution of metal carboxylates in Jackson Pollock's Alchemy (1947) by micro-attenuated total reflection FT-IR spectroscopic imaging. Anal Chem. 2017;89:1283–9.

[17] Capobianco G, Bracciale MP, Sali D, Sbardella F, Belloni P, Bonifazi G, et al. Chemometrics approach to FT-IR hyperspectral imaging analysis of degradation products in artwork cross-section. Microchem J. 2017;132:69–76.

[18] Prati S, Joseph E, Sciutto G, Mazzeo R. New advances in the application of FTIR microscopy and spectroscopy for the characterization of artistic materials. Acc Chem Res. 2010;43:792–801.

[19] Prati S, Sciutto G, Bonacini I, Mazzeo R. New frontiers in application of FTIR microscopy for characterization of cultural heritage materials. Top Curr Chem. 2016;374:129–60.

[20] Bertasa M, Possenti E, Botteon A, Conti C, Sansonetti A, Fontana R, et al. Close to the diffraction limit in high resolution ATR FTIR mapping: demonstration on micrometric multi-layered art systems. Analyst. 2017;142:4801–11.

[21] Van Der Weerd J, Brammer H, Boon JJ, Heeren RM. Fourier Transform Infrared microscopic imaging of an embedded paint cross-section. Appl Spectr. 2002;56:275–83.

[22] Pilc J, White R. The application of FTIR-microscopy to the analysis of paint binders in easel paintings. National Gallery Technical Bull. 1995;16:73–84.

[23] van der Weerd J, Heeren RM, Boon JJ. Preparation methods and accessories for the infrared spectroscopic analysis of multi-layer paint films. Stud Conserv. 2004;49:193–210.

[24] Mazzeo R, Joseph E, Prati S, Millemaggi A. Attenuated total reflection-Fourier transform infrared microspectroscopic mapping for the characterisation of paint cross-sections. Anal Chim Acta. 2007;599:107–17.

[25] Prati S, Rosi F, Sciutto G, Olivieri P, Catelli E, Miliani C, et al. Evaluation of the effect of different paint cross section preparation methods on the performances of Fourier transformed infrared microscopy in total reflection mode. Microchem J. 2013;110:314–9.

[26] Prati S, Rosi F, Sciutto G, Mazzeo R, Magrini D, Sotiropoulou S, et al. Evaluation of the effect of six different paint cross section preparation methods on the performances of Fourier transformed infrared microscopy in attenuated total reflection mode. Microchem J. 2012;103:79–89.

[27] Prati S, Sciutto G, Catelli E, Ashashina A, Mazzeo R. Development of innovative embedding procedures for the analyses of paint cross sections in ATR FITR microscopy. Anal Bioanal Chem. 2013;405:895–905.

[28] Pouyet E, Lluveras-Tenorio A, Nevin A, Saviello D, Sette F, Cotte M. Preparation of thin-sections of painting fragments: classical and innovative strategies. Anal ChimActa. 2014;822:51–9.

[29] de Fonjaudran CM, Nevin A, Piqué F, Cather S. Stratigraphic analysis of organic materials in wall painting samples using micro- FTIR attenuated total reflectance and a novel sample preparation technique. Anal Bioanal Chem. 2008;392:77–86.

[30] Papliaka EZ, Vaccari L, Zanini F, Sotiropoulou S. Improving FTIR imaging speciation of organic compound residues or their degradation products in wall painting samples, by introducing a new thin section preparation strategy based on cyclododecane pre-treatment. Anal Bioanal Chem. 2015;407:5393–403.

[31] Kazarian SG, Chan KL. Micro- and macro-attenuated total reflection Fourier transform infrared spectroscopic imaging. Appl Spectr Focal Point. 2010;64:135–52.

[32] Chan KL, Kazarian SG. Attenuated total reflection Fourier-transform infrared (ATR-FTIR) imaging of tissues and live cells. Chem Soc Rev. 2016;45:1850–64.

[33] Chan KL, Kazarian SG. ATR-FTIR spectroscopic imaging: recent advances and applications to biological systems. Analyst. 2013;138:1940–51.

[34] Joseph E, Prati S, Sciutto G, Ioele M, Santopadre P, Mazzeo R. Performance evaluation of mapping and linear imaging FTIR microspectroscopy for the characterisation of paint cross sections. Anal Bioanal Chem. 2010;396:899–910.

[35] Chan KL, Kazarian SG. New opportunities in micro- and macro-attenuated total reflection infrared spectroscopic imaging: spatial resolution and sampling versatility. Appl Spectr. 2003;57:381–9.

[36] Joseph E, Ricci C, Kazarian SG, Mazzeo R, Prati S, Ioele M. Macro-ATR-FT-IR spectroscopic imaging analysis of paint cross-sections. Vibr Spectrosc. 2010;53:274–8.

[37] Rosi F, Daveri A, Doherty B, Nazzareni S, Brunetti BG, Sgamellotti A, et al. On the use of overtone and combination bands for the analysis of the CaSO4-H2O system by mid-infrared reflection spectroscopy. Appl Spectrosc. 2010;64:956–63.

[38] Spring M, Ricci C, Peggie DA, Kazarian SG. ATR-FTIR imaging for the analysis of organic materials in paint cross sections: case studies on paint samples from the national gallery, London. Anal Bioanal Chem. 2008;392:37–45.

[39] Ricci C, Bloxham S, Kazarian SG. ATR-FTIR imaging of albumen photographic prints. J Cult Herit. 2007;8:387–95.

[40] Kaszowska Z, Malek K, Pańczyk M, Mikołajska A. A joint application of ATR-FTIR and SEM imaging with high spatial resolution: identification and distribution of painting materials and their degradation products in paint cross sections. Vib Spectrosc. 2013;65:1–11.

[41] Rosi F, Cartechini L, Monico L, Gabrieli F, Vagnini M, Buti D, et al. Tracking metal oxalates and carboxylates on painting surfaces by non-invasive reflection mid-FTIR spectroscopy. In: Casadio F, Keune K, Noble P, van Loon A, Hendriks E, Centeno S, Osmond G, editors. Metal soaps in art – conservation& research. Springer Nature book, 2018, in press.

[42] Heeren RM, Boon JJ. Preprint of the 12thTriennal ICOM-CC meeting. Lyon, ICOM Committee for Conservation. 1999;1:228–33.

[43] Keune K, Boon JJ. Analytical imaging studies of cross-sections of paintings affected by lead soap aggregate formation. Stud Conserv. 2007;52:161–76.

[44] van Loon A. Color changes and chemical reactivity in seventeenth-century oil paintings. Amsterdam: University of Amsterdam. PhD Thesis, permalink http://hdl.handle.net/11245/1.292118 2008:–.

[45] Keune K. Binding medium, pigments and metal soaps characterised and localised in paint cross-sections. Amsterdam: University of Amsterdam. PhD Thesis, permalink http://hdl.handle.net/11245/1.248900 2005:–.

[46] van der Weerd J. Microspectroscopic analysis of traditional oil paint. Amsterdam: University of Amsterdam. PhD Thesis. http://hdl.handle.net/11245/1.200948 2002:–.

[47] Mazel V, Richardin P, Touboul D, Brunelle A, Walter P, Laprevote O. Chemical imaging techniques for the analysis of complex mixtures: new application to the characterization of ritual matters on African wooden statuettes. Anal Chim Acta. 2006;570:34–40.

[48] Bruni S, Cariati F, Casadio F, Toniolo L. Spectrochemical characterization by micro-FTIR spectroscopy of blue pigments in different polychrome works of art. Vib Spectros. 1999;20:15–25.

[49] Rosi F, Federici A, Brunetti BG, Sgamellotti A, Clementi S, Miliani C. Multivariate chemical mapping of pigments and binders in easel painting cross-sections by micro IR reflection spectroscopy. Anal Bioanal Chem. 2011;399:3133–45.

[50] Sciutto G, Prati S, Bonacini I, Oliveri P, Mazzeo R. FT-NIR microscopy: an advanced spectroscopic approach for the characterisation of paint cross-sections. Microchem J. 2014;112:87–96.

[51] Sciutto G, Oliveri P, Prati S, Quaranta M, Lanteri S, Mazzeo R. Analysis of paint cross-sections: a combined multivariate approach for the interpretation of μATR FTIR hyperspectral data arrays. Anal Bioanal Chem. 2013;405:625–33.

[52] Cotte M, Susini J, Solé VA, Taniguchi Y, Chillida J, Checroun E, et al. Applications of synchrotron-based micro-imaging techniques to the chemical analysis of ancient paintings. J Anal At Spectrom. 2008;23:820–8.

[53] Cotte M, Checroun E, Mazel V, Solé VA, Richardin P, Taniguchi Y, et al. Combination of FTIR and X-rays synchrotron-based micro-imaging techniques for the study of ancient paintings. A practical point of view. e-PS. 2009;6:1–9.

[54] Pouyet E, Fayard B, Salomé M, Taniguchi Y, Sette F, Cotte M. Thin-sections of painting fragments: opportunities for combined synchrotron-based micro-spectroscopic techniques. Herit Sci. 2015;3:3–3.3 and references therein.
[55] Echard JP, Cotte M, Dooryhee E, Bertrand L. Insights into the varnishes of historical musical instruments using synchrotron microanalytical methods. Appl Phys A. 2008;92:77–81.
[56] Salvadó N, Buti S, Tobin MJ, Pantos E, Prag A, Pradell T. Advantages of the use of sr-ft-ir microspectroscopy: applications to cultural heritage. Anal Chem. 2005;77:3444–51.
[57] Saviello D, Pouyet E, Toniolo L, Cotte M, Nevin A. Synchrotron-based ftir microspectroscopy for the mapping of photo-oxidation and additives in acrylonitrile-butadiene-styrene model samples and historical objects. Anal Chim Acta. 2014;843:59–72.
[58] Nasse MJ, Walsh MJ, Mattson EC, Reininger R, Kajdacsy-Balla A, Macias V, et al. High-resolution Fourier-transform infrared chemical imaging with multiple synchrotron beams. Nat Methods. 2011;8:413–6.
[59] Carr JL, Miller LM, Dumas P. Synchrotron radiation as a source for infrared microspectroscopy imaging with 2D multi-element detection. In: Moss D, editor. Biomedical applications of synchrotron infrared spectromicroscopy. Cambridge: Royal Society of Chemistry, 2011:226–59.
[60] Petibois C, Deleris G, Piccinini M, Cestelli-Guidi M, Marcelli A. A bright future for synchrotron imaging. Nat Photonics. 2009;3:179.
[61] Hirschmugl CJ, Gough KM. Fourier transform infrared spectrochemical imaging: review of design and applications with a focal plane array and multiple beam synchrotron radiation source. App Spectros. 2012;66:475–91.
[62] Schmidt Patterson C, Carson D, Phenix A, Khanjian H, Trentelman K, Mass J, et al. Synchrotron-based imaging FTIR spectroscopy in the evaluation of painting cross-sections. e-PS. 2013;10:1–9.
[63] Latour G, Robinet L, Dazzi A, Portier F, Deniset-Besseau A, Schanne-Klein M-C. Correlative nonlinear optical microscopy and infrared nanoscopy reveals collagen degradation in altered parchments. Sci Rep. 2016;6:26344.
[64] Prati S, Quaranta M, Sciutto G, Bonacini I, Litti L, Meneghetti M, et al. Use of nano gold obtained by laser ablation for SEIRA analyses of colorants. Herit Sci. 2014;2:28.
[65] Prati S, Milosevic M, Sciutto G, Bonacini I, Kazarian SG, Mazzeo R. Analyses of trace amounts of dyes with a new enhanced sensitivity FTIR spectroscopic technique: MU-ATR (metal underlayer ATR spectroscopy). Anal Chim Acta. 2016;941:67–79.
[66] Sciutto G, Prati S, Bonacini I, Litti L, Meneghetti M, Mazzeo R. A new integrated TLC/MU-ATR/SERS advanced approach for the identification of trace amounts of dyes in mixtures. Anal Chim Acta. 2017;991:104–12.
[67] Rosi F, Legan L, Miliani C, Ropret P. Micro transflection on a metallic stick: an innovative approach of reflection infrared spectroscopy for minimally invasive investigation of painting varnishes. Anal Bioanal Chem. 2017;409:3187–97.
[68] Miliani C, Rosi F, Brunetti BG, Sgamellotti A. In situ noninvasive study of artworks: the MOLAB multitechnique approach. Acc Chem Res. 2010;43:728–38.
[69] Blümich B, Casanova F, Perlo J, Presciutti F, Anselmi C, Doherty B. Noninvasive testing of art and cultural heritage by mobile NMR. Acc Chem Res. 2010;43:761–70.
[70] Ricci C, Miliani C, Brunetti BG, Sgamellotti A. Non-invasive identification of surface materials on marble artefacts with fiber optic mid-FTIR reflectance spectroscopy. Talanta. 2006;69:1221–6.
[71] Rosi F, Miliani C, Clementi C, Kahrim K, Presciutti F, Vagnini M, et al. An integrated spectroscopic approach for the non-invasive study of modern art materials and techniques. Appl Phys A. 2010;100:613–24.

[72] Miliani C, Rosi F, Borgia I, Benedetti P, Brunetti BG, Sgamelotti A. Fiber-optic Fourier transform mid-infrared reflectance spectroscopy: a suitable technique for in situ studies of mural paintings. Appl Spectrosc. 2007;61:293–9.

[73] Rosi F, Grazia C, Fontana R, Gabrieli F, Pensabene Buemi L, Pampaloni E, et al. Disclosing Jackson Pollock's palette in Alchemy (1947) by non-invasive spectroscopies. Herit Sci. 2016;4:18.

[74] Brunetti B, Miliani C, Rosi F, Doherty B, Monico L, Romani A, et al. Non-invasive investigations of paintings by portable instrumentation: the MOLAB experience. Top Curr Chem (Z). 2016;374:10.

[75] Miliani C, Daveri A, Brunetti BG, Sgamellotti A. CO2 entrapment in natural ultramarine blue. Chem Phys Lett. 2008;466:148–51.

[76] Zumbühl S, Eggenberger U, Scherrer NC, Berger A. Early Viridian Pigment Composition: characterization of a (hydrated) chromium oxide borate pigment. Stud Conserv. 2009;54:149–59.

[77] Daveri A, Malagodi M, Vagnini M. The bone black pigment identification by noninvasive, in situ infrared reflection spectroscopy. J Anal Methods Chem. 2018. Article ID 6595643:8 https://doi.org/10.1155/2018/6595643.

[78] Miliani C, Rosi F, Burnstock A, Brunetti BG, Sgamellotti A. Non-invasive in-situ investigations versus micro-sampling: a comparative study on a Renoir's painting. Appl Phys A. 2007;89:849–56.

[79] Szmelter I, Cartechini L, Romani A, Pezzati L. Multi-criterial studies of the masterpiece 'The Last Judgement' attributed to hans memling at the national museum of Gdańsk (2010–2013). In: Sgamellotti A, Brunetti BG, Miliani C, editors. Science and art. The painted surface. Cambridge: Royal Society in Chemistry, RSC, 2014:230–51.

[80] Rosi F, Daveri A, Miliani C, Verri G, Benedetti P, Pique F, et al. Non-invasive identification of organic materials in wall paintings by fiber optic reflectance infrared spectroscopy: a statistical multivariate approach. Anal Bioanal Chem. 2009;395:2097–106.

[81] Rosi F, Daveri A, Moretti P, Brunetti BG, Miliani C. Interpretation of mid and near-infrared reflection properties of synthetic polymer paints for the non-invasive assessment of binding media in twentieth-century pictorial artworks. Microchem J. 2016;124:898–908.

[82] Poli T, Chiantore O, Nervo M, Piccirillo A. Mid-IR fiber-optic reflectance spectroscopy for identifying the finish on wooden furniture. Anal Bioanal Chem. 2001;400:1161–71.

[83] Ploeger R, Chiantore O, Scalarone D, Poli T. Mid-infrared fiber-optic reflection spectroscopy (FORS) analysis of artists' alkyd paints on different supports. Appl Spectrosc. 2011;65:429–35.

[84] Saviello D, Toniolo L, Goidanich S, Casadio F. Non-invasive identification of plastic materials in museum collections with portable FTIR reflectance spectroscopy: reference database and practical applications. Microchem J. 2016;124:868–77.

[85] Picollo M, Bartolozzi G, Cucci C, Galeotti M, Marchiafava V, Pizzo B. Comparative study of Fourier transform infrared spectroscopy in transmission, attenuated total reflection, and total reflection modes for the analysis of plastics in the cultural heritage field. Appl Spectros. 2014;68:389–96.

[86] Stacey RJ, Dyer J, Mussell C, Lluveras-Tenorio A, Colombini MP, Duce C, et al. Ancient encaustic: an experimental exploration of technology, ageing behaviour and approaches to analytical investigation. Microchem J. 2018;138:472–87.

[87] Miliani C, Sgamellotti A, Verri G, Benedetti P, Brunetti BG, Daveri A, et al. Noninvasive, in situ identification of binding media using reflection mid-FTIR. In: Piqué F, Verri G, editor(s). Organic materials in wall paintings: project report —. Los Angeles: The Getty Conservation Institute, 2015

[88] Caséfalvayová L, Strlič M, Karjalainen H. Quantitative NIR chemical imaging in heritage science. Anal Chem. 2011;83:5101–6.

[89] Cucci C, Delaney JK, Picollo M. Reflectance hyperspectral imaging for investigation of works of art: old master paintings and illuminated manuscripts. Acc Chem Res. 2016;49:(2070-79 and references therein).

[90] Wu T, Li G, Yang Z, Zhang H, Lei Y, Wang N, et al. Shortwave infrared imaging spectroscopy for analysis of ancient paintings. Appl Spectrosc. 2016;71:977–98.
[91] Ricciardi P, Delaney JK, Facini M, Zeibel JG, Picollo M, Lomax SQ, et al. Near infrared reflectance imaging spectroscopy to map paint binders in situ on illuminated manuscripts. Angew Chem Int Ed. 2012;51:5607–10.
[92] Dooley KA, Lomax S, Zeibel JG, Miliani C, Ricciardi P, Hoenigswald A, et al. Mapping of egg yolk and animal skin glue paint binders in early renaissance paintings using near infrared reflectance imaging spectroscopy. Analyst. 2013;138:4838–48.
[93] Dooley KA, Coddington J, Krueger J, Conover DM, Loew M, Delaney JK. Standoff chemical imaging finds evidence for Jackson Pollock's selective use of alkyd and oil binding media in a famous 'drip' painting. Anal Methods. 2017;9:28–37.
[94] Delaney JK, Conover DM, Dooley KA, Glinsman L, Janssens K, Loew M. Integrated X-ray fluorescence and diffuse visible-to-near-infrared reflectance scanner for standoff elemental and molecular spectroscopic imaging of paints and works on paper. Herit Sci. 2018;6:31.
[95] Dooley KA, Conover DM, Deming Glinsman L, Delaney JK. Complementary standoff chemical imaging to map and identify artist materials in an early Italian renaissance panel painting. Angew Chem Int Ed. 2014;53:13775–9.
[96] Legrand S, Alfeld M, Vanmeert F, De Nolf W, Janssens K. Macroscopic Fourier transform infrared scanning in reflection mode (MA-rFTIR), a new tool for chemical imaging of cultural heritage artefacts in the mid-infrared range. Analyst. 2014;139:2489–98.
[97] Rosi F, Miliani C, Braun R, Harig R, Sali D, Brunetti BG, et al. Noninvasive analysis of paintings by mid-infrared hyperspectral imaging. Angew Chem Int Ed. 2013;52:5258–61.
[98] Sabbah S, Harig R, Rusch P, Eichmann J, Keens A, Gerhard JH. Remote sensing of gases by hyperspectral imaging: system performance and measurements. Opt Eng. 2012;51:111717.
[99] Rosi F, Harig R, Miliani C, Braun R, Sali D, Daveri A, et al. Mid-infrared hyperspectral imaging of painting materials. In: Pezzati L, Targowski P, editor(s). Proceedings of SPIE Optics for arts, architecture, and archaeology IV Vol. 8790. DOI 10.1117/12.2020477 2013.
[100] Gabrieli F, Dooleya KA, Zeibel JG, Howe JD, Delaney JK. Standoff mid-infrared emissive imaging spectroscopy to identify and map materials in polychrome objects. Angew Chem Int Ed. 2018;57:7341–5.
[101] Daveri A, Paziani S, Marmion M, Harju H, Vidmand A, Azzarelli M, et al. New perspectives in the non-invasive, in situ identification of painting materials: the advanced MWIR hyperspectral imaging. Trends Anal Chem. 2018;98:143–8.
[102] Polak A, Kelman T, Murray P, Marshall S, Stothard DJ, Eastaugh N, et al. Use of infrared hyperspectral imaging as an aid for paint identification. J Spectral Imaging. 2016;5:a2. DOI: 10.1255/jsi.2016.a2.
[103] Polak A, Kelman T, Murray T, Marshalla S, Stothard DJ, Eastaugh N, et al. Hyperspectral imaging combined with data classification techniques as an aid for artwork authentication. J Cul Herit. 2017;26:1–11.

Maria Cristina Caggiani and Philippe Colomban

7 Raman microspectroscopy for Cultural Heritage studies

Abstract: The Raman effect is at the basis of Raman scattering and microspectrometry: in the first part of the chapter, it is very shortly exposed together with differences with infrared (IR) spectroscopy, and advantages and drawbacks of the technique. The importance of the choice of the excitation wavelength, of the spectrometer (fixed, portable and handheld) and of the optics is underlined, while the information provided by the technique for inorganic and organic materials is considered. The surface-enhanced Raman spectroscopy (SERS) theory and principle applications are also taken into account. In the second part of the chapter, all the different applications of Raman and SERS to cultural heritage materials are contemplated: minerals, gemstones, rocks, patinas and corrosion products, glass, pottery, mortars, dyes, binders, resins, paper, parchment, inks and human remains. For each category of objects, the answers that Raman microspectrometry and SERS can give to the archaeometric and conservation-related questions, the *in situ* investigations, the search of specific spectral parameters and the use of chemometrics are shown, together with the most recent advances in the field.

Keywords: Raman microspectroscopy, SERS, Mobile Raman

7.1 Raman microspectrometry: advantages and drawbacks

The inelastic light scattering that results from the interaction of a monochromatic, coherent beam of light produced by a laser (light amplification by stimulated emission of radiation) with (liquid, solid or gas) matter is called Raman scattering, from the name of Sir Chandrashekhara Venkata Râman who obtained alone The Nobel Prize for that in 1930, even though other Russian, French and German contributors published significant works before [1, 2]. Theoretical prediction had been made by L. Brillouin in France and A. Smekal in Russia, a few years before. Actually, before the availability of lasers in the 1970s and then of charge-coupled device detectors (CCDs) in the 1980s, the technique remained confidential.

Briefly, in the Raman effect light is absorbed by electronic levels and the final de-excitation leads to transitions between vibrational levels [3, 4]: the Raman spectrum is thus rather similar to that obtained by infrared (IR) absorption; both can be predicted from analysis of the functional groups present in the matter and from the symmetry of

This article has previously been published in the journal *Physical Sciences Reviews*. Please cite as: Colomban, P., Caggiani, M.-C. Raman Microspectroscopy for Cultural Heritage Studies. *Physical Sciences Reviews* [Online] **2018**, 3. DOI: 10.1515/psr-2018-0007.

https://doi.org/10.1515/9783110457537-007

the normal modes. The process is very fast, at the femtosecond scale (1 fs = 10^{-15} s). Contrarily to IR bands, Raman peaks are generally narrow, because of the absence of strong dipolar coupling. If the matter is colourless or does not absorb the exciting light beam, the interaction with electronic levels is virtual. Raman spectra are generally expressed in cm^{-1} as IR ones, but THz or meV unities can be also used. Vibrational transition can proceed from the fundamental level (Stokes Raman spectrum, not very sensitive to temperature because fundamental levels are always occupied) or from the excited one (anti-Stokes spectrum). A Raman spectrum displays a peak fingerprint symmetric with respect to the elastic scattering (Rayleigh) signal, located at the same wavenumber/energy as the laser. The intensity of the anti-Stokes peaks decreases very rapidly as the distance from the Rayleigh peak increases, because the population of the upper vibrational level decreases to zero. Usually, only Stokes side is considered, and the cm^{-1} scale is measured with respect to the excitation energy.

Understanding a Raman spectrum requires the identification of the vibrational unit and its symmetry as it happens for the unit-cell identification by crystallography. This is often obvious for molecular compounds (IR and Raman techniques are called molecular spectroscopies) but more complicated for solids [5].

In the last 10 years, the miniaturization of laser sources and the increasing performances of CCD detector and of optical filters has boosted the development and use of the methods, especially in the field of Cultural Heritage. The reasons are the following: (i) it is an optical technique, which continues and maximizes the expertise gained by visual examination of artefacts and allows the use of advanced optics (microscope, long working distance (WD) optics, etc.); (ii) it is a scattering technique; in other words in most of the cases no preparation of the analysed artefact is required; (iii) the primary interaction occurs with electronic levels, which leads to a specific Raman phenomenon when the matter is coloured, the resonance Raman effect, able to detect traces of chromophores; furthermore, the interaction with metal particles/surfaces may enhance the Raman signal, this is the so-called surface-enhanced Raman spectroscopy (SERS) [3, 4, 6]; (iv) the Raman probe being very local (i. e. the modulation of the electronic cloud by atomic vibrations), the technique informs about the structure of both ordered/crystalline and disordered/amorphous matter; (v) Raman mapping with variable step resolution [from ca. 0.5 to 100 µm, typically) can be made on flat samples. More details about the description of the Raman scattering can be found in [3, 4, 7, 8]. Mobile Raman setups are available since 2000 with different degrees of performances [9–15].

Some drawbacks directly arise from the physics of the light–matter interaction: (i) the Raman intensity is very variable, almost null for ionic bonded compounds (no electron cloud in between atoms) and depends on the number of electrons involved in covalent bonding (in other words, if the polarizability and Z number are high); (ii) a resonance Raman spectrum depends on the laser lines and the observed spectrum is limited to modes involving the chromophore, which limits the pertinence of databases and makes quantitative measurements difficult without calibration and

(iii) illumination of coloured matter may induce a strong local heating (i. e. phase transition, oxidation and decomposition).

SERS was first discovered in the study of diluted species adsorbed on rough silver electrode [16, 17], but it started to be really used to detect diluted species after the publication by Creighton et al. [18], of an easy preparation method of silver colloids. Actually, the efficiency of the colloids is time dependent and optimized sensitivity is obtained after maturation of the preparation. It is assumed that SERS effect is maximal when molecules are squeezed between metal nanoparticles with optimal size (so-called hot spots). Many works have been made to prepare (commercially available) colloids and substrates for SERS [19, 20].

7.2 Procedure: choice of laser line, optics and spectrometer

7.2.1 Choice of the laser excitation

Light scattering intensity depends on the intensity of the exciting laser but also on its energy with exponent 4. Consequently, the Raman signal is very strong under blue excitation, weak with the red one and extremely weak under the IR one. Therefore, the effect is intrinsically remarkable under UV excitation, but adapted optics, mirrors and detectors cause a loss of efficiency and/or are very expensive. Another phenomenon may arise from the illumination of matter: transitions between electronic levels lead to the emission of fluorescence, a rather slow phenomenon (10^{-3} s) but very strong and broad, which may totally cover the Raman spectrum. Only electronic levels having energy lower than the laser line are excited, which may lead to choose red or IR laser excitation. However, the sources of fluorescence should be considered, and three types can be encountered: (i) transition metal and rare-earth/lanthanide ions diluted in oxide matrix give rise to groups of bands with moderate width (tens to hundreds cm^{-1}) separated by large windows. Group components inform about the symmetry of the site occupied by the ions. Changing the laser excitation, up or down, generally allows recording a Raman spectrum free of fluorescence components; (ii) the more or less degraded biological materials contaminating inorganic materials extracted from the soil lead to broad fluorescence; illumination with a few mW of blue or UV line or tens mW of green laser leads to decomposition/elimination of the film and disappearance/reduction of the fluorescence spectrum after a few (tens) minutes of illumination; (iii) polymers and organic compounds, especially those processed with solvents or natural organic/inorganic mixtures like ivory, exhibit strong fluorescence: in that case IR laser excitation (785 or 1,064 nm) can limit or avoid fluorescence generation and a good Raman spectrum can be recorded [21]; alternatively, a fast detection associated to pulsed laser can take advantage of the long life of fluorescence to suppress it [22]; automatic comparison of the spectra obtained with two energy-shifted excitations allows the computed extraction of the sole Raman spectrum [23]. The availability of different exciting lasers is thus recommended; for example, it has been clearly demonstrated in the study of paintings that only the combination of different laser excitations (with the addition of Fourier Transform Infrared Spectroscopy (FTIR) information) gives a complete view of the ingredients used by an artist in a painting [24]. Blue or violet lasers should be preferred for the study of inorganic materials and near infrared (nIR) for organic ones.

7.2.2 Choice of the spectrometer

Three types of instruments are available [9, 13]:

i. bench top instruments, usually excited with two lasers, sometimes more (sources from UV (255 or 325 nm) to nIR (785 nm) and rarely 1,064 nm). A remote head can be connected by optic fibres to facilitate the analysis of artefacts too big to be put under a vertical standard microscope; in rare cases, the instrument is coupled with a horizontal microscope in order to study vertical surfaces of sculptures, etc. The performances of such instruments can be very good: flat background, high resolution (~0.5 cm^{-1}), access to low wavenumber range (10–100 cm^{-1}) and high sensitivity; the CCD detector is cooled to ~200 (Peltier effect) or to 80 K (N_2) to maximize the detection; note that the efficiency of detectors is not constant over all the spectral window, especially for nIR excitation that requires corrections or limits the working range; a large palette of optics is available; some of these instruments can be moved from place to place and adjustment/reliability is computer controlled; the use of a microscope to focus the laser beam and to collect the scattered light limits the exploration of the polarizability tensor (backscattering); 90^d setting between the laser beam and the collection of the scattering, with additional polarization rotator, is needed to analyse all the Raman signature.
ii. Mobile Raman setups associating a laser source (532 or 785 nm, usually) connected by optic fibres (up to 30 m) to a remote head and to the spectrometer (Figure 7.1); the CCD is cooled at 200 K; generally, these instruments do not work above 3,200 cm^{-1}; resolution is medium (~5 cm^{-1}); a large palette of optics is available, and a laptop is sufficient to visualize and manage the instrument.
iii. Handheld instruments excited with 532, 785 or 1,024 nm laser; the low resolution (~10 cm^{-1}) and the uncertainness of the (computed) background hinders the detection/measurement of broad bands; only (very) low magnification (fixed) optic is possible. Visualization on a large screen and data treatments are often not possible/easy.

As mentioned above, the intensity of the Raman signature is very variable as a function of composition and bonding, and hence only high-sensitivity instruments detect most of the phases. Some characteristic peaks being very close (e. g. sulphates, nitrates and carbonates), a good resolution is needed to distinguish and compare the different phases [25].

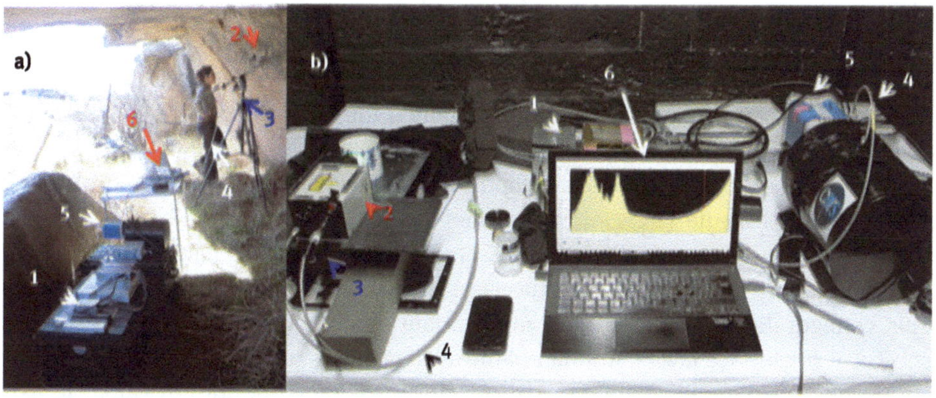

Figure 7.1: Examples of on-site measurements: (a) outdoor analysis of pigments and over layers in a Rock Art shelter (South Africa) and (b) study of soft-paste porcelain in the storage room of a museum: 1: laser; 2: remote head; 3: support; 4: optic fibres; 5: CCD detector and spectrometer; 6: laptop.

7.2.3 Optics

Unfortunately, the proper attention to optics quality is often neglected. The high efficiency of CCD leads to promote the closing of the so-called confocal hole – which is placed symmetrically to the focus point with respect to the objective lens – to decrease the size of the focused laser spot. However, optics are designed to optimize the spatial resolution in the XY focus plan in all the visible range (300–600 nm) at the expense of the *z* direction: the focus volume ranges from an ovoid in air, typically three to four times longer than broader, to a half-diabolo in solids (Figure 7.2(a)). Squeezing the confocal hole limits the volume anisotropy; nevertheless, using a high-magnification optics of

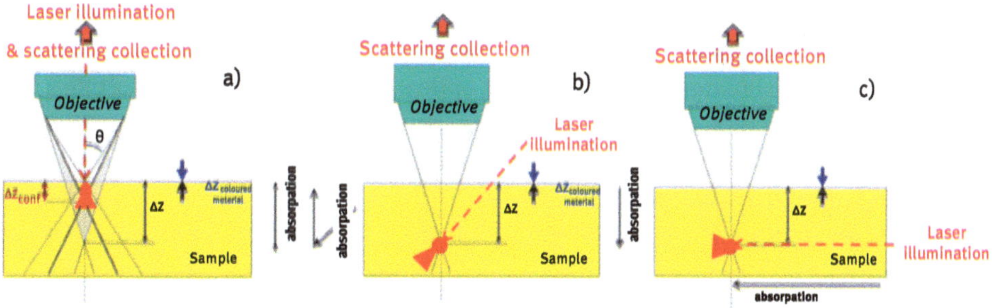

Figure 7.2: Schematic of the volume of matter probed using a microscope objective: (a) usually the laser focus is made at the sample surface and the same optic is used for illumination and collection of the scattered light (backscattering); slits or pinholes allow limiting the light pathway in the optics and reduce the size of the spot, especially along *z*-direction (Δz) (Δz is strongly reduced, up to a few tens nanometer for coloured materials); (b and c) decoupling laser focalization and collection of scattered light is possible; (c) historically the 90^{d} setting was the first to be used; microSORS setting allows decreasing/avoiding the contribution of surface layer.

high quality is much more preferable, though more expansive [13]. The 90-degree configuration, besides requiring an accurate positioning and often a sample preparation, implies a higher heating risk – due to the longer optical path in the sample – connected to the material absorption, especially if the laser colour corresponds to the absorbed one (Figure 7.2(a) and Figure 7.2(c)). The use of optic fibres also controls confocality. In order to overcome the depths reached with conventional confocal microscopy in turbid media [26, 154], a new methodology (SORS: spatially offset Raman spectroscopy) was conceived that consists in collecting the Raman scattering from regions laterally offset away from the laser excitation (surface) spot (Figure 7.2(b)). This configuration, combined with multivariate analysis, allows to retrieve Raman spectra of subsurface layers. Another pathway to gain this result – though less efficient – is the defocusing SORS modality [27] displacing the microscope objective along its axis by a (negative) defocusing distance Δz. Defocusing was also employed in the development of micro-SORS: it allowed the application to thinner layers, whereas in the conventional SORS the collection area diameters and the spatial offsets are at least of the order of a millimetres [28]. Standard microscope optics is often proposed by sellers of Raman instruments. Their WD (i. e. the distance between the front lens and the sample surface) is small, less than mm for an x100 objective (total magnification by conjugation with the spectrometer = 1,000, i. e. a spot waist of ~2–3 µm); this requires a flat/polished sample but maximizes the numerical aperture (NA), up to ~0.9 due to the wide angle of collection of the scattered light. Actually, such objectives are not good to analyse objects with complex shapes: objectives with a long WD up to ~15–20 mm typically for x100 objective, up to 13 mm for the highest magnification x200 (spot waist ~0.5 µm) [13] and much longer WD for x50 or x10 objectives, allow analysing many artefacts without risks of contact. Obviously, NA value is lower, which actually is better when the measurement cannot be made in obscurity; a very good perpendicularity between the analysed surface and the optic is required. Measurements of the exact size of analysed spot generated by the different objectives at the sample surface are recommended and easy [29]. Estimation of the in-depth penetration is much more complicate, especially for coloured matter, and the focused volume enlarges in relation with optical index; besides, the absorption can be very high for coloured material, giving rise to resonance Raman scattering. Penetration can be limited to a few tens of nanometres, which makes Raman scattering a technique of surface analysis.

For UV and nIR excitation (<300 nm, >900 nm), specific optics is recommended or even required.

7.2.4 Sample support and mapping

The use of high magnification objectives has different advantages (high spatial resolution that permits to record the Raman signature of each grain/phase individually, to avoid/limit fluorescence by chosen place where its intensity is minimal, limiting the heating, etc.) but requires the absence of vibrations and/or a strong and vibration-free coupling between the support of the object and that of the remote head (Figure 7.1). The performance of computers allows collection and exploitation of a series of spectra, and bench top instruments usually have a XY motorized stage permitting to collect spectra on large areas. Some instruments offer automatic z-control and z-profiling. Data management is conducted with the software of the instrument, but specific ones able to extract phase distribution, amount, crystallinity, and particle size have been developed [7, 30–34].

7.3 Information provided by Raman scattering

We will consider separately inorganic and organic phases because both recording conditions and extracted information can be very different.

As said before, the first information present in a Raman spectrum concerns the nature of the bond and the symmetry of the vibrational unit cell. A complete description of this approach is provided by [31] and more specifically for natural rocks and minerals by [34]. Some empirical rules can be noted: the stronger peak between ~500 and 1,200 cm^{-1} is generally the symmetric stretching mode, protonated compounds (hydrates and hydroxides) exhibit O-H stretching modes in the 3,100–3,800 cm^{-1} range, lamellar and heavy cation-containing compounds exhibit a strong peak below 200 cm^{-1}, etc. As expected from the Group Theory [5] the peak number of high symmetry cubic or hexagonal phases is small, whereas many peaks are observed for a monoclinic structure. Note, the Raman spectroscopy is more sensitive to orientational disorder and centres of symmetry than X-ray diffraction and the symmetry viewed by each technique can be slightly different [35, 36]. If the laser spot is small and does not include a large distribution of grain, the recorded spectra will be partly polarized, and some bands will not appear or will display a different intensity from that expected for a non-polarized spectrum. This implies that observed spectra could be different from those listed in databases.

7.3.1 Rocks, minerals, gemstones, synthetic compounds and corrosion layers

The identification of minerals is easy, and good reference papers and databases developed by mineralogists are available (see e. g. [34, 37–39]). Good identification is obtained on gems, even with handheld or mobile instruments [40, 41], thanks to their polished surface and the good knowledge of their Raman signature. Note that the composition of minerals is often complex (many partial substitutions) and can vary at the sub-millimetre scale from one side of a crystal to the other one. Raman scattering is also very efficient to detect patinas and corrosion layers as they are made of covalently bond compounds (sulphurs, sulphates, oxides, carbonates and nitrates) with a thickness more than ~ 0.3–0.5 µm. [25].

7.3.2 Glass, glazes and enamels

Usually, due to the large distribution of configuration for the vibrational unit-cell, the spectrum of a glassy phase exhibits broad bands. The case of glassy silicates, the matter of glass but also of glaze and enamels and of phases present in pottery body (paste) is the most important in Cultural Heritage. The basic unit of silicates (glassy or crystalline) is the SiO_4 tetrahedron. Non-distorted symmetry is T_d, and Raman modes expected from Group Theory analyses are as follows: a single (A_1 character) (strong) symmetric mode, 3 (F_2) (very weak) asymmetric stretching modes, 2 symmetric (E) and 3 (F_2) asymmetric medium modes plus medium librational (R') and translational (T', lattice) modes [42–49]. Spectra of glassy phases are rather similar to that of the crystalline ones, except for the strong broadening; in the low wavenumber range, R' and T' modes merge in a single asymmetric band, the so-called Boson peak [43, 49]. As shown in Figure 7.3, the ratio of the bending mode area versus the stretching one varies and is very useful to measure the degree of connectivity – i. e. the polymerization degree of SiO_4 tetrahedron.

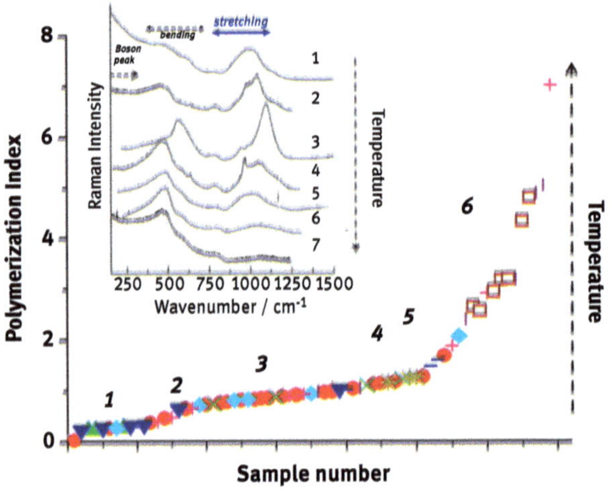

Figure 7.3: Representative spectra of glass and glaze types (from 1, a lead-rich glaze, to 7, a porcelain glaze) showing the evolution of the bending vs. stretching band area, i. e. the polymerization index. The variation of the index is illustrated for a series of glass and glaze types [see [43, 45] for details].

Furthermore, due to the dominant character of A_1 stretching mode, the component of the SiO_4 stretching band, measured between 750 and 1,200 cm^{-1}, typically can be used as the fingerprint of the different types of SiO_4 forming the silicate network: isolated tetrahedron (Q_0), tetrahedral connected by sharing 1 (Q_1), 2 (Q_2), 3(Q_3) or all 4 (Q_4) oxygen atoms [46]. Consequently, with the decrease of O-Si-O bridge number, the melting temperature decreases and the bending-to-stretching area ratio (from ~6 for high-connected silicates to ~0.3 for fully depolymerized ones) is thus directly related to the melting temperature of the glassy phase [43–45, 85]. A very simple abacus built with the maximal (or band centre of gravity) wavenumber of bending and stretching SiO_4 bands allows classifying the different types of glass as shown in Figure 7.4.

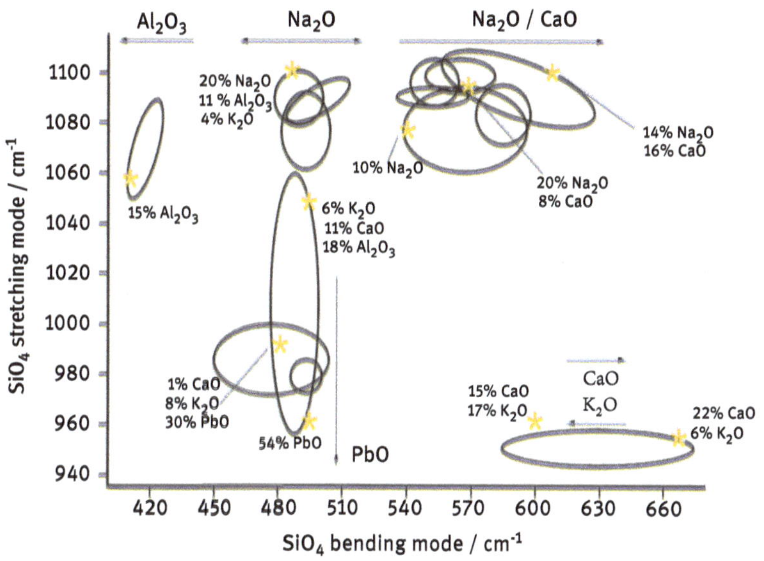

Figure 7.4: Plot of SiO_4 stretching vs. bending wavenumber constituting an abacus of glass types [see [85] for details].

7.3.3 Organic phases, polymers and inorganic/organic composites

IR spectroscopy is largely used to characterize and identify synthetic and natural polymers and organic derivative due to the easiness of getting a spectrum, especially by Attenuated Total Reflectance (ATR) and diffuse FTIR [50–52]. The great advantages of Raman scattering versus IR spectroscopy are the higher spatial resolution and that no preparation is needed. However, advanced spectrometers are needed: if the 785 nm excitation is convenient to get a good spectrum of modern polymers and organic compounds, the degradation of the surface of these materials leads to the choice of 1,064 nm excitation to get spectra of ancient natural resins (amber, copal, gums, glues and shellac) and lacquer or varnishes [53–57]. Thus, 785 and especially 1,064 nm Raman instruments are the most efficient to analyse both inorganic and organic phases of ivory [21], bones [58] and coral [59, 60].

7.4 Surface-enhanced Raman spectroscopy

For a long time, chromatographic techniques have been preferred to identify dyes after extraction from their support. The small amount of dye in the extracted mixture makes their identification difficult by FTIR. SERS technique was mainly developed in biochemistry because its specificity can be much better than fluorescence spectroscopy [61]. The use of SERS is advantageous in many ways: it is convenient for dilute media and it can allow the quenching of fluorescence backgrounds. Another striking point is that it may provide vibration fingerprints of individual molecules even in mixtures. This could help the understanding of the related chemical reactions by monitoring of spectral changes arising from the substrate.

The first application to the study of Cultural Heritage materials dates from 1987, when Guineau and Guichard [62] identified alizarin in extract from an eighth-century textile sample on a silver electrode, a procedure similar to that used for the first observation of pyridine SERS signal 13 years before by Fleischmann et al. [16]. Actually, the SERS procedure really starts to be used after 2002 to study alizarin, purpurin, lakes and indigo [63, 64]. Leona et al. [65] show how to use silver film deposited on glass [Tollens mirrors) or colloids prepared following Lee and Meisel [66]. From our experience using Creighton et al. silver colloids [18], the method is also good [61]. Murcia-Mascaros et al. [67] demonstrate that absorption on ormosil (also called xerogel [68]) is an alternative method. Canamares et al. and Pozzi and Leona [19, 69] reviewed the topics and preparation methods of the SERS promoter: films, colloids dispersed in water or centrifugated paste and substrates prepared by physical techniques (laser ablation, photo-reduction, electron beam, etc.). *In situ* technique (i. e. without extraction) with photo-reduced silver nanoparticles was also tested positively [70]. A much simpler technique is the manual deposit of a drop of a few microliters of colloids onto the sample with a micropipette [71–73]. Paste deposit was also used [74]. The development of peelable gels retains attention in order to be non-destructive [75–78]. Inkjet nanoparticle delivery was also tested [79]. Finally, tip-enhanced Raman spectroscopy (TERS) was also tested [80]. The Raman signature obviously depends on the exciting wavelength and on interactions with the substrate. Therefore spectral libraries and/or reference studies are most of the time very helpful to identify the molecules [19]. Chemometric techniques are very useful.

7.5 Cultural Heritage applications

7.5.1 Answers to the archaeometric questions

Some of the principal questions arising from an interdisciplinary study in which archaeometry – constituting an important interaction between liberal arts and scientific disciplines – is called to answer are “How?”, “Who?”, “When?” and

"Where?". Raman microspectrometry very often can (partly) answer these questions, since it can give information concerning characterization, authenticity check, relative dating and provenance.

Characterization can aim at the knowledge of the structure of natural compounds such as gemstones and minerals or at the discovery of raw materials and technological production used for artificial material manufacturing. The raw materials can in some cases spread light on the provenance, but it must be considered that for certain materials, the source, working and selling places do not necessarily coincide. The evolution and diffusion of production technologies often allows individuating the progresses of a culture or the exchanges of know-how between different human groups; this is the case, for example, of the glazed pottery. Raman microspectrometry intervention in this field may concern the identification of significative minerals in the pottery or porcelain body, in the glaze or at their interface [81–83] or in the composition (and firing temperature) of glazes [83], glass [84] or enamels [85].

The chronological determination essentially answers the need of dating an object. One of the most common applications is the authentication. This can be carried out also in an indirect way, not giving an absolute date but a relative one: a typical example is the Raman identification of pigments or glass compositions that were synthesized or used only after a certain date (*terminus post quem*) [86] or of compounds banished after a certain date (*terminus ante quem*). Attempts of attributions can also be based on the recognition of materials and expedients typical of certain artists or workshops [87].

Concerning the provenance question, apart from the information that in some cases can be given by the raw materials, another much-used approach, also applied to Raman spectroscopy, is that of multivariate statistical analysis. It is useful for distinguishing groups of objects characterized by differences or similarities that can be attributed to provenance, especially in the cases of geological sources of natural materials such as obsidian [88] or of natural resins [55, 89], or to firing process [90].

Apart from archaeometric problems, another important and always increasing contribution of Raman spectroscopy to Cultural Heritage must be reminded, involving all the conservation issues, from the investigation of the causes of alteration and degradation [91, 92] to the subsequent conservation and restoration recommendations [53].

All these issues will be discussed more in detail in Section 5.3, considering each different material (Table 7.1).

7.5.2 Approaches

The more and more widespread *in situ* Raman measurement campaigns for cultural heritage starting from the first years of the 2000 are mainly due to the increasing availability of air-cooled lasers and the miniaturization of detectors and their

Table 7.1: The analysed materials in literature, divided by category, the questions asked ([a] Characterization, [b] Conservation issues, [c] Relative dating, [d] Authenticity check, [e] Provenance), the SERS usage, the best laser choice, and useful references.

	Materials	Questions	SERS	Best laser	Useful references
Inorganic					
Natural	Minerals	a b e		Visible – Resonance	[39, 114–116, 193]
	Gemstones	a d e		Visible – Resonance	[38, 39]
	Rocks	a b e		Visible	[112]
	Corrosion	a b		Visible	[25, 116, 125–131]
Natural/ Crafted	Pigments	a b c d e	X	Visible – Resonance	[34, 39, 114–116]
Crafted	Pottery	a b c d e		Visible	[2, 45, 47, 54, 144]
	Mortars & Plasters	a b		Visible	[152]
	Glass	a b c d		UV-blue	[110]
	Patinas	a b		Visible	[25, 135, 138]
Organic					
Natural	Human remains	a b		IR	[186]
Natural/	Dyes	a b c d e	X	nIR-IR	[156–160]
Crafted	Binders, glues, varnishes	a b d	X	nIR-IR	[54, 114, 163]
	Resins	a b c d e	X	nIR-IR	[55, 164, 194]
Crafted	Parchment	a b d		IR	[175, 176]
	Textiles	a b	X	IR	[162]
	Inks	a b d	X	nIR-IR	[180]

electronics as well as the laptop computer, allowing the use of transportable and then portable instrumentation (Figure 7.1); a review registers this evolution with its drawbacks and advantages [9]. This approach is nowadays employed in all the fields of the archaeometric research, involving both indoor and outdoor campaigns; some examples will be described in the following paragraphs.

The as-acquired raw spectra can undergo a series of mathematic treatments to support their interpretation and to make them comparable to each other. The removal of cosmic spikes and the background correction by means of baseline subtraction are the first steps. If needed, the spectra can undergo further treatments, such as normalization or decomposition of bands [49]. Band intensities (or those of the deconvoluted signals) can be related to each other with the aim of calculating ratios that are specific spectroscopic indicators of some characteristic of the materials (see for example Section 3.2 and Figure 7.3 and Figure 7.4 for glass polymerization degree, and Section 5.3.5 for resin maturation index). Furthermore, the spectroscopic parameters, such as position, intensity and width (Full width at half height: FWHM), can act as variables and be subjected to statistical treatments like cluster analysis or principal component analysis with the aim of distinguishing groups of objects [91], which is a much-used procedure in the provenance problems.

Despite the possibility of gaining more knowledge from the combined use of Raman spectroscopy and chemometrics, since cultural heritage objects are complex and since the questions involved are being multiple, a multi-technique approach, combining the vibrational spectroscopy with elemental analyses/microscopic observations, is almost always preferred for a thorough study of cultural heritage.

7.5.3 Materials

7.5.3.1 Minerals, gemstones, rocks

A recent review paper addresses the identification of minerals in the field of Cultural Heritage studies [39]. In most cases, the Raman analysis of minerals is part of wider studies involving pottery, glazes, glass, enamels and especially pigments. In some cases, the general interest concerning one phase constituting a pigment or ceramic bodies or decorating pottery is so widespread and specific that must be deepened. This is the case, for example, of all that concerns blue pigments. The chromophores of lazurite (the sodalite group blue feldspathoid-colouring lapis lazuli rock) [93–96], ultramarine blue (the synthetic pigment) [97, 98] and haüyne (another mineral belonging to the sodalite group) [99] consist of S_2^- and S_3^- ions trapped in the feldspathoid (mineral) or zeolite (synthetic) host matrix. Since the Raman spectrum of coloured minerals depends from the used exciting laser line, the spectra of S_n^- chromophore recorded under UV, blue, green, red and IR excitation have variable intensities (Figure 7.5). The plot of the normalized intensity shows that the resonance is maximal between 514 and 785 nm; therefore, visible excitation should be used to induce the resonance spectrum of the chromophore, whilst UV/IR excitation should be chosen to identify the host matrix [99, 100]. Another group of minerals that, for its usage covering all the times and all the planet, and for the implication in corrosion mechanisms, deserves specific mention is that of iron oxides and hydroxides, for which different works list and study the Raman spectra [101–103].

The Raman analyses of gemstones, now a routine investigation, is not only used to identify fakes, but – thanks to its characteristics of non-destructivity and confocality – it is a very suitable technique for the analysis of the inclusions, which can be significative of provenance and genesis [39]. This is the case, for example, of the information gained, together with the employment of X-ray maps, on deep mantle diamond formation [104]. Concerning gemstones, a specific branch is that involving cultural heritage-related objects [105]. The latter being usually very precious, such as silver or golden ecclesiastic objects richly decorated with gemstones, *in situ* investigations with portable instrumentation are recommended, which is today common practice, also taking advantage of different excitation wavelengths [40, 106].

Lithic raw materials constituting archaeological prehistoric tools can be compared with geological samples by Raman spectroscopy with the aim of identifying their origin [107]. These are made of silica rocks in which recently moganite Raman

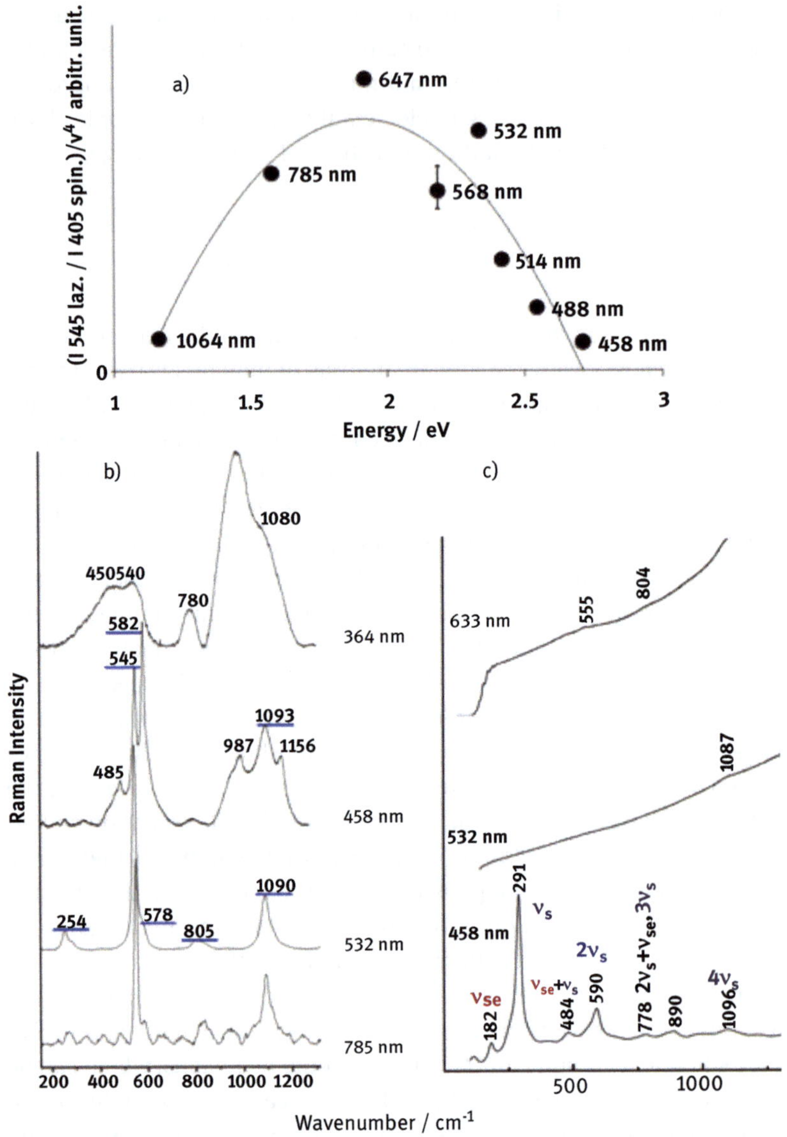

Figure 7.5: (a) Curve plotting the intensity of 545 cm^{-1} S_3^- ion in lazurite peak divided by that of 405 cm^{-1} spinel reference peak and normalized to ν^4, against the incident laser energy [87]; (b) representative spectra recorded on lazurite under different wavelength excitations: S_n^- ion signature (peaks underlined in blue) is dominant for IR and green excitation, whereas only the hosting glass matrix signature is recorded under UV excitation [99]; (c) comparison of the Raman spectra recorded under different excitations for a red-stained glass window coloured by $CdS_{1-x}Se_x$ nanoparticles; fundamental Se and S modes and their overtones/combinations are labelled [142].

signature has been revealed [108] and the ratio between quartz and moganite peaks is considered as spectroscopic parameter [109]. Also, Raman spectroscopy coupled with statistical multivariate analysis can be used to distinguish the major obsidian sources [88, 110] .

It has been underlined that Raman microspectrometry is the best technique for rapidly and non-destructively distinguishing the different varieties of jade: jadeite-jade, nephrite jade and other counterfeiting materials [111, 112].

7.5.3.2 Pigments

One of the most widespread applications of Raman spectroscopy within conservation science undoubtedly concerns pigments of different sorts, of all ages, dispersed in different mediums and decorating various materials. It is by now established that, when possible, portable Raman instrumentation can be successfully used for pigment identification. Databases exist useful as references [2, 113–116]. Notwithstanding, it must be taken into account that they are often not sufficient for the identification, due, for example, to pigment degradation caused by the uprising of reactions in mixtures [92] or by the high incident laser power and/or energy [117]; degradation can be simulated in order to verify the spectral changes [118]. A recent review [39] considers the different kinds of substrates, such as rock art; manuscripts; wall, panel and canvas paintings and even gemstones. Concerning blue pigments, excluding all that is linked with lapis lazuli (see Section 5.3.1), the most studied ones are azurite, cobalt-containing phases [119] and indigo/maya blue (indigo in palygorskite structure [120]). In a very recent paper, the effects of X-ray and ion irradiation on azurite were evaluated with a multi-analytical approach [121]. Blue minerals very rarely used as pigments such as aerinite, veszelyite, vivianite and glaucophane are also considered [39], whose reference Raman spectra cannot be found in pigment database, but rather in mineral ones [193] or in specific studies. Not considering copper-based minerals (see Section 5.3.3), the most widespread green mineral pigments are green earths. Their differentiation (glauconite, celadonite and others) can be based on Raman spectral features; furthermore, the use of the appropriate excitation wavelength can cause resonance effects [122]. Iron oxides and hydroxides yellow and red minerals used as pigments are discussed in Section 5.3.1. Lead- (lead white $PbCO_3$ and red lead Pb_3O_4) and sulphur (realgar, pararealgar As_4S_4 and orpiment As_2S_3)-based yellow and red pigments are much studied due to their strong tendency to photodegradation [39]. Black pigments include graphite and amorphous carbon black, which can be distinguished according to the source of carbon [123]; other dark pigments such as manganese oxides, cobalt oxides and magnetite must be analysed with high-sensitivity instrumentation and very low-power values, because the easy heating of the strongly absorbing phases can hinder the identification of the resultant spectra [117]. On the contrary, white pigments – mainly constituted of carbonates – are usually easily identifiable.

Pigments in painted multilayers can be non-invasively investigated by means of micro-SORS application, which is suitable for the micrometric dimensions of layers commonly encountered in cultural heritage [28]. A combined approach of micro-SORS and mapping is recent, and it was applied for the first time on a real sample by [124], for the study of an inhomogeneous painted stratigraphy on porcelain.

7.5.3.3 Alloys patinas and corrosion products

Whereas the metallic bonding cannot be studied by standard Raman spectroscopy, the corrosion products are identifiable phases. The main aim of their study is understanding the degradation processes and evaluating the best conservation and restoration procedures. Various collections of Raman spectra of different corrosion products, especially of iron [30] [125] or copper-based alloys, can be found [25, 116, 126, 127], with bronze disease being one of the most urgent problems to be understood and faced [126, 128–130] . Non-invasive *in situ* bronze and brass corrosion patinas investigation has also been successfully tested, thanks to the combination of portable Raman and X-Ray Fluorescence (XRF) analyses [25]. Semi-quantitative Raman mapping on the cross-sections is very useful to visualize the distribution pattern of the mixture of phases and to determine if the corrosion is active or not [30, 32, 125, 131]. Silver corrosion products were also considered studying the best Raman setup for their identification [132].

So-called artistic (or artificial) patinas - as opposed to the natural ones - are also studied. They are intentionally applied by the artist to finalize the aesthetic appearance of the work [25]. They can be organic compounds such as drying oils and waxes, more difficult to be identified especially if *in situ* [133], or inorganic coloured layers. Few Raman-based studies were carried out on the latter, especially concerning copper-based alloys, most of them involving preliminary phases of laboratory patination and artificial weathering with the aim of differentiating intentional application from natural patinas formation and of implementing the best strategies of conservation and restoration [134–138]. In fact, the influence that the artificial patina may have on the formation of the corrosion products is interestingly underlined [137].

7.5.3.4 Synthetic inorganic materials: pottery, glass, mortars

The Raman analysis application to pottery and glass is mainly aimed at the investigation of raw materials and production procedures, with the final objective of studying technological and cultural models' evolution within the same context and contamination and circulation among different cultural groups. The two production cycles can be connected, especially for what concerns enamelling and glazing [139], which are the most studied elements of these objects with Raman spectroscopy [82, 83, 85, 140, 141] . Thanks to what is explained in Section 3.2, in fact, glass type differentiation – and thus firing temperature hypothesis – can be attempted with this technique, especially when glass, glazes or enamels are coloured by means of an ionic chromophore, because on the contrary, the scattering of a pigment, especially if resonant, can hide the glass signature revelation [38, 60, 85, 99, 142]. A type of glass application much studied with Raman is that of mosaics tesserae [143] and in general opaque glass-ceramics [84, 143], where the attention is focussed on the pigments and on the opacifying phases. Opaque red and yellow glass coloured with metal – copper and silver, respectively – has deserved special in-depth analyses because of the huge absorption of the light by metal nanoparticles. Main spectral features are observed at very low wavenumber (10–50 cm^{-1}) [144, 145] . Furthermore, the Raman spectrum of the host matrix is perturbated and limited to the near environment of the metal particle [146, 147]. Glass Raman signature inform about the local silicate network conformation (controlled by the Raman spectroscopic parameter "polymerization index") with respect to the colourless glass matrix [144]. These studies were carried out also on stained glass [146]; it was demonstrated that stained glass windows can be analysed *in situ* with portable instrumentation, also differentiating original and restored pieces [91]. Concerning pottery, glazes, interface between body and coating and – to a lesser extent – ceramic bodies are studied for ceramic types ranging from soft paste to porcelain; since about the first years of 2000, *in situ* investigations are commonly carried out, as for example for Iznik pottery [148], Kuthaya fritwares [90], Meissen porcelain and stoneware [47, 149] as well as Limoges enamels on metal [150]. Note that the identification of clays is difficult and XRD or FTIR techniques are more efficient.

Apart from wall paintings studies [151], Raman spectroscopy is successfully applied also to plasters and mortars, with the aim of gaining information about the substrate preparation methodology [152] or of identifying degradation products such as black crusts and efflorescence [153]. Micro-Raman mapping is also very useful to test the penetration depth of treatments within plasters cross-sections [154, 155].

7.5.3.5 Natural and synthetic dyes, organic media and resins

Organic colorant identification by means of Raman spectroscopy can be challenging, but thanks to red or, better, IR excitation wavelengths and the help of databases [156–158] and specific flowcharts [159], it can be achieved. When fluorescence is too strong, SERS can successfully be used to enhance the Raman scattering, requiring very little material [160]. Actually, dye identification is one of the most common applications of SERS to cultural heritage [19], ranging from the first experiments on madder-dyed textiles [62] to the most recent attempts of quantitative measurements [161]. Ancient dyes SERS and Raman study are often associated to the textile one, but very rarely the fibre constituting the textile itself is analysed [162]. Even though binding media and glues are more often studied by means of other techniques such as IR spectroscopy (FTIR) or (pyrolysis)-gas chromatography-mass spectrometry ((Py)-GC-MS), it has been demonstrated that the combined use of FT-Raman, SERS and IR micro-ATR can give complementary information about objects with complex multi-layered decorations

[73]. Databases of 785 nm-acquired or 1,064 nm (FT) Raman spectra of binding medias and varnishes used in cultural heritage are available [114, 163]. Daher et al. [56] proposed a flowchart useful for the discrimination of different classes of organic media (resins, glues, gums and oils) by means of both Raman and IR spectroscopic features. Specific studies on fossil resins such as copal and amber were conducted due to the interest arising from their archaeological importance and therefore giving rise to the study of geological samples of different provenances, such as the Baltic [164], Romanian [53, 89], Mexican [165], Czech and Moravian [166]. Other Raman work in the literature has concentrated on the determination of the extent of resin maturation, hence the age of the resins, investigating the relative intensity of the two strong bands in amber at about 1,646 and 1,450 cm^{-1}, respectively, arising from C = C conjugated stretching and CH_2 deformation [164, 167–169]. Edwards et al. [170] have investigated the potential of Raman spectroscopy for the observation of insect inclusions in amber resins in order to understand their state of preservation.

7.5.3.6 Other organic materials: paper, parchment, inks

Paper, parchment and ink studies are strictly connected with each other [171], even though Raman-based works on paper are much rarer than those concerning parchment, the latter being the constituting support of illuminated manuscripts; even when Raman spectroscopy is used for pigments or ink identification, paper [172] or parchment [173, 174] may be analysed by means of other techniques such as NMR or FTIR. On the other hand, minimally invasive tip-enhanced Raman spectroscopy (TERS) was successfully tested on nineteenth-century paper and ink [80]. FT-Raman can be used for parchment characterization [175], also with the aim of detecting traces of preparation procedures [176] or of investigating the effects of biological or chemical deterioration [177]. Furthermore, it has been shown with 785 nm laser excitation that Raman spectroscopy is more efficient than ATR-FTIR in the parchment substrate analysis in the presence of Ca salts, since the latter Raman signals are narrow, whereas ATR-FTIR spectra result fully overwhelmed by $CaCO_3$ ones [178]. Objects made of protein materials exhibit a strong fluorescence under the laser and 785 or better 1,064 nm excitations are required. Alternatively, the use of very high magnification (x200) microscope objective, close to the diffraction limit, allows minimizing the contribution of the support and collecting the pigment signature [117].

Much wider is the Raman spectroscopy application to ink analysis, which is reviewed and summarized in previous literature [179–181]. Iron gall inks often draw the attention of analysts due both to their copious use for centuries and to their characteristic corrosive effect: *in situ* portable Raman investigations and FT-Raman ones were carried out on them [182–184]. The Vinland Map controversy is famous: during the Raman investigation of the inks, TiO_2 anatase, which is not a medieval pigment but a common impurity of clay or kaolin, was found, hence the claimed attribution to modern age (post 1923) [185].

7.5.3.7 Human archaeological remains

Tissues and biomaterials analysed by Raman spectroscopy associated to forensic archaeology are summarized by [186], showing that this technique may not only give information about these material compositions, but also help distinguishing genuine and fake/counterfeit ones. nIR (785 nm) or IR (1,064 nm) (FT)-Raman excitation wavelengths are used due to the need of fluorescence overcoming. Both inorganic and organic components can be detected, as for example in teeth, for which the intensity ratio of the stretching modes of the organic and the inorganic components can be correlated with the duration of the burial [187]. Apart from teeth [188], bones and skin are the most studied, the former for the evaluation of their mineralization during burial [189], even when they are decorated [190], the latter for the study of the changes in molecular structure [191] also due to the microclimate [192].

Acknowledgements: The authors thank all institutions and their members providing opportunity to study artefacts.

References

[1] Krishnan RS, Shankar RK. Raman effect: history of the discovery. J Raman Spectrosc. 1981;10:1–8.
[2] Colomban P. The destructive/non-destructive identification of enameled pottery, glass artifacts and associated pigments—a brief overview. Arts. 2013;2:77–110.
[3] Long DA. Raman spectroscopy. New York: McGraw-Hill International Book Company, 1977.
[4] Long DA. The Raman effect: a unified treatment of the theory of Raman scattering by molecules. Chichester: John Wiley & Sons Ltd, 2002.
[5] Poulet H, Mathieu JP. Spectres de vibration et symétrie des cristaux. idem, Vibrational spectra and symmetry of crystal, Gordon & Breach Sci. Publ. Ltd (1976) New-York: Gordon & Breach Sci. Publ. Ltd, 1970.
[6] Moskovits M. Surface-enhanced Raman spectroscopy: a brief retrospective. J Raman Spectrosc. 2005;36:485–96.
[7] Gouadec G, Bellot-Gurlet L, Baron D, Colomban P. Raman mapping for the investigation of nanoPHased materials (Ch. 4). In: Zoubir A, editor(s). Raman imaging, techniques & applications Vol. 168. Berlin: Springer Series in Optical Sciences, 2012:85–118.
[8] Colomban P, Gouadec G. Raman scattering theory and interpretation (Ch. 2). In: Amer MS, editor(s). Raman spectroscopy for soft matter applications. Hoboken: John Wiley & Sons, Inc., 2009:11–29.
[9] Colomban P. The on-site/remote Raman analysis with portable instruments - A review of drawbacks and success in Cultural Heritage studies and other associated fields. J Raman Spectrosc. 2012;43:1529–35.
[10] Miliani C, Rosi F, Brunetti BG, Sgamellotti A. In situ non-invasive study of artworks: the MOLAB multitechnique approach. Acc Chem Res. 2010;43:728–38.
[11] Vandenabeele P, Edwards HGM, Jehlička J. The role of mobile instrumentation in novel applications of Raman spectroscopy: archaeometry, geosciences, and forensics. Chem Soc Rev. 2014;43:2628–49.
[12] Colomban P, Milande V, Lucas H. On-site Raman analysis of medici porcelain. J Raman Spectrosc. 2004;35:68–72.

[13] Colomban P. On-site Raman study of artwork: procedure and illustrative examples. J Raman Spectrosc. 2018;49:921–34. DOI: 10.1002/jrs.4042.
[14] Madariaga JM. Analytical methods in the field of cultural heritage. Anal Methods. 2015;7:4848–76.
[15] Vandenabeele P, Edwards HGM, Moens L. A decade of Raman spectroscopy in art and archaeology. Chem Rev. 2007;107:675–86.
[16] Fleischmann M, Hendra PJ, Mcquillan AJ. Raman-spectra of pyridine adsorbed at a silver electrode. Chem Phys Lett. 1974;26:163–6.
[17] Jeanmaire DL, Van Duyne RP. Surface Raman spectroelectrochemistry. 1. heterocyclic, aromatic, and aliphatic-amines adsorbed on anodized silver electrode. J Electroanal Chem. 1977;84:1–20.
[18] Creighton JA, Blatchford CG, Albrecht MG. Plasma resonance enhancement of Raman-scattering by pyridine adsorbed on Silver or Gold sol particles of size comparable to the excitation wavelength. J Chem Soc Faraday Trans. 1979;75:790–8.
[19] Pozzi F, Leona M. Surface-enhanced Raman spectroscopy in art and archaeology. J Raman Spectrosc. 2016;47:67–77.
[20] Banholzer MJ, Millestone JE, Qin L, Mirkin CA. Rationally designed nanostructures for surface-enhanced Raman spectroscopy. Chem Soc Rev. 2008;37:885–97.
[21] Mancini D, Tournié A, Caggiani MC, Colomban P. Testing of Raman spectroscopy as a non-invasive tool for the investigation of glass-protected miniature portraits. J Raman Spectrosc. 2012;43:294–302.
[22] Matousek P, Towrie M, Ma C, Kwok WM, Phillips D, Toner WT, et al. Fluorescence suppression in resonance Raman spectroscopy using a high-performance picosecond Kerr gate. J Raman Spectrosc. 2001;32:983–8.
[23] Culka A, Jehlička J. Sequentially shifted excitation: A tool for suppression of laser-induced fluorescence in mineralogical applications using portable Raman spectrometers. J Raman Spectrosc. 2018;49. DOI: 10.1002/jrs.5320.
[24] Stanzani E, Bersani D, Lottici PP, Colomban P. Analysis of artist's palette on a sixteenth century wood panel painting by portable and laboratory Raman instruments. Vibr Spectrosc. 2016;85:62–70.
[25] Colomban P, Tournié A, Meynard P, Maucuer M. On-site Raman and XRF analysis of Japanese/Chinese Bronze/Brass Patina – the search of specific Raman signatures. J Raman Spectrosc. 2012b;43:799–808.
[26] Matousek P, Clark IP, Draper ERC, Morris MD, Goodship AE, Everall N, et al. Subsurface probing in diffusely scattering media using spatially offset Raman spectroscopy. Appl Spectrosc. 2005a;59:393–400.
[27] Eliasson P, Claybourn M, Matousek P. Deep subsurface Raman spectroscopy of turbid media by a defocused collection system. Appl Spectrosc. 2007;61:1123–7.
[28] Conti C, Colombo C, Realini M, Zerbi G, Matousek P. Subsurface Raman analysis of thin painted layers. Appl Spectrosc. 2014;68:686–91.
[29] Havel M, Colomban P. Smart Raman and Rayleigh spectroscopy for the analysis of nanomaterials. Spectroscopy Europe-Microsc Anal. 2006;20:11–4.
[30] Monnier J, Bellot-Gurlet L, Baron D, Neff D, Guillot I, Dillmann P. A methodology for Raman structural quantification imaging and its application to iron indoor atmospheric corrosion products. J Raman Spectrosc. 2011;42:773–81.
[31] Nakamoto K. Infrared and Raman spectra of inorganic and coordination compounds: part a: theory and applications in inorganic chemistry, 6th ed. Hoboken: J. Wiley & Sons, 2009. DOI: 10.1002/9780470405840.
[32] Veneranda M, Aramendia J, Arrizabalaga I, Bellot-Gurlet L, Colomban P, Castro K, et al. FTIR spectroscopic semi-quantification of iron corrosion phases: a new method to evaluate the protective ability index of rust layers affected by chlorides. Corrosion Sci 2018;133:68–77.

[33] Havel M, Baron D, Colomban P. Smart Raman/Rayleigh imaging of nanosized sic materials using the spatial correlation model. J Mater Sci. 2004;39:6183–90.
[34] Karr C, Jr, editor. Infrared and Raman spectroscopy of lunar and terrestrial minerals. New-York: Academic Press-Elsevier, 1975. DOI: 10.1016/B978-0-12-399950-4.50001-X.
[35] Colomban P. Raman study of the inorganic polymer – superionic nasicon transformation: dynamic, static orientational disorder and superionic conductivity. J Mol Struct. 1986; 143:191–4.
[36] Colomban P. Imagerie Raman de matériaux et dispositifs nano/microhétérogènes, Paris: Techniques de l'Ingénieur. RE 5v2/13 2013b
[37] Colomban P, Sagon G, Faurel X. Differentiation of antique ceramics from the Raman spectra of their colored glazes and paintings. J Raman Spectrosc. 2001;32:351–60.
[38] Pinet M, Smith DC, Lasnier B. La microsonde Raman en gemmologie, revue de gemmologie. Paris: Association Française de Gemmologie, 1992. (Hors Série: Paris).
[39] Bersani D, Lottici PP. Raman spectroscopy of minerals and mineral pigments in archaeometry. J Raman Spectrosc. 2016;47:4999–530.
[40] Jehlička J, Culka A, Bersani D, Vandenabeele P. Comparison of seven portable Raman spectrometers: beryl as a case study. J Raman Spectrosc. 2017;48:1289–99.
[41] Zieman MA. *In situ* micro-Raman spectroscopy on minerals on-site in the grotto hall of the new palace, Park Sanssouci, in Potsdam. J Raman Spectrosc. 2006;37:1019–25.
[42] Colomban P. Polymerisation degree and raman identification of ancient glasses used for jewellery, ceramics enamels and mosaics. J Non-Crystalline Solids. 2003a;323:180–7.
[43] Colomban P, Prinsloo LC. Optical spectroscopy of silicates and glasses. In: Yarwood J, Douthwaite R, Duckett S, editor(s). Spectroscopic properties of inorganic and organometallic chemistry Vol. 40. Cambridge: RSC Publishing, 2009:128–50.
[44] Colomban P, Slodzyck A. Raman intensity: an important tool to study the structure and phase transitions of amorphous/crystalline materials. Opt Mater. 2009;31:1759–63.
[45] Colomban P. Pottery, glass and enamelled artefacts: how to extract information on their manufacture technology, origin and age? ch. 8. In: Edwards H, Vandenabeele P, editor(s). Analytical archaeometry: selected topics. Cambridge: The Royal Society of Chemistry, 2012b:247–70.
[46] Colomban P, Paulsen O. Non-destructive determination of the structure and composition of glazes by Raman spectroscopy. J Am Ceram Soc. 2005;88:390–5.
[47] Colomban P, Milande V. On-site Raman analysis of the earliest known Meissen porcelain and stoneware. J Raman Spectrosc. 2006;37:606–13.
[48] Labet V, Colomban P. Vibrational properties of silicates: a cluster model able to reproduce the effect of "SiO4" polymerization on Raman intensities. J Non-Crystalline Solids. 2013;370:10–7.
[49] Colomban P. On-site Raman identification and dating of ancient glasses: procedures and tools. J Cult Her. 2008;9:e55–60.
[50] Joseph E, Prati S, Sciutto G, Ioele M, Santopadre P, Mazzeo R. Performance evaluation of mapping and linear imaging FTIR microspectroscopy for the characterisation of painting cross-section. Anal Bioanal Chem. 2010;396:899–910.
[51] Manfredi M, Barberis E, Rava A, Robotti E, Gosetti F, Marengo E. Portable diffuse reflectance infrared Fourier transform (DRIFT) technique for the non-invasive identification of canvas ground: IR reflectance collection. Anal Methods. 2015;7:2313–22.
[52] Ricci C, Bloxham S, Kazarian SG. ATR-FTIR imaging of albumen photographic prints. J Cult Herit. 2007;8:387–95.
[53] Badea GI, Caggiani MC, Colomban P, Mangone A, Teodor ED, Teodor ES, et al. FT-Raman and statistical analysis on thermally altered samples of amber. Appl Spectrosc. 2015;69:1457–63.

[54] Asquier M, Colomban P, Milande V. Raman and Infrared analysis of glues used for pottery conservation treatment. J Raman Spectrosc. 2009;40:1641–4.
[55] Daher C, Paris C, Le Hô A-S, Bellot-Gurlet L, Reggert M. Advanced discriminating criteria for natural organic substances of cultural heritage interest: spectral decomposition and multivariate analyses of FT-Raman and FT-IR signatures. Talanta. 2013;115:540–7.
[56] Daher C, Paris C, Le Hô A-S, Bellot-Gurlet L, Echard J-P. A joint use of Raman and infrared spectroscopies for the identification of natural organic media used in ancient varnishes. J Raman Spectrosc. 2010;41:1204–9.
[57] Colomban P, Mancini D. Lacquerware pigments identification with fixed and mobile Raman microspectrometer: a potential technique to differentiate original/fake artworks. Arts. 2013;2:111–23.
[58] Gourrier A, Chadefaux C, Lemaitre E, Bellot-Gurlet L, Reynolds M, Burghammer M, et al. Nanoscale modifications in the early heating stages of bone are heterogeneous at the microstructural scale. PLoS One. 2017;12:e0176–9.
[59] Maia LF, Fleury BG, Lages BG, Barbosa JP, Pinto AC, Castro HV, et al. Identification of reddish pigments in octocorals by Raman spectroscopy. J Raman Spectrosc. 2011; 42:653–8.
[60] Fuerst S, Mueller K, Gianni L, Paris C, Bellot-Gurlet L, Pare CFE, et al. Raman investigations to identify Corallium rubrum in iron age jewelry and ornaments. Minerals. 2016;6:56.
[61] Soualmia F, Touhar SA, Guo L, Xu Q, Garland M, Colomban P, et al. Amino-methyl coumarin as a potential SERS@Ag probe for the evaluation of protease activity and inhibition. J Raman Spectrosc. 2017;48:82–8.
[62] Guineau B, Guichard V., ICOM Committee for Conservation: 8th Triennial Meeting, Preprints, Vol. 2, The Getty Conservation Institute, Marina del Rey, CA, (1987) p. 659.
[63] Canamares MV, Garcia-Ramos JV, Domingo C, Sanchez-Cortes SJ. Surface-enhanced Raman scattering study of the adsorption of the anthraquinone pigment alizarin on Ag nanoparticles. J Raman Spectrosc. 2004;35:921–7.
[64] Shadi IT, Chowdhry BZ, Snowden MJ, Withnall R. Semi-quantitative analysis of indigo by surface enhanced resonance Raman spectroscopy (SERRS) using silver colloids. Spectrochim Acta Part. 2003;59:2213–20.
[65] Leona M, Stenger J, Ferloni E. Application of surface-enhanced Raman scattering techniques to the ultrasensitive identification of natural dyes in works of art. J Raman Spectrosc. 2006;37:981–92.
[66] Lee PC, Meisel D. Adsorption and surface-enhanced raman of dyes on Silver and Gold sols. J Phys Chem. 1982;86:3391–5.
[67] Murcia-Mascaros S, Domingo C, Sanchez-Cortes S, Canamares MV, Garcia-Ramos JV. Spectroscopic identification of alizarin in a mixture of organic red dyes by incorporation in Zr-Ormosil. J Raman Spectrosc. 2005;36:420–6.
[68] Colomban P. Structure of Oxide gels and glasses by IR and Raman scattering: I. Aluminas. J Mater Sci. 1989;24:3002–10.
[69] Canamares MV, Garcia-Ramos JV, Sanchez-Cortes S, Castillejo M, Oujja M. Comparative SERS effectiveness of silver nanoparticles prepared by different methods: A study of the enhancement factor and the interfacial properties. J Colloid Interface Sci. 2008;326:103–9.
[70] Jurasekova Z, Del Puerto E, Bruno G, Garcia-Ramos JV, Sanchez-Cortes S, Domingo C. Extractionless non-hydrolysis surface-enhanced Raman spectroscopic detection of historical mordant dyes on textile fibers. J Raman Spectrosc. 2010;41:1455–61.
[71] Gui OM, Falamas A, Barbu-Tudoran L, Aluas M, Giambra B, Cinta PS. Surface-enhanced Raman scattering (SERS) and complementary techniques applied for the investigation of an Italian cultural heritage canvas. J Raman Spectrosc. 2013;44:277–82.

[72] Van Elsande E, Lecomte S, Le Ho A-S. Micro-Raman spectroscopy (MRS) and surface-enhanced Raman scattering (SERS) on organic colourants in archaeological pigments. J Raman Spectrosc. 2008;39:1001–6.

[73] Daher C, Drieu L, Bellot-Gurlet L, Percot A, Paris C, Le Ho A-S. Combined approach of FT-Raman, SERS and IR micro-ATR spectroscopies to enlighten ancient technologies of painted and varnished works of art. J Raman Spectrosc. 2014;45:1207–14.

[74] Brosseau CL, Casadio F, Van Duyne RP. Revealing the invisible: using surface-enhanced Raman spectroscopy to identify minute remnants of color in Winslow Homer's colorless skies. J Raman Spectrosc. 2011;42:1305–10.

[75] Doherty B, Brunetti BG, Sgamellotti A, Miliani C. A detachable SERS active cellulose film: a minimally invasive approach to the study of painting lakes. J Raman Spectrosc. 2011;42:1932–8.

[76] Doherty B, Presciutti F, Sgamellotti A, Brunetti BG, Miliani C. Monitoring of optimized SERS active gel substrates for painting and paper substrates by unilateral NMR profilometry. J Raman Spectrosc. 2014;45:1153–9.

[77] Leona M, Decuzzi P, Kubic TA, Gates G, Lombardi JR Nondestructive identification of natural and synthetic organic colorants in works of art by surface enhanced Raman scattering. Anal Chem. 2011;83:3990–3.

[78] Lofrumento C, Ricci M, Platania E, Becucci M, Castelluci E. SERS detection of red organic dyes in Ag-agar gel. J Raman Spectrosc. 2013;44:47–54.

[79] Benedetti DP, Zhang J, Tague TJ, Jr, Lombardi JR, Leona M. In situ microanalysis of organic colorants by inkjet colloid deposition surface-enhanced Raman scattering. J Raman Spectrosc. 2014;45:123–7.

[80] Kurouski D, Zaleski S, Casadio F, Van Duyne RP, Shah NC. Tip-enhanced Raman spectroscopy (TERS) for in situ identification of indigo and iron gall ink on paper. J Am Chem Soc. 2014;136:8677–84.

[81] Colomban P, Treppoz F. Identification and differentiation of ancient and modern European porcelains by Raman macro and micro-spectroscopy. J Raman Spectrosc. 2001;32:93–102.

[82] Raskovska A, Minceva-Sukarova B, Grupce O, Colomban P. Characterization of pottery from Republic of Macedonia II. Raman and infrared analyses of glazed pottery finds from Skopsko Kale. J Raman Spectrosc. 2010;41:431–9.

[83] Tanevska V, Colomban P, Minceva-Sukarova B, Grupce O. Characterization of pottery from the republic of Macedonia I: Raman analyses of byzantine glazed pottery excavated from Prilep and Skopje (12th–fourteenth century). J Raman Spectrosc. 2009;40:1240–8.

[84] Caggiani MC, Mangone A, Mastrorocco F, Taccogna C, Laviano R, Giannossa LC. The tetris game of scientific investigation. increase the score embedding analytical techniques. Raw materials and production technology of Roman glasses from Pompeii. Microchem J. 2017;131:21–30.

[85] Caggiani MC, Colomban P, Valotteau C, Mangone A, Cambon P. Mobile Raman spectroscopy analysis of ancient enamelled glass masterpieces. Anal Meth. 2013;5:4345–54.

[86] Ricciardi P, Colomban P, Tournié A, Milande V. Nondestructive on-site identification of ancient glasses: genuine artefacts, embellished pieces or forgeries? J Raman Spectrosc. 2009b;40:604–17.

[87] Caggiani MC, Valotteau C, Colomban P. Inside the glassmaker technology: search of Raman criteria to discriminate between Emile Gallé and Philippe-Joseph Brocard enamels and pigment signatures. J Raman Spectrosc. 2014b;45:456–64.

[88] Carter EA, Hargreaves MD, Kononenko N, Graham I, Edwards HGM, Swarbrick B, et al. Raman spectroscopy applied to understanding prehistoric obsidian trade in the pacific region. Vibr Spectrosc. 2009;50:116–24.

[89] Teodor ES, Teodor ED, Virgolici M, Manea MM, Truica G, Litescu SC. Non-destructive analysis of amber artefacts from the prehistoric Cioclovina hoard (Romania). J Archaeolog Sci. 2010;37:2386–96.

[90] Colomban P, De Laveaucoupet R, Milande V. On-site Raman spectroscopic analysis of Kutahya fritwares. J Raman Spectrosc. 2005;36:857–63.

[91] Colomban P, Tournié A. On-site Raman identification and dating of ancient/modern stained glasses at the Sainte-Chapelle, Paris. J Cult Herit. 2007;8:242–56.

[92] Caggiani MC, Colomban P. Testing of Raman spectroscopy as a non-invasive tool for the investigation of glass-protected pastels. J Raman Spectrosc. 2011a;42:790–8.

[93] Schmidt CM, Walton MS, Trentelman K. Characterization of Lapis Lazuli pigments using a multitechnique analytical approach: implications for identification and geological provenancing. Anal Chem. 2009;81:8513–8.

[94] Bacci M, Cucci C, Del Federico E, Ienco A, Jerschow A, Newman JM, et al. An integrated spectroscopic approach for the identification of what distinguishes Afghan lapis lazuli from others. Vib Spectrosc. 2009;49:80–3.

[95] Bornhauser P, Bougeard D. Intensities of the vibrational spectra of siliceous zeolites by molecular dynamics calculations. II—Raman spectra. J Raman Spectrosc. 2001;32:279–85.

[96] Ledé B, Demortier A, Gobeltz-Hautecoeur N, Lelieur J-P, Picquenard E, Duhayon C. Observation of the v_3 Raman band of S_3^- inserted into sodalite cages. J Raman Spectrosc. 2007;38:1461–8.

[97] Del Federico E, Shofberger W, Schelvis J, Kapetanaki S, Tyne L, Jerschow A. Insight into framework destruction in ultramarine pigments. Inorg Chem. 2006;45:1270–6.

[98] Gobeltz N, Demortier A, Lelieur JP, Duhayon C. Correlation between EPR, Raman and colorimetric characteristics of the blue ultramarine pigments. J Chem Soc Faraday Trans. 1998;94:677–81.

[99] Caggiani MC, Acquafredda P, Colomban P, Mangone A. The source of blue colour of archaeological glass and glazes: the Raman spectroscopy/SEM-EDS answers. J Raman Spectrosc. 2014a;45:1251–9.

[100] Colomban P. Lapis lazuli as unexpected blue pigment in Iranian Lâjvardina ceramics. J Raman Spectrosc. 2003b;34:420–3.

[101] Defaria DLA, Silva SV, deOliveira MT. Raman microspectroscopy of some iron oxides and oxyhydroxides. J Raman Spectrosc. 1997;28:873–8.

[102] Froment F, Tournié A, Colomban P. Raman identification of natural red to yellow pigments: ochre and iron-containing ores. J Raman Spectrosc. 2008;39:560–8.

[103] Bikiaris D, Daniilia S, Sotiropoulou S, Katsimbiri O, Pavlidou E, Moutsatsou AP, et al. Ochre-differentiation through micro-Raman and micro-FTIR spectroscopies: application on wall paintings at Meteora and Mount Athos, Greece. Spectrochimica Acta Part. 1999; 56:3–18.

[104] Smith EM, Shirey SB, Nestola F, Bullock ES, Wang J, Richardson SH, et al. Large gem diamonds from metallic liquid in Earth's deep mantle. Science. 2016;354:1403–5.

[105] Domínguez-Bella S. Archaeometry and cultural heritage: the contribution of mineralogy. In: Herrero JM, Vendrell M, editors. Seminarios SEM. Madrid: Sociedad Española de Mineralogía, 2012:5–28.

[106] Barone G, Bersani D, Jehlička J, Lottici PP, Mazzoleni P, Raneri S, et al. Nondestructive investigation on the 17-eighteenth centuries Sicilian jewelry collection at the Messina regional museum using mobile Raman equipment. J Raman Spectrosc. 2015;46:989–95.

[107] Olivares M, Tarrino A, Murelaga X, Baceta JI, Castro K, Etxebarria N. Non-destructive spectrometry methods to study the distribution of archaeological and geological chert samples. Spectrochimica Acta Part. 2009;73:492–7.

[108] Schmidt P, Bellot-Gurlet L, Slodczyk A, Fröhlich F. A hitherto unrecognised band in the Raman spectra of silica rocks: influence of hydroxylated Si–O bonds (silanol) on the Raman moganite band in chalcedony and flint (SiO_2). PHys Chem Miner. 2012;39:455–64.
[109] Capel Ferrón C, León-Reina L, Jorge-Villar S, Compaña JM, Aranda MAG, López Navarrete JT, et al. Combined Raman spectroscopic and Rietveld analyses as a useful and nondestructive approach to studying flint raw materials at prehistoric archaeological sites. Archaeol Anthropol Sci. 2015;7:235–43.
[110] Colomban P, Tournié A, Bellot-Gurlet L. Raman identification of glassy silicates used in ceramic, glass and jewellry: a tentative differentiation guide. J Raman Spectrosc. 2006;37:841–52.
[111] Chen T-H. A Raman spectroscopic study of heat-treated nephrite. PHase Transitions. 2008;81:205–16.
[112] Smith DC. A review of the non-destructive identification of diverse geomaterials in the cultural heritage using different configurations of Raman spectroscopy. Geolo Soc London Specl Pubs. 2006;257:9–32.
[113] Caggiani MC, Cosentino A, Mangone A. Pigments Checker version 3.0, a handy set for conservation scientists: a free online Raman spectra database. Microchem J. 2016;129:123–32.
[114] Burgio L, Clark RJH. Library of FT-Raman spectra of pigments, minerals, pigment media and varnishes, and supplement to existing library of Raman spectra of pigments with visible excitation. Spectrochim Acta Part. 2001;57:1491–521.
[115] Bell IM, Clark RJH, Gibbs PJ. Raman spectroscopic library of natural and synthetic pigments (pre-1850 AD). Spectrochimica Acta Part. 1997;53:2159–79.
[116] Bouchard M, Smith DC. Catalogue of 45 reference Raman spectra of minerals concerning research in art history or archaeology, especially on corroded metals and coloured glass. Spectrochim Acta Part. 2003;59:2247–66.
[117] Caggiani MC, Colomban P. Raman identification of strongly absorbing phases: the ceramic black pigments. J Raman Spectrosc. 2011b;42:839–84.
[118] Miguel C, Claro A, Pereira GA, Muralha VSF, Melo MJ. A study on red lead degradation in a medieval manuscript Lorvão Apocalypse (1189). J Raman Spectrosc. 2009;40:1966–73.
[119] Bouchard M, Gambardella A. Raman microscopy study of synthetic cobalt blue spinels used in the field of art. J Raman Spectrosc. 2010;41:1477–85.
[120] Sánchez Del Río M, Picquart M, Haro-Poniatowski E, Van Elslande E, Uc VH. On the Raman spectrum of Maya blue. J Raman Spectrosc. 2006;37:1046–53.
[121] Carrasco E, Oujja M, Sanz M, Marco JF, Castillejo M. X-ray and ion irradiation effects on azurite, malachite and alizarin pictorial models. Microchem J. 2018;137:381–91.
[122] Ospitali F, Bersani D, Di Lonardo G, Lottici PP. 'Green earths': vibrational and elemental characterization of glauconites, celadonites and historical pigments. J Raman Spectrosc. 2008;39:1066–73.
[123] Coccato A, Jehlička J, Moens L, Vandenabeele P. Raman spectroscopy for the investigation of carbon-based black pigments. J Raman Spectrosc. 2015;46:1003–15.
[124] Rousaki A, Botteon A, Colombo C, Conti C, Matousek P, Moens L, et al. Development of defocusing micro-SORS mapping: study of a nineteenth century porcelain card. Anal Methods. 2017;9:6435–42.
[125] Aramendia J, Gomez-Nubla L, Bellot-Gurlet L, Castro K, Paris C, Colomban P, et al. Protective ability index measurement through Raman quantification imaging to diagnose the conservation state of weathering steel structures. J Raman Spectrosc. 2014;45:1076–84.
[126] Martens W, Frost RL, Kloprogge JT, Williams PA. A Raman spectroscopic study of the basic copper sulPHates-implications for copper corrosion and 'bronze disease'. J Raman Spectrosc. 2003;34:145–51.

[127] Frost RL. Raman spectroscopy of selected copper minerals of significance in corrosion. Spectrochim Acta Part A: Mol Biomol Spectrosc. 2003;59:1195–204.
[128] Mccann LI, Trentelman K, Possley T, Golding B. Corrosion of ancient Chinese bronze money trees studied by Raman microscopy. J Raman Spectrosc. 1999;30:121–32.
[129] Ospitali F, Chiavari C, Martini C, Bernardi E, Passarini F, Robbiola L. The characterization of Sn-based corrosion products in ancient bronzes: a Raman approach. J Raman Spectrosc. 2012;43:1596–603.
[130] Bertolotti G, Bersani D, Lottici PP, Alesiani M, Malcherek T, Schlüter J. Micro-Raman study of copper hydroxychlorides and other corrosion products of bronze samples mimicking archaeological coins. Anal Bioanal Chem. 2012;402:1451–7.
[131] Bellot-Gurlet L, Neff D, Reguer S, Monnier J, Saheb M, Dillmann P. Raman studies of corrosion layers formed on archaeological irons in various media. J Nano Res. 2009;8:147–56.
[132] Martina I, Wiesinger R, Jembrih-Simburger D, Schreiner M. Micro-Raman characterisation of silver corrosion products: instrumental set up and reference database. e-PS. 2012;9:1–8.
[133] Bongiorno V, Grosso P, Piccardo P, Magnani LG, Carnasciali MM. Virtues of Giambologna from Grimaldi chapel archaeometric characterisation part ii: 'artistic' and natural patinas. J Miner Met Mater Soc. 2016;68:2222–32.
[134] Hayez V, Segato T, Hubin A, Terryn H. Study of copper nitrate-based patinas. J Raman Spectrosc. 2006;37:1211–20.
[135] Kosec T, Ropret P, Legat A. Raman investigation of artificial patinas on recent bronze-part II: urban rain exposure. J. Raman Spectrosc. 2012;43:1578–86.
[136] Bongiorno V, Campodonico S, Caffara R, Piccardo P, Carnasciali MM. Micro-Raman spectroscopy for the characterization of artistic patinas produced on copper-based alloys. J Raman Spectrosc. 2012;43:1617–22.
[137] Hayez V, Costa V, Guillaume J, Terryna H, Hubin A. Micro Raman spectroscopy used for the study of corrosion products on copper alloys: study of the chemical composition of artificial patinas used for restoration purposes. Analyst. 2005;130:550–6.
[138] Ropret P, Kosec T. Raman investigation of artificial patinas on recent bronze – part I: climatic chamber exposure. J Raman Spectrosc. 2012;43:1578–86.
[139] Freestone IC. The relationship between enamelling on ceramics and on glass in the Islamic world. Archaeometry. 2002;44:251–5.
[140] Colomban P, Tournié A, Caggiani MC, Paris C. Pigments and enamelling/gilding technology of Mamluk mosque lamps and bottle. J Raman Spectrosc. 2012a;43:1975–84.
[141] Greiff S, Schuster J. Technological study of enamelling on Roman glass: the nature of opacifying, decolourizing and fining agents used with the glass beakers from Lübsow (Lubieszewo, Poland). J Cult Herit. 2008;9:e27–32.
[142] Fornacelli C, Colomban P, Turbanti Memmi I. Toward a Raman/FORS discrimination between art nouveau and contemporary stained glasses from CdSxSe1-x nanoparticles signatures. J Raman Spectrosc. 2015;46:1129–39.
[143] Gedzevičiūtė V, Welter N, Schüssler U, Weiss C. Chemical composition and colouring agents of Roman mosaic and millefiori glass, studied by electron microprobe analysis and Raman microspectroscopy. Archaeolog Anthropol Sci. 2009;1:15–29.
[144] Colomban P. The use of metal nanoparticles to produce yellow, red and iridescent colour, from Bronze Age to present times in lustre pottery and glass: solid state chemistry, spectroscopy and nanostructure. J Nano Res. 2009;8:109–32.
[145] Simon G, Courty A, Colomban P, Meziane L, Lisiecki I. Low wavenumber Raman scattering of cobalt nanoparticles self-organized in 3D superlattices far from surface plasmon resonance. J Raman Spectrosc. 2016;47:248–51.

[146] Colomban P, Tournié A, Ricciardi P. Raman spectroscopy of copper nanoparticle-containing glass matrices: ancient red stained-glass windows. J Raman Spectrosc. 2009;40:1949–55.
[147] Colomban P, Schreiber H. Raman signature modification induced by copper nanoparticles in silicate glass. J Raman Spectrosc. 2005;36:884–90.
[148] Colomban P, Milande V, Le Bihan L. On-site Raman analysis of Iznik pottery glazes and pigments. J Raman Spectrosc. 2004a;35:527–35.
[149] Simsek G, Colomban P, Casadio F, Bellot-Gurlet L, Faber KT, Zelleke G, et al. On-site identification of early Meissen Böttger red stoneware made at Meissen using portable XRF/ Raman instruments: 2, glaze and gilding analyses. J Am Cer Soc. 2015;98:3006–13.
[150] Kirmizi B, Colomban P, Blanc M. On-site analysis of Limoges enamels from sixteenth to nineteenth centuries: an attempt to differentiate between genuine artefacts and copies. J Raman Spectrosc. 2010;41:1240–7.
[151] Maguregui M, Knuutinen U, Castro K, Madariaga JM. Raman spectroscopy as a tool to diagnose the impact and conservation state of Pompeian second and fourth style wall paintings exposed to diverse environments (house of marcus lucretius). J Raman Spectrosc. 2010;41:1400–9.
[152] Edwards HGM, Farwell DW. The conservational heritage of wall paintings and buildings: an FT-Raman spectroscopic study of prehistoric, Roman, mediaeval and Renaissance lime substrates and mortars. J Raman Spectrosc. 2008;39:985–92.
[153] Pérez-Alonso M, Castro K, Martinez-Arkarazo I, Angulo M, Olazabal MA, Madariaga JM. Analysis of bulk and inorganic degradation products of stones, mortars and wall paintings by portable Raman microprobe spectroscopy. Anal Bioanal Chem. 2004;379:42–50.
[154] Matousek P, Morris MD, Everall N, Clark IP, Towrie M, Draper ERC, et al. Numerical simulations of subsurface probing in diffusely scattering media using spatially offset Raman spectroscopy. Appl Spectrosc. 2005b;59:1485–92.
[155] Conti C, Colombo C, Matteini M, Realini M, Zerbi G. Micro-Raman mapping on polished cross sections: a tool to define the penetration depth of conservation treatment on cultural heritage. J Raman Spectrosc. 2010;41:1254–60.
[156] Vandenabeele P, Moens L, Edwards HGM, Dams R. Raman spectroscopic database of azo pigments and application to modern art studies. J Raman Spectrosc. 2000a;31:509–17.
[157] Schmidt CM, Trentelman KA. 1064 nm dispersive Raman micro-spectroscopy for the in-situ full paper identification of organic red colorants. e-PS. 2009;6:10–21.
[158] Schulte F, Brzezinka K-W, Lutzenberger K, Stege H, Panne U. Raman spectroscopy of synthetic organic pigments used in twentieth century works of art. J Raman Spectrosc. 2008;39:1455–63.
[159] Scherrer NC, Zumbuehl S, Delavy F, Fritsch A, Kuehnen R. Synthetic organic pigments of the 20th and twenty-first century relevant to artist's paints: raman spectra reference collection. Spectrochim Acta Part. 2009;73:505–24.
[160] Whitney AV, Van Duyne RP, Casadio F. An innovative surface-enhanced Raman spectroscopy (SERS) method for the identification of six historical red lakes and dyestuffs. J Raman Spectrosc. 2006;37:993–1002.
[161] Ricci M, Trombetta E, Castellucci E, Becucci M. On the SERS quantitative determination of organic dyes. J Raman Spectrosc. SI 2018;49:997–1005.
[162] Edwards HGM, Farwell DW, Webster D. FT Raman microscopy of untreated natural plant fibres Spectrochimica Acta Part A. Mol Biomol Spectrosc. 1997;53:2383–92.
[163] Vandenabeele P, Wehling B, Moens L, Edwards H, De Reu M, Van Hooydonk G. Analysis with micro-Raman spectroscopy of natural organic binding media and varnishes used in art. Anal Chim Acta. 2000b;407:261–74.
[164] Brody RH, Edwards HGM, Pollard AM. A study of amber and copal samples using FT-Raman spectroscopy. Spectrochim Acta Part. 2001;57:1325–38.

[165] Vandenabeele P, Grimaldi DM, Edwards HGM, Moens L. Raman spectroscopy of different types of Mexican copal resins. Spectrochimica Acta Part. 2003;59:2221–9.

[166] Jehlička J, Jorge VSE, Edwards HGM. Fourier transform Raman spectra of Czech and Moravian fossil resins from freshwater sediments. J Raman Spectrosc. 2004;35:761–7.

[167] Moreno YM, Christensen DH, Nielsen OF. A NIR-FT-Raman spectroscopic study of amber. Asian J Spectrosc. 2000;4:49–56.

[168] Winkler W, Musso M, Kirchner EC. Fourier transform Raman spectroscopic data on the fossil resin siegburgite. J Raman Spectrosc. 2003;34:157–62.

[169] Shen ZX, Yee SL, Tang TS, Qin L, Tang SH. Amber identification using micro-Raman spectroscopy. Asian J Spectrosc 1. 1997;1:127–33.

[170] Edwards HGM, Farwell DW, Jorge Villar SE. Raman microspectroscopic studies of amber resins with insect inclusions. Spectrochimica Acta Part. 2007a;68:1089–95.

[171] Chiriu D, Ricci PC, Cappellini G. Raman characterization of XIV–XVI centuries Sardinian documents: inks, papers and parchments. Vibr Spectrosc. 2017;92:70–81.

[172] Castro K, Pessanha S, Proietti N, Princi E, Capitani D, Carvalho ML, et al. Noninvasive and nondestructive NMR, Raman and XRF analysis of a blue coloured map from the seventeenth century. Anal Bioanal Chem. 2008;391:433–41.

[173] El Bakkali A, Lamhasni T, Haddad M, Ait Lyazidi S, Sanchez-Cortes S, Del Puerto Nevado E. Non-invasive micro Raman, SERS and visible reflectance analyses of coloring materials in ancient Moroccan Islamic manuscripts. J Raman Spectrosc. 2013;44:114–20.

[174] Bruni S, Caglio S, Guglielmi V, Poldi G. The joined use of n.i. spectroscopic analyses – FTIR, Raman, visible reflectance spectrometry and EDXRF – to study drawings and illuminated manuscripts. Appl Phys A. 2008;92:103–8.

[175] Mannucci E, Pastorelli R, Zerbi G, Bottani CE, Facchini A. Recovery of ancient parchment: characterization by vibrational spectroscopy. J Raman Spectrosc. 2000;31:1089–97.

[176] Edwards HGM, Farwell DW, Newton EM, Rull Perez F, Villar SJ. Application of FT-Raman spectroscopy to the characterisation of parchment and vellum, I; novel information for paleograPHic and historiated manuscript studies. Spectrochimica Acta Part. 2001b; 57:1223–34.

[177] Edwards HGM, Perez RF. Application of FT-Raman spectroscopy to the characterisation of parchment and vellum, II; effects of biodeterioration and chemical deterioration on spectral interpretation. J Raman Spectrosc. 2004;35:754–60.

[178] Bicchieri M, Monti M, Piantanida G, Pinzari F, Sodo A. Non-destructive spectroscopic characterization of parchment documents. Vibr Spectrosc. 2011;55:267–72.

[179] Bitossi G, Giorgi R, Mauro M, Salvadori B, Dei L. Spectroscopic techniques in cultural heritage conservation: a survey. Appl Spectrosc Rev. 2005;40:187–228.

[180] Centeno SA. Identification of artistic materials in paintings and drawings by Raman spectroscopy: some challenges and future outlook. J Raman Spectrosc. 2016;47:9–15.

[181] Casadio F, Daher C, Bellot-Gurlet L. Raman spectroscopy of cultural heritage materials: overview of applications and new frontiers in instrumentation, sampling modalities, and data processing. Top Curr Chem. 2016;374 :62. DOI: 10.1007/s41061-016-0061–z.

[182] Bicchieri M, Monti M, Piantanida G, Sodo A. All that is iron-ink is not always iron-gall!. J Raman Spectrosc. 2008;39:1074–8.

[183] Lee AS, Otieno-Alego V, Creagh DC. Identification of iron-gall inks with near-infrared Raman microspectroscopy. J Raman Spectrosc. 2008;39:1079–84.

[184] Lee AS, Mahon PJ, Creagh DC. Raman analysis of iron gall inks on parchment. Vib Spectrosc. 2006;41:170–5.

[185] Brown KL, Clark RJH. Analysis on the pigmentary material on the vinland map and tartar relation by Raman Microprobe Spectroscopy. Anal Chem. 2002;74:3658–61.

[186] Edwards HGM, Munshi T. Diagnostic Raman spectroscopy for the forensic detection of biomaterials and the preservation of cultural heritage. Anal Bioanal Chem. 2005; 382:1398–406.
[187] Bertoluzza A, Brasili P, Castrì L, Facchini F, Fagnano C, Tinti A. Preliminary results in dating human skeletal remains by Raman spectroscopy. J Raman Spectrosc. 1997;28:185–8.
[188] Kirchner MT, Edwards HGM, Lucy D, Pollard AM. Ancient and modern specimens of human teeth: a Fourier transform Raman spectroscopic study. J Raman Spectrosc. 1997;28:171–8.
[189] Mclaughlin G, Lednev IK. Potential application of Raman spectroscopy for determining burial duration of skeletal remains. Anal Bioanal Chem. 2011;401:2511–8.
[190] Edwards HGM, Farwell DW, De Faria DLA, Monteiro AMF, Afonso MC, De Blasis P, et al. Raman spectroscopic study of 3000-year-old human skeletal remains from a sambaqui, Santa Catarina, Brazil. J Raman Spectrosc. 2001a;32:17–22.
[191] Gniadecka M, Wulf HC, Faurskov Nielsen O, Christensen DH, Hart Hansen JP. Fourier transform Raman spectroscopy of fifteenth century mummies from Qilakitsoq Greenland. J Raman Spectrosc. 1997;28:179–84.
[192] Edwards HGM, Wilson AS, Nik Hassan NF, Davidson A, Burnett A. Raman spectroscopic analysis of human remains from a seventh century cist burial on Anglesey, UK. Anal Bioanal Chem. 2007b;387:821–8.
[193] RRUFF.INFO http://rruff.info (accessed 27th February 2018)
[194] Lau D, Livett M, Prawer S. Application of surface-enhanced Raman spectroscopy (SERS) to the analysis of natural resins in artworks. J Raman Spectrosc. 2008;39:545–52.

Ken Sutherland

8 Gas chromatography/mass spectrometry techniques for the characterisation of organic materials in works of art

Abstract: The power of GC/MS to resolve, characterise and quantify complex mixtures of organic compounds with high sensitivity has made it an indispensable analytical tool to address detailed questions about the chemical constituents of works of art. This paper provides an overview of the technique and its particular suitability to material studies of art and historical artefacts, and reviews its diverse research applications concerning the organic composition of artists' and conservation materials. Options with regard to sample preparation by chemical derivatisation, pyrolysis techniques, and methods for the analysis of volatile organic compounds are discussed, as well as various approaches to the treatment and interpretation of data. The greatest value is gained from GC/MS when it is used as a complementary technique, informed by and in synergy with other methods of analysis.

Keywords: gas chromatography/mass spectrometry, works of art, organic materials, derivatisation, pyrolysis, solid phase microextraction, multivariate analysis, peak deconvolution

8.1 Introduction

The organic constituents of works of art have been a source of fascination since chemical tests were first applied to elucidate artists' methods in the eighteenth century, driven in part by an infatuation with the "lost" techniques and media of ancient and Old Master painters [1]. It was not until recent decades, however, that instrumental analytical techniques became sophisticated and accessible enough to allow the diverse organic materials employed by artists – paint media, varnishes, adhesives, etc. – to be characterised with any degree of reliability. Among these, separation techniques such as gas chromatography (GC) showed early promise for the resolution of complex mixtures of chemical species that are typically encountered in samples from art objects, including terpenes in plant resins [2] and fatty acids derived from oils and fats [3]. But it was particularly the combination of gas chromatography with mass spectrometry (MS), with MS providing a means to identify the compounds separated by GC, that emerged as a powerful and essential technique for

This article has previously been published in the journal *Physical Sciences Reviews*. Please cite as: Sutherland, K. Gas chromatography/mass spectrometry techniques for the characterisation of organic materials in works of art. *Physical Sciences Reviews* [Online] **2019**, 4. DOI: 10.1515/psr-2018-0010

https://doi.org/10.1515/9783110457537-008

organic analysis in art technical studies, and has remained a standard tool for museum and cultural heritage laboratories to date. Figure 8.1 illustrates the expansion in literature discussing GC/MS analyses of art materials: the first published studies appeared in the 1970s-1980s [4–6], and following a lag of some years (attributable to the limited number of specialised scientists and accessible instrumentation) there was a rapid increase of applications in the 1990s. By the first years of the twenty-first century the technique had become a firmly established and indispensable component of art research.

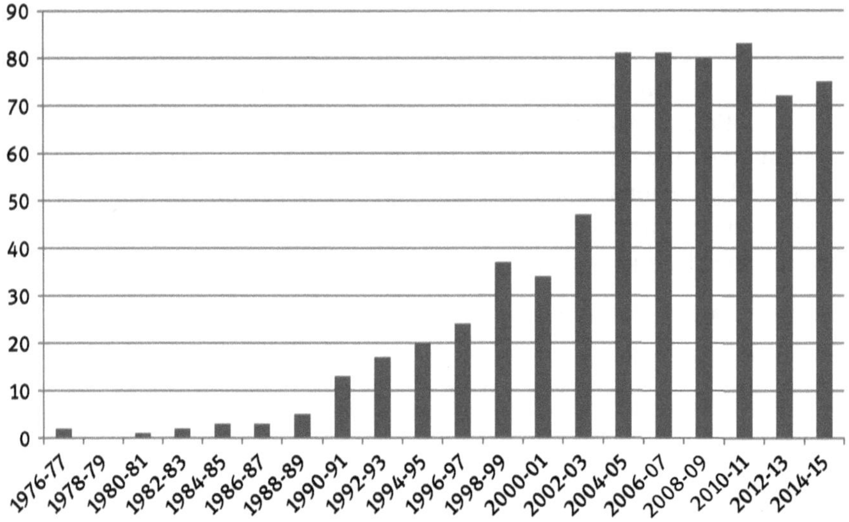

Figure 8.1: Graph showing growth in published GC/MS studies of works of art. Data derived from the Art and Archaeology Technical Abstracts and Web of Science databases (with entries selected that relate specifically to GC/MS and the study of art). The graph is intended to illustrate a trend only; actual numbers are greatly underestimated due to the specificity of search terms used and the large numbers of published studies that are not indexed in these sources.

This paper provides an overview of GC/MS and its value for material studies of works of art, the diversity of analytical protocols available with regard to sample preparation and introduction, the increasing variety of applications in the cultural heritage field, strategies for the treatment and interpretation of large and complex datasets, and its complementary nature with respect to other methods of instrumental analysis. The extensive body of literature on the closely related topic of GC/MS analysis of archaeological organic residues will not be addressed here since it has been the subject of previous reviews by several authors [7, 8]. Furthermore, this is not intended as an exhaustive review of GC/MS-based studies of art materials, but rather a selective and critical survey to illustrate key trends. A more comprehensive summary of recent published applications of GC/MS in art and archaeology was provided by Bonaduce et al. [9], and detailed accounts that deal with analysis of specific material classes (lipids, proteins, natural and synthetic resins) can be found in Colombini and Modugno [10]. While now several decades old, and dated in some technical aspects, the essential reference work of Mills and White [11], remains a lucid and engaging introduction to the organic analysis, including GC and GC/MS, of art materials.

8.2 Instrumentation and applicability to analysis of art materials

GC/MS is referred to as a hyphenated method, combining the benefits of two distinct analytical techniques: GC to separate organic compounds in a complex sample, and MS to facilitate their characterisation, identification and/or quantification[1] (Figure 8.2). As the name implies, separation by GC occurs in the gas phase: for typical applications most samples from art objects are obtained in solid form, and some kind of preparation is therefore required. This may involve chemical digestion and modification (derivatisation) on a micro-scale in order to produce volatile compounds that are amenable to analysis, following injection of a liquid solution into the heated GC inlet. Alternatively, heating in an inert atmosphere (pyrolysis) can effect a controlled decomposition into small molecules that are transferred directly to the inlet. Sample derivatisation and pyrolysis are considered in more detail in the following section.

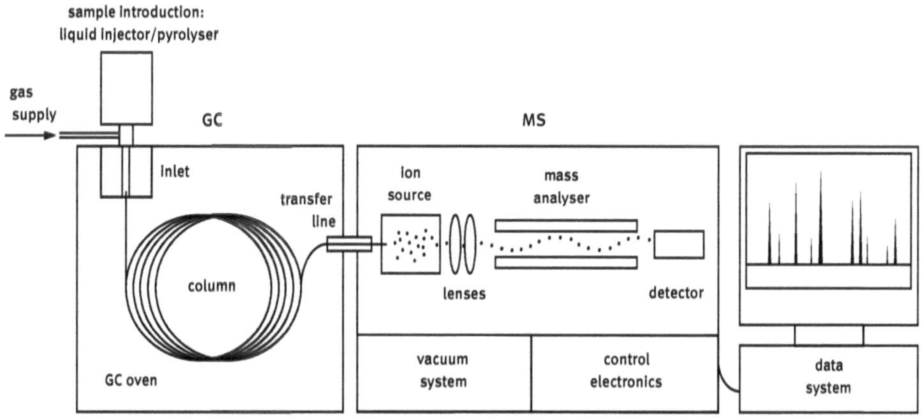

Figure 8.2: Schematic diagram of GC/MS instrument.

1 The acronyms GC and MS are used, here and generally, to refer both to the techniques (gas chromatography, mass spectrometry) and instruments (gas chromatograph, mass spectrometer)

Separation occurs on a capillary column, c. 15–60 m in length, coated on the inside with a thin film of viscous liquid, the *stationary phase*, which is typically a modified polysiloxane or polyethylene glycol. The volatilised sample is passed through the column in a stream of inert carrier gas such as helium, the *mobile phase*, and it is the relative affinity of any given compound for the mobile and stationary phases – as well as its boiling point – that determines the length of time it resides on the column before being eluted. The column is housed inside a precisely temperature-controllable oven, since increasing the temperature during the analytical run greatly improves the speed of separation. Sample compounds are thus separated on the basis of their *retention time*, which itself is a distinctive parameter that is useful for identification, in combination with the information subsequently derived from MS [12].

The compounds are eluted from the column directly into the *ion source* of the mass spectrometer, operating under vacuum, where they are exposed to an energetic beam of electrons that cause ionisation and fragmentation of the molecules. The resultant ionic molecules and fragments can then be separated according to their mass-to-charge ratio (*m/z*) by the *mass analyser*. The most common mass analysers for GC/MS are quadrupole and ion trap types, both of which use electromagnetic fields to filter or separate the ions prior to detection by an electron multiplier. Each compound will produce a characteristic and repeatable pattern of fragments under given ionisation conditions, and its *mass spectrum* – a plot of the abundance of ions recorded at each *m/z* value – is therefore a diagnostic fingerprint for that compound.[2] It can be compared with mass spectral databases of known compounds, or interpreted first-hand based on structural information indicated by the patterns of ions and fragments [13]. Certain types of instrument can perform multiple stages of fragmentation and analysis of selected ions (a process called tandem MS or MS/MS) to give an enhanced level of structural information [14].

GC/MS analyses are most often performed in *scan mode*, i.e. measuring ions within a predetermined range of *m/z* values, which are analysed sequentially by the MS multiple times per second. The data generated are commonly presented as a *total ion chromatogram* (TIC), a plot of the sum of all ions detected in each full scan. The TIC, in common with other types of chromatogram, displays the separated compounds as peaks and gives some indication of their relative abundance. From the TIC, a mass spectrum can be viewed for any individual scan or for the sum of scans over a peak or region (Figure 8.3). Because of the complex, multidimensional nature of GC/MS data – combining axes of time (or scan number), abundance and *m/z* ratios – it can be treated in a multitude of ways to extract an enormous amount of information about the sample and its components. A common example is an *extracted ion chromatogram* (EIC) which

2 An alternative process to electron ionisation (EI) is chemical ionisation (CI), in which a reagent gas such as isobutane or ammonia is present in the ion source during ionisation. This results in a "soft" ionisation with less fragmentation, and can be useful for molecular weight determination of unknown compounds. See for example [32].

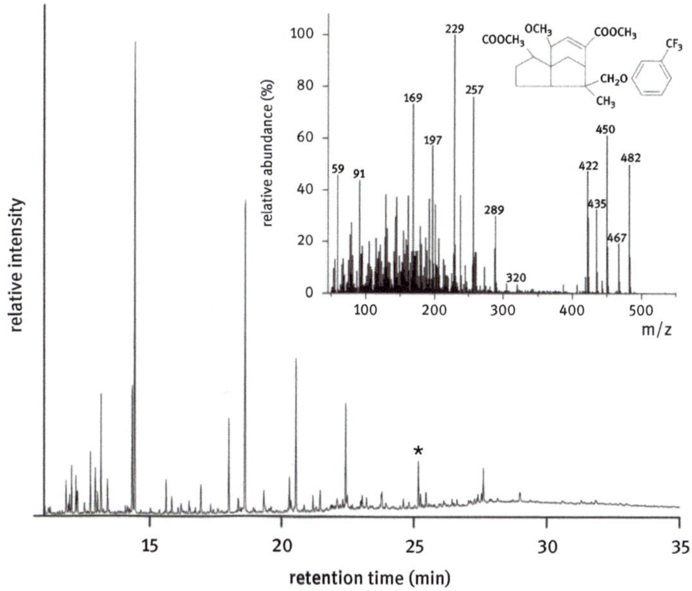

Figure 8.3: Top: total ion chromatogram from GC/MS analysis of a varnish sample from a painting by William Merritt Chase, with inset mass spectrum (from peak in TIC marked *) for a derivative of a cyclic terpene characteristic of shellac. Bottom: William Merritt Chase, *Portrait of Anna Traquair Lang*, 1911, 151 × 121 cm, Philadelphia Museum of Art, 1928-63-4; photograph Joe Mikuliak. For details see K. Sutherland, "Bleached shellac picture varnishes: characterization and case studies", Journal of the Institute of Conservation (2010) 33:129–145, copyright © Icon, The Institute of Conservation, reprinted by permission of Taylor & Francis Ltd, www.tandfonline.com on behalf of Icon, The Institute of Conservation.

can be isolated from the total ion data to display the abundance of ions of a single m/z value during the analytical run. This can be valuable for enhancing the visibility of a specific compound or class of compounds within a complex chromatogram, such as hopanoid compounds in bituminous artists' paints [15] or dye marker compounds in textile samples [16] (Figure 8.4). Further strategies for data treatment are discussed later in this chapter.

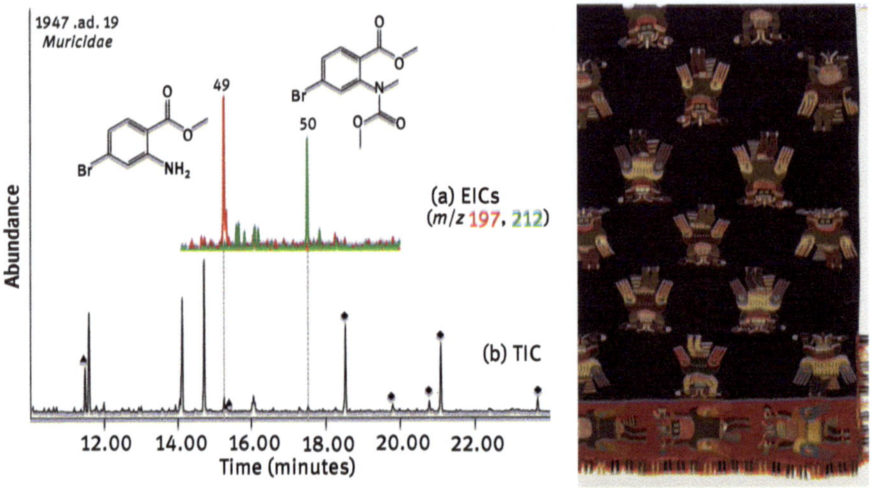

Figure 8.4: Left: total ion chromatogram from GC/MS analysis of a green textile thread from a 2500 year old Peruvian burial cloth with (inset) extracted ion chromatograms indicating the presence of characteristic compounds derived from Tyrian purple (6,6′-dibromoindigotin). Right: Paracas burial cloth, ca. 2500 BP, Peru, Montreal Museum of Fine Arts 1947.ad.19; photograph Jacques Beardsell, Centre de conservation du Québec. For details see J. Poulin, "A new methodology for the characterisation of natural dyes on museum objects using gas chromatography-mass spectrometry", Studies in Conservation (2018) 63:36–61, reprinted by permission of the publisher (Taylor & Francis Ltd, http://www.tandfonline.com).

An alternative to the use of scan mode and EICs, if analysis is aimed at detecting specific compounds at high sensitivity rather than providing a general characterisation, is *selected ion monitoring* (SIM). With SIM, the MS is programmed to measure only a small group of ions that are characteristic of the compounds of interest, rather than a full scan range: it has been used, for example, to enhance the detection of amino acid derivatives in the analysis of proteinaceous paint binders [17, 18]. Due to the great increase in signal-to-noise, SIM improves the sensitivity for the defined compounds, but removes the possibility to detect and/or quantify other materials that may be present in the sample.

The value of GC/MS for the study of art materials lies in its high sensitivity and specificity, which are key factors because of the particular challenges associated with such analyses. Firstly, the possibility to remove samples from valuable works of art is necessarily limited, and those taken are typically microscopic in size (often in the order of micrograms or tens of micrograms). Secondly, the material is often complex: a paint medium or adhesive, for example, may be formulated from multiple materials (such as egg/oil/plant resin) and its characterisation further complicated by the presence of inorganic pigments or fillers. A further source of complexity is contamination, which can be defined in this context as any accretions the object has accumulated over time, as well as artefacts of sampling and analysis: these may include organic materials from conservation treatments, storage or display environments, and handling, in addition to tools/containers and reagents used to extract and prepare the sample. Finally, ageing must be taken into account when interpreting a sample's composition: many organic materials used by artists are prone to oxidation, cross-linking or other forms of chemical alteration over time, and data from modern reference materials may be misleading or entirely useless as a basis for assessing the composition of samples hundreds of years old. This is a particular issue in the analysis of terpenic plant resins, which are especially susceptible to these types of alteration, especially when aged in a thin film such as a varnish or paint layer [19–21]; it can even be a problem for materials considered to be stable, notably beeswax, which will show a depleted content of certain characteristic compounds in ancient artefacts such as Roman Egyptian paintings [22].

8.3 Derivatisation and pyrolysis

Many of the chemical species that constitute art materials are too large or polar to be volatilised and resolved using standard GC conditions. This necessitates some kind of sample preparation, commonly involving chemical treatment on a micro-scale to break down (hydrolyse) large molecules and polymers and to modify (derivatise) active groups such as carboxyl, hydroxyl, or amines. Reagents that generate methyl and silyl derivatives are most commonly used, although a great number of alternatives are available for specific purposes. In addition to improving the chromatographic properties of the analyte molecules, derivatisation may also produce compounds with distinctive mass spectra that can facilitate their detection in complex samples [23, 24].

The choice of sample preparation procedure will depend on the nature of the material and the specific question to be addressed: there is no single "universal" method, and the increasing variety of protocols described in the literature can be bewildering to a novice analyst. Some pre-knowledge of the sample, derived from complementary analysis techniques such as Fourier transform infrared spectroscopy (FTIR), can be invaluable to guide the selection of an appropriate procedure, based on a determination of the general class(es) of organic material present (protein, carbohydrate, wax, etc.) Similarly important is a clear definition of the research question prior to analysis, based on consultation among the various collaborators (scientists, conservators, curators): a thorough characterisation of a complex sample may be informative in some instances, time-consuming and unhelpful in others. Analytical protocols have been developed for GC/MS that address a broad range of material types, involving multiple stages of sample extraction/derivatisation and analysis [25–27] (Figure 8.5), and while these have the benefit of being comprehensive they are also relatively laborious and require larger sample sizes (up to 1 mg). In contrast, other methods use reagents, such as the quaternary ammonium methylating agent TMTFTH (trimethyl[α,α,α-trifluoro-*m*-tolyl] ammonium hydroxide or *m*-[trifluoromethyl]phenyltrimethylammonium hydroxide), that will perform both hydrolysis and derivatisation in a single solution [16, 23, 28–30], a great advantage for microscopic samples from works of art as it avoids analyte loss that may result from extraction or other pre-treatments. However, TMTFTH is suited for the characterisation of a more limited range of material types (primarily oils, waxes, natural resins and dyes).

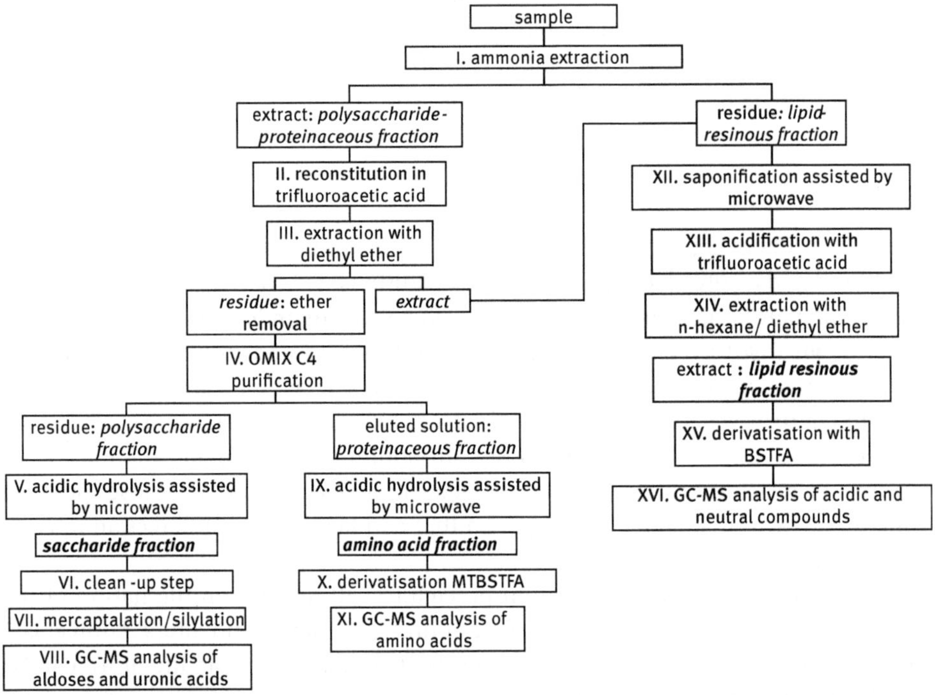

Figure 8.5: Multi-step analytical procedure for the characterisation of several classes of material within a single sample. Reprinted with permission from Lluveras et al., "GC/MS analytical procedure for the characterization of glycerolipids, natural waxes, terpenoid resins, proteinaceous and polysaccharide materials in the same paint microsample avoiding interferences from inorganic media", Analytical Chemistry (2010) 82:376–386, copyright 2010 American Chemical Society.

Using the appropriate sample preparation protocol, the majority of natural organic materials that are encountered as constituents of works of art – oils, waxes, natural resins, proteins, polysaccharides, organic dyes/pigments, etc. – can be characterised with a high degree of specificity, based on detection of diagnostic marker compounds. The major classes of organic materials, and their characteristic components, are summarised in Table 8.1.

An alternative to chemical methods for breaking down macromolecules into small, volatile species is pyrolysis, which is performed at 400–800°C in a small furnace, purged with the inert carrier gas and connected directly to the GC inlet. Pyrolysis-GC/MS (Py-GC/MS) initially found applications in the study of polymeric materials that are resistant to chemical decomposition, including fossil or semi-fossil resins such as amber and copal [19, 60, 61], Asian lacquers [62] and synthetic resins [63, 64].[3] Certain polymers will undergo pyrolysis in a controlled way to produce molecules that are readily separated by GC, most notably the synthetic resins (acrylics, styrenes, polyurethanes, etc.) in plastics and modern paint media, for which this is a very effective and straightforward method of analysis [65–67], for more details see [68, 69]. Plasticisers and other additives in polymer formulations are also effectively released and separated by Py-GC/MS [66, 70–72], as well as some classes of synthetic organic pigments [73, 74]. For other materials pyrolysis reactions are far more complex and unpredictable, and may produce numerous compounds that are either small and undiagnostic, or polar and not easily separated and detected. This is particularly true for aged/oxidised natural materials such as plant oils and resins [75]. The use of pyrolysis in combination with a derivatising reagent – termed reactive pyrolysis or (with the use of certain types of methylating reagent) thermally-assisted hydrolysis and methylation (THM) – overcomes this limitation, combining the advantages of both analytical strategies, and allowing effective and reproducible analysis of a wide range of material classes in a single sample with minimal preparation [76]. The most commonly used reagent for this purpose is tetramethylammonium hydroxide (TMAH) [19, 21, 32, 45, 50, 77], although pyrolysis with the silylating reagent hexamethyldisilazane (HMDS) has shown value for certain applications [58, 78, 79]. Examples of studies using pyrolysis with methylation and silylation are included in Table 8.1.

The application of Py-GC/MS techniques to the study of works of art has grown considerably in recent years, in a large part because of their versatility and as a result of a greater understanding of the pyrolysis reactions and products. Multi-stage or "double-shot" techniques, using different temperatures to effect thermal desorption or pyrolysis of specific sample components, have further enhanced the level of information obtained from analyses [19, 80]. But another significant factor is improvements

---

3 Certain synthetic resins are amenable to chemical derivatisation, including alkyds, which have a chemical similarity to drying oils. See [30].

Table 8.1: Classes of natural organic materials used by artists and conservators that are amenable to analysis by GC/MS, and marker compounds used for their characterisation.

Material class	Examples	Marker compounds	References
Vegetable oils, animal fats	Drying oils (linseed, poppy, walnut, tung oil), non- or semi-drying oils (soy, safflower, castor oil), tallow, stearine	Fatty acids and their oxidation products, glycerol	[31–36]
Waxes	Beeswax, spermaceti, carnauba, paraffin wax	Long-chain fatty acids and alcohols, hydroxy acids (or esters of these), alkanes	[22, 37–39]
Bituminous materials	Bitumen, asphalt	Hopanes, steranes, alkanes	[15, 40]
Plant resins (diterpene)	Pine resin (rosin), larch resin (Venice turpentine), sandarac, copals, amber	Diterpenes (abietanes, pimaranes, labdanes) and their oxidation products	[14, 19, 21, 41–43]
Plant resins (triterpene)	Dammar, mastic, elemi	Triterpenes (dammaranes, ursanes, euphanes, lupanes, oleananes, amyrins) and their oxidation products	[20, 44–46]
Insect resins	Shellac	Hydroxy aliphatic acids, cyclic sesquiterpene acids	[47, 48]
Proteins	Gelatine (collagen), keratin (wool, horn), albumins (egg white, yolk), casein	Amino acids	[17, 49–51]
Carbohydrates	Plant gums (gum arabic, tragacanth, fruit tree gums), starch, cellulose	Monosaccharides (aldopentoses, aldohexoses, uronic acids)	[52–56]
Organic dyes, pigments	Flavonoids, anthraquinoids, tannins, indigoids, orceins	Various compound types depending on colorant source	[16, 57–59]

that have been made in the design of commercially available pyrolysers – with small internal volumes and rapid, even heating – that greatly improve the sensitivity and reproducibility of analyses. A new pyrolysis method has recently been reported that employs a thermal separation probe (TSP) inserted directly into a special GC inlet that can be heated rapidly to 450°C. The combination of controlled heating and the direct transfer of pyrolysis products to the GC column makes it possible to analyse extremely small samples (as little as a few micrograms) and provides structural information, for example on the composition of ambers, that is not readily obtained using more conventional pyrolysis techniques [42, 81].

8.4 Analysis of VOCs

An increasing number of GC/MS studies in the cultural heritage field are concerned with the measurement of volatile organic compounds (i.e. those that show high vapour pressure at normal, ambient temperatures). This is due in a large part to the growing availability of solid phase microextraction (SPME) techniques, which allow VOC analysis to be carried out easily, inexpensively, and with high sensitivity. SPME is carried out using a small length (c. 1 cm) of fused silica fibre that is coated with an adsorbent or absorbent material; a variety of coatings and coating thicknesses are available and are selected based on the polarity and molecular weight range of the analytes. VOCs from the material or environment to be tested are sorbed by the fibre coating, which is then inserted into the heated GC inlet using a device similar to a syringe, and the compounds are thus driven off the fibre for analysis in the usual way [82]. Derivatisation can be performed in several ways, including the treatment of the SPME fibre with a reagent prior to exposure to the sample material, to improve the stability and detection of the absorbed compounds. SPME can be performed in the headspace above a solid sample in a vial (HS-SPME), from direct immersion in a liquid solution (DI-SPME), or – increasingly commonly for museum applications – *in-situ* to study VOCs in various environments, such as a display case or storage area [83], or between the pages of a book [24]. The technique has been applied for the assessment of building or art storage environments [83–85], to monitor museum artefacts for pesticides [86] and fungal growth [87], to measure degradation products of paper and other cellulose-based materials [24, 88–91], or for the characterisation of materials such as waxes [92], polymers and plastics [93–95]. While sampling (for HS- or DI-SPME) may be necessary to achieve the most reproducible, quantitative measurements, SPME has the advantage that it can be used to characterise museum objects and materials in a fully non-contact and non-sampling way [92, 94].

Another technique that has been applied recently for VOC analysis is evolved gas analysis (EGA). This is more commonly used as a complement to Py-GC/MS, to provide a rapid characterisation of the composition and thermal degradation properties of materials such as plastics [96] or proteins [51]. However, Samide et al. [97], proposed a new approach in which the pyrolyser was operated at a low temperature (115°C), allowing potentially harmful VOCs in museum construction and packing materials to be analysed effectively by GC/MS, and providing a straightforward alternative to SPME and other thermal desorption techniques for laboratories equipped with a conventional pyrolyser.

8.5 Data treatment and interpretation

Although the essential technology of GC/MS has remained unchanged for decades, incremental improvements (especially in the design of columns, and of MS components to improve efficiency in the generation and detection of ions) have led to progressive enhancements in the instruments' resolution and sensitivity. Coupled with advances in the speed of electronics and software, this means that a single analytical run may now contain hundreds or even thousands of component peaks. Such vast datasets are potentially rich sources of chemical information about the sample, but present new challenges that necessitate different approaches to data treatment and interpretation.

For some analytical questions qualitative data interpretation may still be sufficient – that is, based on the detection of marker compounds that are diagnostic for specific materials – and the enhanced sensitivity of modern instrumentation has improved the ability to detect such markers at trace levels. Examples include characteristic compounds that result from pyrolysis of the polymeric components of sandarac and copals, to indicate the presence of these resins in artists' paints and varnishes [19]; and 12-hydroxy stearic acid as evidence for the use of castor wax (hydrogenated castor oil) as an additive in modern oil paints [31]. Semi-quantitative approaches to interpretation have traditionally included the measurement of peak area ratios, such as those of saturated (palmitic and stearic) fatty acids that are widely used to infer the type of drying oil in a paint medium [3, 28, 33], and the ratios of oxidised abietane molecules as indicators of the state of oxidation of a paint or adhesive sample containing Pinaceae resin [21]. For more fully quantitative interpretations, correlation coefficients of the relative molar compositions (derived from peak area ratios) of a group of compounds have been used, for example to interpret patterns of amino acids diagnostic of certain types of protein or their mixtures in artists' materials [49].

An alternative approach for datasets with large numbers of peaks is the use of multivariate statistical techniques such as principal component analysis (PCA), which are increasingly used to highlight patterns and relationships among samples analysed by GC/MS. PCA is a method that reduces redundant or irrelevant information in a large dataset and derives a small number of variables (principal components) that account for most of the variance. These can be displayed in two-dimensional "scores plots" that indicate groupings of related samples (Figure 8.6). Applications to the study of works of art by GC/MS have included analysis of proteinaceous [25, 27] and gum-based paint

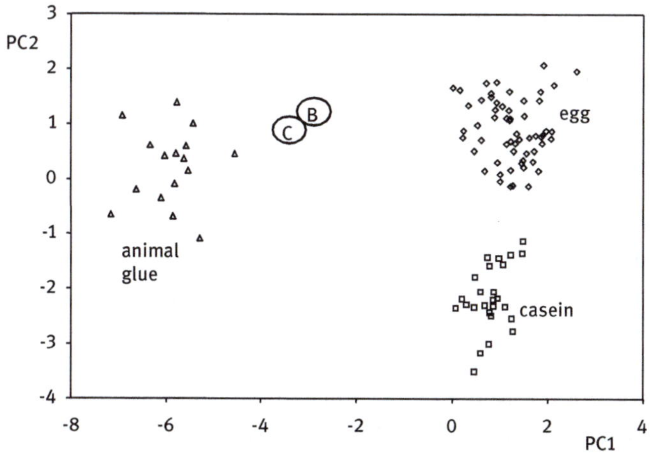

Figure 8.6: Top: PCA score plot from GC/MS analysis of the amino acid fraction of proteinaceous reference materials, along with data for paint samples from an eighteenth century painting (B) and a fourth century sarcophagus (C). The position of the paint samples between the clusters for animal glue and egg suggests a mixture of these materials. Bottom: Sarcofago delle Amazzoni (source of sample C). Reprinted with permission from Andreotti et al., “Combined GC/MS analytical procedure for the characterization of glycerolipid, waxy, resinous, and proteinaceous materials in a unique paint microsample”, Analytical Chemistry (2006) 78:4490–4500, copyright 2006 American Chemical Society; image of sarcophagus reproduced by permission of the Ministero per i beni e le attività culturali – Polo Museale della Toscana – Firenze.

media [56], shellac resins/varnishes [48], Asian papers [98] and modern oil paints [35]. Multivariate techniques such as PCA are effective tools to extract and visualise information that may not be evident from more straightforward or empirical approaches to data analysis, but they require considerable expertise to determine the most appropriate ways to input and process the GC/MS data, using specialised software programmes, and to interpret the results.

Another powerful strategy in data interpretation is peak deconvolution, which facilitates detection of marker compounds in complex GC/MS data, particularly those present at trace levels. Peak deconvolution is a process that extracts "pure" mass spectra (i.e. free of extraneous peaks from background or co-eluting compounds) for single compounds, based on an automated analysis of the profiles of individual *m/z* values in the TIC data. A widely used system is AMDIS (Automated Mass Spectral Deconvolution and Identification System), a freeware programme developed by the National Institute of Standards and Technology [99], in which peak deconvolution is combined with automated searching of extracted spectra against a user-created library to identify "target" compounds in a GC/MS analysis. This system has been adapted notably by scientists at the Getty Conservation Institute, Los Angeles, and the Cultural Heritage Agency of the Netherlands, Amsterdam, as an interpretation tool that was initially developed for the study of Asian lacquers [77] but has since been broadened to encompass a variety of natural and synthetic organic materials.

Another critical aspect of data interpretation is the exchange of information and expertise among the growing community of specialists involved in the study of cultural heritage materials. This is exemplified by the activities of MaSC (the Users' Group for Mass Spectrometry and Chromatography) and in particular the series of meetings and hands-on training workshops in GC/MS and related analytical methods that the Group has organised since 2004 [18, 100]. Practical workshops with a focus on GC/MS have also been held in conjunction with the above-mentioned RAdICAL (Recent Advances in Characterizing Asian Lacquer) project [101].

8.6 Complementary analytical techniques

While GC/MS is clearly a powerful tool for the precise characterisation of organic materials in works of art, it is most informative when used in combination with other types of instrumental analysis. Since there is always some degree of selectivity in the sample preparation procedure, with regard to its optimisation for certain compound types, pre-knowledge of the sample is beneficial. As mentioned above, FTIR in particular is widely used in the cultural heritage field and provides an effective means of "screening" samples, to indicate the type(s) of organic material present and to guide the selection of a suitable GC/MS protocol. FTIR can also supplement information derived from GC/MS, for example, by indicating in which form fatty acids are present (as glyceride or wax esters, free acids, or metal carboxylates) in an oil- or wax-based paint sample, a distinction that is not possible with some common GC/MS procedures, and which may provide insights into the ageing or manufacture of the paint [102].[4] Knowledge of the inorganic components of a sample (obtained from FTIR

4 Several research groups have developed GC/MS protocols that allow discrimination of the different forms of fatty acid: see [26, 32, 34, 39].

or Raman spectroscopy, x-ray fluorescence spectroscopy, or scanning electron microscopy with energy dispersive spectroscopy) can be valuable, since pigments or other inorganics may influence the chemical alteration of organic materials detected by GC/MS [102], or may interfere with chemical derivatisation procedures [17, 55].

A variety of MS techniques other than GC/MS are finding applications in the study of organic artists' materials, as summarised in a recent review by Calvano et al. [103]. These include "soft" ionisation methods such as matrix-assisted laser desorption ionisation (MALDI) and electrospray ionisation (ESI) MS that allow the analysis of larger molecules than is generally possible with GC/MS, such as peptides, oligosaccharides and triglycerides. Liquid chromatography with different types of detection, including MS, has long been used for the precise characterisation of organic colorants [104]. Several MS techniques allow the determination of both organic and inorganic constituents in a sample, including time-of-flight secondary ion MS (TOF-SIMS), which has been used to perform chemical imaging of discrete features and layers within samples such as paint cross-sections. Direct temperature-resolved MS (DTMS) is a type of in-source pyrolysis-MS technique that provides a rapid characterisation of organic and certain inorganic components in a complex sample, with some resolution of material fractions achieved by the heating profile, but without the separation of discrete compounds that is possible with chromatography. More recently, direct analysis in real time (DART) MS, a technique that can be used directly on surfaces without sampling, has been applied to the study of historic papers, and may have further applications for non-invasive analysis of works of art and historic artefacts [103].

With regard to GC/MS, several specialised techniques are available that to date have had little application in the study of art. Comprehensive two-dimensional gas chromatography MS (GC×GC/MS) is a method in which the sample is subjected to two independent chromatographic separations that, when used with a fast MS such as TOF, provide high resolving power for trace compounds in complex mixtures. Promising results have been obtained using GC×GC/MS to analyse natural resins and waxes [105] and Asian papers [98] (Figure 8.7). Laser pyrolysis GC/MS (La-Py-GC/MS), a technique more commonly used for geological samples, allowed layer-specific characterisation of resins and synthetic colorants in an exploratory study of cross-sections made from paint mock-ups [106]. GC with combustion isotope ratio MS (GC-C-IRMS) has shown great value in archaeological studies for the determination of plant and animal sources of organic residues, based on the isotopic signatures of individual organic compounds [107], and will likely find utility for questions regarding the origins of organic materials in works of art.

8.7 Outlook and conclusions

While new analytical techniques continue to be developed and adapted for applications to study museum artefacts, GC/MS is likely to remain a standard and essential

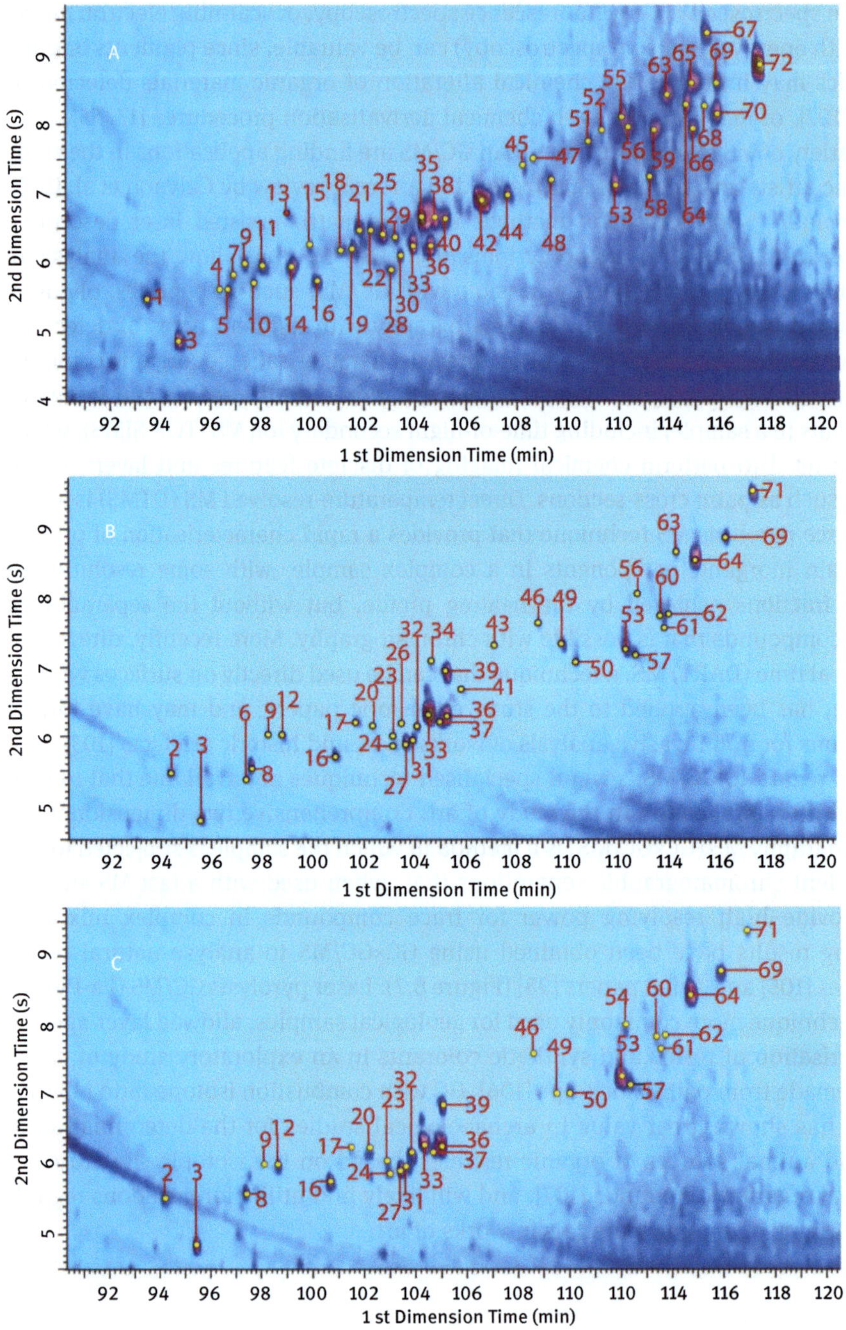

Figure 8.7: Py-GC×GC/MS data showing separation in two dimensions of chemical markers for different types of Asian paper: kozo (A), mitsumata (B) and gampi (C). Reprinted from Han et al., "Analytical characterization of East Asian handmade papers: A combined approach using Py-GC×GC/MS and multivariate analysis", Journal of Analytical and Applied Pyrolysis (2017) 127:150–158, copyright (2017), with permission from Elsevier.

tool in the cultural heritage field. Its use has expanded with the development of an increasing variety of accessories and sample preparation protocols that allow for highly versatile analyses, along with the reduced cost and increased accessibility of instrumentation. The value of GC/MS analysis has grown in synergy with the accompanying increased understanding of the organic constituents of works of art, their interactions and degradation mechanisms. With an enhanced knowledge of the fundamental chemistry, it has been possible to address increasingly sophisticated questions, besides the basic identification of materials, relating to issues with their stability and conservation, manufacturing processes, adulteration, and authenticity. These avenues of investigation have also been facilitated by a concurrent growth of research into the historical technical literature and other documentary sources that provide information, for example, on the processing and preparation of raw materials for artists' paints and varnishes. Collaboration among scientists of different specialties, and with experts in other areas (conservators, artists, historians), will continue to be a key factor in advancing the use of GC/MS and finding new applications in the art museum.

Acknowledgements: I am grateful to Jennifer Poulin, Canadian Conservation Institute, and Mark Ormsby, National Archives and Records Administration, for their helpful comments on the manuscript.

References

[1] Nadolny J. The first century of published scientific analyses of the materials of historical painting and polychromy, circa 1780–1880. Rev Conservation. 2003;4:39–51.

[2] Masschelein-Kleiner L. Perspectives de la chimie des liants picturaux anciens. Bulletin (Institut Royal du Patrimoine Artistique/Koninklijk Instituut voor het Kunstpatrimonium). 1963;6:109–26.

[3] Mills JS. The gas chromatographic examination of paint media. Part I, Fatty acid composition and identification of dried oil films. Stud Conserv. 1966;11:92–108.

[4] Arpino P, Moreau J-P, Oruezabal C, Flieder F. Gas chromatographic-mass spectrometric analysis of tannin hydrolysates from the ink of ancient manuscripts (XIth to XVIth century). J Chromatogr. 1977;134:433–9.

[5] Mills J, White R. Organic mass-spectrometry of art materials: work in progress. Natl Gallery Tech Bull. 1982;6:3–18.

[6] Weiss NR, Biemann K. Application of mass spectrometric techniques to the differentiation of paint media. In: Brommelle N, Smith P, editors. Conservation and restoration of pictorial art. London: Butterworths, 1976: 88–92.

[7] Kałużna-Czaplińska J, Rosiak A, Kwapińska M, Kwapiński W. Different analytical procedures for the study of organic residues in archeological ceramic samples with the use of gas chromatography-mass spectrometry. Crit Rev Anal Chem. 2016;46:67–81.

[8] Evershed R. Organic residue analysis in archaeology: the archaeological biomarker revolution. Archaeometry. 2008;50:895–924.

[9] Bonaduce I, Ribechini E, Modugno F, Colombini MP. Analytical approaches based on gas chromatography mass spectrometry (GC/MS) to study organic materials in artworks and archaeological objects. Top Curr Chem. 2016;374:291–327.

[10] Colombini MP, Modugno F, editors. Organic mass spectrometry in art and archaeology. Chichester: John Wiley & Sons, Ltd., 2009.

[11] Mills JS, White R. The organic chemistry of museum objects, 2nd ed. Oxford: Butterworth-Heinemann, 1994.

[12] Grob RL, Barry EF, editors. Modern practice of gas chromatography, 4th ed. Hoboken: John Wiley & Sons, Inc., 2004.

[13] McLafferty FW, Tureček F. Interpretation of mass spectra, 4th ed. Sausalito: University Science Books, 1993.

[14] Azemard C, Menager M, Vieillescazes C. Analysis of diterpenic compounds by GC-MS/MS: contribution to the identification of main conifer resins. Anal Bioanal Chem. 2016; 408:6599–612.

[15] Kokkori M, Boon JJ, Burnstock A, McGlinchey C, Mandt P. From the streets to the canvas: on the use of bituminous asphalt in early Soviet Constructivist paintings. In: Wallert A, editor. Painting techniques: history, materials and studio practice. Amsterdam: Rijksmuseum, 2017: 280–7.

[16] Poulin J. A new methodology for the characterisation of natural dyes on museum objects using gas chromatography-mass spectrometry. Stud Conserv. 2018;63:36–61.

[17] Gautier G, Colombini MP. GC-MS identification of proteins in wall painting samples: a fast clean-up procedure to remove copper-based pigment interferences. Talanta. 2007;73:95–102.

[18] Schilling M, Mazurek J. Identification of proteinaceous, oil, and plant gum binding media by quantitative GC-MS analysis: training course outline and notes. In: Sutherland K, editor. MaSC 2007 GCMS workshop handbook. Philadelphia Museum of Art, 2007:18 pp.

[19] Van den Berg KJ, van der Horst J, Boon JJ. Recognition of copals in aged resin/oil paints and varnishes. In: Bridgland J, editor. ICOM-CC 12th Triennial Meeting, Lyon, 29 August–3 September 1999, Preprints. London: Earthscan Ltd., 1999:855–61.

[20] Van der Doelen G, van den Berg KJ, Boon JJ. Comparative chromatographic and mass-spectrometric studies of triterpenoid varnishes: fresh material and aged samples from paintings. Stud Conserv. 1998;43:249–64.

[21] Van den Berg KJ, Boon JJ, Pastorova I, Spetter L. Mass spectrometric methodology for the analysis of highly oxidized diterpenoid acids in Old Master paintings. J Mass Spectrom. 2000;35:512–33.

[22] Regert M, Colinart S, Degrand L, Decavallas O. Chemical alteration and use of beeswax through time: accelerated ageing tests and analysis of archaeological samples from various environmental contexts. Archaeometry. 2001;43:549–69.

[23] Sutherland K. Derivatisation using m-(trifluoromethyl)phenyltrimethylammonium hydroxide of organic materials in artworks for analysis by gas chromatography-mass spectrometry: unusual reaction products with alcohols. J Chromatogr A. 2007;1149:30–7.

[24] Lattuati-Derieux A, Ramalho O, Egasse C, Thao-Heu S, Dupont A-L. Evaluation of solid-phase microextraction on-fiber derivatization for the analysis of paper degradation compounds. e-Preservation Sci. 2015;12:38–49.

[25] Andreotti A, Bonaduce I, Colombini MP, Gautier G, Modugno F, Ribechini E. Combined GC/MS analytical procedure for the characterization of glycerolipid, waxy, resinous, and proteinaceous materials in a unique paint microsample. Anal Chem. 2006;78:4490–500.

[26] Koller J, Baumer U. An investigation of the 'Lacquers of the West': A methodical survey. In: Kühlenthal M, editor. Japanese and European Lacquerware. München: Karl M. Lipp Verlag, 2000:339–49.

[27] Lluveras A, Bonaduce I, Andreotti A, Colombini MP. GC/MS analytical procedure for the characterization of glycerolipids, natural waxes, terpenoid resins, proteinaceous and polysaccharide materials in the same paint microsample avoiding interferences from inorganic media. Anal Chem. 2010;82:376–86.
[28] White R, Pilc J. Analyses of paint media. Natl Gallery Tech Bull. 1996;17:91–103.
[29] Pitthard V, Finch P, Bayerová T. Direct chemolysis–gas chromatography–mass spectrometry for analysis of paint materials. J Sep Sci. 2004;27:200–8.
[30] Schilling MR, Keeney J, Learner T. Characterization of alkyd paint media by gas chromatography-mass spectrometry. In: Roy A, Smith P, editors. Modern Art, New Museums: contributions to the 2004 IIC Congress, Bilbao. London: IIC, 2004: 197–201.
[31] Izzo FC, van den Berg KJ, van Keulen H, Ferriani B, Zendri E. Modern oil paints – formulations, organic additives and degradation: some case studies. In: van den Berg KJ, Burnstock A, de Keijzer M, Krueger J, Learner T, de Tagle A, et al., editors. Issues in contemporary oil paint. Cham: Springer, 2014:75–104.
[32] Van den Berg JDJ, van den Berg KJ, Boon JJ. Identification of non-cross-linked compounds in methanolic extracts of cured and aged linseed oil-based paint films using gas chromatography-mass spectrometry. J Chromatogr A. 2002;950:195–211.
[33] Schilling MR, Mazurek J, Learner TJS. Studies of modern oil-based artists' paint media by gas chromatography/mass spectrometry. In: Learner TJS, editor. Modern paints uncovered. Los Angeles: Getty Conservation Institute, 2007:129–39.
[34] Banti D, La Nasa J, Lluveras Tenorio A, Modugno F, van den Berg KJ, Lee J, et al. A molecular study of modern oil paintings: investigating the role of dicarboxylic acids in the water sensitivity of modern oil paints. RSC Adv. 2018;8:6001–12.
[35] Lee J, Bonaduce I, Modugno F, La Nasa J, Ormsby B, van den Berg KJ. Scientific investigation into the water sensitivity of twentieth century oil paints. Microchemical J. 2018;138:282–95.
[36] White R, Kirby J. Rembrandt and his circle: seventeenth-century Dutch paint media re-examined. Natl Gallery Tech Bull. 1994;15:64–78.
[37] Bonaduce I, Colombini MP. Characterisation of beeswax in works of art by gas chromatography-mass spectrometry and pyrolysis-gas chromatography-mass spectrometry procedures. J Chromatogr A. 2004;1028:297–306.
[38] Regert M, Langlois J, Colinart S. Characterisation of wax works of art by gas chromatographic procedures. J Chromatogr A. 2005;1091:124–36.
[39] Stacey R, Dyer J, Mussell C, Lluveras-Tenorio A, Colombini MP, Duce C, et al. Ancient encaustic: an experimental exploration of technology, ageing behaviour and approaches to analytical investigation. Microchemical J. 2018;138:472–87.
[40] Languri GM, Boon JJ. Between myth and reality: on the mummy pigment from the Hafkenscheid Collection. Stud Conserv. 2005;50:161–78.
[41] Baumer U, Dietemann P, Koller J. Identification of resinous materials on 16th and 17th century reverse-glass objects by gas chromatography/mass spectrometry. Int J Mass Spectrom. 2009;284:131–41.
[42] Poulin J, Helwig K. Inside amber: the structural role of succinic acid in Class Ia and Class Id resinite. Anal Chem. 2014;86:7428–35.
[43] Van der Werf ID, van den Berg KJ, Schmitt S, Boon JJ. Molecular characterization of copaiba balsam as used in painting techniques and restoration procedures. Stud Conserv. 2000;45:1–18.
[44] White R, Kirby J. A survey of nineteenth- and early twentieth-century varnish compositions found on a selection of paintings in the National Gallery collection. Natl Gallery Tech Bull. 2001;22:64–84.
[45] Watts S, de la Rie R. GCMS analysis of triterpenoid resins: in situ derivatization procedures using quaternary ammonium hydroxides. Stud Conserv. 2002;47:257–72.

[46] Dietemann P, Higgitt C, Kälin M, Edelmann MJ, Knochenmuss R, Zenobi R. Aging and yellowing of triterpenoid resin varnishes – influence of aging conditions and resin composition. J Cult Heritage. 2009;10:30–40.
[47] Colombini MP, Bonaduce I, Gautier G. Molecular pattern recognition of fresh and aged shellac. Chromatographia. 2003;58:357–64.
[48] Sutherland K, del Río JC. Characterisation and discrimination of various types of lac resin using gas chromatography mass spectrometry techniques with quaternary ammonium reagents. J Chromatogr A. 2014;1338:149–63.
[49] Schilling M, Khanjian H. Gas chromatographic analysis of amino acids as ethyl chloroformate derivatives. III. Identification of proteinaceous binding media by interpretation of amino acid composition data. In: Bridgland J, editor. ICOM-CC 11th Triennial Meeting Edinburgh, 1–6 September 1996, Preprints. London: James and James, 1996: 220–7.
[50] Chiavari G, Gandini N, Russo P, Fabbri D. Characterization of standard tempera painting layers containing proteinaceous binders by pyrolysis(/methylation)-gas chromatography-mass spectrometry. Chromatographia. 1998;47:420–6.
[51] Orsini S, Parlanti F, Bonaduce I. Analytical pyrolysis of proteins in samples from artistic and archaeological objects. J Anal Appl Pyrolysis. 2017;124:643–57.
[52] Bleton J, Mejanelle P, Sansoulet J, Goursaud S, Tchapla A. Characterization of neutral sugars and uronic acids after methanolysis and trimethylsilylation for recognition of plant gums. J Chromatogr A. 1996;720:27–49.
[53] Pitthard V, Finch P. GC-MS analysis of monosaccharide mixtures as their diethyldithioacetal derivatives: application to plant gums used in art works. Chromatographia Suppl. 2001;53: S317–21.
[54] Ormsby BA, Townsend JH, Singer BW, Dean JR. British watercolour cakes from the eighteenth to the early twentieth century. Stud Conserv. 2005;50:45–66.
[55] Lluveras-Tenorio A, Mazurek J, Restivo A, Colombini MP, Bonaduce I. Analysis of plant gums and saccharide materials in paint samples: comparison of GC-MS analytical procedures and databases. Chem Cent J. 2012; 6. DOI: 10.1186/1752-153X-6-115.
[56] Riedo C, Scalarone D, Chiantore O. Multivariate analysis of pyrolysis-GC/MS data for identification of polysaccharide binding media. Anal Methods. 2013;5:4060–7.
[57] Fabbri D, Chiavari G, Ling H. Analysis of anthraquinoid and indigoid dyes used in ancient artistic works by thermally assisted hydrolysis and methylation in the presence of tetramethylammonium hydroxide. J Anal Appl Pyrolysis. 2000;56:167–78.
[58] Casas-Catalán MJ, Doménech Carbó MT. Identification of natural dyes used in works of art by pyrolysis-gas chromatography/mass spectrometry combined with in situ trimethylsilylation. Anal Bioanal Chem. 2005;382:259–68.
[59] Degani L, Riedo C, Gulmini M, Chiantore O. From plant extracts to historical textiles: characterization of dyestuffs by GC-MS. Chromatographia. 2014;77:1683–96.
[60] Shedrinsky AM, Grimaldi DA, Boon JJ, Baer NS. Application of pyrolysis-gas chromatography and pyrolysis-gas chromatography/mass spectrometry to the unmasking of amber forgeries. J Anal Appl Pyrolysis. 1993;25:77–95.
[61] Galletti GC, Mazzeo R. Pyrolysis/gas chromatography/mass spectrometry and Fourier-transform infrared spectroscopy of amber. Rapid Commun Mass Spectrom. 1993;7:646–50.
[62] Niimura N, Miyakoshi T, Onodera J, Higuchi T. Characterization of Rhus vernicifera and Rhus succedanea lacquer films and their pyrolysis mechanisms studied using two-stage pyrolysis-gas chromatography/mass spectrometry. J Anal Appl Pyrolysis. 1996; 37:199–209.
[63] De Witte E, Terfve A. The use of a Py-GC-MS technique for the analysis of synthetic resins. In: Brommelle NS, Thomson G, editors. Science and technology in the service of conservation:

preprints of the contributions to the Washington Congress, 3–9 September 1982. London: International Institute for Conservation of Historic and Artistic Works, 1982: 16–8.
[64] Learner T. The analysis of synthetic resins found in twentieth century paint media. In: Wright MM, Townsend JH, editors. Resins: ancient and modern. Edinburgh: Scottish Society for Conservation and Restoration, 1995: 76–84.
[65] Learner T. The analysis of synthetic paints by pyrolysis-gas chromatography-mass spectrometry (PyGCMS). Stud Conserv. 2001;46:225–41.
[66] La Nasa J, Orsini S, Degano I, Rava A, Modugno F, Colombini MP. A chemical study of organic materials in three murals by Keith Haring: a comparison of painting techniques. Microchemical J. 2016;124:940–8.
[67] Defeyt C, Schilling M, Langenbacher J, Escarsega J, Rivenc R. Polyurethane coatings in twentieth century outdoor painted sculptures. Part II: comparative study of four systems by means of Py-GC/MS. Heritage Sci. 2017;5. DOI: 10.1186/s40494-017-0129-2.
[68] Moldoveanu SC. Analytical pyrolysis of synthetic organic polymers. Amsterdam: Elsevier B.V., 2005.
[69] Tsuge S, Ohtani H, Watanabe C. Pyrolysis-GC/MS data book of synthetic polymers. Oxford: Elsevier, 2011.
[70] Giachet MT, Schilling M, McCormick K, Mazurek J, Richardson E, Khanjian H, et al. Assessment of the composition and condition of animation cels made from cellulose acetate. Polym Degrad Stab. 2014;107:223–30.
[71] Fremout W, Verboven M, Saverwyns S. Accelerated tobacco smoke staining on waterborne acrylic paintings caused by exuding surfactants: a study with Py-GC/MS and THM-GC/MS. e-Preservation Sci. 2014;11:47–53.
[72] Sutherland K, Schwarzinger C, Price BA. The application of pyrolysis gas chromatography mass spectrometry for the identification of degraded early plastics in a sculpture by Naum Gabo. J Anal Appl Pyrolysis. 2012;94:202–8.
[73] Ghelardi E, Degano I, Colombini MP, Mazurek J, Schilling M, Learner T. Py-GC/MS applied to the analysis of synthetic organic pigments: characterization and identification in paint samples. Anal Bioanal Chem. 2015;407:1415–31.
[74] Sonoda N. Characterization of organic azo-pigments by pyrolysis-gas chromatography. Stud Conserv. 1999;44:195–208.
[75] Bonaduce I, Andreotti A. Py-GC/MS of organic paint binders. In: Colombini MP, Modugno F, editors. Organic mass spectrometry in art and archaeology. Chichester: John Wiley & Sons, Ltd. 2009:303–26.
[76] Chiavari G, Bocchini P, Galletti GC. Rapid identification of binding media in paintings using simultaneous pyrolysis methylation gas chromatography. Sci Technol Cult Heritage. 1992;1:153–8.
[77] Schilling MR, Heginbotham A, van Keulen H, Szelewski M. Beyond the basics: A systematic approach for comprehensive analysis of organic materials in Asian lacquers. Stud Conserv. Supplementary Issue 3; 2016;61:3–27.
[78] Tamburini D, Pescitelli G, Colombini MP, Bonaduce I. The degradation of Burmese lacquer (thitsi) as observed in samples from two cultural artefacts. J Anal Appl Pyrolysis. 2017;124:51–62.
[79] Chiavari G, Prati S. Application of pyrolysis to the diagnostic in the artistic field. In: van Grieken R, Janssens K, Van't dack L, Meersman G, editors. Art 2002: 7th International Conference on Non-destructive Testing and Microanalysis for the Diagnostics and Conservation of the Cultural and Environmental Heritage. Antwerp: University of Antwerp, 2002:8
[80] Wei S, Pintus V, Schreiner M. Photochemical degradation study of polyvinyl acetate paints used in artworks by Py-GC/MS. J Anal Appl Pyrolysis. 2012;97:158–63.

[81] Moffatt E, Salmon A, Poulin J, Fox A, Hay J. Characterization of varnishes on nineteenth-century Canadian furniture. J Can Assoc Conserv. 2015;40:3–18.
[82] Pawliszyn J. Solid-phase microextraction: theory and practice. New York: Wiley-VCH, 1997.
[83] Dyer J, Ward C, Rode N, Hacke M, Shashoua Y. Reassessment of anoxic storage of ethnographic rubber objects. In: Bridgland J, editor. ICOM-CC 16th Triennial Conference Preprints, Lisbon, 19-23 September 2011. Lisbon: Critério-Produção Grafica, Lda., 2011: 10.
[84] Tsukada M, Rizzo A, Granzotto C. A new strategy for assessing off-gassing from museum materials: air sampling in Oddy test vessels. AIC News. 2012;37:1–7.
[85] Desauziers V, Bourdin D, Mocho P, Plaisance H. Innovative tools and modeling methodology for impact prediction and assessment of the contribution of materials on indoor air quality. Heritage Sci. 2015;3:1–8.
[86] Ormsby M, Johnson JS, Heald S, Chang L, Bosworth J. Investigation of solid phase microextraction sampling for organic pesticide residues on museum collections. Collection Forum. 2006;20:1–12.
[87] Sawoszczuk T, Syguła-Cholewińska J, del Hoyo-Meléndez JM. Optimization of headspace solid phase microextraction for the analysis of microbial volatile organic compounds emitted by fungi: application to historical objects. J Chromatogr A. 2015;1409:30–45.
[88] Banou P, Alexopoulou A, Chranioti C, Tsimogiannis D, Terlixi A-V, Zervos S, et al. The effect of oil binders on paper supports via VOC analysis. J Cult Heritage. 2016;20:589–98.
[89] Pedersoli Júnior JL, Ligterink FJ, van Bommel M. Non-destructive determination of acetic acid and furfural in books by solid-phase micro-extraction (SPME) and gas chromatography-mass spectrometry (GC/MS). Restaurator. 2011;32:110–34.
[90] Weston RJ, Smith GJ, Scheele SM, Williams SH. Accelerated hydrothermal degradation of fibres of Phormium tenax (New Zealand flax). J Cult Heritage. 2012;13:413–8.
[91] Nguyen TP, le Bourg E, Bouvet S, Rottier V. Optimisation of the quality control of cellulosic materials used for the conservation and storage of paper-based cultural heritage, based on the emissions of volatiles. In: Bridgland J, editor. ICOM-CC 16th Triennial Conference Preprints, Lisbon, 19–23 September 2011. Lisbon: Critério-Produção Grafica, Lda., 2011: 9.
[92] Lattuati-Derieux A, Thao S, Langlois J, Regert M. First results on headspace-solid phase microextraction-gas chromatography/mass spectrometry of volatile organic compounds emitted by wax objects in museums. J Chromatogr A. 2008;1187:239–49.
[93] Curran K, Možir A, Underhill M, Gibson LT, Fearn T, Strlič M. Cross-infection effect of polymers of historic and heritage significance on the degradation of a cellulose reference test material. Polym Degrad Stab. 2014;107:294–306.
[94] Lattuati-Derieux A, Egasse C, Thao-Heu S, Balcar N, Barabant G, Lavédrine B. What do plastics emit? HS-SPME-GC/MS analyses of new standard plastics and plastic objects in museum collections. J Cult Heritage. 2013;14:238–47.
[95] Ormsby M. Analysis of laminated documents using solid-phase microextraction. J Am Inst Conserv. 2005;44:13–26.
[96] Schilling MR, Learner T. Evolved gas analysis as a tool for characterizing plastics. In: Bridgland J, editor. ICOM-CC 16th Triennial Conference Preprints, Lisbon, 19–23 September 2011. Lisbon: Critério-Produção Grafica, Lda., 2011: 11 pp.
[97] Samide MJ, Liggett MC, Mill J, Smith GD. Relating volatiles analysis by GC-MS to Oddy test performance for determining the suitability of museum construction materials. Heritage Sci. 2018;6:1–10.
[98] Han B, Vial J, Inaba M, Sablier M. Analytical characterization of East Asian handmade papers: a combined approach using Py-GC×GC/MS and multivariate analysis. J Anal Appl Pyrolysis. 2017;127:150–8.

[99] D'Arcy P, Mallard WG. AMDIS – user guide. Gaithersburg MD: National Institute of Standards and Technology, 2004.

[100] MaSC (the Users' Group for Mass Spectrometry and Chromatography) https://mascgroup.org/workshops-meetings/past-workshops-and-meetings/. Accessed 21 August 2018.

[101] RAdICAL (Recent Advances in Characterizing Asian Lacquer) http://www.getty.edu/conservation/our_projects/education/radical/. Accessed 21 August 2018.

[102] Higgitt C, Spring M, Saunders D. Pigment-medium interactions in oil paint films containing red lead or lead-tin yellow. Natl Gallery Tech Bull. 2003;24:75–95.

[103] Calvano CD, van der Werf ID, Palmisano F, Sabbatini L. Revealing the composition of organic materials in polychrome works of art: the role of mass spectrometry-based techniques. Anal Bioanal Chem. 2016;408:6957–81.

[104] Pauk V, Barták P, Lemr K. Characterization of natural organic colorants in historical and art objects by high-performance liquid chromatography. J Sep Sci. 2014;37:3393–410.

[105] Frysinger G, Hall G, Gaines R, Sutherland K. Comprehensive two-dimensional gas chromatography with mass spectrometric detection (GC×GC-MS) and its application to the separation and identification of organic compounds in some natural resins and waxes. In: Sutherland K, Maines C, van den Berg KJ, Ferreira E, Higgitt C, editors. MaSC 2007 Meeting Program. Philadelphia Museum of Art: 2007:18 (abstract).

[106] Prati S, Fuentes D, Sciutto G, Mazzeo R. The use of laser pyrolysis-GC-MS for the analysis of paint cross sections. J Anal Appl Pyrolysis. 2014;105:327–34.

[107] Evershed R. Compound-specific stable isotopes in organic residue analysis in archaeology. In: Colombini MP, Modugno F, editors. Organic mass spectrometry in art and archaeology. Chichester: John Wiley & Sons, Ltd. 2009:391–432.

Ilaria Degano

9 Liquid chromatography: Current applications in Heritage Science and recent developments

Abstract: Liquid chromatography has been widely employed in the analysis of materials in Heritage Science, due to its ease of use and relatively low-cost, starting from thin layer chromatography of organic binders in paintings, of archaeological waxes and resins, and finally of natural dyes. High performance systems employing analytical columns containing packed stationary phases gradually supplanted thin layer chromatography (TLC) in the field, since the separation, detection and quantitation of specific species contained in a sample in the field of Cultural Heritage requires selective, sensitive and reliable methods, allowing for analysing a wide range of samples, in terms of analyte types and concentration range. Today, the main applications of High-Performance Liquid Chromatography in this field are related to the separation and detection of dyestuffs in archaeological materials and paint samples by reversed-phase liquid chromatography with suitable detectors. Proteomics and lipidomics are also gaining momentum in the last decade, thanks to the increased availability of instrumentation and procedures. In this chapter, principles and theory of liquid chromatography will be presented. A short review of the instrumentation needed to perform an analysis will be provided and some general principles of sample preparation revised. More details on the detection systems, the chromatographic set-ups and specific sample treatment strategies will be provided in the individual sections dedicated to the applications to Heritage Science of the main types of liquid chromatographic techniques. In particular, the applications of thin layer chromatography will be shortly described in paragraph 4.1. The applications of Reverse Phase High Performance Liquid Chromatography (RP-HPLC) will be discussed in detail in paragraph 4.2, including the analysis of natural and synthetic dyes and pigments and the profiling of lipid materials. The possibility to perform proteomic analysis will be presented and a link to the relevant Chapter in this book provided. The most important and promising applications of ion exchange chromatography (IC) will be discussed in paragraph 4.3. Finally, size exclusion and gel permeation chromatography (GPC) will be presented in paragraph 4.4, including applications to the study of polymeric network formation in paint binders, of the phenomena related to the depolymerisation of cellulose in paper and of cellulose and lignin in wood samples. The possibility to study synthetic polymers as artists' materials and restorers' tools by size exclusion (SEC) or gel permeation (GPC) will also be introduced. In the conclusions, future perspectives of liquid chromatography in Heritage Science will be briefly discussed.

This article has previously been published in the journal *Physical Sciences Reviews*. Please cite as: Degano, I. Liquid chromatography: current applications in Heritage Science and recent developments. *Physical Sciences Reviews* [Online] **2019**, 4. DOI: 10.1515/psr-2018-0009.

https://doi.org/10.1515/9783110457537-009

Keywords: liquid chromatography, TLC, HPLC, size exclusion chromatography/gel permeation chromatography, ion chromatography, detectors

9.1 Principles and theory

Liquid chromatography is a separation technique in which the mobile phase is a liquid and the elution can be carried out in a column filled with stationary phase, a capillary, or on a plane. As the sample is introduced, the components distribute between the mobile and the stationary phase; the introduction of additional mobile phase carries solute molecules in a continuous series of transfers between the two phases. The separation can occur accordingly to one or more of the following interactions: adsorption/partition, ionic exchange and size exclusion. Moreover, if the stationary phase is functionalized with specific species or chemical groups (e. g. chiral active sites), affinity interaction can occur.

Elution types can be isocratic, when the composition of the mobile phase does not change during the separation, or gradient, when the composition of the mobile phase changes during elution time.

Normal and reverse phase (RP) chromatography involve adsorption/partition interactions and the typical analytes are organic compounds. In these cases, the elution order and retention times mostly depend on the polarity of the mobile phase with respect to that of the stationary phase. The solvents, which range from aliphatic hydrocarbons to water, are ordered in "eluotropic series" that take into account the intermolecular forces between eluents and stationary phases and report the so-called "eluting power" for specific solvents, for specific stationary phases.

Direct phase chromatography entails the use of a stationary phase more polar than the mobile phase. The most used polar stationary phases are composed by silica or alumina gels and the eluents are mostly apolar organic solvents.

RP chromatography is performed when the mobile phase is more polar than the stationary phase. The stationary phase is generally composed by silica particles to which apolar groups are chemically bonded, such as octadecyl groups (C18). Methanol, acetonitrile and water are the most used eluents. In RP-HPLC, the retention of the analytes is also strongly influenced by their hydrophobicity, which can be modified by additives such as acids, bases or buffers (mobile phase modifiers).

Ion chromatography (IC) is a technique where ionic interactions are exploited to separate ions in an aqueous mobile phase, according to type and strength of their charge. The stationary phase is typically an ion exchange resin manufactured starting from a macroporous polymer (such as cross-linked polystyrene) functionalized with cation or anion exchanger groups, which can in turn be strong or weak exchangers. The typical analytes are inorganic anions and cations, but IC can also be applied to detect organic ions as carboxylic acids, carbohydrates and amino acids, as well as proteins, peptides and nucleotides.

Finally, size exclusion chromatography (SEC) is based on the separation of the analytes according to their hydrodynamic volume (roughly equivalent to the size of a molecule in solution) depending on their ability to perform preferential paths along the chromatographic column. When size exclusion chromatography is performed using aqueous solvents, it is also called gel permeation chromatography (GPC). The typical analytes are natural and synthetic macromolecules. SEC and GPC are widely used to purify or fraction samples in preparative chromatography, but can also be applied as analytical techniques to estimate the average molecular weight of a sample in polymer and materials science. For each column, it is possible to assess a range where the molecular weight of the analytes and the elution volume are in a logarithmic relationship. Thus, in the presence of a suitable set of standards, the molecular weight of analytes in an unknown sample may be estimated.

9.2 Instrumentation

The instrumentation needed to perform analytical liquid chromatography differs depending on the type of stationary phase used and the type of detector needed for the identification, and possibly quantitation, of the analytes.

9.2.1 Paper and thin layer chromatography

To perform chromatographic analyses on a planar, or thin layer, stationary phase, an elution chamber is needed. Whenever the stationary phase is paper, we refer to it as paper chromatography. The application of this technique for analytical purposes is today rather obsolete. In TLC, the stationary phase is a highly polar solid adsorbent (alumina, silica or cellulose, typically) coated on a disposable aluminum foil, and the elution occurs inside the elution chamber. Before inserting the plate in the elution chamber, the liquid sample is deposed in a small spot on the bottom of the stationary phase, and its components migrate at different characteristic rates; after the separation process, the analytes appear as vertically separated spots. The distance travelled by each compound with respect to that of the solvent can be used to identify components (retention factor, Rf) if suitable standards or reference materials are available. The sample components are then detected by visual inspection if coloured, or after spraying the TLC plate with a suitable visualizing agent. Paper and thin layer chromatography are cheap and easy to use, but suffer from poor performances in terms of analyte quantitation, with high detection limits and low reproducibility.

9.2.2 High-performance liquid chromatography

To perform the analysis on a column or a capillary, the typical set-up for analytical purposes entails a mobile phase supplier, one or more pumps delivering the eluent, the solvent programmer, the sample injection valve, the column (or the capillary) and one or more detectors. A system to acquire and process the data is needed, such as a computer. Figure 9.1 depicts a general scheme for HPLC applications.

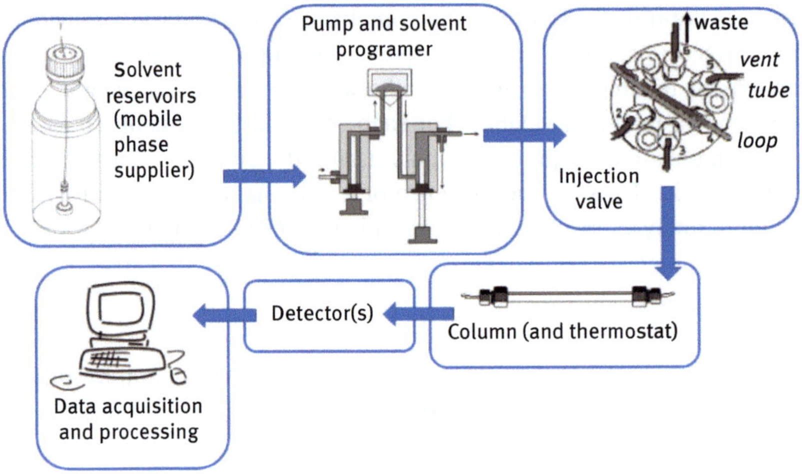

Figure 9.1: General scheme of an HPLC instrument.

The solvent reservoir(s) contains the eluents to be used for the analysis. The pump can be designed in different modes, but has stringent specifications, namely the ability to operate continuously at high pressures, up to 6000 psi for high performance liquid chromatography (HPLC), and up to 10,000 psi for ultra-high-performance liquid chromatography (UPLC). Moreover, a precise and reproducible flow control is required and in the specific range needed for the target application. For most applications, 0.2–10 mL/min flow is used. For nano-LC applications, flow rates down to the tenth of microliter per minute is required. The solvent programmer is necessary to perform the analysis in a gradient mode. It is a crucial part of the equipment, since accurate and reproducible solvent programmes ensure the reproducibility of the analyses and allow high-throughput applications.

The column is the core of the chromatographic separation and several columns are available on the market. Discussing the different types of stationary phases is beyond the scope of the present chapter. Nonetheless, some general considerations apply to all kind of chromatographic analytical columns. More than 80% of the commercial columns are packed in stainless steel to support high pressures and guarantee inertness towards the most common eluents. Tubing, fittings and frits holding the packing in the column are made of materials resistant to solvents and buffers, such as stainless steel or peek. The stationary phase is in most cases made of spherical packings, which might or might not be functionalized, made of silica particles or polymers. Both the pore size and the diameter of the particles influence the back-pressure needed to elute the sample components. Most important, they affect the superficial area of the stationary phase and the coefficient of diffusion of the analytes into the stationary phase, and ultimately have a strong effect on the retention mechanisms and the overall efficiency of the separation. The influence of different experimental parameters on the efficiency of a separation is considered by the van Deemter equation, which relates the efficiency of a separation (in terms of height equivalent to a theoretical plate) to the linear mobile phase velocity by considering the physical, kinetic, and thermodynamic properties of the system [1]. Most columns are in the 10–30 cm length range, with 2-4.6 mm internal diameter and 2–10 µm particle size diameter. UPLC (Ultra High-Pressure Liquid Chromatography) allows the use of sub-2-µm particles for the stationary phase by exploiting more effective solvent pumps and shorter connections, which can resist to relatively higher back-pressures. UPLC allows obtaining a high chromatographic efficiency in shorter runtimes, thus providing faster analyses with lower solvent consumption.

The sample is injected through an injection port, which includes a loop for sample loading and an internal valve. Most recent instruments are equipped with autosamplers, whose functioning is based on the same type of valves and loops as manual injection. Autosamplers allow multiple samples to be analysed in a sequence that can run overnight, thus increasing the throughput of the analytical set up.

The range of detectors used in liquid chromatography is wide and their exhaustive description goes beyond the scopes of the present Chapter. The most generally used detector for organic compounds, which can be applied to most chromatographic setups, is based on UV-Vis absorption. Different types of spectrophotometric detectors can be coupled online to the chromatographic set up by employing conical flow cells to host the eluent as exited from the column. Single wavelength or multi-wavelength detectors can be used; Photodiode Array Detectors (PDA or also known as DAD) are also employed and provide information on the entire UV-Vis absorption spectrum of eluted species. The tridimensional information achieved by DAD allows obtaining chromatograms over a range of wavelengths during a single run, thus providing more information on the sample composition than a single wavelength detector. In particular, the combination of retention time and spectrum allows for the positive identification of the analytes, ruling out coelution of different compounds. Moreover, the acquisition of the UV-Vis spectrum of each separated peak allows for selecting an optimal wavelength for verifying peak purity and peak identity, and to quantify the analytes with optimal detection limits. A scheme for DAD is reported in Figure 9.2. Fluorescence detectors (FD) are also frequently used, providing higher selectivity and sensitivity towards fluorescent analytes. The sensitivity of FD is typically 2 or 3 orders of magnitude higher than that of absorption detectors; the outstanding selectivity is due to the possibility to change both excitation and emission wavelengths during the chromatographic run. In order to improve the absorption or the fluorescence of the analytes, offline or online derivatization can be applied.

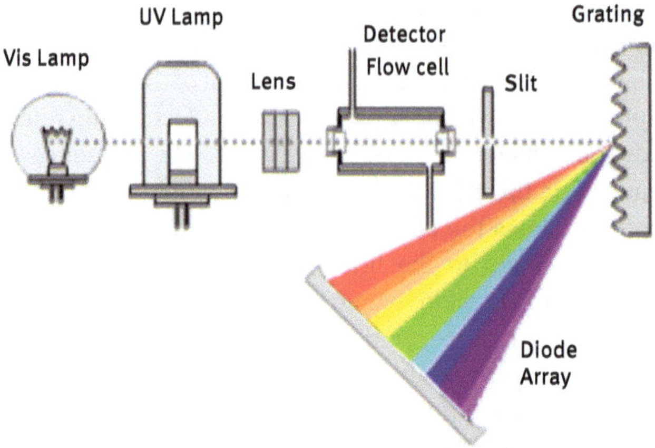

Figure 9.2: Scheme of a Photodiode Array Detector (or Diode Array Detector) in the UV-Vis range.

None of these spectrophotometric techniques is the main choice for detection system in IC applications, mostly employed to separate organic and inorganic ions, amino acids, proteins or nucleic acids. IC applications mainly rely on conductometer detectors.

For SEC applications, besides the spectrophotometric detectors, also molecular-weight-sensitive detectors, such as refractive index, light scattering and viscometry are used.

Finally, many set-ups of liquid chromatography can rely on mass spectrometry as detection system, which has enriched and widened the possible applications in the field of Cultural Heritage. Spray ionization techniques such as electrospray ionization (ESI) and atmospheric pressure chemical ionization (APCI) are the typical interfaces used for coupling mass spectrometry with HPLC. ESI is produced by forcing a liquid to pass through a capillary, then an electric field applied to the capillary tip causes the ionization of the molecules. The desired effect is the formation of liquid-charged droplets from which gas phase ions are formed. ESI mass spectra are characterized by protonated/deprotonated molecules and adducts, i. e. $[M + H]^+$, $[M + Na]^+$, $[M - H]^-$, $[M + HCOO]^-$; multiply charged ions are typically produced in ESI sources. APCI is based on chemical ionization carried out at atmospheric pressure. The solution passes through a pneumatic nebulizer and is desolvated in a heated quartz tube or heating block, thus producing vaporization of solvent and analyte molecules, aided by a nebulizing gas. An electric field applied to a corona discharge needle inside the ion source allows the ionization of the gas forming the primary ions, which in turns ionize the vaporized solvent molecules by proton transfer. These latter ions in turn ionize the analyte molecules. Differently from ESI, APCI produces singly charged ions with fragmentation to some extent [2]. As for the analyser of the spectrometer, the most common are the ones based on quadrupole, ion trap or Time of Flight (ToF). The mass spectra are acquired in sequence by scanning the selected mass interval repetitively in quadrupoles and ion traps, or by measuring the ToF of the ions through the flight tube in ToF detectors. The ions abundance in each spectrum is summed and reported in the Total Ion Current (TIC) chromatogram. Specific information on mass spectrometric detection is available in other Chapters in this book, while more details on the most common detectors are provided along with the specific applications described in the following paragraphs.

9.3 Sample preparation

Sample treatments for liquid chromatography applications are generally aimed at providing a solution free of interferences and enriched in the analytes under study. In the field of Heritage Science, the amount of sample available for the analysis is often limited; thus, research has dealt with different strategies to maximize the information obtainable from one microsample by purifying and pre-concentrating the analytes, and sample treatment has been the focus of several studies. To be specifically analysed by liquid chromatography, a sample needs to be treated and solubilized in a solvent compatible with the method and not harmful for the stationary phase or the detector. In detail, the solvent shall be miscible with the mobile phase, and be characterized by a similar strength with respect to the eluent used at the beginning of the elution, in order not to compete with it (resulting in turn in not ideal peak shapes). Moreover, the solution containing the analytes must be filtered, to remove any particulate that might harm or clog the stationary phase. Standard filtration is performed on micro-syringe filters with 0.45 μm pore size. Whenever extremely small volumes are used to extract tiny samples, centrifugation is a viable alternative strategy to purify the sample before injection.

Keeping these restrictions and best practices in mind, different protocols have been optimized in the last decades, depending on the target analytes and sample characteristics.

Some examples will be provided in the following sections.

9.4 Applications to heritage science

The first applications of liquid chromatography in Heritage Science were based on Thin Layer Chromatography; however HPLC quickly supplanted TLC in diagnostic applications. TLC has regained momentum recently, thanks to the possibility to couple it with Surface Enhanced Raman Spectroscopy.

Nonetheless, the great majority of applications of liquid chromatography in Heritage Science are based on partition HPLC. HPLC reversed-phase chromatography has been applied for the last 30 years to the characterization of organic dyes, which are mixtures of polar, water-soluble compounds. The most common detection systems are based on DAD, but coupling mass spectrometry with RP-HPLC has led to several publications highlighting the suitability of this detector, especially when dealing with unknown degradation/ageing pathways and products. RP-HPLC coupled with mass spectrometry has not only been applied to the study of dyes and pigments, but also to lipid materials and in proteomic applications.

In selected cases, Ion Chromatography (IC) is also used in the field of Cultural Heritage, to obtain qualitative and quantitative information on particulates and deposits containing salts. Also in this case, chromatography is the technique of election, since sensitivity and selectivity are of paramount importance.

Size exclusion or GPC is not often applied in the field of Heritage Science for diagnostic purposes, while it provided very interesting results in the characterization of reference materials needed to mimic the actual behaviour of organic components in works-of-art. A typical example is the evaluation of the physical-chemical properties of preservatives for stone, paintings, etc, which are polymeric in most cases, and whose characteristics are easily observed by size exclusion or GPC. Other examples entail the study of the behaviour in terms of ageing and interaction with different organic and inorganic molecules of natural macromolecules present in archaeological and historical works-of-art, such as wood and cellulosic or protein-based textiles; binding media such as polysaccharides (natural gums), lipids (siccative oils) or proteins (egg, casein, collagen); varnishes such as terpenoid resins.

9.4.1 Thin layer chromatography (TLC)

TLC has been applied in the field of Heritage Science in very early applications. Its use is described by Mills and White [3] and reviewed by Striegel and Hill [4] for paint materials, waxes and resins, and by Schweppe [5] for the analysis of natural dyes. The diffusion of HPLC has much limited the application of TLC, since the former allows more reproducible and most of all quantitative results to be obtained, while lowering the detection limits thanks to sensitive detectors. In the last few years, the interest in TLC has risen again thanks to its possible coupling with SERS detection for dyes and pigments analyses, through the elution of the analytes on a suitable surface able to produce plasmonic effects [6, 7]. On the one hand, the separation of the individual components of the samples ease the interpretation of their Raman spectra; the application of SERS hinders the fluorescence provided by most samples of natural and synthetic dyes, avoiding interference in obtaining the Raman spectra; finally, the extreme sensitivity of SERS detection overcomes the limitations of the TLC approach.

9.4.2 Reverse phase high performance liquid chromatography (RP-HPLC)

RP-HPLC is widely applied in the field of Heritage Science and in particular to the analysis of natural and synthetic dyes [8–17]; in proteomics-related applications; in the characterization of lipids in archaeological samples [18–23] or used as paint binders [24–26]; in the analysis of biomolecular markers in archaeological residues.

9.4.2.1 Analysis of natural and synthetic dyes and pigments

RP chromatography is particularly useful to analyse organic dyes and pigments, which are complex mixtures of organic compounds most often aromatic and polar, characterized by a strong absorption in the UV-Vis range. Generally, the characterization of the dye source is based on the identification of molecular markers, followed by a qualitative and semi-quantitative comparison of analytical results with the profiles of known reference materials. The most commonly employed stationary phases are C8 or C18, most often in the size of 150 × 4.6 or 150 × 2 mm, with 3 or 5 µm particle size, while the elution is performed in gradient mode, with binary systems including water and acetonitrile or methanol. Amongst the organic solvents, acetonitrile is often preferred since it has lower absorbance in the 200–275 nm range, and lower back-pressure due to lower viscosity with respect to methanol [27]. Mobile phase modifiers are used to improve the chromatographic efficiency and peak shape. Being most dyes slightly acidic due to phenolic or carboxylic functional groups, phosphate buffers are the most used. Methanesulfonic acid (MSA) has been used to improve the resolution of indigoid compounds in [28]. When synthetic dyes or pigments are concerned, counter ion chromatography might be used to improve retention and peak symmetry, by using for instance tetrabutylammonium hydroxide as a mobile phase modifier [17]. To separate the most polar coloured compounds or degradation products, such as phenolic acids contained in tannins, or phenols due to the degradation of aromatic dyes, polar embedded reversed-phase columns such as RP-amide can also be applied [29]. As an alternative, or in all cases whenever mass spectrometry is proposed as detector, organic modifiers are employed, such as formic acid or trifluoroacetic or acetic acids. In most recent years, UHPLC was successfully applied, improving the efficiency of the separations and lowering the detection limits [12, 13, 30].

The most used detectors for the analysis of dyestuffs are UV/Vis, Diode array (DAD or PDA), fluorescence (FD), and mass spectrometric detectors, with ESI or APCI interfaces (ESI-MS, APCI-MS). Notably, post-column derivatization might be applied, especially for enhancing fluorescence [31].

To extract natural and synthetic organic dyes and pigments from paint samples or textile fibres, several procedures have been tested in the last 20 years [32, 33]. Although several comparative investigations have been published [9, 14, 34, 35], the method of election has not been identified yet, which is optimal for all the investigated materials, since the range of chemical classes exploited as colouring materials is extremely broad, and the molecular markers of each dye differ in terms of solubility and stability in acidic or alkaline media. The most commonly employed methods rely on the solubilization of the colouring molecules from the solid sample by solid-liquid extraction in a heated bath, sometimes aided by ultrasounds or microwaves. The less stable dyes and pigments, as carotenoids or glycosides, or the hydrophobic ones as indigoids are commonly extracted in dimethylformamide (DMF) or dimethylsulfoxide (DMSO). Organic lakes used as pigments, and mordant dyes are commonly extracted in acidic media to hydrolyze the chemical bonds between the analytes and the substrate. Until very recently, the most widespread method entailed hydrolysis in an acidic (for HCl) methanolic solution. In the last 10 years, mild extraction methods were developed, entailing the use of acidic complexation agents such as hydrofluoric, formic, oxalic, trifluoroacetic, or acetic acid, alone or combined with EDTA. These mild methods were developed to overcome the issues caused by the application of strong acidic conditions, such as cleavage of glycosidic bonds, dehydration or decarboxylation of molecular markers, or esterification of phenolcarboxylic compounds (see Figure 9.3). In some cases, the sample is further purified and the analytes pre-concentrated by introducing a liquid/liquid extraction step with a suitable solvent, such as ethylacetate.

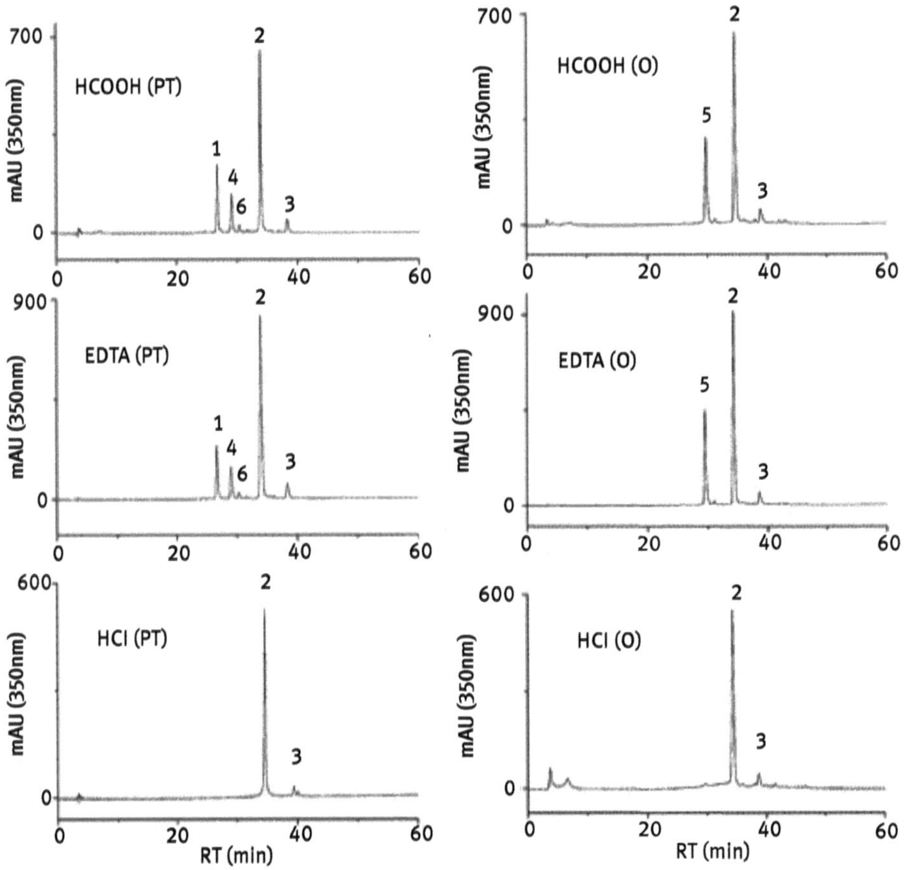

Figure 9.3: HPLC profiles of flavonoid compounds extracted from silk dyed with pagoda tree buds (PT) or onionskins (O), using the formic acid, EDTA, and HCl extraction methods; DAD detection with plot of absorbance at 350 nm. 1 = rutin; 2 = quercetin; 3 = kaempferol; 4 = kaempferol 3, 7-diglucoside; 5 = quercetin 4′-glucoside; 6 = isomer of peak 1 and 4 components (reprinted with permission from [16], ©2005 American Chemical Society).

9.4.2.2 Proteomic applications

Proteomics is the study of proteomes and their functions, being the proteome the complete set of proteins encoded by the genome, accordingly to IUPAC. For proteomic applications in Heritage Science, the researchers have relied on analytical protocols already developed for biological studies. The most used set-ups entail the use of nano-LCs coupled with high-resolution mass spectrometry through electrospray sources, applied to a tryptic digest of the proteins contained in the sample. The columns are most often capillary columns with C18 stationary phase, and the eluents are water and acetonitrile with formic acid as mobile phase modifier.

This extremely complex field of study is gaining momentum for the identification of proteinaceous paint binders in paint samples, fossils or well-preserved archaeological residues. Several examples in the field of Heritage Science are reviewed in [36, 37]. Beyond protein detection, studies related to the biological species of the identified proteins are present in the literature [38]. However, the approaches used in biomedical research cannot be easily applied to historical and archaeological materials because of the extensive protein degradation occurring in these samples and new specific procedures need to be developed to address these analytical challenges [39].

9.4.2.3 Profiling of lipid materials

Lipids are widespread as paint binders and as residues in archaeological objects (as food residues, illuminants, pharmaceuticals). They most often originally consisted in triacylglycerides (TAGs) deriving from plant oils or animal fats. In some cases, lipids found in archaeological residues also derive from vegetal or animal waxes. A relevant lipid fraction is also contained in egg and milk and other dairy products. Finally, lipids can be also found in semi-synthetic binders, such as modified modern oils and alkyd resins.

Although gas chromatography-mass spectrometry has been widely applied to the characterization of lipid materials in works-of-art, after saponification and derivatization of the fatty acids deriving from the alkaline hydrolysis of triglycerides or transesterification, RP-HPLC is used whenever the source of the lipid has to be unambiguously identified based on its profile in terms of surviving and/or oxidized TAGs, diglycerides and monoglycerides. In the analysis of lipids in the field of Heritage Science, the most used stationary phase is C18. Different column sizes and dimensions have been proposed, and good results were achieved by employing core-shell stationary phases, which enable the analyst to obtain good separations in short analytical runs (less than 30 min), using traditional HPLC pumps and providing efficiencies comparable to those of UHPLC systems [18, 40]. The most common eluents are methanol and isopropanol, but also acetonitrile and water are used [41].

Besides UV spectrophotometers and refractive index detectors, the most successful detection system in these studies is mass spectrometry [21, 24, 40–42] since RP-HPLC is generally unable to separate the several isomers in lipid samples. APCI-MS provides information on the nature of the triacylglycerides by yielding mass spectra containing information on the protonated molecular ions $[M + H]^+$ and diacylglycerol ions $[M - RCO_2]^+$. Moreover, the difference in the relative abundance of diacylglycerol fragments $[M - RCO_2]^+$ ion provided by the APCI ionization allows for distinguishing the positional isomers of an individual triacylglycerol [43]. The application of ESI interface produces the molecular ion relative to the analytes, which are often detected as $[M + Na]^+$. To gain structural information, tandem mass spectrometry is then applied, and ESI-MS/MS has featured several applications in the field in recent years. Moreover, high resolution mass spectrometry can provide the raw formula of the analytes and of their tandem mass product ions, thus improving the selectivity of the method and the possibility to ascertain the nature of unknown species. Methods based on HPLC-ESI-MS/MS have been applied to interesting case studies recently, such as the characterization of oil based paint media in modern and contemporary artworks [44–46], to the evaluation of the behaviour of plant oils used as binding media prepared in different mockups and aged in different conditions [24, 47], to the identification of lipids not related to food processing, such as ergot or whale oil, in archaeological findings [21, 23]. Figure 9.4 shows the chromatograms relative to the extracts of Winsor & Newton alkyd paint mock-ups: one was naturally cured for 9 months while the other was subjected to artificial ageing in the presence of acetic acid vapour. The TAGs profile highlights the differences in composition of the paint binder resulting from ageing.

The extraction of TAGs, DAGs and MAGs from a paint or archaeological sample is commonly performed by solid-liquid extraction in a suitable solvent, such as binary mixtures of hexane with dichloromethane or chloroform, possibly aided by microwaves, followed by drying and reconstitution in a solvent miscible with the eluents. The samples are filtered prior to injection.

A new procedure allowing for the simultaneous detection of free fatty acids, MAGs, DAGs and TAGs has recently been published, entailing the derivatization of the free fatty acids with hydoxyquinoline to increase their retention on a C18 column with a water/methanol/isopropanol gradient elution [41].

9.4.3 Ion exchange chromatography

Ion chromatography has been used in the field of Heritage Science mostly as a routine method to identify inorganic salts (mainly anions) and small organic acids (formic, oxalic and acetic acids) in black crusts and generally damaged layers on stone buildings or mortars [48–50]. It has also been exploited for the quantification of indoor pollutants or ions leaching from glass objects [51–59]. In these cases, a conductometer detector is employed, sometimes along with an UV–Vis spectrophotometer detector. Since the eluent is usually a strong electrolyte, such as a strong base or acid, the contribution to the background of any electrolyte that might interfere is usually removed using an ion suppressor prior to detection. In the process, the solution passes through an ion

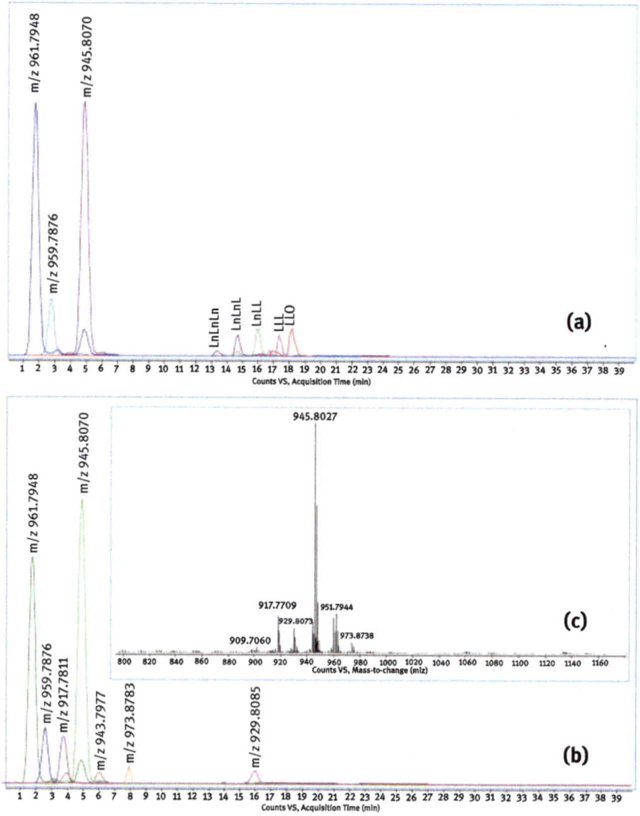

Figure 9.4: (a) HPLC-ESI-Q-ToF chromatogram of Winsor & Newton alkyd paint ("Griffin Fast Drying Oil Colour") (PBk9) after 9 months of natural ageing; (b) FIA mass spectrum; (c) HPLC-MS chromatogram after 6 months of artificial ageing. The chromatograms were obtained by overlapping the extract ion chromatograms of the identified TAG species (reprinted from [46] with permission from Elsevier).

exchange system (the "suppressor") in which the cations are replaced by H^+ to neutralize the alkaline eluent to water in anion chromatography, while the anions are replaced by OH^- in cation chromatography to neutralize the acidic eluent to water.

Anion exchange IC has also been used in a completely different set-up to identify plants gums used in paintings, after hydrolysis assisted by microwaves in the corresponding sugars by trifluoroacetic acid and after purification of the sample from its other components. The detector used by Colombini et al. in [60] was a pulsed amperometry (PAD) using gold working electrodes.

9.4.4 Size exclusion (SEC) or gel permeation chromatography (GPC)

SEC has been applied in the field of Heritage Science to study the behaviour of reference materials upon natural or accelerated ageing, and to the characterization of wood and paper depolymerisation upon degradation. Its application to historical or archaeological samples is indeed limited by the relative high amount of sample needed for the analysis, which is usually ten or hundred times higher than for RP-HPLC or IC applications. The main applications of this technique are related to the study of the degree of cross-linking or depolymerization a consequence of curing, ageing or damage in natural and synthetic polymers, such as: polysaccharides (gums) or protein-based materials used as paint binders; cellulose in paper artefacts; cellulose, hemicellulose and lignin in wood objects; synthetic polymers used as consolidants in restoration or as part of multi-material modern works-of-art. Natural resins have also been studied by SEC. Depending on the stationary phase used, and the type of analyte, different detectors were used, such as Refractive index detector (RID), UV (UV) or Diode array detector (DAD), Fluorescence-multiangle laser light scattering detector (MALS) or Cold Vapour Generation Atomic Fluorescence Spectrometry (CVGAFS).

9.4.4.1 Polymeric network in paint binders

The characterization of the filming properties of paint binders can be performed by size exclusion chromatography, since they entail polymerization, cross-linking and condensation, beside oxidative modifications and shortening of side-chains. The evaluation of the degree of polymerization can help understanding the reactions occurring upon curing of the paint layers or during ageing. Since the amount of sample required for SEC or GPC is relatively high, this type of study can only be performed on reference materials, possibly subjected to artificial ageing. SEC and GPC have been used to gain information on the influence of different parameters on the curing of materials, since the presence of different types of pigments, additives, or even the environmental conditions (temperature, light exposure, relative humidity) can influence the polymerization-depolymerization rates of the paint binders. The choice of column and detector strongly depends on the type of material under study.

Some examples include the evaluation of the effect on the curing of protein binders (casein and ovalbumin) of the presence of azurite ($Cu_3(CO_3)_2(OH)_2$), calcium carbonate ($CaCO_3$), hematite (Fe_2O_3), and red lead (Pb_3O_4) [61], or of cinnabar (HgS) [62], performed on a SEC system coupled to a diode array detector (DAD) or an UV and cold vapour generation atomic fluorescence spectrometry detector (SEC-UV-CVGAFS). The latter technique is based on the pre-column derivatization of protein thiol groups with a mercurial probe to detect mercury-protein complexes by CVGAFS. In all these cases, the separation of the proteins was performed on a Biosep SEC S2000 column in 50 mM Phosphate Buffered Saline pH 7.4.

The curing of oil binders has also been studied by SEC. In particular, the effects due to preprocessing performed by heating, chemical bleaching or treatment with metallic salts on linseed oil were studied by SEC with refractive index and DAD detectors. The separation was performed on a PLGEL 5 µm 1000 Å column and tetrahydrofuran was used both to solubilize the sample and as eluent [63]. Recently, the same analytical set-up was applied to study the behaviour of historical linseed oil/colophony varnishes formulations [47]. The results obtained by SEC quantitation showed that changing the oil/colophony proportions in the varnish strongly affected the polymerization rate of the oil, favoring hydrolysis in presence of higher amount of colophony, while linseed oil inhibited the formation of diterpenic dimers and trimers. The different polymeric fractions are clearly visible in the SEC chromatograms, reported in Figure 9.5.

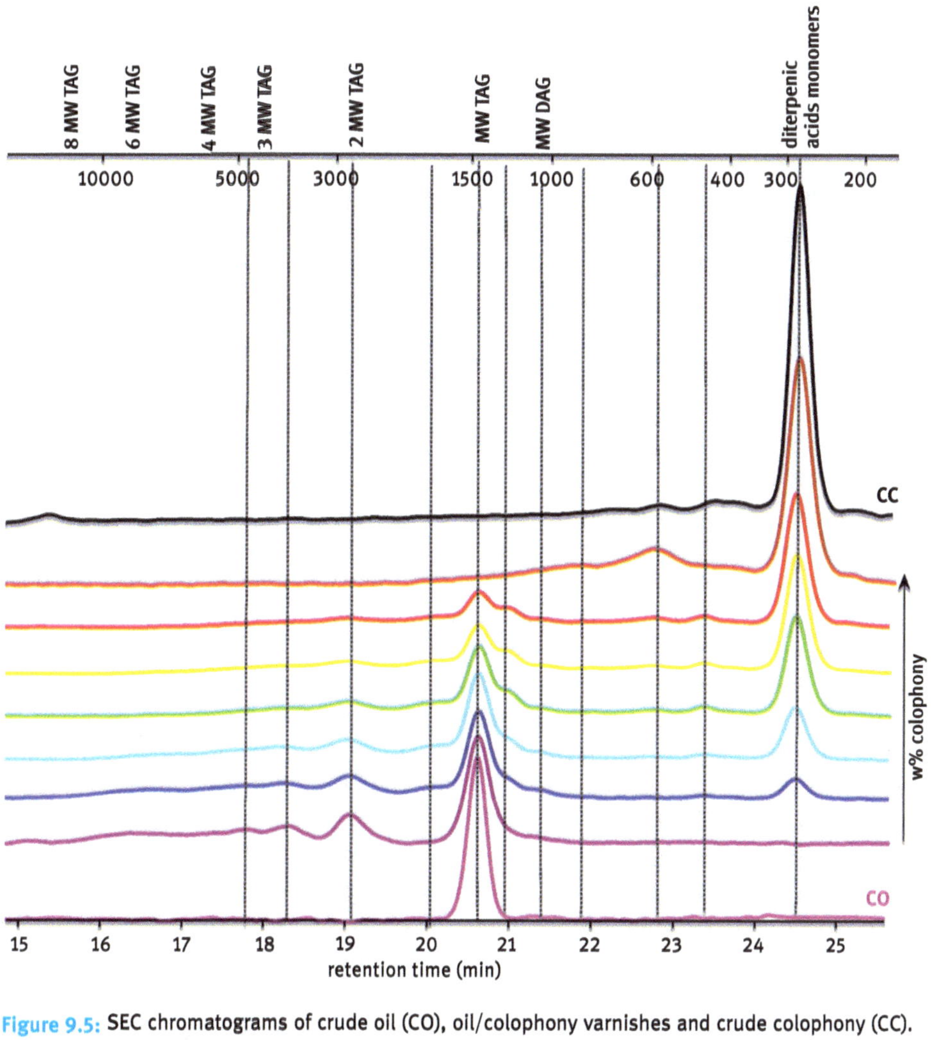

Figure 9.5: SEC chromatograms of crude oil (CO), oil/colophony varnishes and crude colophony (CC). From bottom to top the oil/colophony proportions of the varnishes are 100/0, 80/20, 66/33, 50/50, 42/58, 33/66 and 0/100, wt.%/wt.% (reprinted from [47] with permission from Elsevier).

Dammar and mastic resins were also studied using a high-performance size exclusion chromatography system (HP-SEC) coupled with RID and UV/Vis detectors, to model the polymerization and degradation of triterpenoid resins, which entails radical polymerization, cross-linking and condensation, oxidative modifications and shortening of side-chains [64].

9.4.4.2 Depolymerisation of cellulose and lignin in paper and wood samples

The evaluation of the degree of polymerization in dependence on the conservation parameters and in relation with other variables measured on the samples such as the extent of oxidation or the increase in acidity, or even the crystallinity index, can provide important results in the study of the factors affecting the degradation of paper and wood objects.

The evaluation of the degree of depolymerization in historical paper as a consequence of ageing in general, and conservation conditions in particular, is generally evaluated by either viscosimetry or SEC [65]. Samples need to be solubilized prior to injection; one of the most promising systems entail the dissolution in LiCl–*N,N*-dimethylacetamide (DMAc). In this case, the use of proper standards and a thorough interpretation of the results is needed to perform studies on the depolymerization of cellulose in paper, since the formation of the strong interactions between the solvent and cellulose influences its hydrodynamic volume and thus impede to rely on the commonly applied calibration curves. Note that the eluent shall be a LiCl–DMAc solution. The chromatographic set-up described in the literature comprises a 7.5 × 300 mm crosslinked polystyrene–divinylbenzene gel column and a differential refractometer detector [66, 67]. Derivatization and subsequent dissolution in tetrahydrofuran is another option [68].

In the case of wood samples, solubilization is much more difficult due to the poor solubility of both lignin, cellulose and hemicellulose in all solvents [69]. Samples are thus generally subjected to a derivatization reaction prior analysis. GPC analysis applied to lignin extracted from archaeological waterlogged wood has been used to determine the molecular weight distribution of lignin [70–72]. In the most recent among these studies [72], lignin was analysed after extraction and acetylation, while wood as a whole was solubilized in ionic liquids and both cellulosic and lignin fractions were analysed after acetylation and benzoylation. For lignin GPC, the wood powder was refluxed under nitrogen in a 0.1 M HCl dioxane-water solution (85:15 %); the suspension was centrifuged, and the supernatant added dropwise into a 0.01 M HCl aqueous solution, then kept at + 4 °C overnight for complete lignin precipitation. The precipitate was collected by centrifugation, washed with acidified distilled water (pH 2), freeze-dried, and finally acetylated with acetic anhydride: pyridine 1:1 v/v. Wood benzoylation and acetylation reactions were both performed on the wood powder in 1-allyl-3-methylimidazolium chloride, with benzoyl chloride and acetyl chloride, respectively. A sequence of an Agilent PL gel 5 µm, 500 Å and a PL gel 5 µm, 10^4 Å columns in series was used, and the solvent for sample dissolution and elution was tetrahydrofuran. The detection was performed with an ultraviolet diode array (UV) detector set at 240 or 280 nm for benzoylated and acetylated samples, respectively [72]. GPC provided data on the depolymerization of lignin related to the conservation state of wood in archaeological context.

9.4.4.3 Synthetic polymers as artists' materials and restorers' tools

Synthetic materials have been exploited by artists and restorers since their appearance on the market. The variety of formulations of synthetics resins in terms of composition and rheologic properties has allowed on the one hand their extensive use in paint tubes and sprays, plastic design materials and varnishing or coatings for artistic purposes; on the other hand, their adhesive properties have made them perfect as consolidants, adhesives and protective varnishes for conservation. Even if SEC and GPC are very well-known analytical tools employed by synthetic chemists to evaluate the behaviour of different polymeric resins upon ageing, their long term stability and, as a consequence, suitability as conservation or restoration materials have only been evaluated in few case studies [33].

9.5 Conclusions

In conclusion, liquid chromatography has been widely applied in Heritage Science to analyse organic and inorganic materials, by performing qualitative and semi-quantitative analyses.

For decades, the most used techniques where those based on reverse-phase HPLC with UV-Vis detection for the study of organic dyes and pigments, and those exploiting IC to detect inorganic salts or small organic acids in damage layers and patinas, mostly on stone building.

As more efficient stationary phases were introduced on the market, and mass spectrometry became a viable detector for liquid chromatography in terms of ease-of-use, robustness and availability in research laboratories, new applications started to be developed. These include for instance the analysis of lipid materials by RP-HPLC-ESI-MS/MS and proteomic applications, in the field of both paint binders and archaeological amorphous residues characterisation.

Size exclusion chromatography has been used mostly on reference materials, due to the relatively big sample size required; nonetheless, SEC and GPC applications allowed for studying in detail peculiar cases such as the degradation of paper and wood objects.

In the future, the applicability of liquid chromatography will increase for several reasons.

First, since most of the analyses rely on the comparison with reference materials prepared *ad hoc*, the number of applications of HPLC to lipid and protein analysis will increase thanks to the creation of dedicated databases of acyl glycerol and protein/peptides profiles (already available since a long time for the study of organic dyes and pigments), based on the analysis of reference mock-ups, also artificially aged, and of actual paintings and archaeological samples.

By a technical point of view, novel methods will be developed thanks to the continuous production of stationary phases useful for the simultaneous analysis of different classes of analytes. The application of two-dimensional liquid chromatography is also promising, as demonstrated by the paper by Pirok et al. [73] who developed a LC × LC method for the separation of authentic aged synthetic dyes, with ion-exchange chromatography in the first dimension and fast ion-pair liquid chromatography in the second.

Finally, the further development of different interfaces between liquid chromatography and mass spectrometric detection will also allow for increasing the number of detectable species, and thus the range of application of HPLC; this is particularly interesting for SEC and GPC analyses, which would benefit from a MS detection of the products.

References

[1] van Deemter JJ, Zuiderweg FJ, Klinkenberg A. Longitudinal diffusion and resistance to mass transfer as causes of nonideality in chromatography. Chem Eng Sci. 1956;5:271–89.

[2] Giorgi G. overview of mass spectrometric based techniques applied in the cultural heritage field. In: Colombini MP, Modugno F, editors. Organic mass spectrometry in art and archaeology. 2009.

[3] Mills JS, White R. Analytical methods. In: Mills JS, White, editors. The Organic Chemistry of Museum Objects. Elsevier. 1987:13–25.

[4] Striegel MF, Hill J. Thin-Layer chromatography for binding media analysis. Sci Tools Conserv. CA: Getty Conservation Institute. 1996;1–174.

[5] Schweppe H. Handbuch Der Naturfarbstoffe: vorkommen, Verwendung, Nachweis. Landsberg/Lech: Ecomed, 1993.

[6] Brosseau CL, Gambardella A, Casadio F, Grzywacz CM, Wouters J, Van Duyne RP. Ad-Hoc surface-enhanced raman spectroscopy methodologies for the detection of artist dyestuffs: Thin layer chromatography-surface enhanced Raman Spectroscopy and in situ on the fiber analysis. Anal Chem. 2009;81:3056–62.

[7] Cañamares MV, Reagan DA, Lombardi JR, Leona M. TLC-SERS of Mauve, the first synthetic dye. J Raman Spectrosc. 2014;45:1147–52.

[8] Lech K, Wilicka E, Witowska-Jarosz J, Jarosz M. Early synthetic dyes - A challenge for tandem mass spectrometry. J Mass Spectrom. 2013;48:141–7.

[9] Manhita A, Ferreira T, Candeias A, Barrocas Dias C. Extracting natural dyes from wool-an evaluation of extraction methods. Anal Bioanal Chem. 2011;400:1501–14.

[10] Sanyova J, Reisse J. Development of a mild method for the extraction of anthraquinones from their aluminum complexes in madder lakes prior to hplc analysis. J Cult Herit. 2006;7:229–35.

[11] Surowiec I, Quye A, Trojanowicz M. Liquid chromatography determination of natural dyes in extracts from historical scottish textiles excavated from peat bogs. J Chromatogr A. 2006;1112:209–17.

[12] Taujenis L, Olšauskaite V. Identification of main constituents of historical textile dyes by ultra performance liquid chromatography with photodiode array detection. Chemija. 2012;23:210–5.

[13] Troalen LG, Phillips AS, Peggie DA, Barran PE, Hulme AN. Historical textile dyeing with *genista tinctoria* L.: A comprehensive study by UPLC-MS/MS analysis. Anal Methods. 2014;6:8915–23.

[14] Valianou L, Karapanagiotis I, Chryssoulakis Y. Comparison of extraction methods for the analysis of natural dyes in historical textiles by high-performance liquid chromatography. Anal Bioanal Chem. 2009;395:2175–89.

[15] Vanden Berghe I, Gleba M, Mannering U. Towards the identification of dyestuffs in early iron age scandinavian peat bog textiles. J Archaeol Sci. 2009;36:1910–21.

[16] Zhang X, Laursen RA. Development of mild extraction methods for the analysis of natural dyes in textiles of historical interest using LC-Diode array detector-MS. Anal Chem. 2005;77:2022–5.

[17] van Bommel MR, Vanden Berghe I, Wallert AM, Boitelle R, Wouters J. High-performance liquid chromatography and non-destructive three-dimensional fluorescence analysis of early synthetic dyes. J Chromatogr A. 2007;1157:260–72.

[18] Saliu F, Degano I, Colombini MP. Identification of triacylglycerols in archaelogical organic residues by core-shell reversed phase liquid chromatography coupled to electrospray ionization-quadrupole-time of flight mass spectrometry. J Chromatogr A. 2014;1346:78–87.

[19] Kimpe K, Jacobs PA, Waelkens M. Mass spectrometric methods prove the use of beeswax and ruminant fat in late roman cooking pots. J Chromatogr A. 2002;968:151–60.

[20] Romanus K, Van Neer W, Marinova E, Verbeke K, Luypaerts A, Accardo S, et al. Brassicaceae seed oil identified as illuminant in Nilotic Shells from a First Millennium AD Coptic Church in Bawit, Egypt. Anal Bioanal Chem. 2008;390:783–93.

[21] Blanco-Zubiaguirre L, Ribechini E, Degano I, La Nasa J, Carrero JA, Iñañez J, et al. GC–MS and HPLC-ESI-QToF characterization of organic lipid residues from ceramic vessels used by basque whalers from 16th to seventeenth centuries. Microchem J. 2018;137:190–203.

[22] La Nasa J, Lucejko JJ, Humpf H-U, Ribechini E. Advancements in the chemical structures of ergot acyl glycerides by high performances liquid chromatography coupled with high resolution mass spectrometry. Microchem J. 2018;141:229–39.

[23] Lucejko JJ, La Nasa J, Porta F, Vanzetti A, Tanda G, Mangiaracina CF, et al. Long-Lasting ergot lipids as new biomarkers for assessing the presence of cereals and cereal products in archaeological vessels. Sci Rep. 2018;8:3935.

[24] Degano I, La Nasa J, Ghelardi E, Modugno F, Colombini MP. Model study of modern oil-based paint media by triacylglycerol profiling in positive and negative ionization modes. Talanta. 2016;161:62–70.

[25] La Nasa J, Degano I, Modugno F, Colombini MPMP. Alkyd paints in art: Characterization using integrated mass spectrometry. Anal Chim Acta. 2013;797:64–80.

[26] La Nasa J, Degano I, Modugno F, Colombini MPMP. Industrial Alkyd Resins: Characterization of Pentaerythritol and Phthalic acid esters using integrated mass spectrometry. Rapid Commun Mass Spectrom. 2015;29:225–37.

[27] Halpine SM. An improved dye and lake pigment analysis method for high-performance liquid chromatography and diode-array detector. Stud Conserv. 1996;41:76–94.

[28] Nowik W, Marcinowska R, Kusyk K, Cardon D, Trojanowicz M. High performance liquid chromatography of slightly soluble brominated indigoids from tyrian purple. J Chromatogr A. 2011;1218:1244–52.

[29] Restivo A, Degano I, Ribechini E, Colombini MPMP. Development and OPTIMISATION of an HPLC-DAD-ESI-Q-ToF method for the determination of phenolic acids and derivatives. PLoS ONE. 2014;9:e88762.

[30] Serrano A, Van Bommel M, Hallett J. Evaluation between ultrahigh pressure liquid chromatography and high-performance liquid chromatography analytical methods for characterizing natural dyestuffs. J Chromatogr A. 2013;1318:102–11.

[31] Surowiec I, Orska-Gawryś J, Biesaga M, Trojanowicz M, Hutta M, Halko R, et al. Identification of natural dyestuff in archeological coptic textiles by HPLC with fluorescence detection. Anal Lett. 2003;36:1211–29.

[32] Degano I, Ribechini E, Modugno F, Colombini MPP. Analytical methods for the characterization of organic dyes in artworks and in historical textiles. Appl Spectrosc Rev. 2009;44:363–410.

[33] Degano I, La Nasa J. Trends in high performance liquid chromatography for cultural heritage. Top Curr Chem. 2016;374:20.

[34] Wouters J, Grywacz C, Claro A. A comparative investigation of hydrolysis methods to analyze natural organic dyes by HPLC-PDA. Stud Conserv. 2011;56:231–49.

[35] Sanyova J. Mild extraction of dyes by hydrofluoric acid in routine analysis of historical paint micro-samples. Microchim Acta. 2008;162:361–70.

[36] Dallongeville S, Garnier N, Rolando C, Tokarski C. Proteins in art, archaeology, and paleontology: from detection to identification. Chem Rev. 2016;116:2–79.

[37] Vinciguerra R, De Chiaro A, Pucci P, Marino G, Birolo L. Proteomic strategies for cultural heritage: from bones to paintings. Microchem J. 2016;126:341–8.

[38] Lluveras-Tenorio A, Vinciguerra R, Galano E, Blaensdorf C, Emmerling E, Perla Colombini M, et al. GC/MS and proteomics to unravel the painting history of the Lost Giant Buddhas of Bāmiyān (Afghanistan). PLoS ONE. 2017;12:e0172990.

[39] Orsini S, Yadav A, Dilillo M, McDonnell LA, Bonaduce I. Characterization of degraded proteins in paintings using bottom-up proteomic approaches: new strategies for protein digestion and analysis of data. Anal Chem. 2018;90:6403–8.

[40] La Nasa J, Ghelardi E, Degano I, Modugno F, Colombini MP. Core shell stationary phases for a novel separation of triglycerides in plant oils by high performance liquid chromatography with electrospray-quadrupole-time of flight mass spectrometer. J Chromatogr A. 2013;1308:114–24.

[41] Nasa JL, Degano I, Brandolini L, Modugno F, Bonaduce I. A Novel HPLC-ESI-Q-ToF Approach for the determination of fatty acids and acylglycerols in food samples. Anal Chim Acta. 2018;1013:98–109.

[42] Byrdwell WC, Neff WE. Dual parallel electrospray ionization and atmospheric pressure chemical ionization mass spectrometry (MS), MS/MS and MS/MS/MS for the analysis of triacylglycerols and triacylglycerol oxidation products. Rapid Commun Mass Spectrom. 2002;16:300–19.

[43] Saliu F, Modugno F, Orlandi M, Colombini MP. HPLC-APCI-MS analysis of triacylglycerols (TAGs) in historical pharmaceutical ointments from the eighteenth century. Anal Bioanal Chem. 2011;401:1785–800.

[44] La Nasa J, Zanaboni M, Uldanck D, Degano I, Modugno F, Kutzke H, et al. Novel application of liquid chromatography/mass spectrometry for the characterization of drying oils in art: elucidation on the composition of original paint materials used by Edvard Munch (1863–1944). Anal Chim Acta. 2015;896:98–109.

[45] Stella EM, Bracci S, Iannaccone R, La Nasa J, Colombini MP. Violon. Céret by Pablo Picasso: The case of a lost painting. A methodological approach. J Cult Herit. 2018. 10.1016/j.culher.2018.05.012.

[46] La Nasa J, Degano I, Modugno F, Colombini MP. Effects of acetic acid vapour on the ageing of alkyd paint layers: Multi-analytical approach for the evaluation of the degradation processes. Polym Degrad Stab. 2014;105:257–64.

[47] Tirat S, Degano I, Echard J-P, Lattuati-Derieux A, Lluveras-Tenorio A, Marie A, et al. Historical linseed oil/colophony varnishes formulations: study of their molecular composition with micro-chemical chromatographic techniques. Microchem J. 2016;126:200–13.

[48] Gobbi G, Zappia G, Sabbioni C. Anion determination in damage layers of stone monuments. Atmos Environ. 1995;29:703–7.
[49] Sabbioni C, Ghedini N, Bonazza A. Organic anions in damage layers on monuments and buildings. Atmos Environ. 2003;37:1261–9.
[50] Bonazza A, Sabbioni C, Ghedini N. Quantitative data on carbon fractions in interpretation of black crusts and soiling on european built heritage. Atmos Environ. 2005;39:2607–18.
[51] Smolík J, Mašková L, Zíková N, Ondráčková L, Ondráček J. Deposition of suspended fine particulate matter in a library. Herit Sci. 2013;1:7.
[52] Ghedini N, Ozga I, Bonazza A, Dilillo M, Cachier H, Sabbioni C. Atmospheric aerosol monitoring as a strategy for the preventive conservation of urban monumental heritage: the florence baptistery. Atmos Environ. 2011;45:5979–87.
[53] Niklasson A, Johansson L-G, Svensson J-E. Atmospheric corrosion of historical organ pipes: influence of acetic and formic acid vapour and water leaching on lead. In: Metal 04: proceedings of the International Conference on Metals Conservation = actes de la conférence internationale sur la conservation des métaux, Canberra, Australia, 4–8 October 2004, 2004:273–80.
[54] Kontozova-Deutsch Deutsch F, Bencs L, Krata A, Van Grieken R, De Wael KV. Optimization of the ion chromatographic quantification of airborne fluoride, acetate and formate in the Metropolitan museum of art, New York. Talanta. 2011;37:105–13.
[55] Hodgkins RE, Grzywacz CM, Garrell RL. An improved ion chromatography method for analysis of acetic and formic acid vapours. e-Preservation Sci. 2011;8:74–80.
[56] Kontozova-Deutsch V, Krata A, Deutsch F, Bencs L, Van Grieken R. efficient separation of acetate and formate by ion chromatography: Application to air samples in a cultural heritage environment. Talanta. 2008;75:418–23.
[57] Franzoni E, Sassoni E, Graziani G. Brushing, Poultice or Immersion? the role of the application technique on the performance of a novel hydroxyapatite-based consolidating treatment for limestone. J Cult Herit. 2015;16:173–84.
[58] Tittarelli F, Moriconi G, Bonazza A. atmospheric deterioration of cement plaster in a building exposed to a urban environment. J Cult Herit. 2008;9:203–6.
[59] Verhaar G, van Bommel MR, Tennent NH. Weeping glass: the identification of ionic species on the surface of vessel glass using ion chromatography. In: Roemich H, Fair L, editors. ICOM-CC Glass and Ceramics working group interim meeting. Paris: International Council of Museums; 2016. p. 123–33.
[60] Colombini MP, Ceccarini A, Carmignani A. Ion chromatography characterization of polysaccharides in ancient wall paintings. J Chromatogr A. 2002;968:79–88.
[61] Duce C, Bramanti E, Ghezzi L, Colombini MP, Spepi A, Biagi S, et al. Interactions between inorganic pigments and proteinaceous binders in reference paint reconstructions. Dalt Trans. 2013;42:5945–6236.
[62] Duce C, Ghezzi L, Onor M, Bonaduce I, Colombini MP, Tine' MR, et al. Physico-Chemical characterization of protein-pigment interactions in tempera paint reconstructions: Casein/Cinnabar and Albumin/Cinnabar. Anal Bioanal Chem. 2012;402:2183–93.
[63] Van der Berg JDJ, Carlyle L, Holcapek M, Boon JJ, Vermist ND. Effects of traditional processing methods of linseed oil on the composition of its triacylglycerols. J Sep Sci. 2004;27:181–99.
[64] Theodorakopoulos C, Boon JJ. A high performance size exclusion chromatographic study on the depth-dependent gradient in the molecular weight of aged triterpenoid varnish films. Prog Org Coatings. 2011;72:778–83.
[65] Kolar J, Strlič M. Ageing and stabilisation of paper. Ljubljana: National and University Library; 2004.

[66] Strlič M, Kolar J, Žigon M, Pihlar B. Evaluation of size-exclusion chromatography and viscometry for the determination of molecular masses of oxidised cellulose. J Chromatogr A. 1998;805:93–9.
[67] Strlič M, Kolar J. Size exclusion chromatography of cellulose in LiCl/N,N-dimethylacetamide. J Biochem Biophys Methods. 2003;56:265–79.
[68] Stol R, Pedersoli JL, Poppe H, Kok WT. Application of size exclusion electrochromatography to the microanalytical determination of the molecular mass distribution of celluloses from objects of cultural and historical value. Anal Chem. 2002;74:2314–20.
[69] Lucejko JJ, Modugno F, Ribechini E, Tamburini D, Colombini MP. Analytical instrumental techniques to study Archaeologicalwood Degradation. Appl Spectrosc Rev. 2015;50:584–625.
[70] Salanti A, Zoia L, Tolppa EL, Giachi G, Orlandi M. Characterization of waterlogged wood by NMR and GPC Techniques. Microchem J. 2010;95:345–52.
[71] Colombini MP, Orlandi M, Modugno F, Tolppa EL, Sardelli M, Zoia L, et al. Archaeological wood characterisation by PY/GC/MS, GC/ MS,NMR and GPC techniques. Microchem J. 2007;85:164–73.
[72] Zoia L, Tamburini D, Orlandi M, Łucejko JJ, Salanti A, Tolppa EL, et al. chemical characterisation of the whole plant cell wall of archaeological wood: an integrated approach. anal bioanal chem. 2017;409:4233–45.
[73] Pirok BWJ, Knip J, van Bommel MR, Schoenmakers PJ. Characterization of synthetic dyes by comprehensive two-dimensional liquid chromatography combining ion-exchange chromatography and fast ion-pair reversed-phase chromatography. J Chromatogr A. 2016;1436:141–6.

Stepanka Kuckova, Pavel Cejnar, Jiri Santrucek and Radovan Hynek

10 Characterization of proteins in cultural heritage using MALDI–TOF and LC–MS/MS mass spectrometric techniques

Abstract: Identification of proteinaceous materials in artworks is of high interest to restorers-conservators, art historians and archaeologists, because it helps to shed light on the used painting techniques, to attribute unknown artworks, to make conclusions on prehistoric diets, etc. The chapter is devoted to the mass spectrometry instrumentation, evaluation of obtained data and it is showing several examples of the application of matrix-assisted laser desorption/ionization–time-of-flight and liquid chromatography–electrospray ionization–quadrupole–time-of-flight mass spectrometers on cultural heritage samples.

Keywords: MALDI-TOF MS, LC-MS/MS, mass spectrometry instrumentation, statistical analysis, proteinaceous and non-proteinaceous materials, cultural objects

10.1 Introduction to mass spectrometry of proteins

Identification of proteinaceous materials in artworks is of high interest to restorers-conservators, art historians and archaeologists, because it helps to shed light on the used painting techniques, to attribute unknown artworks, to make conclusions on prehistoric diets, etc. This chapter is devoted to usage of mass spectrometry (MS) techniques and is based on our experience with proteins and analyses of natural organic dyes of cultural heritage objects.

Mass spectrometric analyses are based on determination of molecular weights of analysed compounds. These methods are very straightforward and unambiguous because molecular weight is one of the basic and easily understandable characteristics of a molecule. Basically, there are two ways of proteins analysis by MS. The first one, very rare, is based on determination of molecular weights of intact proteins. A much more common way of protein analysis is to digest a protein polypeptide chain specifically by proteolytic enzyme, typically by trypsin, which cleaves peptide bonds with high specificity behind residues of arginine and lysine if they are not followed by proline. Thus, each protein is cleaved into a set of specific peptide fragments with characteristic molecular weights in the range of 700–4,000 Da (typically). These weights can be

This article has previously been published in the journal *Physical Sciences Reviews*. Please cite as: Kuckova, S., Cejnar, P., Santrucek, J, Hynek, R. Characterisation of Proteins in Cultural Heritage Using MALDI-TOF and LC-MS/MS Mass Spectrometric Techniques. *Physical Sciences Reviews* [Online] **2019**, 4. DOI: 10.1515/psr-2018-0011.

https://doi.org/10.1515/9783110457537-010

determined with high accuracy. The set of specific peptide fragments characterizes each protein similar to how a fingerprint does an individual person. A protein then can be identified by comparing experimentally obtained molecular weights with theoretical weights derived from protein sequence stored in sequence databases.

Nowadays, a more common, modern and effective way of protein identification is based on determination of amino acid sequence of peptides by inducing their fragmentation. The peptides are identified with the aid of sequence database and used to produce a list of proteins present in the sample. In this way up to several thousand proteins can be identified in one sample.

All the methods of proteins identification described above were developed for work with liquid samples in early 1990s. However, in 2004, an innovative approach has been proposed by Hynek and Kuckova [1]. This method led to a revolution in the analysis of proteins contained in insoluble materials, and demonstrated its great potential for the analysis of proteins in artworks. The technique is actually based on peptide mass mapping. This does not require problematic and timely extraction of proteins from artworks which, in some cases, can hardly be possible. Instead, a tiny sample, e. g. a small piece of colour layer of artwork, is submersed into a drop of solution containing trypsin. This enzyme has a relatively small molecule, which is able to penetrate insoluble materials and cleave the entrapped polypeptide chains of proteins. The small specific peptide fragments are released into the solution unlike the large intact proteins. Then the solution can be analysed by an appropriate MS method (Figure 10.1).

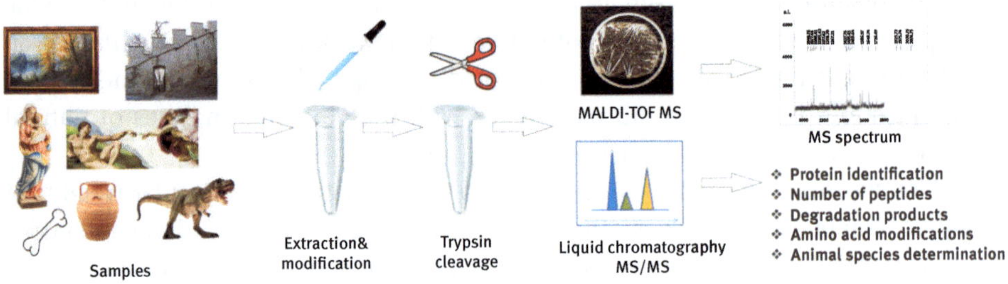

Figure 10.1: Proteomic approach applied on cultural heritage objects.

Identification of proteins in the cultural heritage objects was always considered to be one of the most important, because revealing of the protein composition provides information on painting techniques, explains unusual mortar properties, helps recognizing animal species, etc. Nowadays, mass spectrometric techniques like MALDI (matrix-assisted laser desorption/ionization) and LC–MS/MS (liquid chromatography coupled to tandem MS) are used most often for the protein identification. Several other protein identification methods exist, however their use is somewhat limited [2, 3]. The use of Fourier transform infrared spectroscopy is limited to samples with relatively high-protein concentration [4]. Gas chromatography–mass spectrometry (GC–MS), based on observing characteristic ratio of amino acids released after hydrolysis, is generally not reliable for identification of complex mixtures of proteins [4, 5]. Enzyme-linked immunosorbent assay is very sensitive, but specific antibody usage is necessary for each targeted protein [4, 6]. Other methods like GALDI (graphite-assisted laser desorption/ionization mass spectrometry) [7, 8], immuno-surface-enhanced Raman scattering [9, 10], time-of-flight (TOF)–secondary ion mass spectrometry [11] and chemiluminescent imaging techniques are used in specific cases [12].

10.2 Mass spectrometry instrumentation

Basic working principle of a MS is to transform sample molecules into the form of ions in gas phase, separate them according to their m/z and then detect and measure their abundance by detector. Every MS consists of several parts: a sample inlet for introduction of analysed compounds (capillary from GC/LC, a sample plate with solid samples); an ion source, which converts the analysed molecules into the form of ions (either positively or negatively charged); one or several analysers to separate ions according to their m/z ratios; a detector producing a signal proportional to the amount of ions falling on it; and a data processing system producing mass spectra in the suitable form for further analysis. It is also possible to fragment the separated ions and measure m/z ratio of fragments; so called tandem mass spectrum (MS/MS) is obtained in such case, which allows for determining amino acid sequence of individual peptides.

10.2.1 Ion sources

For ionization of non-volatile biomolecules (peptides, proteins, polysaccharides) MALDI and electrospray ionization (ESI) are used. Introduction of these ionization techniques in the 1980s started a new era of MS analysis culminating in the modern proteomics. Both ESI and MALDI are considered to be soft ionization methods producing mainly intact protonated pseudomolecular ions $(M+H)^{n+}$.

Matrix plays a key role in the MALDI ionization process [13]. Commonly used matrices are weak acids, e. g. α-cyano-4-hydroxycinnamic acid (HCCA), 2,5-dihydroxybenzoic acid (DHB) or 3,5-dimethoxy-4-hydroxycinnamic acid (sinapinic acid). DHB is used in analysis of peptides, proteins and synthetic polymers but it is of particular use for analysis of glycans and glycoproteins. HCCA is typically used for measurement of peptides and small proteins (<20 kDa) whereas sinapinic acid is used for larger proteins.

In any case, the matrix molecules must have a strong absorption at the wavelength of the laser used for ionization. UV lasers have been used with pulse frequency up to 10 kHz. Thanks to that, a spectrum of a sample can be obtained within few seconds. During sample preparation, a microlitre of a sample is mixed with several microlitres of the matrix solution. The sample-matrix mixture is deposited on a target (usually a steel plate). After air drying process, a spot containing crystals of matrix and sample molecules remains on the target. Upon irradiation with laser beam (in deep vacuum), matrix molecules absorb laser-emitted photons, the spot is warmed up, and the irradiated matrix and sample molecules are desorbed. They create a cloud containing neutral matrix molecules, photonically ionized matrix molecules and the molecules of the sample. Great molar excess of matrix over sample protects sample from any harm caused by laser. In the dense plume matrix and sample molecules undergo multiple collisions. It is assumed that during collisions protons are transferred from matrix molecules to sample molecules thus creating protonated pseudomolecular ions which are subsequently accelerated in high-voltage direct electric field for the mass measurement. MALDI typically provides monocharged protonated ions $(M+H)^+$.

Whereas MALDI produces ions from the solid state samples, ESI can be used for liquids [14]. Compounds to be analysed are dissolved in a mixture of solvents (usually mixtures of water, acetonitrile, methanol or isopropanol). Also small amount of acid (often 0.1 % solution of formic acid) is added to promote ionization if the measurement is performed in the positive mode; in the negative mode a base is added (trimethylamine, piperidine). Non-volatile buffers should be avoided as they would crystalize inside the ESI source. Sample solution is pumped through a capillary. A several kilovolts potential difference between the capillary tip and counter-electrode produces a very intense field which causes the solution to spray in the form of fine charged droplets. The ESI source is heated to 150–200 °C so the liquid from droplets evaporates, the droplets shrink and surface charge increases. When surface charge limit is reached the droplets undergo fissions to create smaller droplets. As the evaporation continues, the droplets undergo further fissions until there is only one sample molecule in the droplet. At this time the rest of the solution is evaporated, and the charges carried by the droplet are transferred to the analyte. This, so called charge-residue model of ionization, is valid for proteins [15]. Alternative theory, called ion evaporation model, proposes that, as the droplet shrinks to a certain limit and reaches high enough surface intensity of electric field, the solvated ions can escape from the surface of the droplet and carry charges off with them; this model is probably involved in the ionization of small molecules (peptides) [15]. The characteristic property of ESI is, in contrast with MALDI, creation of multiply charged (protonated) ions. In ESI, peptides carry two to four protons, and proteins can carry up to several tens of protons. ESI is a softer technique compared to MALDI, it allows for direct measurement of noncovalent protein complexes (native ESI). The definite advantage of ESI is the fact that its response is proportional to the concentration of analytes in the solution; therefore, ESI is routinely used for quantitative analysis unlike MALDI where the measured ion intensity depends on many variables which are difficult to control.

10.2.2 Mass analysers

When the analytes are ionized and transferred to gas phase, they can be separated according to their *m*/*z* ratio in an analyser. There are five types of mass analysers used in protein MS–TOF, quadrupole, quadrupole ion trap (QIT), orbitrap and Fourier transform ion cyclotron resonance (FTICR). One of their most important characteristics is resolution, i. e. the ability to separate ions with close *m*/*z* ratios. TOF analyser, orbitrap and FTICR exhibit high resolution whereas quadrupole and quadrupole ion trap demonstrate low resolution.

Time-of-flight (TOF) analyser separates ions by their speeds. Ions are accelerated by electric field outside the analyser, and then they drift in field-free region of the flight tube (Figure 10.2) [16]. All ions of the same charge acquire same kinetic energy, however they differ in their masses and consequently in their speeds. *M*/*z* ratios are determined from the flight times of ions, i. e. how long the ion flies from ion source to the detector. TOF analysers can operate in two modes – linear mode and reflector mode. In linear mode electric field accelerates ions after ionization and directs them to the flight tube through which they get to a detector. This mode is generally used for the analysis of large molecules (proteins). Compared to the reflector mode it has lower resolution, lower accuracy of measured *m*/*z* ratios but virtually no upper mass limit.

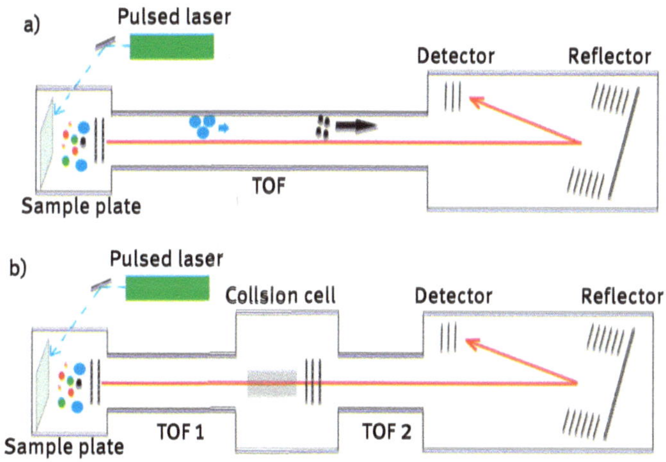

Figure 10.2: Schematics of MALDI–TOF and MALDI–TOF/TOF instruments. (a) After ionization ions obtain high-kinetic energies and are separated according to their speeds in the field-free region of the flight tube. The reflector compensates for spread in kinetic energies of ions with the same *m*/*z* ratio. (b) In TOF/TOF instruments, a collision cell is placed between two TOF analysers. Ions of particular *m*/*z* ratio are selected by the first TOF analyser, fragmented in the collision cell and the fragment ions separated by the second TOF analyser.

In reflector mode the trajectory of ions in the flight tube is more complicated. The ions are accelerated and directed to the flight tube similarly to linear mode. However, at the far end of the flight tube a reflector (ion mirror) is placed. The reflector is an assembly of circular electrodes with increasing electric potential. As the ions fly through the reflector they are gradually slowed down, eventually stopped and then accelerated at a slight angle back to the flight tube where they hit detector. The purpose of the reflector is to equalize kinetic energies of the ions with the same m/z ratio and thus increase resolution of acquired mass spectra and accuracy of measured masses. MALDI–TOF represents an ideal match of two techniques which benefits from both the pulse nature of MALDI source and the ability of TOF analyser to separate incoming bundles of ions in time and to send them to the detector. It is possible to connect two TOF analysers in a series thus enabling acquisition of tandem mass spectra (Section 10.2.4) [17].

Quadrupole mass analyser [18] consists of four parallel metal rods (Figure 10.3(a)). When an ion is introduced to the quadrupole it starts oscillating between the rods under the influence of combined static and alternating electric fields. For particular values of these electric parameters only ions of certain m/z ratio have stable trajectory through the quadrupole. All other ions hit one of the rods, discharge and never reach the detector. In order to obtain a mass spectrum, electric parameters are progressively changed (scanned) while keeping their ratio constant, thus ions of different m/z ratio can successively pass through the quadrupole. Quadrupoles are widely used in the modern MSs despite their low resolution. When only alternating field is applied no separation by m/z ratio occurs, and all ions can pass through the quadrupole being focused in its centre. This is useful for ion passage through MS. Quadrupoles can also serve as collision cells when filled with collision gas (nitrogen, argon). Ions passing through the quadrupole with only alternating voltage on collide with molecules of the

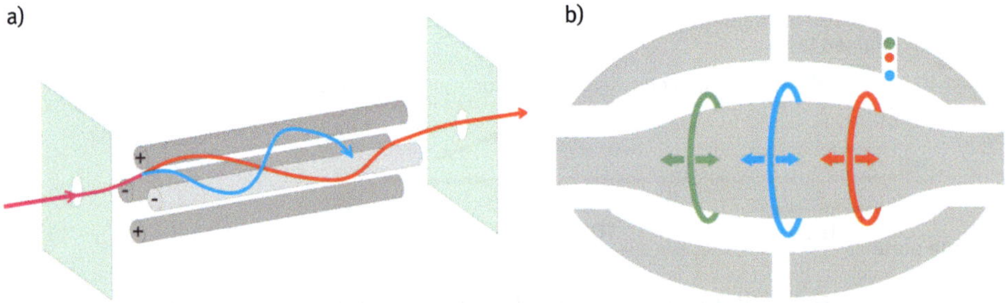

Figure 10.3: Schematics of quadrupole and orbitrap. (a) Quadrupole consists of metal rods. Ions flying through the quadrupole oscillate under the influence of electric fields. Only ions of certain m/z ratio have stable trajectory for a given value of electric fields. (b) Ions enter orbitrap through the opening in outer electrode. Under the influence of an electric field they start to move around and along the central electrode creating concentric rings. Motion of ions induce image current in the outer electrode which is recorded and used for determination of m/z ratios of ions by Fourier transform.

collision gas, which increases their inner energy until the bonds in the molecules start to break creating fragments which are useful for structural studies.

Quadrupole ion trap (QIT) is another type of analyser with low resolution [19], although analysers of this type are very useful for detailed structural studies. QIT consists of a ring electrode and two caps electrodes. Ions can be trapped inside QIT by a combination of alternating and direct electric fields. Helium gas present in QIT takes away the excess energy from the trapped ions by collisions. In order to obtain a mass spectrum, the trapped ions need to be expelled from the trap. This is achieved by increasing amplitude of the alternating field voltage applied to the ring electrode while keeping its frequency constant. The ions then leave the trap through the holes in the cap electrodes and impinge on the external detector.

Orbitrap is the youngest member of mass analysers' family being commercially introduced in 2005 [19]. Since its introduction it has become widely used thanks to its high resolution, high-mass accuracy, affordable price and low operational costs. Orbitrap consists of an inner spindle-like electrode and an outer cylindrical electrode which is cut in halves, which are electrically insulated from each other (Figure 10.3(b)). Orbitrap is an electrostatic trap with electric potential of nearly −3 kV (for analysis of positive ions) applied to the central electrode. Ions are injected into the orbitrap tangentially through a small opening in the outer electrode. The applied electric field causes the ions to oscillate around and along the central electrode. The frequency of oscillations along the central electrode depends on m/z ratio. The oscillating ions create electric current in the outer electrode, and the mass spectrum is obtained by Fourier transform of the recorded current. The longer the current recording period the better is the resolution. For a given scanning time the resolution is inversely proportional to $\sqrt{m/z}$ thus decreasing when mass of ions increases.

The best mass resolution is offered by FTICR [20] type of analyser. Although the extreme resolution is tempting, its high-acquisition price, operational costs and complicated operation procedures are primary reasons of why FT-ICR is not widely used. The ICR cell itself consists of a pair of excitation electrodes and a pair of detection electrodes. The cell is placed inside a superconducting magnet with the magnetic field intensity of several Tesla. The magnetic field causes ions to move inside the cell along circular trajectories near the centre of the cell with the unique frequencies which depend on m/z ratio of ions. In order to carry out detection of ions present in the cell the voltage on the excitation electrodes is set to oscillate at various frequencies. The radius of the circular motion of excited ions increases, and they fly in tight packets along the detecting electrodes creating current in them. Fourier transform is then applied to deconvolve the resulting signal. Similar to orbitrap, the longer is the measurement of the induced current the higher is the resolution of obtained spectra. Compared to orbitrap, the decrease of resolution with increasing m/z ratio is faster since the resolution is inversely proportional to the increasing m/z ratio.

10.2.3 Detectors

Finally, the separated ions have to be detected to provide a measurable signal [21]. Orbitrap and FT-ICR analysers function as detectors because they record image current created by moving ions which is then converted into mass spectra by Fourier transform. Other types of analysers require external detectors. The most common detectors are electron multipliers either with discrete or continuous dynodes. Multipliers are made of material which can easily give off electrons. When the incoming ion hits the conversion dynode several electrons are released and directed by electric potential to another dynode where each electron knocks out a few electrons. Finally, all electrons reach the anode. A special form of an electron multiplier is the multichannel plate detector which is made of a flat plate with a dense array of tubes (channels) that pass through the plate at a slight angle. Each channel represents a small electron multiplier.

10.2.4 Peptide fragmentation

As mentioned above, MS does not only provide molecular weights, but can also aid in structural analysis when molecular ions are fragmented and the spectrum of fragments is collected. Knowing how molecules dissociate provides clues to their structure [22]. The most common method for fragmenting ions is collision-induced dissociation (CID). As the name suggests, the dissociation is caused by collisions of ions with molecules of inert gas (nitrogen, argon). Multiple collisions increase the inner energy of a molecule till the weakest bonds start breaking. The extent of the molecule dissociation can be influenced by the energy of collisions.

From the MS/MS spectra of molecules with known structure it is possible to find the rules for certain classes of molecules. These rules then help to interpret MS/MS spectra of unknown molecules. For example, it is known that CID induces fragmentation of peptides at their peptides bonds with minimal dissociation of amino acids side chains. The fragmentation is fairly random and produces a series of ions from which it is possible to “read” the peptide amino acid sequence. Usually a sequence is not recognized directly from MS/MS spectrum (*de novo* sequencing). Instead, it is determined by matching measured spectrum to theoretical spectrum during protein sequence database search [23].

Electron transfer/capture dissociation (ETD, ECD) [24] is a less common method of ion fragmentation. The dissociation is caused by transfer of electron from the anion radical (ETD) or by capture of electron from a heated cathode (ECD). In proteomics these methods are especially useful for study of labile posttranslational modifications like phosphorylation.

Although it is possible to fragment all ions with various m/z ratios at once, it is much more common to select precursor ions to be fragmented first, then to perform the actual fragmentation and finally to acquire the spectra of fragment (product) ions. From the instrumental point of view these steps can all be performed in one analyser, e. g. QIT. The sequence of the steps in acquiring MS/MS spectrum is as follows: first all ions are expelled from the trap but the ions of interest which are fragmented by collisions with helium gas molecules, and then the fragments are analysed as described above. Alternatively it is possible to isolate and fragment one of the fragment ions to obtain MS^3 spectrum. In this way it is possible to carry out detailed structural studies of molecules and to distinguish molecules with similar structures. Another way of performing MS/MS measurement is to connect several analysers in series where every step of the MS/MS analysis is performed by a different analyser. A classical setup represents triple quadrupole instruments where the first quadrupole separates precursors, the second one is filled with the collision gas and acts as collision cell and the third one separates fragments according to their m/z ratio. More novel setup is based on connection of two quadrupoles with TOF analyser in Q-TOF instruments. Replacement of the third quadrupole with TOF analyser allows for acquiring both MS and MS/MS spectra with high accuracy and resolution. Although orbitrap is similar to QIT trapping device, the MS/MS fragmentation is not carried out inside it; the precursors are fragmented outside in a linear ion trap or a quadrupole, and then the fragments are sent into orbitrap for acquisition of high-resolution MS/MS spectra.

10.3 Identification of proteins and statistical analysis of complex samples

Depending on the type of acquired data there are two ways of identifying proteins [23], and both approaches use databases of protein or nucleic acid sequences. There are many regularly updated sequence databases available on the internet (e. g. Swissprot, NCBInr). Since genomes of many species have been sequenced by now, complete proteomes are available for download and can be used for protein identification.

When only masses of peptides are measured, peptide mass fingerprinting (PMF) is employed. In PMF, it is assumed that the set of peptides obtained after protease digestion is unique for each protein, thus it can be used for protein identification. Before actual identification a few parameters are set such as the protease used for digestion, possible modifications of amino acids (e. g. alkylation of cysteines, phosphorylation, oxidation of methionines), assumed uncertainties of peptide masses, etc. It is also possible to select biological taxon if multispecies database is to be searched. During the search, sequences of protein are digested *in silico*, masses of resulting peptides calculated and compared to those collected in experiment. The result is the list of proteins potentially present in the analysed sample. Each protein found is given an identification reliability score, which must exceed certain threshold for the identification to be trustworthy. PMF can identify pure proteins or mixtures containing up to three proteins.

For samples with tandem mass spectra acquired (MS/MS, see Section 10.2.4), more sophisticated algorithms are required. As in the case of PMF, sequences in the database are digested *in silico*, masses of peptides calculated and compared to experimentally obtained precursor masses. When a match is found, a theoretical fragmentation spectrum is created and compared to the measured MS/MS spectrum. As well, the reliability score is calculated. It is important to note that MS/MS spectra identify peptides only. Thus, at the last step the identified peptides are used to produce a list of possible proteins present in the sample. In general, at least two independent peptides are required for a reliable identification of a protein. When an identified peptide is unique, particular protein recognition is straightforward; however, some peptides might be present in multiple proteins due to the fact that homologous proteins from the same organism (or closely related organisms) have similar sequences.

Statistical analysis is applied afterwards to identify sources of variance in collected spectra and/or to find differences between tested sample groups. For this purpose, the experiment conditions should be set to eliminate as much variation obtained from selected sample acquisition method as possible including any differences in used extraction protocols, enzyme cleavage conditions, storage conditions and suppressing any processing delays. In practice though, the samples are often processed in identical replicates to identify the effect of variation for the same group of samples and to focus on identification of variation among groups only.

Before further analysis, the whole preprocessed (smoothed) spectra or the lists of peaks identified in spectra are standardized (normalized) in order to suppress differences caused by sample weight, volume, age and other factors. Normalization can be performed in a number of ways. The most basic way is normalization of the spectrum intensities in respect to integral spectrum mass (integral of intensity over a given m/z interval). This approach is also known as normalization by total ion count, and is used as a common standard.

Most variation of the data is usually determined by several distinct peaks spread over a wide m/z range. In various spectra, intensities of these peaks may vary from high to very low. To identify these peaks, high-dimensional exploratory data analysis like principal components analysis (PCA) or factor analysis is employed. Such techniques transform original spectrum samples into points in multidimensional space, where each dimension is associated to a different source of variance. The so called component loadings or factor loadings then identify sensitive m/z positions associated with particular dimension. When samples are properly treated, mass spectra of replicated samples are visualized as points located close together, while the spectra of samples treated under different conditions which change their protein composition would generally form a separate group. In the field of cultural heritage, the technique is used for example for distinguishing of animal glues [25] or to study changes due to ageing and the influence of organic/inorganic components on protein identification [26]. When the samples are not separated into distinct groups according to the main sources of variance, other statistical techniques can be utilized for identification and confirmation of changes among particular sample groups. These techniques include two-sample statistical *t*-tests, analysis of variance or discriminant analysis.

10.4 Protein binders identification

Identification of protein composition is required in many fields of cultural heritage like archaeology, art restoration, polychrome sculptures and historical mortars. Protein analyses can be performed with fragments, powders and cross-sections taken from cultural objects. Particularly important is the possibility to analyse cross-sections of multi-layered samples, although this is still quite rare in the field of organic materials analyses of cultural heritage objects. Cross-sections often contain valuable samples that are impossible to re-acquire again, and the only possibility to do the analyses is by using archived samples. Moreover, in combination with application of fluorescent dyes specific for proteins (e. g. Sypro Ruby), direct analysis of cross-sections allows for simultaneous identification and localization of protein binders in particular layer [27].

Both mass spectrometric methods (MALDI–TOF and LC–MS/MS) can help solving the basic problem of analytical methods (e. g. GC and LC [28, 29]) used for protein binders identification that are based on observing of typical ratios of amino acids characteristic for pure proteinaceous binders – they allow for identifying binders but not their mixtures which are common among artwork samples. MALDI–TOF can determine the mixtures only in some cases while the more sophisticated LC–MS/MS allows for this unambiguously (including their rich mixtures).

Nowadays, we can find a lot of different ways of sample pretreatments and clean-up strategies. One of the most popular samples pretreatment is the extraction of proteins before enzymatic digestion by agents that differ according to the analysed samples. For example in palaeontology there are many demineralizing and extraction protocols. Schroeter et al. [30] compared 22 unique sub-extractions (after using different extraction solutions) of chicken bone as well as and compare their total protein yield, diversity and coverage of bone proteins. After the extraction (using, e. g. NH_3, NH_4HCO_3) the proteins can be denatured (e. g. by 6 M guanidine, 6 M urea), the disulphide bonds in protein can be reduced by dithiothreitol and then to prevent the re-creating of the disulphide bonds the alkylation reaction is carried out with iodoacetamide [31]. The final extract is subjected to enzymatic digestion that can be accelerated using microwave oven (15 min digestion at 700 W instead of several hours for the classic procedure) similar to how it is done for analyses of paintings or wall paints [31, 32]. The newest sample pretreatment is removing of post-translation modification (e. g. glycosylation). This deglycosylation step by a specific enzyme, *N*-glycosidase F, preceding the proteolytic digestion is used for improving the quality and reliability of identification of egg proteins [33, 34]. Finally, it is possible to modify surface of standard stainless steel MALDI target e. g. by coating of colloidal graphite (GALDI) that increases (up to five times) the MS signal [35].

Our experience shows that high quality results can be obtained using following simple procedure. Before the analysis the samples should be immersed in 50 mM NH_4HCO_3 solution containing 10 µg/mL of trypsin: sample powders and fragments are to be immersed in trypsin solution in Eppendorf microtubes, and cross-sections samples are to be covered by a drop of this solution (Figure 10.4). Immersed samples should be left for 2 h at room temperature for cleaving. After the trypsin digestion, the solution is purified on reverse phase ZipTip (C18): after equilibrating, binding and washing steps the resulting peptides are desorbed from the stationary phase. The final solution is used for mass spectrometry analyses.

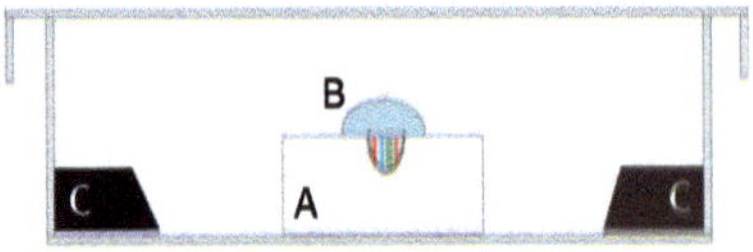

Figure 10.4: Scheme of a cross-section covered with a drop of trypsin solution in a Petri dish. (A) Cross-section of sample, (B) trypsin solution and (C) wet papers [27].

MALDI–TOF mass spectrometry has first been applied in the field of cultural heritage in 2004, when trypsin was applied for protein digestion to fragments of artworks for identifying proteinaceous binders [1]. At this time it was possible to only identify protein binders by using libraries of reference materials (egg, milk, animal glues, blood, etc.). Today, another type of mass spectrometry, LC–MS/MS, is more popular for the protein analyses. Publicly available PCA software for MALDI–TOF was also developed providing an easy and efficient way for mass spectrum analyses [26] making the MALDI–TOF a powerful analysis tool.

Evaluation of MALDI–TOF spectra by PCA allows for detecting small differences in spectra that are usually hard to be seen with the naked eye. Although it is was possible to distinguish individual protein binders before, in some cases this type of evaluation allows one to distinguish materials containing milk proteins (e. g. milk, curd, whey and casein; Figure 10.5) and from different but very similar animal species (e. g. hen, duck, goose and quail eggs) which helps getting more detailed information on local materials usage by different artistic workshops in the past and verify the hypothesis that different types of eggs could be used in Old Masters' recipes of materials [36].

10.4.1 Analyses of historical mortars

Analysis of historical mortars and plasters is important for revealing old technological approaches, understanding their unusual properties and, as the result, choosing

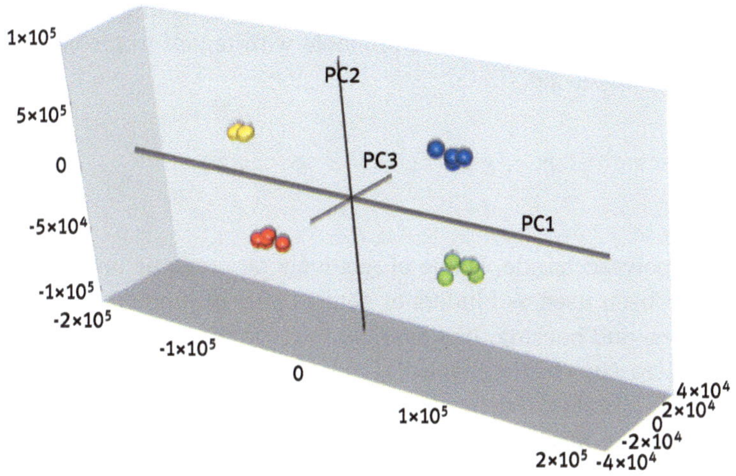

Figure 10.5: Results of PCA analysis of milk products used as additives into historical mortars: red pellets – casein, green – milk, blue – whey and yellow – curd.

a proper technological process for restoration. Since ancient times, highly proteinaceous materials were added to mortars and plasters in order to enhance their physical properties – e. g. air entrainment, compressibility, hardness, water resistance, etc. [37, 38]. The analyses are complicated and time consuming due to extremely low amounts of protein additives, aggressive processes during hardening of fresh mortar, and unpredictable changes of the materials caused by their ageing.

The analysis of Gothic mortar taken from remains of a bridge pier in Roudnice nad Labem, the third oldest stone bridge in Bohemia (1333), is an example of successful identification of protein additive by MALDI–TOF–MS [39]. The presence of milk proteins was confirmed by LC–ESI–Q-TOF, where beta-casein was identified by two peptides (Table 10.1).

Table 10.1: Results from LC–ESI–Q-TOF analysis of the bridge mortar where the milk protein was identified by two peptides from beta-casein. The peptides were not oxidized or deamidated.

Accession	Protein	Sequence	Range
CASB_BUBBU	Beta-casein	K.AVPYPQR.D	192–198
CASB_BUBBU	Beta-casein	R.GPFPIIV.-	218–224

The measured MS/MS spectra were searched for through entire database of bony vertebrates. According to the search engine, the best match was found with beta-casein coming from Asian water buffalo (*Bubalus bubalis*, BUBBU), however, the same sequences can be found in bovine beta-casein (but not in goat or sheep). The presence of bovine beta-casein would be more probable with regard to the central European location of the bridge.

10.4.2 Identification of starch plant origin by proteins connected with its metabolism

Starch, a plant storage polysaccharide, is one of relatively rare organic material in artworks. Starches have been used as binders or components of binders, fillers of pigments in colour layers, and backing. It is possible to recognize the starch origin with the aid of optical microscopy by the shape of starch granules which is specific for each plant. However, it is possible to detect and identify them in very low amounts simultaneously with protein materials contained in samples using LC–MS/MS, because although starch is a polysaccharide, its grains contain proteins that are connected to the production or degradation of polysaccharide chains. In bachelor thesis of Svajcrova the most abundant proteins are described that were found by LC–MS/MS in starches from wheat, corn, rice and potatoes [40]. Although the biological functions of the detected proteins are very similar, they differ in their amino acid sequences and can be attributed to the plant of the starch origin.

Sometimes it is possible to find starch in unexpected places (in the field of cultural heritage), like in samples from paintings in Swedish church in Härnevi created by "Albertus Pictor" workshop around 1480. The paintings contain caseins (milk proteins) and have never been overpainted/limewashed. They have suffered from dirt, soot and mould. Surprisingly, there is no archive information on its treatments/cleaning etc., although cleaning is an essential procedure because of old heating systems used in old churches. Cleaning was often performed by using soft bread as a sponge to remove the dirt, and this process resulted in leaving starch traces on the surface.

The result of LC–MS/MS analyses is shown in Table 10.2. The unusual finding can be explained by the cleaning procedure performed by restorers in 80s of last century. At that time wall frescoes were cleaned using bread. Since the whole *Viridiplantae* part of Swissprot database was searched, proteins from other plants (soybean, corn) have also been identified – most likely because homologous wheat proteins were not present in the database. Another explanation could be that bread contains mostly wheat, barley and, possibly, traces of other plant grains (e. g. soybean, corn).

10.4.3 Distinguishing between whole egg and egg yolk tempera

Until today, no analytical method could distinguish between different egg binders (egg yolk and whole egg binder) in artwork samples, especially in cross-sections. However, using LC–MS/MS the binders can be unambiguously distinguished. The

Table 10.2: The result of proteomic analysis of sample from Swedish church in Härnevi from fourteenth century. The proteins were identified by searching *Viridiplantae* part of Swissprot database.

Plant	Protein	# Peptides
Wheat	Serpin-Z1A	5
Corn	Heat shock 70 kDa protein	5
Soybean	Heat shock 70 kDa protein	5
Barley	Sucrose synthase 2	4
Wheat	Alpha-amylase/trypsin inhibitor CM3	3
Corn	Luminal-binding protein 2	3
Wheat	Beta-amylase	2
Wheat	Glutenin, high-molecular weight subunit DY10	2
Barley	Protein synthesis inhibitor II	2
Barley	Elongation factor 1-alpha	2
Wheat	Avenin-like b8	2
Wheat	Glutenin, high-molecular weight subunit PW212	2
Wheat	Gamma-gliadin B	2
Barley	Beta-amylase	2
Wheat	Alpha-amylase inhibitor 0.28	2
Wheat	Glucose-1-phosphate adenylyltransferase small sub., chloroplastic	2
Wheat	Serpin-Z2B	2

determination of presence of egg yolk and whole egg is based on analysis of the ratio between peptides coming from proteins contained in egg white and egg yolk. It is important to focus on the ratio because in real situations the egg white is impossible to be completely separated from egg yolk.

Distinguishing of egg yolk from whole egg binder is demonstrated on two paintings of Czech painter Antonin Calkovský (1933–2005) – Rectangle and night landscape (1982) and Desert landscape (1981) (Figure 10.6). The results of LC–MS/MS are shown in Tables 10.3 and 10.4. The presence of collagens from animal species like mammoth (MAMAE) or dog (CANFA) is highly doubtful, but because collagens of many animal species are very similar (they have almost the same amino acid sequences) they were automatically assigned by software to these animals as it was described in part 3 of this chapter. Although, this does not imply that the artworks contain animal glue were prepared from these animals. The animal species recognition procedure using proteomics is discussed in Section 10.4.4, but it has not been applied in this case. In a sample from "Rectangle and night landscape" 3 peptides were attributed to egg white and 5 to egg yolk proteins (Table 10.3). The ratio between these peptides is 3:5 = 0.60. This, according to our results (whole egg: 0.5–0.8 and egg yolk: 0.0–0.3) obtained from a 23-year-old model samples, confirms the presence of whole egg [41]. A sample from "Desert landscape" contains 2 peptides from egg white and 9 from egg yolk proteins (Table 10.4). The ratio is 2:9 = 0.22, which

(a) (b)

Figure 10.6: Paintings of Antonin Calkovský: Rectangle and night landscape (1982) and Desert landscape (1981) on right.

Table 10.3: LC–MS/MS results of sample from Rectangle and night landscape (1982). The sample contains predominantly collagens (italic), egg yolk (bold-italic) and egg white (bold).

Accession	Protein	# Peptides
CO1A1_BOVIN	*Collagen alpha-1(I) chain*	*53*
CO1A1_MAMAE	*Collagen alpha-1(I) chain*	*42*
CO1A1_RAT	*Collagen alpha-1(I) chain*	*38*
CO1A2_BOVIN	*Collagen alpha-2(I) chain*	*23*
CO1A2_CANFA	*Collagen alpha-2(I) chain*	*16*
CO1A2_MAMAE	*Collagen alpha-2(I) chain*	*13*
CO3A1_BOVIN	*Collagen alpha-1(III) chain*	*11*
CO1A2_MOUSE	*Collagen alpha-2(I) chain*	*9*
CO3A1_RAT	*Collagen alpha-1(III) chain*	*7*
OVAL_CHICK	**Ovalbumin**	**3**
VIT2_CHICK	***Vitellogenin-2***	***3***
APOV1_CHICK	***Apovitellenin-1***	***1***
VIT1_CHICK	***Vitellogenin-1***	***1***

indicates whole egg binder. The presence of collagens in both samples is the result of contamination of both samples with animal glue from the ground layer. The correct recognition of both used binders was confirmed by the painter's notices on reverse side of paintings where the used techniques were described.

10.4.4 Animal species identification

Alternatively to DNA methods, proteomics coupled with MS techniques can be used for species identification, studies of evolutionary history, phylogenetic relationships as well as tracing of biological origins of proteinaceous materials used in culture heritage

Table 10.4: LC–MS/MS results of sample taken from Desert landscape (1981). The sample contains collagens (italic), egg yolk (bold-italic) and egg white (bold).

Accession	Protein	# Peptides
VIT2_CHICK	***Vitellogenin-2***	***5***
CO1A2_BOVIN	*Collagen alpha-2(I) chain*	*3*
CO1A1_BOVIN	*Collagen alpha-1(I) chain*	*3*
VIT1_CHICK	***Vitellogenin-1***	***2***
APOV1_CHICK	***Apovitellenin-1***	***2***
OVAL_CHICK	**Ovalbumin**	**1**
LYSC_TRASA	**Lysozyme C**	**1**

[42–46]. The animal species are recognized by a method to discriminate unidentifiable bone fragments, using a new technology called ZooMS (ZooArchaeology by MS). It is based on the persistence and slow evolution of collagen that is used as a molecular fingerprint for bones identification. Bones are identified by differences in the mass of the peptides which arise as a result of sequence differences between species [42].

The animal species source of animal glues can be determined based on the detection patterns of 12 type I collagen-derived marker peptides [44]. By this method, a painting (The Harvest by Camille Pissarro, 1882) was analysed. It was found that that glues used in the ground and size layers of the canvas originate from cattle (similar to 65 %) and sheep (similar to 35 %).

Not only collagens but also milk proteins can be assigned to particular animal species. Ancient food remains from Subeixi Cemeteries (500–300 years BC) of the Turpan Basin in Xinjiang, China, were analysed by an improved method based on LC coupled with MALDI–TOF/TOF–MS to further identify protein origin. The specific sequence of bovine casein and the homology sequence of goat/sheep casein were identified [46].

10.5 Organic dyes, lipids, plant gums and resins identification

MALDI–TOF–MS can not only be used for identification of large molecules like peptides (900–4,000 m/z) but also for small molecules like organic dyes with masses within 200–500 m/z (e. g. red anthraquinoid dyes, flavones, indigo and some synthetic dyes such as phthalocyanines), lipids, plant gums and natural and synthetic resins.

10.5.1 Natural and synthetic organic dyes and pigments

Determination of dyes in samples can be performed either directly (e. g. on threads; Figure 10.7) or after extraction with hydrofluoric acid [47] (fragments from paintings/

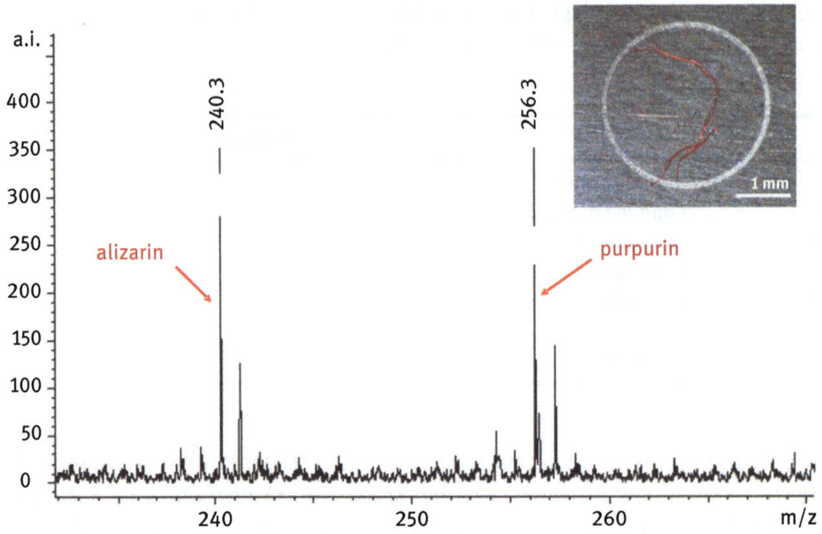

Figure 10.7: Laser desorption/ionization–time-of-flight MS analysis of a red thread from twentieth century red-coloured shoes. The spectrum indicates presence of two anthraquinone dyes alizarin and purpurin.

sculptures) – both in positive and negative acquisition modes by MALDI or ESI MS [48]. Dyes can be analysed without the presence of an organic matrix, which is a major advantage of analysis of small molecules. Since no matrix is used, the technique is called laser desorption/ionization–time-of-flight MS (LDI–TOF). LDI was applied on the analysis of synthetic pigments in L. Fontana [49] and J. Pollock [50] paintings, and in other modern works of art by artists like M. Rothko, B. Newman and J. de Rivera [51].

10.5.2 Lipids

MALDI-FTICR MS has successfully been used for characterization of lipids in egg tempera [52] and for studying the effects of the traditional processing methods (heating, addition of siccative) of drying oils [53] by detection of triacylglycerols, their oligomers and phosphatidylcholines. The method can unambiguously discriminate between egg- and drying oil-based binders [54] and also allow a simultaneous identification of lipids and proteins in microsamples using an extraction process described in reference [55]. In another study, colloidal graphite (GALDI) was used for a development of a rapid identification of all marker lipids of egg-based binder and for simultaneous identification of protein binders [8]. An LC–ESI–MS method provides reliable separation of non-polar triglycerides in drying oil as well as in polar oxidized and hydrolysed tri-, di- and monoglycerides and free fatty acids. The applicability of the method was shown on 62-year-old paints from Winsor and Newton [56].

10.5.3 Plant gums

Recently MALDI–TOF was used for identification of arabic, cherry and locust-bean gums by their mass fingerprints in cultural heritage samples for the first time [57]. Granzotto et al. [57] successfully identified a wide range of oligosaccharidic combinations of pentose, hexose, deoxyhexose and hexuronic acid in arabic gum whereas cherry and locust-bean gums showed, respectively, $Pent_xHex_y$ and Hex_n profiles [57]. In another work, Granzotto and Sutherland [58] extracted unambiguous MS profile of Acacia gum (gum Arabic, sp. *A. senegal*) characterized by specific and recognized oligosaccharides, from as little as 0.1 µg of material (naturally aged gum and gum mixed with lead white). The same authors also distinguished *A. senegal* from *A. seyal* by their mass patterns of specific oligosaccharides produced by enzymatic digestion [59].

10.5.4 Natural and synthetic resins

FTICR–MS using MALDI ion source was used for analyses of aged natural dammar resin [60]. The collected mass spectra revealed the clusters corresponding to oxygenated derivatives of terpenoids which differ by the number of oxygen atoms in their molecules MALDI–TOF directly used on cellulose-coated thin layer chromatography plates detected highly oxidized terpenoids, which are hard to detect by other techniques [61]. The validity of the method was proved on natural di- and triterpenoid resins used as paint varnishes by Old Masters. Also MALDI and nano-ESI–MS were also employed for the characterization of synthetic resins using poly(ethylene glycol)-based additives present in water extracts of acrylic emulsion paints [62]. ESI–Q-TOF was used for molecular characterization of three low-molecular-weight resins (ketone resin MS2A, the aldehyde resin Laropal A81, and the hydrocarbon resin Regalrez 1094) which are now commonly used in restoration [63].

10.6 Concluding remarks

In comparison to other analytical methods, the analyses by mass spectrometric techniques are very sensitive and allow for identifying proteins, including unexpected kinds. They are capable of identifying complex mixtures of proteins, studying of protein post-translation modifications and amino acid modifications caused by ageing or environmental conditions. Obtaining of accurate sequence of peptides provides the possibility to determine animal and plant species origins of proteinaceous materials.

Like no other method in use, MS is capable of direct determination of proteins, lipids, resins, plant gums and dyes, by employing the analysis of molecular weight. Among well distinguishable artistic materials are egg temperas, starches and milk materials. Other advantages of the mass spectrometric techniques are the affordability of the methodology for many different materials and also the possibility to analyse cross-sections of microsamples.

The main drawback of the described MS techniques is their relatively high cost, particular demand of maintenance and necessity of skilled operators. Nowadays, it is possible to analyse proteins in extremely small and old samples. However, the protein sequence databases are not complete, because not all animal species and animal breeds' proteomes have been sequenced yet and identified proteins sequences entered in the databases. Therefore, some proteins present in samples might not be identified.

The MS techniques require a minimum amount of sample (e. g. a fragment from painting), however it must be taken from the cultural heritage object itself. The sampling protocol often does not leave any visible traces on the analysed material and thus the MS analyses are considered to be semi-destructive.

The MS techniques are utilized for analyses of any type of materials that contain proteins – e. g. proteinaceous binders in artworks, protein additives in historical mortars, archaeological findings of prehistoric food, leathers, parchments, bones and teeth. Applications of these methods reveal painting techniques, recipes of Old Master builders, composition of prehistoric diets, animal species origin of cultural heritage objects, evolutionary history and phylogenetic relationships of animals (including extinct species). Besides proteins, other materials such as lipids, resins, polysaccharides in plant gums, organic and synthetic dyes and pigments, can also be identified.

Acknowledgements: The authors cordially thank Igor Fogas for providing samples and photographs from Calkovský paintings, Misa Asp for providing sample from Swedish Härnevi, the Alexander Humboldt Foundation for financial support, and professor Milan Kodicek from Laboratory of Applied Proteomics at University of Chemistry and Technology in Prague, for his valuable comments and chapter editing. We also thank our Canadian friend Dmitry Gavrilov, Ph.D., for helping with the article proofreading.

References

[1] Hynek R, Kuckova S, Hradilova J, Kodicek M. Matrix-assisted laser desorption/ionization time of flight mass spectrometry as a tool for fast identification of protein binders in color layers of paintings. Rapid Commun Mass Spectrom. 2004;18:1896–900. DOI: 10.1002/rcm.1570.

[2] Calvano CD, van der Werf ID, Palmisano F, Sabbatini L. Revealing the composition of organic materials in polychrome works of art: the role of mass spectrometry-based techniques. Anal Bioanal Chem. 2016;408:6957–81. DOI: 10.1007/s00216-016-9862-8.

[3] Dallongeville S, Garnier N, Rolando C, Tokarski C. Proteins in art, archaeology, and paleontology: from detection to identification. Chem Rev. 2016;116:2–79. DOI: 10.1021/acs.chemrev.5b00037.

[4] Krizova I, Schultz J, Nemec I, Cabala R, Hynek R, Kuckova S. Comparison of analytical tools appropriate for identification of proteinaceous additives in historical mortars. Anal Bioanal Chem. 2018;410:189–200. DOI: 10.1007/s00216-017-0709-8.

[5] Colombini MP, Andreotti A, Bonaduce I, Modugno F, Ribechini E. Analytical strategies for characterizing organic paint media using gas chromatography/mass spectrometry. Acc Chem Res. 2010;43:715–27. DOI: 10.1021/ar900185f.
[6] Lee HY, Atlasevich N, Granzotto C, Schultz J, Loike J, Arslanoglu J. Development and application of an ELISA method for the analysis of protein-based binding media of artworks. Anal Methods. 2015;7:187–96. DOI: 10.1039/c4ay01919a.
[7] Colombini MP, Modugno F. Organic mass spectrometry in art and archaeology. Wiley: New York, 2009. ISBN 978-0-470-51703-1.
[8] Calvano CD, van der Werf ID, Palmisano F, Sabbatini L. Identification of lipid- and protein-based binders in paintings by direct on-plate wet chemistry and matrix-assisted laser desorption ionization mass spectrometry. Anal Bioanal Chem. 2015;407:1015–22. DOI: 10.1007/s00216-014-8359-6.
[9] Perets EA, Indrasekara AS, Kurmis A, Atlasevich N, Fabris L, Arslanoglu J. Carboxy-terminated immuno-SERS tags overcome non-specific aggregation for the robust detection and localization of organic media in artworks. Analyst. 2015;140:5971–80. DOI: 10.1039/c5an00817d.
[10] Kuckova S, Hamidi-Asl E, Sofer Z, Marvan P, de Wael K, Sanyova J, et al. A simplified protocol for the usage of new immuno-SERS probes for the detection of casein, collagens and ovalbumin in the cross-sections of artworks. Anal Methods. 2018;10:1054–62. DOI: 10.1039/c7ay01864a.
[11] Atrei A, Benetti F, Potenza M, Dei L, Carretti E, Niccolucci V, et al. Characterization of organic binders in a thirteenth century painted wooden panel: comparison of ToF-SIMS and Dot-ELISA results. Int J Mass Spectrom. 2018;430:63–8. DOI: 10.1016/j.ijms.2018.04.006.
[12] Sciutto G, Dolci LS, Buragina A, Prati S, Guardigli M, Mazzeo R, et al. Development of a multiplexed chemiluminescent immunochemical imaging technique for the simultaneous localization of different proteins in painting micro cross-sections. Anal Bioanal Chem. 2011;399:2889–97. DOI: 10.1007/s00216-010-4258-7.
[13] Karas M, Hillenkamp F. Laser desorption ionization of proteins with molecular masses exceeding 10,000 Daltons. Anal Chem. 1988;60:2299–3001.
[14] Fenn JB, Mann M, Meng CK, Wong SF, Whitehouse CM. Electrospray ionization for mass spectrometry of large biomolecules. Science. 1989;246:64–71. DOI: 10.1126/science.2675315.
[15] Konermann L, Ahadi E, Rodriguez AD, Vahidi S. Unraveling the mechanism of electrospray ionization. Anal Chem. 2013;85:2–9. DOI: 10.1021/ac302789c.
[16] Boesl U. Time-of-flight mass spectrometry: introduction to the basics. Mass Spectrom Rev. 2017;36:86–109. DOI: 10.1002/mas.21520.
[17] Tripković T, Charvy C, Alves S, Lolić AĐ, Baošić RM, Nikolić-Mandić SD, et al. Identification of protein binders in artworks by MALDI–TOF/TOF tandem mass spectrometry. Talanta. 2013;113:49–61. DOI: 10.1016/j.talanta.2013.03.071.
[18] Savaryn JP, Toby TK, Kelleher NL. A researcher's guide to mass spectrometry-based proteomics. Proteomics. 2016;16:2435–43. DOI: 10.1002/pmic.201600113.
[19] Zubarev RA, Makarov A. Orbitrap mass spectrometry. Anal Chem. 2013;85:5288–96. DOI: 10.1021/ac4001223.
[20] Nikolaev EN, Kostyukevich YI, Vladimirov GN. Fourier transform ion cyclotron resonance (FT ICR) mass spectrometry: theory and simulations. Mass Spectrom Rev. 2016;35:219–58. DOI: 10.1002/mas.21422.
[21] Koppenaal DW, Barinaga CJ, Denton MB, Sperline RP, Hieftje GM, Schilling GD, et al. MS detectors. Anal Chem. 2005;77:418–27. DOI: 10.1021/ac053495p.
[22] Paizs B, Suhai S. Fragmentation pathways of protonated peptides. Mass Spectrom Rev. 2005;24:508–48. DOI: 10.1002/mas.20024.
[23] Cottrell JS. Protein identification using MS/MS data. J Proteomics. 2011;74:1842–51. DOI: 10.1016/j.jprot.2011.05.014.

[24] Qi Y, Volmer DA. Electron-based fragmentation methods in mass spectrometry: an overview. Mass Spectrom Rev. 2017;36:4–15. DOI: 10.1002/mas.21482.

[25] Fremout W, Kuckova S, Crhova M, Sanyova J, Saverwyns S, Hynek R, et al. Classification of protein binders in artist's paints by matrix-assisted laser desorption/ionisation time-of-flight mass spectrometry: an evaluation of principal component analysis (PCA) and soft independent modelling of class analogy (SIMCA). Rapid Commun Mass Spectrom. 2011;25:1631–40. DOI: 10.1002/rcm.5027.

[26] Kuckova S, Rambouskova G, Hynek R, Cejnar P, Oltrogge D, Fuchs R. Evaluation of mass spectrometric data using principal component analysis for determination of the effects of organic lakes on protein binder identification. J Mass Spectrom. 2015;50:1270–8. DOI: 10.1002/jms.3699.

[27] Kuckova S, Sandu IC, Crhova M, Hynek R, Fogas I, Muralha VS, et al. Complementary cross-section based protocol of investigation of polychrome samples of a sixteenth century Moravian sculpture by optical, vibrational and mass spectrometric techniques. Microchem J. 2013;110:538–44. DOI: 10.1016/j.microc.2013.07.002.

[28] Fico D, Margapoti E, Pennetta A, Benedetto GE. An enhanced GC/MS procedure for the identification of proteins in paint microsamples. J Anal Methods Chem. Article Number: 6032084. 2018. DOI: 10.1155/2018/6032084.

[29] Zangrando R, Piazza R, Cairns WR, Izzo FC, Vianello A, Zendri E, et al. Quantitative determination of un-derivatised amino acids in artistic mural paintings using high-performance liquid chromatography/electrospray ionization triple quadrupole mass spectrometry. Anal Chim Acta. 2010;675:1–7. DOI: 10.1016/j.aca.2010.06.045.

[30] Schroeter ER, DeHart CJ, Schweitzer MH, Thomas PM, Kelleher NL. Bone protein "extractomics": comparing the efficiency of bone protein extractions of *Gallus gallus* in tandem mass spectrometry, with an eye towards paleoproteomics. PEERJ. 2016;4:e2603. doi:10.7717/peerj.2603.

[31] Leo G, Cartechini L, Pucci P, Sgamellotti A, Marino G, Birolo L. Proteomic strategies for the identification of proteinaceous binders in paintings. Anal Bioanal Chem. 2009;395:2269–80. DOI: 10.1007/s00216-009-3185-y.

[32] Pramanik BN, Mirza UA, Ing YH, Liu YH, Bartner PL, Weber PC, et al. Microwave-enhanced enzyme reaction for protein mapping by mass spectrometry: a new approach to protein digestion in minutes. Protein Sci. 2002;11:2676–87. DOI: 10.1110/ps.0213702.

[33] Vinciguerra R, Galano E, Vallone F, Greco G, Vergara A, Bonaduce N, et al. Deglycosylation step to improve the identification of egg proteins in art samples. Anal Chem. 2015;87:10178–82. DOI: 10.1021/acs.analchem.5b02423.

[34] Orsini S, Yadav A, Dilillo M, McDonnell LA, Bonaduce I. Characterization of degraded proteins in paintings using bottom-up proteomic approaches: new strategies for protein digestion and analysis of data. Anal Chem. 2018;90:6403–8. DOI: 10.1021/acs.analchem.8b00281.

[35] Li XP, Wilm M, Franz T. Silicone/graphite coating for on-target desalting and improved peptide mapping performance of matrix-assisted laser desorption/ionization-mass spectrometry targets in proteomic experiments. Proteomics. 2005;5:1460–71. DOI: 10.1002/pmic.200401023.

[36] Thompson DV. The practice of tempera painting. New Haven: Yale University Press, 1936. Reprinted by Dover Publications (1962).

[37] Sickels LB. Organic additives in mortars. Edinburgh Archit Res. 1981;8:7–20.

[38] Jasiczak J, Zielinski K. Effect of protein additive on properties of mortar. Cem Concr Compos. 2006;28:451–7. DOI: 10.1016/j.cemconcomp.2005.12.007.

[39] Kuckova S, Santrucek J, Adamec M, Hynek R, Zeman A. Chemical analysis of Gothic mortar from a bridge pier in Roudnice nad Labem (Czech Republic). J Liq Chromatogr Related Technol. 2016;16:739–44. DOI: 10.1080/10826076.2016.1238394.

[40] Svajcrova L: Příprava polysacharidových pojiv používaných v uměleckých dílech. Bachelor thesis, Faculty of Education, Charles University in Prague, Czech Republic (2014).

[41] Kuckova S, Baumer U, Dietemann P: Proteomics distinguishing between whole egg and egg yolk tempera. Paper under preparation (2018).

[42] Buckley M, Collins M, Thomas-Oates J, Wilson JC. Species identification by analysis of bone collagen using matrix-assisted laser desorption/ionisation time-of-flight mass spectrometry. Rapid Commun Mass Spectrom. 2009;23:3843–54. DOI: 10.1002/rcm.4316.

[43] Buckley M. Species identification of bovine, ovine and porcine type 1 collagen; comparing peptide mass fingerprinting and LC-based proteomics methods. Int J Mol Sci. 2016;17:445. DOI: 10.3390/ijms17040445.

[44] Kumazawa Y, Taga Y, Takashima M, Hattori S. A novel LC–MS method using collagen marker peptides for species identification of glue applicable to samples with multiple animal origins. Heritage Sci. 2018;6:43. DOI: 10.1186/s40494-018-0209-y.

[45] Brandt LØ, Schmidt AL, Mannering U, Sarret M, Kelstrup CD, Olsen JV, et al. Species identification of archaeological skin objects from Danish bogs: comparison between mass spectrometry-based peptide sequencing and microscopy-based methods. PLoS ONE. 2014;9:e106875. DOI: 10.1371/journal.pone.0106875.

[46] Hong C, Jiang H, Lü E, Wu Y, Guo L, Xie Y, et al. Identification of milk component in ancient food residue by proteomics. PLoS ONE. 2012;7:e37053. DOI: 10.1371/journal.pone.0037053.

[47] Sanyova J, Reisse J. Development of a mild method for the extraction of anthraquinones from their aluminum complexes in madder lakes prior to HPLC analysis. J Cult Heritage. 2006;7:229–35. DOI: 10.1016/j.culher.2006.06.003.

[48] Maier MS, Parera SD, Seldes AM. Matrix-assisted laser desorption and electrospray ionization mass spectrometry of carminic acid isolated from cochineal. Int J Mass Spectrom. 2004;232:225–9. DOI: 10.1016/j.ijms.2003.12.008.

[49] Gottschaller P, Khandekar N, Lee LF, Kirby DP. The evolution of Lucio Fontana's painting materials. Stud Conserv. 2012;57:76–91. DOI: 10.1179/2047058411Y.0000000002.

[50] Khandekar N, Mancusi-Ungaro C, Cooper H, Rosenberger C, Eremin K, Smith K, et al. A technical analysis of three paintings attributed to Jackson Pollock. Stud Conserv. 2010;55:204–15. DOI: 10.1179/sic.2010.55.3.204.

[51] Lomax SQ, Lomax JF, De Luca-Westrate A. The use of Raman microscopy and laser desorption ionization mass spectrometry in the examination of synthetic organic pigments in modern works of art. J Raman Spectrom. 2014;45:448–55. DOI: 10.1002/jrs.4480.

[52] van Den Brink OF, Boon JJ, O'Connor PB, Duursma MC, Heeren RM. Matrix-assisted laser desorption/ionization Fourier trans-form mass spectrometric analysis of oxygenated triglycerides and phosphatidylcholines in egg tempera paint dosimeters used for revealing organic materials in works of art for environmental monitoring of museum display conditions. J Mass Spectrom. 2001;36:479–92. DOI: 10.1002/jms.145.

[53] van Den Berg JD, Vermist ND, Carlyle L, Holcapek M, Boon JJ. Effects of traditional processing methods of linseed oil on the composition of its triacylglycerols. J Sep Sci. 2004;27:181–99. DOI: 10.1002/jssc.200301610.

[54] Calvano CD, van der Werf ID, Palmisano F, Sabbatini L. Fingerprinting of egg and oil binders in painted artworks by matrix-assisted laser desorption ionization time-of-flight mass spectrometry analysis of lipid oxidation by-products. Anal Bioanal Chem. 2011;400:2229–40. DOI: 10.1007/s00216-011-4919-1.

[55] van der Werf ID, Calvano CD, Palmisano F, Sabbatini L. A simple protocol for matrix assisted laser desorption ionization-time of flight-mass spectrometry (MALDI–TOF-MS) analysis of lipids and proteins in single microsamples of paintings. Anal Chim Acta. 2012;718:1–10. DOI: 10.1016/j.aca.2011.12.056.

[56] van Dam EP, van Den Berg KJ, Gaibor AN, van Bommel M. Analysis of triglyceride degradation products in drying oils and oil paints using LC–ESI–MS. Int J Mass Spectrom. 2017;413:33–42. DOI: 10.1016/j.ijms.2016.09.004.

[57] Granzotto C, Arslanoglu J, Rolando C, Tokarski C. Plant gum identification in historic artworks. Sci Rep. 2017;20:44538. DOI: 10.1038/srep44538.

[58] Granzotto C, Sutherland K. Matrix assisted laser desorption ionization mass fingerprinting for identification of Acacia gum in microsamples from works of art. Anal Chem. 2017;89:3059–68. DOI: 10.1021/acs.analchem.6b04797.

[59] Granzotto C, Sutherland K, Arslanoglu J, Ferguson GA. Discrimination of Acacia gums by MALDI–TOF MS: applications to micro-samples from works of art. Microchem J. 2019;144: 229–41.

[60] Vahur S, Teearu A, Haljasorg T, Burk P, Leito I, Kaljurand I. Analysis of dammar resin with MALDI-FT-ICR-MS and APCI-FT-ICR-MS. J Mass Spectrom. 2012;47:392–409. DOI: 10.1002/jms.2971.

[61] Scalarone D, Duursma MC, Boon JJ, Chiantore O. MALDI–TOF mass spectrometry on cellulosic surfaces of fresh and photo-aged di- and triterpenoid varnish resins. J Mass Spectrom. 2005;40:1527–35. DOI: 10.1002/jms.893.

[62] Hoogland FG, Boon JJ. Development of MALDI-MS and nano-ESI-MS methodology for the full identification of poly(ethylene glycol) additives in artists' acrylic paints. Int J Mass Spectrom. 2009;284:66–71. DOI: 10.1016/j.ijms.2009.03.001.

[63] Bonaduce I, Colombini MP, Degano I, Di Girolamo F, Nasa J, Modugno F, et al. Mass spectrometric techniques for characterizing low-molecular-weight resins used as paint varnishes. Anal Bioanal Chem. 2013;405:1047–65. DOI: 10.1007/s00216-012-6502-9.

Part II: **Selected Case studies**

Susanna Bracci and Giovanni Bartolozzi

11 Wall paintings – diagnostic and archaeometric studies

Abstract: The paper deals with the techniques and protocols used for studying wall paintings. A brief introduction about the more recent literature dealing with archaeometric and diagnostic analyses of wall paintings is reported. After that, the illustration of three case studies, spanning from Roman to contemporary wall painting are described.

Keywords: wall paintings, archeometry, non-invasive analyses, conservation

11.1 Introduction

Wall paintings were made in all ages, from many centuries B.C.E. to the present day. Wall painting techniques are many and only summarily divisible into three categories: *fresco, mezzo-fresco, secco paint* [1]. Actually, most of the time, conservators and scientists are faced with mixed techniques, used by individual artists or entire workshops, according to the customer, the period of execution, the environmental characteristics and other variables.

The study and the knowledge of the original materials, the techniques and the executive procedures are essential prerequisites for the evaluation of the type of artwork, for its geochronological collocation, the possible attribution, and, not less important, for the identification of the ongoing degradation phenomena. The knowledge of the materials of the artwork and its state of conservation is also at the basis of a correct and conscious restoration intervention.

In the past, the studies of wall paintings were exclusively based on stylistic analyses and on the personal evaluation of scholars, although some analyses were conducted on Roman wall paintings already in the XIX century by a famous chemist [2].

To date, several analytical techniques are employed, albeit, in practice those applied depend largely on specific fields of interest and circumstances.

The first analytical approaches were based on the analyses of samples taken from the work of art, to be conducted in the laboratory by means of benchtop devices. If the samples were big enough to be further sampled, techniques such as X-ray Diffraction (XRD), X-ray Fluorescence (XRF) or Infrared Spectroscopy (FT-IR) could be also applied. Usually, cross sections were obtained from the samples, subsequently observed with Optical Microscope (OM) and analyzed with Scanning Electron Microscope equipped

This article has previously been published in the journal *Physical Sciences Reviews*. Please cite as: Bracci, S., Bartolozzi, G. Wall paintings – diagnostics and archeometric studies. *Physical Sciences Reviews* [Online] **2019**, 4. DOI: 10.1515/psr-2018-0013.

https://doi.org/10.1515/9783110457537-011

with elemental analysis (SEM-EDS). Other benchtop techniques such as infrared spectroscopy in different configurations (ATR, micro-ATR, micro-imaging), and Raman spectroscopy were used [3–8]. For the identification of organic materials such as binders, in a *secco* painting, laboratory techniques based on Pyrolisis-Gas Chromatography-Mass Spectroscopy (PY-GC-MS) and Gas Chromatography-Mass Spectroscopy (GC-MS) [9] were used.

All the techniques quoted before, along with other laboratory techniques nowadays available, are still widely used and they are very important especially when a detailed stratigraphic analysis or the identification of organic materials is required. These techniques have proven their significant value and have contributed to breakthroughs in archaeometric research [10–24].

In spite of this, the microinvasive approach had some drawbacks. First, these rather complex techniques require highly skilled analysts and high-tech instruments, which basically means that the measurements have to be performed in specialized laboratories. A second drawback is, for some of them, their time-consuming and expensive nature. This makes it difficult to analyze a high number of samples. Third, these techniques required a sample, even if micro, and some of them destroy the whole or part of the sample analyzed.

Starting from the first pioneering approach with spectroscopic techniques [25, 26] the use of noninvasive analyses was exploited in the last fifteen/twenty years. This increasing attention to the use of noninvasive techniques was fostered by the arising awareness to preserve as much as possible the integrity of the work of art and, in parallel, by the effort of the industries to develop portable devices with good performances.

In general, noninvasive analyses can be divided into two main groups on the basis of the data collected: imaging and single spot techniques.

Among imaging techniques [27] the most used are: RGB imaging that is commonly applied for documenting the surfaces under analysis [21, 28]; raking light (RL) for the evaluation of detachments or surface defects [29]; UV fluorescence (UVf), which is commonly used for mapping of the organic materials such as lakes [30, 31] or binders in *secco* paintings [21, 32]; infrared reflectance (IR) [33, 34] for unveiling the original drawing; Visible Induced Luminescence (VIL), a very powerful tool for mapping the presence of a specific pigment, the Egyptian Blue [35–38].

Among single spot techniques the main role today is played by portable X-ray fluorescence (XRF) [39] and reflectance spectroscopy [40]. pXRF is a relatively fast, easy-to-use and nondestructive analytical tool [41, 42] able to yield information about key chemical elements.

Reflectance spectroscopies include several tools depending on the wavelength range chosen: UV-VIS-NIR (range 350–400 nm, 400–780 nm, 780–2500 nm) or NIR and mid-infrared (such as Total Reflectance Fourier Transform Infrared spectroscopy – TR FT-IR) or Raman spectroscopy.

Fiber-optic reflectance spectroscopy (FORS) is a technique that makes it possible both to identify pigments, colorants and alteration products, and to acquire parameters that are useful for monitoring colorimetric or compositional variations of artefacts [30, 43].

Fourier Transform Infrared (FT-IR) spectroscopy is an analytical technique capable of providing a molecular characterization of the various materials of artistic interest. Portable instruments in total-reflectance configuration make it possible to obtain measurements in a noninvasive way and without any contact with the surface of the work [30, 44].

Raman spectrometry reveals information about composition of pigments and preparation layers [45]. It determines the nature and the arrangement of the chemical bonds in the material. Sometimes it suffers from the presence of fluorescent materials. In such cases, the fluorescence signal may saturate the Raman spectrum but this problem may be overcome by using different lasers sources.

References to many other noninvasive techniques can be found in literature, such as XRD coupled to XRF [46], Laser induced fluorescence (LIF) [47], portable Nuclear Magnetic Resonance-pNMR [48–50], Terahertz spectroscopy (THz) [30], or analyses conducted on the synchrotron lines of big facilities [51].

In between a completely noninvasive approach and one based only on the analyses of samples, a very efficient diagnostic protocol is based on the combination of a first noninvasive step followed by a limited, but very focused, sampling campaign. This latter relies on the data acquired during the first step and is devoted to those questions that still remain unsolved. A multidisciplinary approach is now therefore currently privileged as the drawbacks of a single diagnostic technique can be overcome by integrating and implementing the data set acquired [29, 49, 52–59].

The above synthetic overview clearly demonstrates that a single completely exhaustive technique does not exist. Scholars will always have to select certain techniques, being aware that each one of them is characterized by specific pros and cons. The choice has to be based on different aspects such as the type of research question (composition/fabrication condition), the type of material (extensive or limited collection of samples) and the available resources (time, *in-situ*/*ex-situ* availability, budget).

Any extension to the list of possible techniques will lead to improved archaeometric research.

11.2 Case studies

11.2.1 Domitilla's catacombs – *dei fornai* cubicle

The catacomb of Domitilla is one of the most extensive of underground burial networks in Rome and the cubicle called "*dei fornai*" ("of the bakers") is one of the most important environments in this catacomb. This latter was excavated during the fourth century in a peripheral area of the catacomb close to the end of the exploitation of the burial area [60].

The cubicle consists of a passage leading to a room with two large polygonal apses divided by a sort of central corridor ending in a quadrangular niche; four *arcosolia* open alternately along the major sides of the polygon. It represents a unique pictorial decoration due to the painted subjects and topics. These latter include on one side, various scenes of workers unloading sacks of wheat from ships anchored in the Tiber River, in order to transport them by hand-barrow to windmills and ovens. On the opposite apse, Christ, enthroned between two groups of apostles with the Princes of the Apostles, Peter and Paul sitting in the foreground, is represented.

In the cubicle a diagnostic campaign [61] was carried out, exclusively by means of *in situ* non-invasive methods and techniques following the request of the curator of the catacomb. The question was to identify as much as possible the pigments used and to compare areas that appeared different on a stylistic base. The measurements were conducted after the cleaning by means of laser ablation [62] to remove the carbonatic concretions from the surface.

The set of the techniques used included Colorimetry (Minolta CR600), Digital Microscopy (DM) (Scalar 2000), Fiber Optics Reflectance Spectroscopy (FORS, Ocean Optics Mod. HR2000, VIS 380–850 nm), X-ray Fluorescence Spectrometry (XRF, Innov-X ALPHA 4000), Infrared Spectroscopy in total reflection mode (TR FT-IR, Bruker Optics Alpha) and Visible Induced Luminescence (VIL, Sony Cyber Shot DSC F-828 and two Radium Spectralux halogen tubes, model NL T8 58W/840) (Figure 11.1). During the diagnostic campaign a monitoring of the environmental conditions was performed, too. The recorded data showed that the room was in a constant state of humidity saturation (100% R.H.), and the temperature underwent very little fluctuations during the day even in presence of working people (15°C +/– 0.3°C). The harsh environmental conditions, especially the very high relative humidity, made the acquisitions quite challenging with respect to the instrumentation safety.

As a first approach, color measurements and digital microscopy were performed on 95 selected areas, in order to have a preliminary rough screening suitable for splitting the examined areas in few big groups, corresponding to green, blue, yellow, red, orange, black and white colors.

In the green areas, the combined use of FORS and XRF revealed the presence of two different pigments. The typical fluorescence signals of iron (Fe) and calcium (Ca) in the XRF measurements and the absorption bands in FORS spectra unequivocally ascertained the use of green earth in most of the investigated areas. In few cases the shape of FORS spectra and the high counts of copper (Cu) recorded by XRF were in agreement with the use of a copper based pigment.

The XRF measurements carried out on the blue areas lead to infer the presence of a copper based pigment and FORS spectra, recorded on the same areas, validated the identification of Egyptian blue as the unique blue pigment on the mural paintings. Furthermore, the VIL images of the blue areas showed the characteristic white luminescence in the NIR region, attributable to the use of Egyptian blue (Figure 11.2). FORS

Figure 11.1: Some examples of data acquisition with portable instruments in the Domitilla's Catacombs. (a) and (c) FT-IR TR; (b) XRF.
Photos by Giovanni Bartolozzi.

spectra of all investigated yellow areas presented the typical absorption bands of an iron-based pigment, such as yellow ochre. The XRF analyses confirm the presence of iron (Fe) as the principal element in all areas. In red areas, FORS spectra presented the characteristic signals attributable to an iron oxide based pigment and XRF analyses confirmed the presence of iron (Fe).

TR FT-IR spectra (Figure 11.3) showed the absorption bands of Fe_2O_3, namely hematite (573 cm^{-1}, 479 cm^{-1}). Additionally, in some of the investigated areas, lead (Pb) was also detected suggesting that red ochre had been used in mixture with a white pigment (possibly lead white) to obtain different hues.

The presence of lead (Pb) was also detected by XRF in orange areas but, in this case, the FORS spectrum showed the typical S-shape of minium with the inflection point at 564 nm calculated from the first derivative spectrum (Figure 11.3).

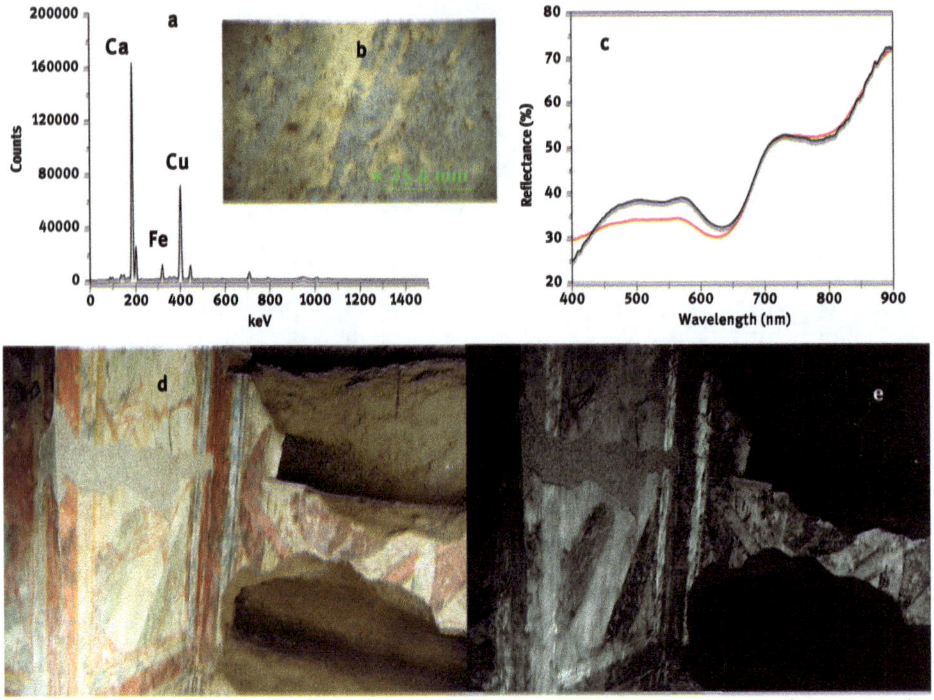

Figure 11.2: (a) XRF spectrum with inserted MD image (b) of one of blue areas painted with Egyptian Blue. (c) FORS spectrum of the same area (black line) compared with a reference spectrum of Egyptian Blue (red line). RGB (d) and VIL (e) images.
Photos by Susanna Bracci.

Notwithstanding the very uncomfortable and harsh experimental conditions, due to narrow and low-ceilinged spaces and, most of all, to the very high humidity in the cubicle, the small dimensions and the manageability of the instruments allowed to obtain almost all the requested information.

According to historical and stylistic studies the cubicle, through its history, had two decorative programs with two different pictorial styles: a first and simple linear decoration, and a second with a more elaborated iconographic program and a richer palette, arising from mixing few pigments in order to obtain different hues (a rare characteristic in painted catacombs). The quality of the latter painted decoration, the presence of some peculiar colors involved (orange lead, copper green), the high cost of some of the identified pigments (Egyptian blue) and the few tombs projected in the last phase (only four *arcosolia*) corroborates the hypothesis that this decoration was a commission promoted by a high-ranking customer with considerable economic capacity.

An important aspect highlighted by this work was the efficacy of the use of a multidisciplinary approach, where scientific competences worked in very close

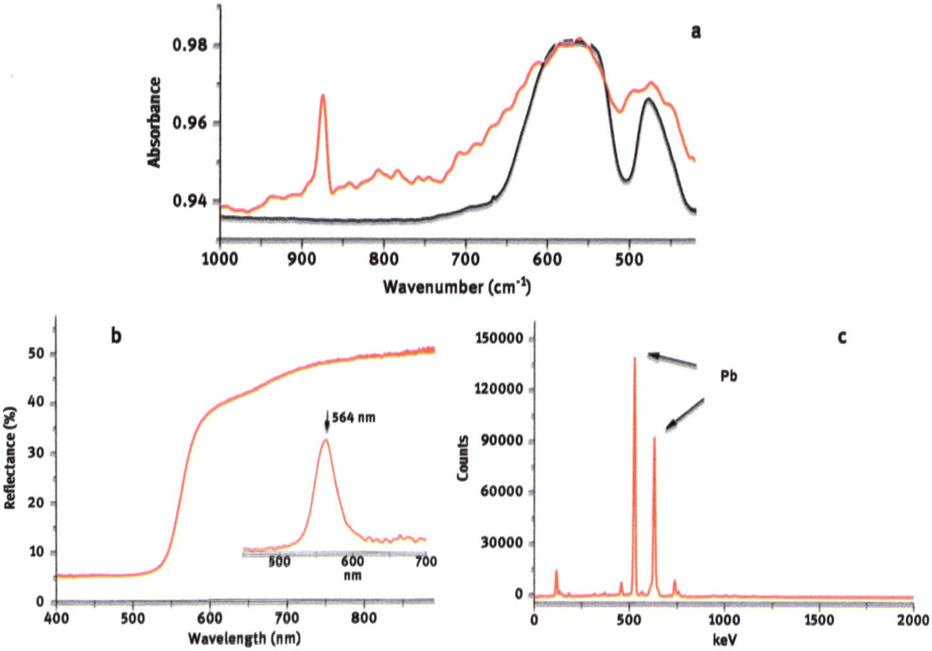

Figure 11.3: (a) FT-IR TR spectrum of a red area (red line) compared with the spectrum of ematite (black line); (b) FORS spectrum of an orange area with inserted the first derivative one showing the inflection point at 564 nm; (c) XRF spectrum of the same area.
Spectra by ICVBC.

contact with historic-artistic ones. Such a collaboration assumed particular importance, since the scientific information resulted very significant when associated to iconographic and historical studies.

11.2.2 Archeological site of Hierapolis

This work was conducted in the framework of the Marmora Phrygiae Project [63], during which many different scientific activities were foreseen both *in situ* during the Italian Archeological Mission [64] in the archeological site of Hierapolis of Phrygiae (Pamukkale, Denizli) [65] and, afterwards, in laboratory.

Here the focus was devoted to the results of the analyses and study of the fragments of painted plaster discovered during the recent archeological excavations inside the byzantine Church of St. Philip in Hierapolis. The site of St. Philip Church, mostly obliterated during the fourteenth century after the Seljuk occupation, yielded many fragments of painted plaster together with a few non-figurative pieces preserved *in situ*. All these fragments constitute a rare document of the early and middle Byzantine painting [66–68]

In this case, most of the fragments collected, were stored in the deposits of the Archeological Mission grouped according to the stratigraphic unit of excavation. Since it was impossible to move the fragments from The Archeological mission in Turkey to the laboratories in Italy and, moreover, due to their high number (hundreds), taking a sample from all would have been too engaging in terms of time and costs. Therefore, the sampling campaign was planned after the first noninvasive *in-situ* approach. This latter was conducted in order to group the fragments also on the basis of the pigments used and the similarities in the decoration pattern, and to identify the more representative fragments of each group. The set of the techniques used comprised imaging techniques such as UVf (Canon EOS 7D, lens EFS 18–135 mm, Flash Quantum T5D) and VIL (Canon EOS 400D, no built-in filter for IR, lens EFS 18–135 mm, Flash Quantum T5D). Single spot techniques used were Digital Microscopy (DM) (Scalar DG-2A), Fiber Optics Reflectance Spectroscopy (FORS, Ocean Optics mod.HR2000, VIS 380–850 nm).

For each box containing the fragments, a survey with UVf was conducted to have a general view about the possible presence and distribution of fluorescent materials such as waxes, oils, and organic treatments. This technique, useful also to recognize red lakes, due to their characteristic pink-orange fluorescence, did not give significant results. With VIL images it was possible to identify and map Egyptian blue (Figure 11.4). After this first wide survey, FORS spectra on each fragment according to the different color present were acquired, following the results obtained by imaging techniques. All fragments were documented with RGB photography and, on the colored surfaces, detailed images at high magnification were acquired.

During the three campaigns in Hierapolis, it was possible to analyze hundreds of fragments from 11 stratigraphic units of excavation, being the fragments from each unit stored in more than one box. By crossing the information obtained with this wide screening it was possible to isolate, from each stratigraphic unit, the most representative fragments in terms of color, stratigraphy, mortars and plaster layers. This selection permitted to reduce the number of samples from hundreds to 36 that were further analyzed in the laboratories. Portable XRF spectroscopy (Bruker, Tracer III-SD) was employed to obtain elemental composition of the pigments. Then, the samples were further sampled under the optical microscope and the pictorial layers, plasters and mortars were analyzed by using microinvasive techniques: infrared spectroscopy (FTIR, Perkin Elmer, System 2000 and Agilent Technologies Cary 660 Series Spectrometer coupled with a microscope) and XRD (PANalytical, X' Pert PRO). The remaining fraction of the samples was used to prepare both cross and thin sections to be studied under the optical microscope (for cross sections in reflectance with Nikon Eclipse E600 and for thin sections in transmission in polarized with a Zeiss, Axioscope A.1). Cross sections were also analyzed by using Scanning Electron Microscope (SEM-EDS, FEI/Philips Electron Optics, ESEM Quanta200 equipped with an X-ray spectrometer) and with μ-Raman (Bruker, Senterra).

Interesting data were acquired on the painted plasters and mortars of the two main phases of the church: the early and middle Byzantine ones. The fragments

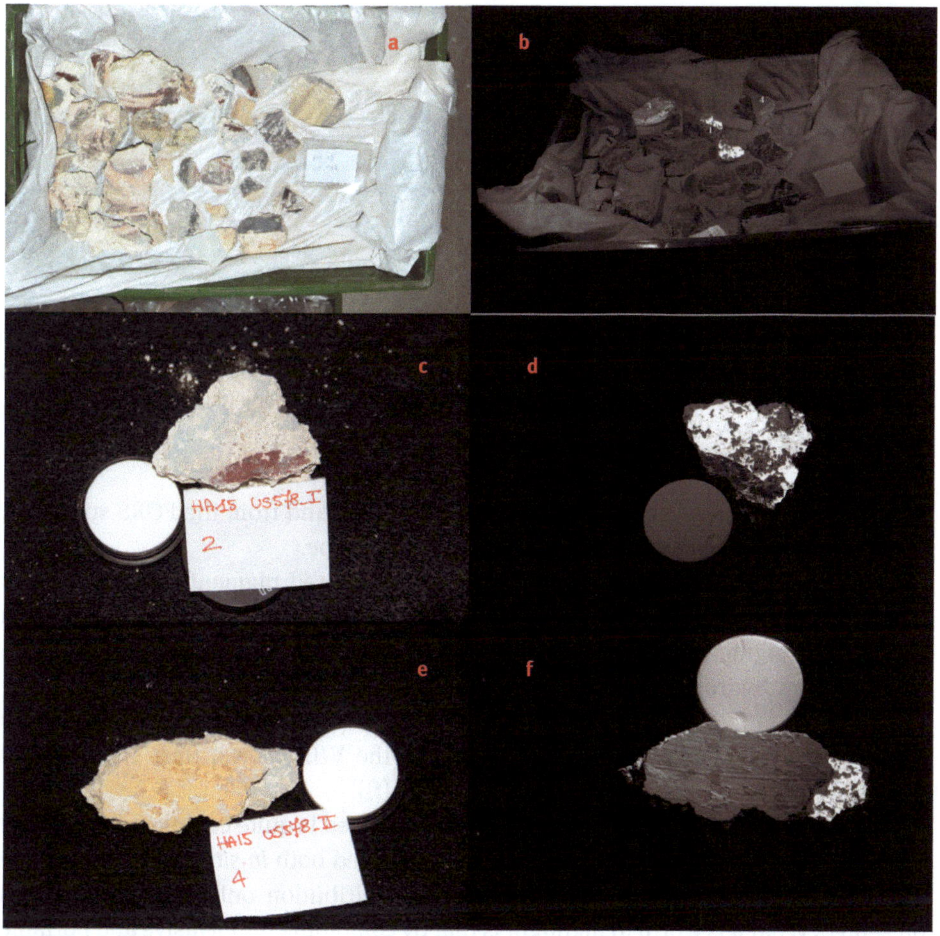

Figure 11.4: (a) RGB and (b) VIL images of one of the boxes containing part of fragments of the stratigrafic unit (US578). (c) RGB and (d) VIL images of one fragment with Egyptian blue on the external layer; (e) RGB and (f) VIL images of one fragment with Egyptian blue in the internal layer below the plaster and yellow painted layers.
Photos by Susanna Bracci.

belonging to the early Byzantine phase were characterized by a single layer of painted plaster while those referring to the middle Byzantine phase were characterized by different superimposed layers.

The blue areas analyzed on fragments belonging to both early and middle Byzantine phases were made with lapislazuli (sodium aluminum silicate). The characteristic minimum of reflectance (R%) at around 600 nm, related to the presence of this pigments, was evident (Figure 11.5).

Egyptian blue was present in light blue/green painted layers. Red-orange pigment, used in few fragments, was identified as minium (Pb_3O_4), as inferable from XRF

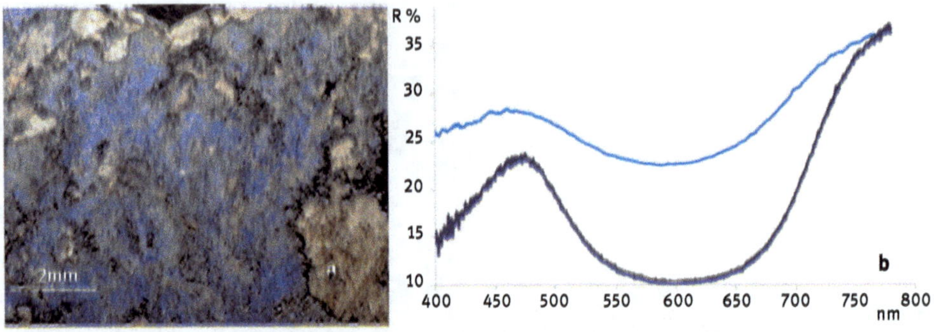

Figure 11.5: (a) Macrophotograph of a blue sample painted with lapislazuli and (b) the corresponding FORS spectrum (light blue line) compared with a lapislazuli reference one (dark blue line). Photo by Susanna Bracci.

spectra, detecting the presence of the lead (Pb) lines, and from the FORS spectra in which this pigment showed the characteristic "S" shape.

On the other hand, XRF and FORS spectra of the red pigments on the samples from the middle Byzantine phase showed only the spectral features related to iron oxide based pigments (Fe_2O_3), such as red ochre. For the yellow hues the pigments used are iron hydroxides ($FeO(OH)$). In some samples, under the painted and the plaster layers, another layer of the light blue color is recognized. This inner layer was obtained using Egyptian blue as evidenced in the VIL images, thus suggesting a successive decorative phase (Figure 11.4 (e) and (f)).

As already mentioned, the case study here reported is quite complex, the number of fragments analyzed *in-situ* and of the data collected both *in-situ* and in the laboratory were very huge. For this reason, in this contribution only part of them was presented. What it is worth to highlight is that by combining noninvasive and portable techniques *in-situ* with laboratory analyses it was possible to get information about almost all the fragments in the deposits (hundreds of fragments). In this complex system, painting materials, plaster and mortars were definitely identified with a focused and limited sampling campaign. The results obtained were useful to archeologists to define the different phases to which the Church was submitted in Byzantine period.

11.2.3 Keith haring's mural painting in Paris

Modern and contemporary wall paintings present a variety of new challenges. Artists nowadays can choose from among a huge variety of products made up of polymeric binders, colorants and inorganic fillers. This entails a different approach with respect to the characterization of a "classical" wall painting. Contemporary artists choose the materials available on the market not on the based on their chemical composition, but only on the final coloring effect not taking into account the possible incompatibilities of materials. Conservation scientists are thus faced with the challenge of identifying the composition of the materials used.

Keith Haring (born in 1958–died in 1990) was a pop American artist. Between 1980 and 1989, he achieved international recognition and participated in numerous group and solo exhibitions. Throughout his career, he devoted much of his time to public works, many of them commissioned, which often carried social messages. He produced more than 50 public artworks between 1982 and 1989, in dozens of cities around the world, many of which were created for charities, hospitals, children's day care centers and orphanages [69].

In the last years, two conservators, Antonio Rava and Will Shank, were entitled for the conservation of some of the Keith Haring's murals. The murals included the *Collingwood Mural* (1984) in Melbourne (Australia) and the *Tuttomondo* (1989) in Pisa (Italy). The latter was studied in the framework of the CoPAC Project (*Conservazione Preventiva dell'Arte Contemporanea*) funded by Tuscany Region (2011–2013) [70, 71]. Another one was the contemporary mural painting of Keith Haring at the Necker Children Hospital in Paris [69] which is the case study here reported. The mural painting was created in 1987, in typical Haring style, with large areas painted with primary colors and characters outlined with the hallmark black. This monumental work of art was painted on the outer walls of a gray concrete fire exit but, due to the weather and the passage of time, the painting was deteriorating. The concrete on which the paint was applied was crumbling and cracking, endangering the work's long-term survival. A number of black lines have degraded and some have detached from the concrete. In this case the conservators had to face the deterioration phenomena of the wall painting and therefore, fragments were collected from the ground below the mural painting assuming these samples to be representative of the whole works of art. The samples were sufficiently large to enable noninvasive and micro-invasive analyses to be performed in laboratory in order to characterize the materials and verify the state of conservation. Samples were studied by a multi-analytical protocol. XRF (Bruker Tracer III-SD) and X-ray diffraction (XRD, PANalytical, X'Pert PRO), were applied directly on the samples. After cross sections were obtained and observed under the optical microscope in reflectance both under VIS and UV light (Nikon Eclipse E600), scanning electron microscope (SEM-EDS, FEI/Philips Electron Optics, ESEM Quanta200 equipped with an X-ray spectrometer), infrared spectroscopy (μ-FT-IR Agilent Technologies Cary 660 Series Spectrometer coupled with a microscope) and μ-Raman spectroscopy (Bruker, Senterra), were applied on them [72] (Figure 11.6).

Observing the stratigraphy from the cross sections, starting from the innermost layer, the concrete contains a matrix where several crystals were embedded. They have various sizes ranging from 10 μm to 300 μm. This concrete layer showed a standard composition based on the presence of calcite ($CaCO_3$), quartz (SiO_2), gypsum ($CaSO_4{\cdot}2H_2O$) and calcium silicates, as highlighted by XRD. Traces of

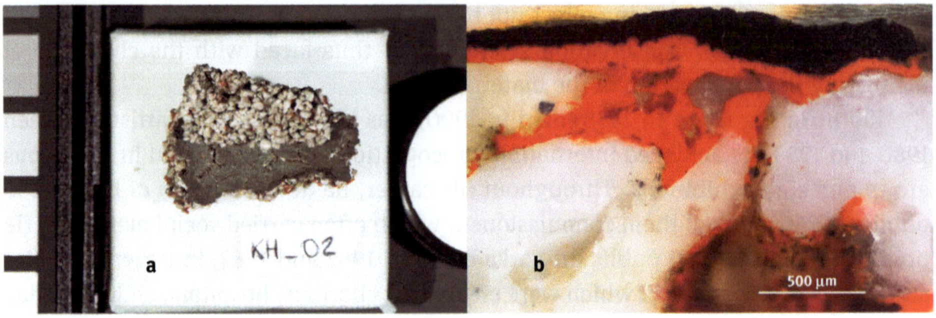

Figure 11.6: (a) Red sample KH_02 showing the black and red layers rolled up on themselves. (b) RGB stratigraphic image under the microscope of the cross section obtained from KH_02 sample. Photos by Susanna Bracci.

iron and aluminum were due to the presence of the corresponding oxides. On the top of this layer, a white ground was visible. It had a nonhomogeneous thickness ranging from 30 µm to 400 µm. XRD analyses identified alcite ($CaCO_3$) and gypsum ($CaSO_4 \cdot 2H_2O$). Titanium was also present in the crystalline form of rutile mineral (TiO_2).

This layer might be a kind of plaster used as a "priming" layer. It covers the substrate thus obtaining a flat and homogeneous surface, which would have provided protection, preventing the support from decaying. The upper coating layer of the support was a white matrix with some gray and white grains. XRF measurements showed high amounts of both calcium (Ca) and iron (Fe). XRD and FT-IR revealed the presence of calcite ($CaCO_3$), gypsum ($CaSO_4 \cdot 2H_2O$) with some traces of magnetite (Fe_3O_4) and aluminous silicates. Preliminary FTIR analysis of the binder of the painting layers, showed the presence of a vinyl resin. In order to confirm and to obtain a more reliable identification of the molecular composition of the paint samples, pyrolysis-GC/MS analysis was performed. This latter confirmed the use of a vinyl resin as binder along with the identification of two plasticizers, the vinyl ester of Versatic acid, the main constituent of VeoVa™, and butyl phthalate, commonly added during the synthesis of vinyl resins and paints [73].

Elemental techniques were not effective in identifying pigments, since no marker elements, (e.g. cadmium, chrome, cobalt) were found. These results however, enabled to exclude some inorganic metal-based pigments. The colored layers were painted directly on the coating with a thickness in between 100 µm and 200 µm, the paint sometimes penetrating into the cracks and porosity of the coating layer (Figure 11.7). The pigments were applied individually, as single layers, though in some fragments a black layer, above the colored one, was present. This latter corresponds to the boundary of the painted figures. The black pigment was identified as bone black by means of µ-FTIR spectra (signals due to PO_4^{3-} and OH^- groups). SEM-EDS analysis confirmed the presence of a high quantity of phosphorus (P) and iron (Fe) while the XRD technique

Figure 11.7: Stratigraphic RGB images of cross sections of blue (a), yellow (b) and green (c) samples under the optical microscope.
Photos by Susanna Bracci.

revealed the presence of magnetite (Fe_3O_4) and hydroxyapatite ($Ca_{10}(PO_4)_6(OH)_2$), the main component of bone black.

μ-Raman spectra acquired on the red layers revealed the presence of a Naphtol AS pigment belonging to the monoazo pigment group. Naphtol AS pigments represent the most important group as regards artistic use and PR112 is one of the most common ones in artists' palettes.

The Naphtol AS subclass pigments are identified by several characteristic and intense bands [74], even if an unequivocal identification is hard to achieve due the high similarity among the Naphtol AS Raman patterns. Within the numerous compounds belonging to Naphtol AS class, PR112 pigment ($C_{24}H_{16}Cl_3N_3O_2$) could have been used by Haring, since its characteristic twin peaks at 1392 cm^{-1} and 1375 cm^{-1} are visible in the Raman spectrum. Indeed, the co-presence of other monoazo Naphtol AS pigments is highly plausible, considering that some strong bands in the Raman spectrum cannot be ascribed to PR112 only. One possible hypothesis involves the use of a mixture of PR112 (the main colorant) and two yellow colorants [75], PY74 and PY73, which were detected in the yellow layers. Green and blue areas were painted with phthalocyanine pigments, as observed from the μ-Raman spectra. Phthalocyanines are a large group of pigments which share the common base structure of porphyrin. The blue varieties were discovered between 1907 and 1929, while the green varieties were discovered during the 1930s but have been available on the art market in the 1950s. In their structures some of the hydrogen atoms are replaced by chlorine or bromine in a variable number to obtain different colors and hues. The green colorant used for this artwork was PG7, as deduced from a comparative analysis with the literature [76], while PB15 is the phthalocyanine used for the blue areas. However, an unambiguous identification of the blue phthalocyanine polymorphs may be difficult and would require more specific methods such as chemometric analysis.

From the data obtained, apart from the extensive knowledge of the structure and materials used, the reasons for the diffuse detachments of fragments were identified. The combined use of a black pigment (carbon black) and the vinyl binder, with VeoVa™ and butyl phthalate as plasticizers, made the surface to be very sensitive to weather conditions. When irradiated by sun the surface temperature increases overcoming the glass transition temperature (Tg) of the polymer. This latter expands and becomes flexible. Then, when the temperature gets colder, the polymer shrinks and pulls out from the concrete. The phenomena are stronger for black pigment because all the energy it receives from the sun is converted into heat. Thus, the black areas are more prone to rolling up of the layer on itself with the consequence of detachment from the cement.

This project was developed in order to support the conservation campaign performed by conservators and therefore the detailed knowledge of the materials and of the ongoing degradation phenomena was a solid base for planning a focused and effective conservation project.

Acknowledgements: The authors gratefully acknowledge the all the colleagues working on the case studies here reported.

Case study 2.1: Marco Realini and Barbara Sacchi from CNR-ICVBC National Research Council, Institute for Conservation and Valorization of Cultural Heritage, Florence (IT); Barbara Mazzei from Pontificia Commissione di Archeologia Sacra, Rome (IT); Maria Gigliola Patrizi, Private Conservator, Rome (IT).

Case study 2.2: Emma Cantisani and Silvia Vettori from CNR-ICVBC National Research Council, Institute for Conservation and Valorization of Cultural Heritage, Florence (IT), Maria Piera Caggia, Francesco D'Andria from CNR-IBAM National Research Council, Institute of Archeological Heritage – Monuments and Sites, Lecce (IT).

Case study 2.3: Donata Magrini, Emma Cantisani, Claudia Conti and Antonio Sansonetti from CNR-ICVBC National Research Council Institute for Conservation and Valorization of Cultural Heritage Florence (IT), Antonio Rava, Private Conservator, Torino (TO); Will Shank from Conservation Resources Management, San Francisco, CA – USA; Maria Perla Colombini from University of Pisa, Department of Chemistry and Industrial Chemistry, Pisa, (IT).

References

[1] Piovesan R, Mazzoli C, Maritan L, Cornale P. Fresco and lime-paint: an experimental study and objective criteria for distinguishing between these painting techniques. Archaeometry. 2012;54:723–36.

[2] Davy H. Some experiments and observations on the colours used in painting by the ancients. Philos Trans R Soc Lond. 1815;105:97–124.

[3] Corzo MA, Afshar M, editors. Art and eternity, the Nefertari wall paintings conservation project 1986–1992. USA: The J. Paul Getty Trust, 1993.

[4] Sotiropoulou S, Papliaka ZE, Vaccari L. Micro FTIR imaging for the investigation of deteriorated organic binders in wall painting stratigraphies of different techniques and periods. Microchem J. 2016;124:559–67.

[5] Perez-Rodriguez JL, Robador MD, Centeno MA, Siguenza B, Duran A. Wall paintings studied using Raman spectroscopy: A comparative study between various assays of cross sections and external layers. Spectrochim Acta A: Mol Biomol Spectrosc. 2014;120:602–9.

[6] Jorge-Villar SE, Rodríguez Temiño I, Edwards HG, Hernández J, Ruiz Cecilia JI, Miralles I. The Servilia tomb: an architecturally and pictorially important Roman building. Archaeol Anthropol Sci. 2018;10:1207–23.

[7] Derrick MR, Stulik D, Landry JM. Infrared spectroscopy in conservation science. Scientific Tools for Conservation, The Getty Conservation Institute, USA. 1999. http://hdl.handle.net/10020/gci_pubs/infrared_spectroscopy. Accessed: Aug 2018

[8] Cather S, editor. The Conservation of wall paintings, proceedings of a symposium organized by the courtauld institute of art and the getty conservation institute, London, July 13–16, 1987. The Getty Conservation Institute, 1997. ISBN: 0-89236-162-X.

[9] Brecoulaki H, Andreotti A, Bonaduce I, Colombini MP, Lluveras A. Characterization of organic media in the wall-paintings of the "Palace of Nestor" at Pylos, Greece: evidence for a secco painting techniques in the Bronze Age. J Archaeol Sci. 2012;39:2866–76.

[10] Egel E, Simon S. Investigation of the paintings materials in Zhongshan Grottoes (Shaanxi, China). Heritage Sci. 2013;1:29.

[11] Fermo P, Piazzalunga A, de Vos M, Andreoli M. A multi-analytical approach for the study of the pigments used in the wall paintings from a building complex on the Caelian Hill (Rome). Appl Phys A. 2013;113:1109–19.

[12] Tomasini E, Castellanos Rodríguez D, Gómez BA, Da Faria DL, Landa CR, Siracusano G, et al. A multi-analytical investigation of the materials and painting technique of a wall painting from the church of Copacabana de Andamarca (Bolivia). Microchem J. 2016;128:172–80.

[13] Valadas S, Candeias A, Dias C, Schiavon N, Cotovio M, Pestana J, et al. A multi-analytical study of the fifteen century mural paintings of the Batalha Monastery (Portugal) in view of their conservation. Appl Phys A. 2013;113:989–98.

[14] Marić-Stojanovic M, Bajuk-Bogdanović D, Uskoković-Mrković S, Holclajtner-Antunović I. Spectroscopic analysis of XIV century wall paintings from patriarchate of Peć monastery, Serbia, spectrochimica acta part A. Mol Biomol Spectrosc. 2018;191:469–77.

[15] Mas S, Miguel C, Melo MJ, Lopes JA, de Juan A. Screening and quantification of proteinaceous binders in medieval paints based on µ-Fourier transform spectroscopy and multivariate curve resolution alternating least squares. Chemom Intell Syst. 2014;134:148–57.

[16] Glavcheva Z, Yancheva D, Velcheva E, Stamboliyska B, Petrova N, Petkova V, et al. Analytical studies of the Alexandrovo Thracian tomb wall painting, Spectrochimica Acta, Part A. Mol Biomol Spectrosc. 2016;152:622–8.

[17] Pelosi C, Agresti G, Andaloro M, Baraldi P, Pogliani P, Santamaria U, et al. Micro-Raman and microstratigrafic analysis of the painting materials in the rock-hewn church of the church of the forty martyrs in Şahinfendi, Cappadocia (Turkey). Archeometry. 2016;58:659–72.

[18] Gil M, Martins MR, Carvalho ML, Souto C, Longelin S, Cardoso S, et al. Microscopy and microanalyses of an extreme case of salt and biodegradation in seventeenth century wall paintings. Microsc Microanalyses. 2015;21:606–16.

[19] Nord GA, Tronner K, Olausson KB. Copper vanadate minerals founds in medieval mural paintings. Stud Conserv. 2012;57:183–6.

[20] Demir S, Şerifaki K, Böke H. Execution technique and pigment characteristics of Byzantine wall paintings of Anaia church in Western Anatolia. J Archeol Sci: Rep. 2018;17:39–46.

[21] Picollo M, Carlesi S, Bartolozzi G, Marchiafava V, Olmi R, Giuntini L, et al. Tecnica e degrado della Crocifissione dell'Angelico attraverso il programma di indagini: analisi, confronto e discussione dei risultati. In: Scudieri M, editor. La Crocifissione dell'Angelico a San Marco quarant'anni dopo l'intervento della salvezza. Indagini, restauri, riflessioni. Quaderni dell'Ufficio e Laboratorio Restauri di Firenze, Polo Museale della Toscana. Livorno, Italia: Sillabe s.r.l., 2016:80–96. ISBN: 978-88-8347-834-5.

[22] Crupi V, Allodi V, Bottari C, D'Amico F, Galli A, Gessini A, et al. Spectroscopic investigation of Roman decorated plaster by combining FT-IR, micro-Raman, and UV-Raman analyses. Vib Spectrosc. 2016;83:78–84.

[23] Paladini A, Toschi F, Colosi F, Rubino G, Santoro P. Stratigraphic investigation of wall painting fragments from Roman villas of the Sabine area. Appl Phys A. 2015;118:131–8.

[24] Gutman M, Lesar Kikelj M, Zupanek B, Kramar S. Wall paintings from the roman Emona (Ljubljana, Slovenia): characterization of the mortar layers and pigments. Archaeometry. 2016;58:297–314.

[25] Bacci M, Baldini F, Carlà R, Linardi R. A colour analysis of the brancacci chapel frescoes. Appl Spectrosc. 1991;45:26–31.

[26] Bacci M. Fiber optics applications to works of art. Sens Actuators B. 1995;29:190–6.

[27] Mairinger F. UV-, IR- and X-ray-imaging. In: Janssens KH, Grieken R, editors. Non-destructive microanalysis of cultural heritage materials, vol. 42, 1st ed. Antwerp: Wilson & Wilson Elsevier, 2004:15–73. ISBN: 9780444507389.

[28] Kanai T, Hirayama H, Fukunaga K. Digital imaging system. In: Tamassia M, editor. Salve mater, the annunciation by Frà Angelico in San Marco, Quaderni del Museo di San Marco, 1, Polo Museale della Toscana. Sillabe SrL Livorno, 2017:99–101. ISBN: 978-88-8347-590-0.

[29] Gil M, Rosado T, Ribeiro I, Pestana JA, Caldeira AT, Carvalho ML, et al. Are they fresco paintings? Technical and material study of Casas Pintadas of Vasco da Gama house in Évora (Southern Portugal). X-Ray Spectrom. 2015;44:154–62.

[30] Bartolozzi G, Catapano I, Croci M, Cucci C, Fukunaga K, Ludeno G, et al. Study of the annunciation fresco by using non-invasive diagnostic and analytical techniques. In: Tamassia M, editor. Salve mater, the annunciation by Frà Angelico in San Marco, Quaderni del Museo di San Marco, 1, Polo Museale della Toscana. Italy: Sillabe SrL Livorno, 2017:103–35.

[31] Buzzegoli E, Keller A. Ultraviolet fluorescence imaging. In: Pinna D, Galeotti M, Mazzeo R, editors. Scientific examination for the investigation of paintings. A Handbook for Conservator-restorers. Firenze: Publisher Centro Di Edifimi srl, 2009:204–6. ISBN: 978-88-7038-474-1.

[32] Mounier A, Dayet L, Daniel F, Belin C. Fluorescence UV des liants employés dans les dorures sur peintures murals médiévales, Archeo-Science. Revue de Archeométrie. 2011;35:19–28. ISBN: 978-2-7535-1847-6.

[33] Saunders D, Liang H, Billinge R, Cupitt J, Atkinson N. A new camera for high-resolution infrared imaging of works of art. Stud Conserv. 2006;51:277–90.

[34] Van Asperen de Boer JR. Infrared reflectography: a method for the examination of paintings. Appl Opt. 1968;7:1711–4.
[35] Nicola M, Aceto M, Gheroldic V, Gobetto R, Chiari G. Egyptian blue in the Castelseprio mural painting cycle. Imaging and evidence of a non-traditional manufacture. J Archaeol Sci: Rep. 2018;19:465–75.
[36] Dyer J, Verri G, Cupitt J. Multispectral Imaging in Reflectance and Photo-induced Luminescence modes: A User Manual. 2013. https://www.britishmuseum.org/pdf/charisma-multispectral-imaging-manual-2013.pdf. Accessed: Aug 2018.
[37] Verri G. The spatially resolved characterization of Egyptian blue, Han blue and Han purple by photo-induced luminescence digital imaging. Anal Bioanal Chem. 2009;394:1011–21.
[38] Accorsi G, Verri G, Bolognesi M, Armaroli N, Clementi C, Miliani C, et al. The exceptional near-infrared luminescence properties of cuprorivaite (Egyptian blue). Chem Commun. 2009;3392–4. DOI: 10.1039/B902563D
[39] Shugar AN, Mass JL, editors. Handheld XRF for art and archeology. Studies in Archeological Science 3. Belgium: Leuven University Press, 2012. ISBN: 978 90 5867 9345.
[40] Bacci M. UV-VIS-NIR, FT-IR, FORS spectroscopies. In: Ciliberto E, Spoto G, editors. Modern analytical methods in art and archaeology. Chemical analysis: a series of monographs on analytical chemistry and its applications, vol. 155. New York: John Wiley & Sons, 2000:321–61. ISBN:978-0-471-29361-3.
[41] Cesareo R, Castellano A, Buccolieri G, Quarta S, Marabelli M, Santopadre P, et al. Portable equipment for energy dispersive X-ray fluorescence analysis of Giotto's frescoes in the Chapel of the Scrovegni. Nucl Instrum Methods Phys Res B. 2004;213:703–6.
[42] Gebremariam KF, Kvittingen L, Banica F-G. Application of a portable XRF analyzer to investigate the medieval wall paintings of Yemrehanna Krestos Church, Ethiopia. X-Ray Spectrom. 2013;42:462–9.
[43] Bacci M, Boselli L, Picollo M, Radicati B. Ultraviolet, visible, near infrared fiber optic reflectance spectroscopy (FORS). In: Pinna D, Galeotti M, Mazzeo R, editors. Scientific examination for the investigation of paintings. A handbook for conservator-restorers. Firenze: Publisher Centro Di Edifimi srl, 2009:197–203. ISBN:978-88-7038-474-1.
[44] Miliani C, Rosi F, Daveri A, Brunetti BG. Reflection infrared spectroscopy for the non-invasive in situ study of artists' pigments. Appl Phys A. 2012;106:295–307.
[45] Barone G, Bersani D, Coccato A, Lauwers D, Mazzoleni P, Raneri S, et al. Nondestructive Raman investigation on wall paintings at Sala Vaccarini in Catania (Sicily). Appl Phys A. 2016;122:1–10.
[46] Pagés-Camagna S, Laval E, Vigears D, Duran A. Nondestructive and *in-situ* analyses of Egyptian wall paintings by X-ray diffraction and X-ray fluorescence portable systems. Appl Phys A. 2010;100:671–81.
[47] Almaviva S, Fantoni R, Colao F, Puiu A, Bisconti F, Fiocchi Nicolai V, et al. LIF/Raman/XRF non-invasive microanalysis of frescoes from St Alexander catacombs in Rome. Spectrochim Acta A: Mol Biomol Spectrosc. 2018;201:207–15.
[48] Blümich B, Haber A, Casanova F, Del Federico E, Boardman V, Wahl G, et al. Noninvasive depth profiling of walls by portable nuclear magnetic resonance. Anal Bioanal Chem. 2010;397:3117–25.
[49] Proietti N, Di Tullio V, Presciutti F, Gentile G, Brunetti BG, Capitani D. A multi-analytical study of ancient Nubian detached mural paintings. Microchem J. 2016;124:719–25.
[50] Di Tullio V, Capitani D, Presciutti F, Gentile G, Brunetti BG, Proietti N. Non-invasive NMR stratigraphy of a multi-layered artefact: an ancient mural painting. Anal Bioanal Chem. 2013;405:8669–75.
[51] Debastiani R, Simon R, Batchelor D, Dellagustin G, Baumbach T, Fiederle M. Synchrotron-based scanning micro-X-ray fluorescence applied to fragments of Roman mural paintings. Microchem J. 2016;126:438–45.

[52] Clementi C, Ciocan V, Vagnini M, Doherty B, Laurenzi Tabasso M, Conti C, et al. Non-invasive and micro-destructive investigation of the *Domus Aurea* wall painting decorations. Anal Bioanal Chem. 2011;401:1815–26.

[53] Gebremariam KF, Kvittingen L, Banica F-G. Physico-chemical characterization of pigments and binders of mural in a church in Ethiopia. Archeometry. 2016;58:271–83.

[54] Paradisi A, Sodo A, Artioli D, Botti A, Cavezzali D, Giovagnoli A, et al. Domus aurea, the 'Sala delle maschere': chemical and spectroscopic investigations on the fresco paintings. Archaeometry. 2012;54:1060–75.

[55] Robador MD, De Viguerie L, Pérez-Rodríguez JL, Rousselière H, Walter P, Castaing J. The structure and chemical composition of wall paintings from islamic and christian times in the Seville Alcazar. Archaeometry. 2016;58:255–70.

[56] Pereira-Pardo L, Gil M, Prieto B, Silva B. Multi-analytical approach to the material characterization of sixteenth century wall paintings from Ribeira Sacra (Galicia, NW Spain): pictorial palette, technique and alterations. Color Res Appl. 2016;41:263–9.

[57] Amato SR, Bersani D, Lottici PP, Pogliani P, Pelosi P. A multi-analytical approach to the study of the mural paintings in the presbytery of Santa Maria Antiqua al foro romano in Rome. Archaeometry. 2017;59:1050–64.

[58] Garofano I, Perez-Rodriguez JL, Robador MD, Duran A. An innovative combination of non-invasive UV-Visible-FORS, XRD and XRF techniques to study Roman wall paintings from Sevilla, Spain. J Cult Heritage. 2016;22:1028–39.

[59] Wooley E, Nadolny J, Shekede L. Tin relief on thirteenth century Cypriot wall painting: technology and conservation. Proc. of Vienna Congress 2012 The Decorative Conservation and the Applied Arts. The International Institute for Conservation of Historic and Artistic Work, 2012: S331–9. DOI: 10.1179/2047058412Y.0000000049.

[60] Mazzei B. Il cubicolo "dei fornai" nelle catacombe di Domitilla a Roma alla luce dei recenti restauri. Acta XVI Congressvs Internationalis Archaeologiae Christianae Costantino e i costantinidi l'innovazione costantiniana, le sue radici e i suoi sviluppi, Pars II, 1927-1942. Rome, 2013.

[61] Bracci S, Realini M, Sacchi B, Bartolozzi G, Patrizi MG, Mazzei B, et al. The cubicle "dei fornai" in Domitilla catacombs (Rome): non-invasive analyses for the characterization of the materials. In: Biscontin G, Driussi G, editors. Proceedings of the international congress "Scienza e Beni Culturali XXIX " Conservazione e Valorizzazione dei Siti archeologici. Approcci scientifici e problemi di Metodo. Venezia, Italy: Arcadia Ricerche srl., 2013:1161–72.

[62] Patrizi MG, Senserrich Espuñes R, Mazzei B, Mascalchi M, Agresti J, Osticioli I, et al. Il cubicolo "dei fornai" nelle catacombe di Domitilla a Roma considerazioni a conclusione della pulitura laser, APLAR4 - Applicazioni laser nel restauro. Saonara (PD): Il Prato Publisher, 2013:143–54. ISBN: 8863362025.

[63] FIRB. Progetto "Marmora Phrygiae", Futuro in Ricerca, MIUR. 2018. http://marmora.ibam.cnr.it/. Accessed: Aug 2018.

[64] MAIER, Missione Archeologica Italiana a Hierapolis. 2018. https://www.hierapolis.unisalento.it/15. Accessed: Aug 2018.

[65] Scardozzi G. The Marmora Phrygiae project in ancient quarries and building sites in Asia minor. In Ismaelli T, Scardozzi G, editors. Research on Hierapolis in Phrygia and other cities in south-western Anatolia: archaeology, archaeometry, conservation. Bari: EDIPUGLIA, 2016: 11–26. ISBN: 978-88-7228-819-1. DOI: 10.4475/819.

[66] Cantisani E, Vettori S, Bracci S, Caggia MP, Neri E, Pedro Leal AS. Fragments of painted plaster from the church of San Philip in Hierapolis: A preliminary archeological and archeometric study. In: Ismaelli T, Scardozzi G, editor(s). Ancient quarries and building sites in Asia minor. Research on Hierapolis in Phrygia and other cities in south-western Anatolia: archaeology,

archaeometry, conservation. Bari: EDIPUGLIA. ISBN: 978-88-7228-819-1. DOI: 2016:523–38. DOI: 10.4475/819.

[67] Cantisani E, Vettori S, Caggia MP. Mortars and plasters of the St Philip Church. In: Ismaelli T, Scardozzi G, editor(s). Ancient quarries and building sites in Asia minor. Research on Hierapolis in Phrygia and other cities in south-western Anatolia: archaeology, archaeometry, conservation. Bari: EDIPUGLIA. ISBN: 978-88-7228-819-1. DOI: 2016:511–21. DOI: 10.4475/819.

[68] Vettori S, Cantisani E, Ricci M, Bracci S, Conti C, Caggia MP. Archaeometric and archaeological study of painted plaster from the Church of St Philip in Hierapolis of Phrygia (Turkey). J Archeol Sci: Rep. 2018. submitted.

[69] Haring. 2018. http://www.haring.com. Accessed: Aug 2018

[70] Cucci C, Bartolozzi G, De Vita M, Marchiafava V, Picollo M, Casadio F. The colors of Keith Haring: A Spectroscopic study on the materials of the mural painting *Tuttomondo* and on reference contemporary outdoor paints. Appl Spectrosc. 2016;70:186–96.

[71] De Vita M, Bartolozzi G, Cucci C, Marchiafava V, Picollo M, Rava A, et al. La colorimetria a supporto degli interventi conservativi sui dipinti murali contemporanei. I murales di Keith Haring, Kermes La rivista del restauro Speciale 98 - Restaurare l'arte contemporanea, XVIII. 2015. ISBN: 978-88-942064-1-8

[72] Magrini D, Bracci S, Cantisani E, Conti C, Rava A, Sansonetti A, et al. A multi-analytical approach for the characterization of wall painting materials on contemporary buildings. Spectrochim Acta A: Mol Biomol Spectrosc. 2017;173:39–45.

[73] La Nasa J, Orsini S, Degano I, Rava A, Modugno F, Colombini MP. A chemical study of organic materials in three murals by Keith Haring: a comparison of painting techniques. Microchem J. 2016;940–8. DOI: 10.1016/j.microc.2015.06.003

[74] Scherrer NC, Zumbuehl S, Delavy F, Fritsch A, Kuehnen R. Synthetic organic pigments of the 20th and twenty-first century relevant to artist's paints: raman spectra reference collection. Spectrochim Acta A: Mol Biomol Spectrosc. 2009;73:505–24.

[75] Ropret P, Centeno SA, Bukovec P. Raman identification of yellow synthetic organic pigments in modern and contemporary paintings: reference spectra and case studies. Spectrochim Acta A: Mol Biomol Spectrosc. 2008;69:486–97.

[76] Buzzini P, Massonnet G. A market study of green spray paints by Fourier transform infrared (FTIR) and Raman spectroscopy. Sci Justice. 2004;44:123–31.

Márcia Vilarigues, Inês Coutinho, Teresa Medici, Luís C. Alves, Bernard Gratuze and Andreia Machado

12 From beams to glass: determining compositions to study provenance and production techniques

Abstract: Combining a stylistic approach with chemical analyses and focusing not only on composition but also on production technology, this research aims to identify possible centres of post-medieval production of glass as well as decorative production techniques used in stained glass, in particular the grisaille.

The use of several combined analytical techniques allowed for the unveiling of the composition of glass and the structure of grisailles. Sorting out the grisaille structure sheds light on the processes employed to produce paints, while compositions of glass objects are used to propose production provenances. Both contribute to the study and critical analysis of how glass heritage has developed over time, including materials and technical practices, uncovering the transfer of technology and movement of people and goods.

Keywords: glass, grisaille, PIXE, LA-ICP-MS

12.1 Sources and provenience of glass raw materials: some general considerations

12.1.1 Silica sources

The glass structural component or vitrifying agent is silica, which can be obtained from sand or crushed quartz pebbles. Concerning the use of sand, despite its possible degree of purity, in addition to silica, several other components, considered impurities, will enter the glass batch. Calcium carbonate, magnesium oxide and alumina are the sand impurities known for giving glass a chemical stability, and iron oxide is responsible for the natural (and sometimes unwanted) green, blue or yellow hues. In addition, the rare-earth elements (REEs) will become part of the glass composition [1, 2]. All these elements present in the sand structure are the main trace elements and trackers for the sand silica source, giving us the possibility to discuss the raw materials provenance and consequently to consider, in a more informed way, the glass provenance through its main component [3–7].

This article has previously been published in the journal *Physical Sciences Reviews*. Please cite as: Vilarigues, M., Coutinho, I., Medici, T., Alves, L., Gratuze, B., Machado, A. From beams to glass: determining compositions to study provenance and production techniques. *Physical Sciences Reviews* [Online] **2019**, 4. DOI: 10.1515/psr-2018-0019.

https://doi.org/10.1515/9783110457537-012

12.1.2 Fluxing agents

Flux agents are added to the batch with the intent of lowering the melting point of the mixture, since pure silica melts at c. 1,700 °C. These oxides, usually alkalis, can be denominated as network modifiers, since their introduction will alter the glass network [2, 8, 9]. The most common fluxing agents used in the past were natron, and soda-rich and potassium-rich ash, being the last ones of vegetal origin. Natron was widely used as a flux in glass production since the first millennium BC until the ninth century from the present era, when it started to be gradually replaced by plant ash [10–14]. In the Mediterranean area, the use of natron was replaced with ashes obtained from plants located near coastlines like the Chenopodiaceae family (comprehending many genera like *Suaeda, Salsola* and *Salicornia*), which were rich in soda [15, 16]. This was the technology used by the prominent Venetian production [11, 13].

In central Europe natron was replaced with ashes from forest plants (for instance fern) and wood, richer in potassium oxide (potash-rich ashes) and with the contents of calcium and magnesium oxides also higher in comparison with soda-rich ashes [17, 18].

During the sixteenth century, this regional division started to blur due to the emigration of Italian glassmakers bringing the soda-rich glass tradition North of the Alps [6, 18–21].

When the produced glass was meant to be very pure and colourless, it was common to import ashes with a higher quality (having fewer impurities) comparing to the ones harvested locally. In Murano, Venice, Levantine ashes were the preferred ones. They arrived in Murano from Egypt, Israel, Syria and Lebanon [15, 22, 23]. Barilla ashes are also rich in sodium and these were imported into Venice from Spain [11, 22], Table 12.1.

From the fifteenth century onwards, Venetians glassmakers started to treat the ashes in order to obtain a special glass known as *cristallo*. This glass that was meant to imitate the precious stone rock crystal was made with purified ashes that needed to be boiled in water, filtered and the remaining residue treated to extract sodium carbonate [22, 25].

At the end of the seventeenth century/beginning of the eighteenth century, raw materials like saltpetre (potassium nitrate), chalk and limestone were being introduced into the batch as sources of potassium and calcium [26, 27]. About the same time in England, lead was being introduced into the batch to act as a stabilizer and to lower the glass melting temperature. In both situations, the main purpose was to achieve a perfectly discoloured glass formulation even when the glass walls had a higher thickness [28–30].

12.1.3 Stabilizers

To chemically and structurally stabilize the glass, some of the alkaline ions that were introduced as flux agents will be replaced with alkaline earth ions (with less

Table 12.1: Chemical composition of the *façon-de-Venise* and the gourd-shaped vessel samples, its majority determined by µ-PIXE, in weight percent of oxides.

Samples	Na_2O	MgO	Al_2O_3	SiO_2	P_2O_5	SO_3	Cl	K_2O	CaO	TiO_2	MnO	Fe_2O_3	CoO	PbO
							Venetian and façon-de-Venise glass							
SCV-V191	16.3	2.5	1.2	67.7	<0.12	0.08	0.69	3.79	6.62	0.06	0.31	0.38	<20 µg/g	0.03
SCV-V193	18.1	2.8	0.8	68.0	0.13	0.12	0.61	2.47	5.91	0.03	0.38	0.33	<20 µg/g	0.03
SCV-V194	19.2	3.2	6.1	57.8	0.28	0.05	0.75	4.69	6.06	0.25	0.55	0.77	150 µg/g	0.25
SCV-V195	18.0	2.7	0.8	68.3	0.11	0.11	0.65	2.39	5.68	0.04	0.45	0.39	<25 µg/g	0.03
SCV-V404	13.7	4.1	1.3	65.5	<0.06	0.05	0.60	6.53	6.76	0.22	0.57	0.53	<20 µg/g	<20 µg/g
SCV-V408*	11.5	3.5	1.5	64.0	0.19	n.m.	0.84	5.46	11.89	0.07	0.34	0.61	<10 µg/g	10 µg/g
PMF0527	16.9	3.2	2.1	60.8	0.27	0.05	0.87	5.26	9.53	0.07	0.51	0.46	50 µg/g	<10 µg/g
PMF0530	15.9	3.0	3.9	66.4	0.33	0.05	0.93	2.01	6.00	0.13	0.49	0.78	0.01	<0.01
PMF0540	18.6	1.9	6.0	62.9	0.42	0.02	1.30	3.76	3.22	0.16	0.92	0.82	0.01	<0.02
PMF0546	18.2	3.0	4.6	59.5	0.31	0.05	0.97	4.40	7.53	0.15	0.48	0.61	0.01	0.03
PMF0550	16.8	2.2	3.2	66.9	0.30	0.04	1.14	2.47	6.19	0.08	0.28	0.30	50 µg/g	<40 µg/g
PMF0556	15.1	3.2	2.6	62.5	0.18	0.06	0.83	5.99	8.79	0.06	0.30	0.27	<30 µg/g	<10 µg/g
PMF0568	15.7	2.5	3.6	66.8	0.21	0.05	0.91	3.36	5.48	0.12	0.50	0.62	60 µg/g	0.01
PMF0569	15.6	2.4	3.6	67.3	0.18	0.06	0.93	3.06	5.48	0.11	0.63	0.58	60 µg/g	<40 µg/g
PMF0570	16.2	3.0	4.0	66.6	0.34	0.05	0.94	1.95	5.79	0.12	0.42	0.67	0.01	<0.01
PMF0996	15.7	2.4	3.6	67.0	0.23	0.06	0.94	3.06	5.56	0.11	0.66	0.61	0.01	<0.01
PMF1010	16.1	2.2	2.3	66.4	0.22	0.06	1.00	4.08	6.63	0.07	0.37	0.53	60 µg/g	0.06
PMF1023	16.9	2.6	3.5	62.8	0.31	0.05	1.22	4.25	7.52	0.12	0.13	0.62	0.08	<0.03
PMF1025	15.0	2.4	3.5	66.9	0.26	0.05	0.92	3.30	5.93	0.14	0.76	0.72	60 µg/g	<0.01

(continued)

Table 12.1 (continued)

Samples	Na_2O	MgO	Al_2O_3	SiO_2	P_2O_5	SO_3	Cl	K_2O	CaO	TiO_2	MnO	Fe_2O_3	CoO	PbO
SJT0011	18.7	2.4	1.8	67.9	0.22	0.07	0.99	2.96	4.05	0.08	0.27	0.52	40 µg/g	0.02
SJT0012	18.6	3.8	2.7	62.3	0.16	0.05	0.91	3.21	7.06	0.08	0.35	0.62	0.01	0.05
SJT0014	14.0	3.2	2.6	61.2	0.21	0.08	0.67	6.91	9.66	0.09	0.47	0.62	50 µg/g	0.05
SJT0038	19.5	2.6	2.1	63.7	0.20	0.08	0.96	4.59	5.31	0.06	0.36	0.48	30 µg/g	0.01
SJT0105	18.0	2.5	1.8	66.5	0.16	0.08	0.85	4.17	4.63	0.10	0.43	0.71	50 µg/g	0.01
SJT0112	14.7	3.5	2.3	62.1	0.14	0.06	0.69	8.22	7.31	0.07	0.36	0.50	0.01	0.02
SJT0122	18.0	2.5	2.6	65.9	0.29	0.05	0.98	2.93	5.27	0.10	0.49	0.80	0.01	0.03
SJT0123	19.7	2.1	2.9	62.9	0.23	0.07	1.01	3.54	6.12	0.10	0.41	0.70	0.01	0.01
SJT0135_ body glass	12.7	3.0	2.3	61.3	0.28	0.05	0.64	9.74	8.04	0.11	0.49	0.82	0.01	0.01
CPU0006	18.4	2.2	1.9	65.7	0.26	0.06	1.14	2.95	6.29	0.07	0.28	0.48	50 µg/g	0.03
							Gourd-shaped vessels							
SCV-V079	19.9	5.6	6.8	55.5	0.45	0.03	0.86	2.08	6.67	0.28	0.35	1.26	<60 µg/g	<20 µg/g
SCV-V082	15.0	4.8	4.8	64.0	0.43	0.02	0.74	2.36	5.09	0.23	1.22	0.94	<40 µg/g	<10 µg/g
SCV-V115	17.6	3.1	2.3	66.5	0.09	0.07	0.76	2.53	6.07	0.09	0.02	0.64	<30 µg/g	0.01
SCV-V177	18.9	4.2	7.7	55.3	0.18	0.04	0.56	4.32	7.40	0.17	0.58	0.59	<30 µg/g	<30 µg/g
SCV-V210	18.3	2.9	1.4	62.9	0.00	0.24	0.69	2.12	5.58	0.04	0.46	0.39	70 µg/g	1.89
SCV-V352	20.6	5.5	7.7	56.7	0.33	0.02	0.91	1.77	3.26	0.31	1.13	1.66	<0.01	<30 µg/g
SCV-V365	18.6	6.1	6.1	60.8	0.35	0.06	1.00	1.43	3.22	0.31	0.43	1.26	<20 µg/g	<10 µg/g
SCV-V390	16.4	6.6	9.2	54.2	0.73	0.03	0.78	1.89	5.03	0.36	2.08	2.37	<20 µg/g	<0.01
SCV-V423	16.9	6.6	9.2	54.3	0.73	0.02	0.82	1.85	4.85	0.34	1.93	2.25	0.01	<30 µg/g

*SJT0128	17.2	6.9	6.5	58.0	0.54	0.02	0.88	1.51	6.10	0.24	0.06	1.87	0.01	<40 µg/g
CMOG B (µ-PIXE)														
Measured	15.5	0.9	4.3	64.5	0.46			1.05	8.48	0.10	0.28	0.38		0.51
Ref. [24]	17.0	1.0	4.4	62.3	0.82			1.00	8.56	0.09	0.25	0.34		0.61
CMOG A (LA-ICP–MS)														
Measured	13.5	2.4	0.9	68.0	0.11			2.85	5.39	0.75	0.98	1.06	0.17	0.07
Ref. [24]	14.3	2.7	1.0	66.6	0.13			2.87	5.03	0.79	1.00	1.09	0.17	0.12

*The composition of this fragment (SCV-V408) was obtained only by means of LA-ICP–MS.
n.m. Stands for "not measured".

mobility). These elements will create a bridge between two or more oxygen ions, re-establishing this way the glass matrix and enhancing its thermal, chemical and mechanical resistances [8, 9, 22].

The alkaline earth oxides (Ca, Mg) acting as natural glass stabilizers were introduced into the batch together with silica and flux raw materials (e. g. sand and plant ashes) as their impurities; alumina was introduced through sand rich in feldspars and in less quantities through non-purified vegetable ashes and contamination from crucibles [31–33].

It is generally accepted that, until the nineteenth century, alkaline earth oxides and alumina were added to the batch without the glassmakers knowing their role on the glass matrix stability [31, 32]. Nevertheless, the possibility that skilled glassmakers might have realized that adding some lime to the batch, in the form of shells for instance (a fact mentioned already by Pliny in the first century), improved the glass resistance when in contact with water should be considered [1, 33, 34].

Concerning lead oxide, it was primarily used as a glass stabilizer and it was commonly introduced into the batch as litharge or minium; however, it was later used as a flux agent. Depending on its concentration, lead oxide can be a glass stabilizer, fluxing or vitrifying agent [9, 32].

12.1.4 Fining agents, opacifiers and colouring agents

A fining agent is used to reduce or eliminate bubbles from the glass. The first intentionally used fining agent was arsenic, during the seventeenth century [9, 22].

Colourants are metallic oxides that, depending on the firing conditions (atmosphere and temperature), will determine the final colour of the glass. Some of these elements were intentionally added to the batch and others were naturally present in the raw materials, as is the case of iron present in the sand. Iron can give colours from bluish-green to yellowish-green depending on whether the firing conditions are reductive or oxidizing. Cobalt was used to attain a deep blue, while copper was introduced in the glass batch on purpose to give the glass a turquoise, green or red colours. Manganese oxide was employed in order to obtain a purple colour; however, it was also used with the purpose of decolourizing the glass, counteracting the colouring effect of the iron. This element also acts as a fining agent [31, 32].

Looking finally to the opacifying agents, these are not many and remain almost unchanged throughout history. The glass opacity is the result of the precipitation of crystalline or colloidal compounds during the glass cooling phase. These precipitates will block the transmission of light [35]. One of the first opacifying agents was antimony, and later compounds based on tin were preferred [36–38]. The white opaque colour can be the result of two different techniques: the precipitation of calcium antimonate, used since the fifteenth century BC and throughout all the roman period, being later reintroduced during the second half of the sixteenth century in Venice, or the tin dioxide (also known as cassiterite), which was preferred in Venice in the renaissance period [1, 39, 40]. Opaque white glass was known in Venice as *lattimo* and was used in decorative elements as the filigree canes but also for blowing objects [35]. Opaque yellow glass was obtained by lead stannate or lead antimonate [31, 35, 38, 41]. Opaque red glass was obtained by addition of copper oxide to a base glass in reducing conditions [40, 42, 43].

Still considering the opacifying agents, it is worth mentioning calcium phosphate obtained by bone ash, in use from the fifth century onwards [38, 41, 44, 45], and arsenic oxide, employed in Venice from the sixteenth century onwards to opacify lead glass white enamels and *girasole* glass: [11, 22].

12.1.5 Raw materials used for the production of grisailles

Grisaille paint is the oldest and one of the most significant painting materials used in stained glass. Grisailles are made of a lead–silica glass and a colorant agent (usually Fe_2O_3, CuO mixed or isolated).

Different recipes for lead–silica glass can be found in the written sources. The main difference between them is the proportion of lead oxide to silica, which varied from 1.5 to 3, as well as the change of the source of lead in the nineteenth century, when the calcined lead was replaced by minium, normally found in nature and used as pigment in illuminations and paintings [46]. Jewish glass, *Pasta* and *Fondant A* are just some of the expressions used for the lead–silica glass. The most common one is *Rocaille* mentioned for the first time in the seventeenth-century Pierre Lebrun's manuscript, *Recueuil des essaies des merveilles de la peinture* [Merrifield, vol. II, [47], but its first recipe appeared in the André Felibien's work [48], 40 years later. This lead–silica glass has been used until the nineteenth century not only in grisailles but also in sanguine paint [49].

Iron and copper oxides were obtained from the burned beaten pieces of iron and/or copper as described in the historical written sources, such as Theophilus, Antonio da Pisa or Johannes Kunckel. The colouring agents were the same along the centuries up to the nineteenth century, when the recipes started to refer to the use of pigments such as hematite or umber. Manganese also appears as a colouring agent in two recipes from the Kunckel's treatise (seventeenth century). These recipes, similar to black enamel ones, could be an experience from the author to substitute the iron and copper grains by manganese oxides.

12.2 Analytical technique for glass characterization

The research methodology was designed on the basis of an interdisciplinary approach, relating archaeometry, history and conservation science. These different but complementary approaches allowed for the characterization of the chemical composition of the glass in terms of major, minor and trace elements, and the discussion of the raw materials employed for the different glass composition types. This will also allow for a better definition of a possible provenance for the glass, and for the disclosure of the technology involved in the making of the grisaille paintings. Finally, a study of the historical materials and techniques used for the production of objects and grisailles and hypothesis on commercial routes of the raw materials and objects will also be proposed.

The technique of laser ablation inductively coupled plasma mass spectrometry (LA-ICP–MS) has been widely employed in the study of cultural heritage objects such as glass [3, 7, 50] and metals [51], and also in the study of the natural glass obsidian [52–54], because of its capability to analyse solid samples. This technique has the great advantage of its very high detection limits that can go to ng/g levels. The disadvantage of LA-ICP–MS is the hole that the laser ablation leaves behind, and the size of this crater can range between 20 and 200 µm. The ablated material (ca. 80–90 % of it) is then deposited around the craters and can be easily removed [55]. However, it is important to point out that in the specific case of glass analysis with the aim of studying provenances, the advantages of this method, that allows one to quantify not only the major, minor but also trace and REE elements, overwhelm the disadvantage of its being a destructive technique, in a micrometer scale.

Ion-beam analytical techniques such as the particle-induced X-ray emission (µ-PIXE) are increasingly being employed in the study and characterization of cultural heritage. One of the great advantages of these techniques, and probably what makes them so attractive for use with historical objects, is the practically negligible damage caused by the ion beam bombardment [56]. This technique allows for the quantification of major, minor and some trace elements. It also allows for elemental mapping of selected areas, being possible to study and quantify the elements present in corrosion layers, to observe and study the elemental diffusion among different layers, among others.

PIXE

Being an X-ray spectroscopy analytical technique, the physical fundamental differences and properties of PIXE from the more widely used XRF (X-ray fluorescence) or SEM–EDS (scanning electron microscopy with energy dispersive X-ray spectroscopy) are consequence of the fact that the ionization of the atomic inner shell electrons that is followed by the emission of elemental characteristic X-rays is due to the interaction of MeV charged particles instead of keV photons or electrons. Using solid-state detectors (Si(Li), silicon drift detectors [SDD], High Purity Ge) for the X-ray detection provides the technique with multielemental analytical capabilities.

The sensitivity of the technique depends on several radiative contributions to the obtained spectra starting with the characteristics elemental X-ray yield. This yield is dependent on the bombarding particle X-ray production cross-section which is higher for protons than for any other heavier ion with the same energy. This is the main reason why protons are usually used in PIXE and also why the acronym is most of the time referred to as proton-induced X-ray emission. Apart from the characteristic elemental X-rays, the radiative contributions to a PIXE spectrum derive from bremsstrahlung radiation due both to the projectile and the kicked out electrons (continuum radiation on the low-energy part of the spectrum with higher contribution from the electrons) and to the interaction of gamma rays resulting from nuclear reactions with light elements (continuum background due to Compton scattering in the detector). Although X-ray production cross-section increases with energy, the bremsstrahlung and nuclear reaction contribution to the spectrum background (both increasing with bombarding particle energy) make that best sensitivity can be attained with protons of energy up to 3 MeV, thus enabling the use of small accelerators. The X-ray production cross-section is also found to decrease quite fast with the atomic number of the excited element what should favour the detection of light elements. However, the final elemental sensitivity depends on the physical characteristics of the detector used such as the active material, its width and the detector window thickness. Common PIXE experimental set-ups allow to detect Na and heavier elements with a sensitivity reaching the µg/g level. Detector energy resolution is also an important factor for the technique sensitivity when the X-ray energy from several elements is similar.

One of the most important characteristics of the technique is its non-destructive character which is essential for the analysis of some CH materials or allows the subsequent analysis by some other complementary techniques. Notice however that this non-destructive characteristic may be hampered by its surface analysis performance that may require special sample preparation for removing the surface altered layer in order to obtain confident bulk sample analysis.

Quantitative analysis

Quantitative determination is performed on the basis that the registered elemental X-ray yield is a function of the number of atoms or mass of the corresponding element present in the sample and of the number of particles that impinge on the target. That function can be written as [57, 58]

$$Y_{Zi} = K(\Omega/4\pi)\sigma_{Zi}(E_0)\ \varepsilon_{Zi} Q \int_{E}^{E0} C_{Zi}\sigma_{Zi}(E)/\sigma_{Zi}(E_0)\ T_{Zi}(E)(\mathrm{d}x/\mathrm{d}E)\mathrm{d}E \tag{12.1}$$

where $T_{Zi}(E)$ is given by

$$T_{Zi}(E) = \mathrm{EXP}\left\{ - \left[\left(\frac{\cos\alpha}{\cos\theta} \right) \left(\mu_{T_{Zi}} \int_{E}^{E0} S(E)^{-1} \mathrm{d}E \right) \right] \right\} \tag{12.2}$$

where the index Zi refers to the element with atomic number Z and characteristic X-ray line i (K_α, K_β, L_α, L_β, etc.), $\sigma_{Zi}(E)$ being the X-ray production cross-section at proton energy E, $S(E)$ the stopping-power of the sample matrix (translates the proton energy loss inside the sample), C_{Zi} is the elemental concentration, ε_{Zi} is the detector efficiency combined with the absorption between sample and detector front window (e. g. experimental vacuum chamber window, air, extra X-ray filter), $\Omega/4\pi$ is the solid angle fraction subtended between sample and detector and T_{Zi} is the depth dependent sample absorption of the i radiation. This T_{Zi} depends on geometric parameters such as the angles with the normal to the sample surface of the incoming beam (α) and of the outgoing X-ray (θ) and also on μ_{Zi} the sample X-ray absorption coefficient for the i radiation. Finally, K is a constant involving only physical constants (such as the Avogadro number and elemental charge).

From these two equations, the following conclusions can be drawn: the reverse problem of finding concentration values once knowing the X-ray yield can only be properly solved if C_{Zi} is considered to be a constant and taken out of the integral (there is no sample depth information and then the sample should be uniform and homogeneous); the results are geometry dependent (α and θ angles) meaning that the sample surface should be as flat as possible (higher influence for low energy X-rays); Q, the accumulated proton charge, proportional to the number of protons impinging the sample, is the spectra normalization factor and should be well measured. Else, we must be confident that the detected elements and the associated chemical compound or bonding are all there is in the sample (sample only contains oxides of elements heavier than Na, for instance) and perform a concentration normalization to 100 %.

All other equation parameters are tabulated, can be found by calibration or obtained through an iterative procedure. Calculations can then be made using a fundamental parameters approach without the need of standard reference materials. Also notice that the limit E of the integral can be extended from $E0$ (initial proton beam energy, then allowing to simplify the equation and using the so-called thin sample approximation) to 0 (zero) where the sample is sufficiently thick to stop the beam (thick target procedure). For a 2-MeV proton beam, although dependent on sample composition, this thickness is around 20–30 µm which makes PIXE a surface analytical technique.

Several software packages are available for performing both PIXE X-ray spectra deconvolution and quantitative analysis such as GeoPIXE [59], GUPIXWIN [60], PIXEKLM [61] and OMDAQ2007 [62].

Nuclear Microprobe

Complementary to the so-called macroPIXE set-up were the lateral dimensions of the used beam that are defined by a set of collimators, the possibility of focusing charge particles by means of electro or electromagnetic lenses as well as performing beam scanning adds beam spatial resolution down to the µm level as well as imaging capabilities. This allows µ-PIXE to become a microscopy technique with quantitative capabilities. Typical in-vacuum chamber nuclear microprobe analysis allows a 2 MeV proton beam to probe a sample surface area up to 2.6 × 2.6 mm^2 while obtaining PIXE elemental distribution maps with at least 256 × 256 pixels image resolution [63, 64]. In-vacuum experimental chamber, however, limits the size of the samples that can be analysed. On account of its size or delicacy, samples can still be analysed using an external beam set-up maintaining the technique capabilities and sensitivity when flooding the proton beam path from the exit nozzle to the sample and the incoming path of X-rays from sample to the detector with helium then reducing beam energy loss and X-ray attenuation.

Comparison with XRF and SEM–EDS

Advantages and drawbacks can always be taken into account when comparing with similar or competitive techniques.

Comparison with SEM–EDS is quite unfavourable for µ-PIXE in what accounts to beam spatial resolution, but quite favourable if elemental sensitivity is considered. Even though 20 or 30 keV electrons have a higher X-ray production cross-section than 2 or 3 MeV protons [65], the much higher bremsstrahlung radiation spectrum contribution of a SEM–EDS renders the technique quite less sensitive than PIXE (~0.1% against µg/g level). Both techniques allow beam scanning and 2D elemental distribution maps. SEM–EDS also probes a sample depth much shallower (up to 2 µm) than MeV proton beams (up to ~50 µm).

In what concerns, the comparison to XRF traditional systems, the X-ray production cross-section increases with Z (atomic number) while for protons strongly decreases with Z, which combined with the typical background radiation from both techniques makes PIXE more sensitive to light elements and XRF typically more sensitive to heavy elements. Notice however that the increment in Z of the XRF cross-section is dependent on the kV accelerating voltage and secondary target used once no fluorescence can be observed for transitions with energies greater than the excitation source. Still being a surface analysis, XRF beams can probe sample depths over 100 µm. Although X-ray beams can be focused by using polycapillary lens (typically down to 50 µm), beam scanning cannot be performed. However, elemental mapping is still possible but only by using sample or X-ray system scanning. On the other hand, the whole capabilities of an XRF energy dispersive spectrometer can be arranged in small and portable systems that may then be crucial when museum or other *in-situ* analyses are mandatory.

PIXE analysis of glass samples

According to the PIXE technique characteristics, some care should be taken when analysing historical glass samples. As previously mentioned, PIXE is a surface analysis and then quantitative results are too much dependent on the samples surface corrosion carried by the strong leaching of some elements and in particular of Na. If an accurate glass bulk composition is required, the corroded glass surface should be removed. Whenever possible, this should be achieved by polishing either a small zone of the entire object or small glass samples embedded in resin that were previously cut from selected glass fragments. Confident glass bulk composition can also be obtained if analysing the cross-section of freshly broken fragments.

Glasses are insulating samples and when analysed in vacuum both impinging charged particles and kicked off electrons frequently promote a large charge build up that degrades the collected X-ray spectrum. The most common solution is to apply a thin conductive carbon coating layer (by thermal evaporation, sputtering or spray means) for samples analysed in vacuum or their analysis in an external beam set-up.

In what accounts to data spectra collection, a single spectrum could be obtained containing information for all possible elements. However, using a 2 MeV or higher proton beam energy would imply that protons would enter a common 8-µm-thick Be windowed detector and degrade the detector crystal. On the other hand, as PIXE yield is high for low Z elements (that including Si the major element in glass), the beam current should be kept low (in order to keep a reasonable total X-ray count rate inferior to 3,000 counts/s) that would account for long irradiation time if heavy element trace amounts are to be determined. The ideal solution would be to have two detectors simultaneously: a low solid angle detector with magnetic backscattered protons deflection and an high solid angle detector with an X-ray filter in front of it for providing efficient detection of heavy elements X-rays and at the same time reducing the large light elements contribution also preventing backscattered protons from entering into the detector. As an alternative, information can be obtained with one detector in two steps: the first one using low proton beam energy (0.7–1.0 MeV) for preventing backscattered proton to cross the common 8-µm-thick Be detector window and at the same time providing good sensitivity for light elements; the second step is to use a filter in front of the detector (Mylar filter with 50–350 µm thickness) and increase the beam energy and current for detecting heavy elements in trace amounts.

Using the microprobe elemental distribution capabilities also allows to ascertain the homogeneity of the analysed sample or define analysis of representative areas.

Summary

PIXE analysis using MeV proton beams is a powerful technique for the study of different type of samples in several science fields such as Material Science, biology/biomedicine and Cultural Heritage due to its fine sensitivity (down to the µg/g level), versatility, non-destructive characteristics as well as for its multi-elemental analytical capabilities. The possibility of beam focusing and beam scanning adds spatial resolution down to the µm level and imaging capabilities then allowing PIXE to become a microscopy technique. The needed sample preparation is usually simple and quantitative results are precise and accurate for uniform and homogeneous samples. PIXE is also versatile in what accounts to the size and type of samples that can be analysed due to the combination of in-vacuum and external beam set-up possibilities. However, PIXE does not provide chemical compounds data (only detects elements) neither delivers sample depth dependent information, at least in a straightforward way. In spite of its low availability, once it requires the use of small accelerators, PIXE is still a very cost-effective and competitive or complementary analytical technique.

UV–vis reflectance spectroscopy was used to study the natural hues present in the glass in order to understand what was affecting the colour of the glass. This technique is very useful for *in situ* analysis, for being easily transportable, non-invasive and quick.

Optical microscopy is a useful tool for a first assessment of the overall morphology of the paint layers, in particular the dark field allows the perception of surface details which are not visible with a bright field, since the illumination is oblique in relation with the surface.

Scanning electron microscopy (SEM) is a valuable resource to analyse glass samples with a greater detail when compared to optical microscopy. SEM equipped with an energy dispersive spectroscopy detector (EDS) produces higher resolution images, which enables the morphological characteristics of the samples to be highlighted [66]. X-ray elemental maps are also very useful especially for heterogeneous surfaces such as glass paintings. Backscattered imaging backscattered electrons (BSE) allows visualizing the main differences on the composition of the surface to be analysed.

12.2.1 Experimental methods and conditions

12.2.1.1 μ-PIXE

Quantitative results were achieved with μ-PIXE technique using an Oxford Microbeams OM150 type scanning nuclear microprobe set-up, either with the in-vacuum or with the external beam configuration. To allow efficient detection of low energy X-rays such as the ones of Na, all the glass fragments were irradiated in vacuum with a focused 1 MeV proton beam. The X-rays were collected by an 8-μm-thick Be windowed SDD detector with 145 eV resolution. In order to avoid or detect possible local glass heterogeneities, X-ray imaging (2D elemental distribution) and spectra were obtained from an irradiated sample area of 750 × 750 μm^2. For trace element quantification (typically elements with atomic number above the one of Fe), a higher energy was needed and a 2-MeV proton beam was used. In this case, the external beam set-up was chosen in order to prevent sample beam-charging and consequently X-ray spectra degradation. X-rays were collected with a SDD detector with 145 eV resolution from a sample area of 800 × 800 μm^2. Operation and basic data manipulation, including elemental distribution mapping, were achieved through the OMDAQ software code and quantitative analysis with GUPIX program. The results are expressed in weight percentage of oxides and were normalized to 100 %. In order to validate the obtained concentration results, a glass reference standard, Corning B, was also analysed. Those values are presented in Table 12.1.

12.2.1.2 LA-ICP–MS

A selection of fragments being part of a larger research project were also analysed by means of LA-ICP–MS. The LA-ICP–MS analysis was carried out on the resin embedded glass cross-section. The ablation system used here is located at the National Centre of Scientific Research in Orleans, France. It consists of a Nd:YAG laser working at 266 nm (quadrupled frequency) operating at a maximum energy of 2 mJ and at a maximum pulse frequency of 15 Hz. The beam diameter can be adjusted from 20 to 100 µm. The glass analysis was performed at 8 Hz with a beam diameter of 80 µm. A pre-ablation time of 20 s is set in order to eliminate the transient part of the signal which is then acquired for 55 s corresponding to 20 mass scans from lithium to uranium (the signal in count/s is measured in low resolution mode for 58 different isotopes). Calibration for glass was undertaken by employing NIST610, and Corning B, C and D glass reference material [55]. The detection limits range from 0.1 % to 0.01 % for major element and from 20 to 500 ng/g for other. The composition is calculated from the average of two ablations carried out in different areas of the sample.

To validate the obtained results, glass reference standard Corning A is also analysed as an unknown sample and results are presented in Table 12.1. The results for the trace elements and REE obtained for the analysed samples from the SCV assemblage are reported in Ref. [67], for the *façon-de-Venise* samples, and in Ref. [68], for the gourd-shaped vessels.

12.2.1.3 UV–vis reflectance spectroscopy

A MAYA 200 PRO from Ocean Optics spectrophotometer with a single beam dispersive optic fibre was used, together with a 2048 CCD Si detector that allows operating in the 200–1,050 nm range. The light source is a HL-200-HP 20 W halogen from Ocean Optics, with a single optical path between 360 and 2,500 nm. The spectra were acquired directly on the glass surface of the objects, in reflectance (R) mode, with a 45°/45° configuration (illumination angle/acquisition) and ca. 2 mm of diameter of analysed area. Spectra were obtained between 360 and 1,050 nm, with integration time of 8 ms/scan and 15 scans. A Spectralon® surface was used as a reference. The obtained data were converted and presented as an absorbance $A' = \log10(1/R)$.

12.2.1.4 Optical microscopy and SEM

The particle morphology of the grisaille powders was observed by transmitted light optical microscopy under plane and cross-polarized light. The microscope, Axioplan 2ie Zeiss, was equipped with a halogen light HAL100 and a digital camera (Nikon DMX1200F), and by SEM at the MicroLab Electron Microscopy Laboratory, Instituto Superior Técnico, Lisbon. The SEM system complies a Hitachi S2400 model equipped with a Rontec standard EDS detector, operating at a beam acceleration voltage of 20 kV, also using a backscattered electron detector for imaging. The powder samples analysed by optical microscopy were prepared in dispersion, on a Cargille Meltmount™ resin with a refractive index of 1.662 and an Abbe V dispersion of 26, between two glass slides.

12.2.2 Sampling

The studied glass has an archaeological provenance, apart from the stained glass samples that belong to private collection. In order to avoid erroneous results by analysing and quantifying corrosion layers instead of the uncorroded bulk glass, it was decided to sample all selected objects. Small glass samples of 2–4 mm^2 were dry cut from the selected glass fragments with a steel wire with diamond powder. Samples were embedded in an epoxy resin (Araldite 2020) from Huntsman™ and polished with SiC sandpapers down to 4,000 mesh. This sampling procedure was performed only in broken objects and on individual fragments without possible connections.

The glass samples painted with grisailles were cut and mounted in epoxy resin using the methodology descibed above, for analysis of cross-sections by μ-PIXE and SEM–EDS.

12.3 Case studies

The case studies were selected to demonstrate the complementarity of the developed research methodology both for the study of the glass objects and for the grisailles paintings.

12.3.1 Seventeenth-century archaeological glass found in Portugal: *façon-de-Venise* and gourd-shaped vessels

In this section, the raw materials used to produce glass, and the trace elements that enter the glass matrix with them, are explored. For the past few years, trace elements have been pointed out as potential tracers for the different sources of employed raw materials. To illustrate this potential, an important case study in the Portuguese panorama concerning archaeological glass dated to the seventeenth century unearthed in the Portuguese territory is addressed. Two well-defined categories of artefacts are considered: *façon-de-Venise* objects and the gourd-shaped vessels. The interest in studying these two specific types of finds relies on the fact that on one hand *façon-de-Venise* glass, being a very well-studied type of glass, offers the possibility to compare our results with published data on glass with known provenances, while, on the other hand, for the gourd-shaped vessels, a different scenario is faced, since, as far as the authors know, this object has rarely been found outside Portugal, thus being a potential candidate for a national glass production. The compositional characteristics of these glasses found in Portugal are discussed and the interconnection between them is explored, uncovering the transfer of technology and movement of people and goods.

12.3.1.1 Archaeological contexts

The 29 fragments considered here came from 4 different archaeological contexts (Figure 12.1) that will be presented in a brief way. To have more information about the archaeological contexts and its history, please see Refs. [69] and [70].

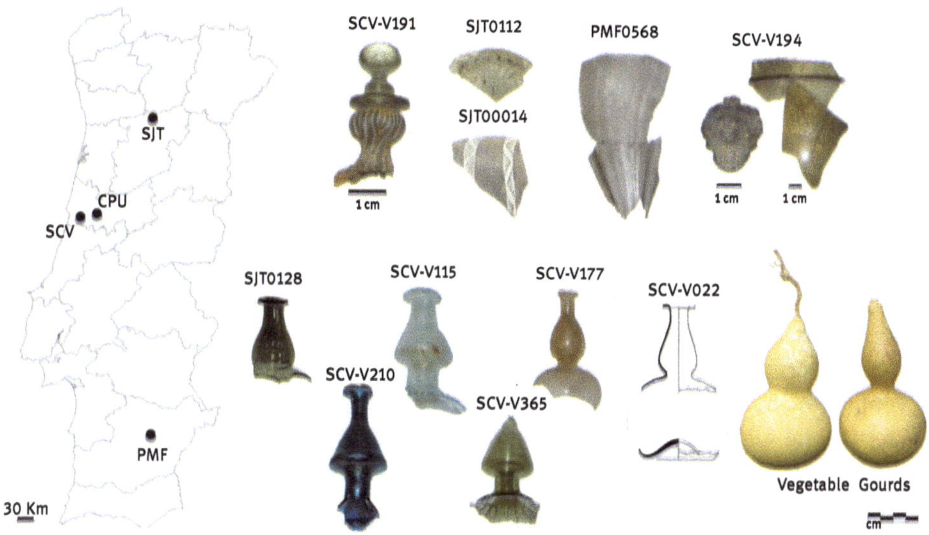

Figure 12.1: Portugal's map, showing the geographical locations of the archaeological sites of the four façon-de-Venise glass sets and of the two gourd-shaped vessel's sets (SJT: São João de Tarouca Monastery; CPU: Courtyard of the Coimbra University; SCV: Santa Clara-a-Velha Monastery; PMF: Praça Miguel Fernandes in Beja). On the left, some examples of fragments under study are presented. Pictures taken by I. Coutinho and archaological drawing by T. Medici.

Santa Clara-a-Velha Monastery (SCV) – This monastery belonged to the Poor Clares (Order of Saint Clare), a contemplative order of nuns in the Roman Catholic Church. The fragments here studied are all from the seventeenth century, which corresponds to the final period of occupation of the monastery. From the thousands of glass fragments found at SCV Monastery [70, 71], a set of six *façon-de-Venise* was selected for the study. Three of them show a perfectly discoloured glass, with different key features: one presents filigree, another shows engraved decoration and the third one has an applied prunt in raspberry shape. A set of fragments, probably belonging to a drinking *tazza* with gilded decoration and a lion's head on the stem, were also chosen. The body glass of this object presents a dark grey hue (SCV-V194). The other two objects are the folded base of a vessel and the mouth of a small flask with mould blown decoration. Both present a high degree of corrosion where original colour of the glass is not perceptible at a naked eye. Also from this assemblage, nine gourd-shaped vessels were selected.

University Courtyard, Coimbra (CPU) – Archaeological excavations were undertaken in the CPU (Coimbra, Portugal), exposing structures with material dated to the seventeenth century From the glass fragments, only one object (the one with *façon-de-Venise* characteristics) was selected from this assemblage to be included in this study. This fragment is possibly the neck of a small flask and is decorated with white filigree canes.

Miguel Fernandes Square, Beja (PMF) – Archaeological excavations at Miguel Fernandes Square in Beja exposed a set of 137 underground silos, used as storage containers for food, mainly cereals, during the fourteenth and fifteenth centuries. Based on more than 1,400 glass fragments retrieved, it was possible to infer the presence of an estimated 366 objects. From these, 13 *façon-de-Venise* fragments dated between the sixteenth and the seventeenth centuries were selected. Most of the fragments are part of beakers and goblets made with colourless glass with natural bluish and greyish hues.

São João de Tarouca Monastery (SJT) – Tarouca is located in northern Portugal, and the monastery was founded between 1,152 and 1,154. It is believed to be the earliest Cistercian monastery in the country [72]. The fragments are dated to the seventeenth century. They comprise glass with filigree and mould blown decoration patterns, one gilded fragment also with engraved decoration, and two objects with the application of strings. The body glass from all fragments shows a greyish or bluish natural hue.

12.3.1.2 *Façon-de-Venise* glass

From the sixteenth century onwards, *façon-de-Venise* glass was the manifestation of the Venetian taste, spreading in Europe throughout several production centres, mainly as a consequence of the Venetian glassmakers' emigration. A number of papers devoted to the analysis of Venetian and *façon-de Venise* glass objects discussed the raw materials employed in the manufacture of these items, and their probable origin [6, 11, 19, 21, 32, 73]. Chemical analysis and research on ancient treatises lead to the understanding that the raw materials employed in Venetian and *façon-de Venise* glass production were carefully selected, in order to accomplish the high-quality glass that made it one of the most wanted luxury items throughout Europe [11]. For example, during the fourteenth century, Venetian glassmakers started to use quartz pebbles from the Ticino River as a silica source, avoiding the use of impure sands.

Considering glass production in Portugal, documentary sources report that glassmakers from Italy (Venice and Altare) and from Antwerp were settled in the Portuguese territory from the seventeenth century onwards [74–76].

The study conducted on *façon-de-Venise* objects from different Portuguese archaeological excavations – SCV, SJT, PMF and CPU – was expected to bring new information on the commercial relationships with other European centres and on glass from possible national provenance.

Twenty-nine fragments were selected and analysed for this study, major and minor composition are presented in Table 12.1. For the great majority of them, glass colour and the way some decorations were applied prevented from considering the glass to have a genuine Venetian origin. For a more exhaustive description of the fragments, see Refs. [67, 69, 70].

Having in consideration what was stated above about raw materials and how it is possible to take advantage of the impurities that enter the glass matrix and use them as important provenance trackers, the silica trace components were the firsts to be explored. Analysing the relation between aluminium and titanium oxides, both entering the glass through silica, makes it possible to formulate some considerations on the employed silica sources. Regarding the division of alumina contents as defined by Ref. [40], the samples can be divided in four groups with low, medium, high and very high alumina contents (Figure 12.2(a)). Venetian and *façon-de-Venise* glass are characterized by their choice of pure raw materials with low contents of impurities [11]. The results (Figure 12.2(a)) show that some of these objects have a chemical composition that is unusual for *façon-de-Venise* glass that do not match with any of the known European *façon-de-Venise* production centres and prevents from attributing them to a genuine Venetian provenance (for typical Venetians compositions see Ref. [13], for values of glass in several *façon-de-Venise* production centres see Ref. [40], Table 4).

It is known that the choice of raw materials was a primary phase of glass manufacture, since these influenced the quality of the final result, particularly in

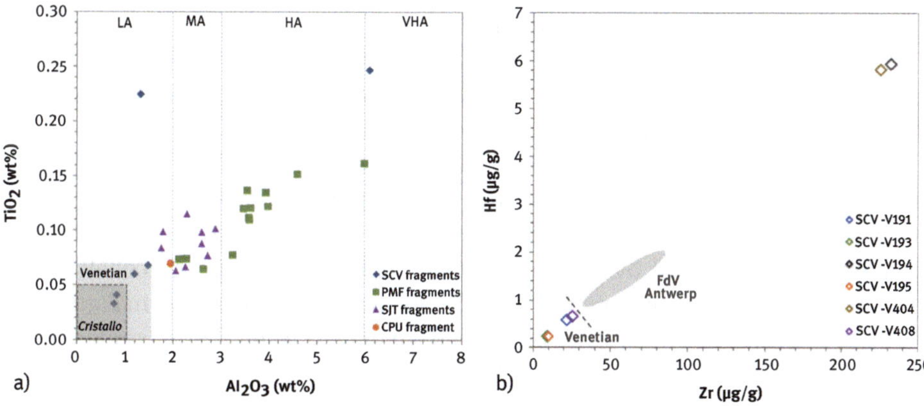

Figure 12.2: Binary charts of (a) Al_2O_3 versus TiO_2 with boundaries for general Venetian glass and *cristallo* glass following published compositions in Ref. [13], and with marked zones for the low alumina (LA), medium alumina (MA), high alumina (HA) and very high alumina (VHA) contents based on Ref. [40], and (b) Zr versus Hf concentrations in µg/g determined by means of LA-ICP–MS for the SCV objects. The dashed lines are general Antwerp *façon-de-Venise* and general Venetian regions taken from the literature [19].

terms of transparency and of the possibility of obtaining colourless glass. The purest material was quartz pebbles and was used by the Venetian glassmakers who imported it from other Italian regions, namely Ticino River [11]. However, the most commonly employed raw material was sand and it is generally accepted that it came from places near the glass production locations [9]. For this reason, it is considered a possibility that the high alumina levels derive from the use of local sands found in the Portuguese territory and rich in feldspar.

According to values published in literature, Venetian glass (both *cristallo* and *vitrum blanchum*) is characterized by low contents of TiO_2 (around 0.05 wt.%), as represented in Figure 12.2(a) by the two marked regions [13]. Three SCV objects, with alumina contents below 2.0 wt.% and titanium dioxide content below 0.1 wt.%, are compatible with genuine Venetian compositions; from these, two are within the *cristallo* boundaries. Comparing the obtained results with coeval genuine Venetian and *façon-de-Venise* glass [11] confirms that these fragments from SCV are probably genuine Venetian objects. To validate the Venetian provenance of the selected SCV fragments, these were analysed by LA-ICP–MS. This allowed quantifying the trace and the REE that are considered the fingerprint of the employed raw materials. In order to compare with values and boundaries present and defined in the literature, the contents of zirconium and hafnium (Figure 12.2(b)) were studied. Both these elements derive from the silica source, and looking to the values of these elements in the samples being studied, four of the SCV fragments are within the Venetian boundaries [19]. This confirms that these objects are genuine imports from Venice.

The fragments from the remaining assemblages do not fall within the Venetian boundaries and are considered to originate from different production centres. The PMF fragments form one group and a majority of samples from SJT form another one.

Regarding the flux properties, the study of these allowed one to confirm that the fragments previously identified as Venetian are also compatible in terms of flux with being made out in this region [67].

Another particularity of the group was the natural hue of the glass featuring shades of grey and blue. Only the SCV fragments imported from Venice do not present such visible hues. UV–vis reflectance spectroscopy allows one to determine that this grey/blue hue is caused by the simultaneous presence of iron and cobalt oxides (Fe(III) – 380, 420 and 440 nm; Co(II) – 530, 590 and 640 nm) as shown in Figure 12.3. As far as it is known, the presence of cobalt in colourless *façon-de-Venise* glass is not usual and has rarely been reported in the literature.

12.3.1.3 Gourd-shaped vessels

This peculiar shape is characterized by a neck that can be globular, pear-shaped or truncated conical, while the body is usually globular [77], often decorated with vertical ribs made by using a dip mould, starting from the bottom of the neck [70, 77]. Many of the gourds discovered to date are of brownish red, green (several shades

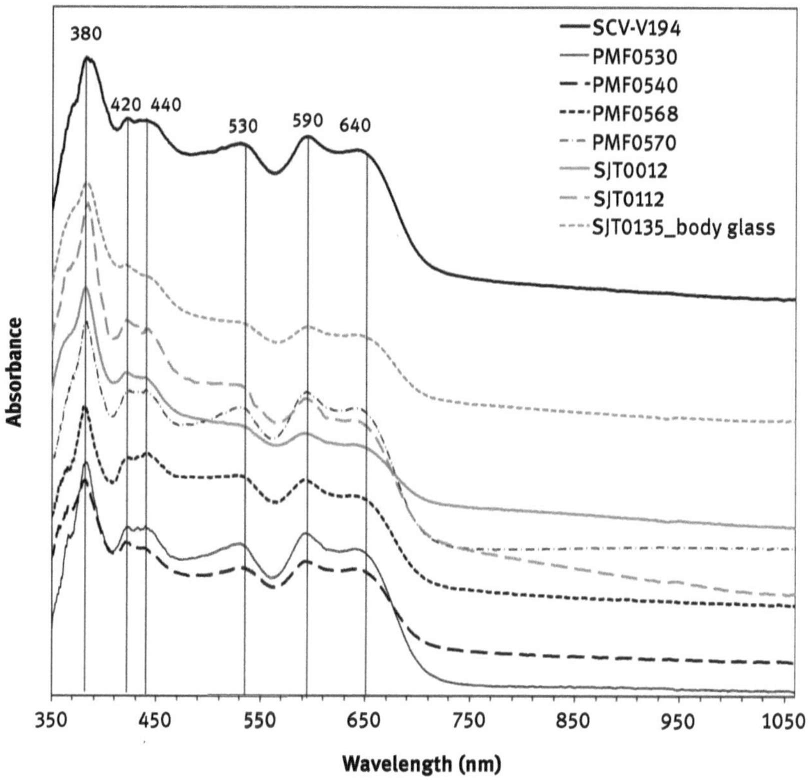

Figure 12.3: Absorbance spectra of several *façon-de-Venise* objects from SCV, PMF and SJT assemblages.

can be observed) or blue glass. Less often, colourless glass gourds have been found, some of them with a greenish or yellowish tinge.

Considering the number of gourd-shaped vessels unearthed in Portugal, these bottles were probably very common in Portugal from the seventeenth century onwards. Glass bottles of this form have been found in archaeological assemblages at SJT, Porto, Coimbra, Santarém, Lisbon, Almada and Moura, all of them dated to the seventeenth century, demonstrating that this form was widely disseminated throughout the country [70]. A painting by the famous seventeenth -century Portuguese painter Josefa de Ayala e Cabrera (also known as Josefa de Óbidos, 1630–1684) depicts a gourd-shaped bottle of colourless glass, with an outline that matches the gourds with a truncated conical neck (picture of the painting in Ref. [78]).

Elsewhere in the Iberian Peninsula, gourd-shaped glass vessels are found in Spain, especially in the territories that belonged to the Kingdom of Aragon. They are mentioned in several inventories in Catalonia (fifteenth century) and Majorca (sixteenth century) [79, 80]. However, none of the examples of gourds mentioned above matches the forms of the gourds discovered in Portugal. When comparing gourds found in Portugal with those discovered elsewhere in Europe, certain morphological differences can be detected. The Portuguese gourds have a pear-shaped or truncated conical neck, as was described above and is illustrated in Figure 12.1. These neck shapes contrast with the most common globular necks in the gourds found in Spain and other European countries.

Because of the complexity of the form, the techniques by which the gourds were manufactured were not fully understood until replicas were made. The master glass artist Robert Wiley conducted some experiments at the research unit Vidro e Cerâmica para as Artes at the Universidade NOVA de Lisboa in Caparica, Portugal. The experiment suggested that the original glassmakers exercised considerable control over temperature while employing the Venetian technique of *mezza-stampatura* (half-moulding). For a complete description of the remake of this shape, see Ref. [68].

Following the same line of thought that was used to approach the *façon-de-Venise* data, the impurities that came along with the raw materials were explored (see Table 12.1). The alumina and titanium oxide contents of the gourds were plotted (Figure 12.4(a)), using the division in alumina content employed before. At least three different silica sources can be identified. The two gourds in the left corner of the chart (SCV-V115 and SCV-V210) have the lowest amounts of alumina, which implies the use of a purer source, or sources, of silica. The remaining gourds, which have an alumina content higher than 5 wt.%, were produced using silica sources much higher in impurities. As we saw before, analysing the reported compositions of contemporaneous glass objects from such European centres, such as Antwerp, Amsterdam, London and Spain (alumina values equal to or less than 2.0 wt.%), we observe that alumina contents this high are unusual [19, 73, 82].

These results were compared with those from the study of *façon-de-Venise* glass found in Portugal discussed above and reported by Coutinho et al. [67], and also with data obtained for six samples of production remains recovered from a coeval glass furnace in Majorca, Spain. The alumina values of the Majorcan samples vary between 3.5 and 5.3 wt.% [81].

The composition of the body glass of a small gourd with *millefiori* decoration from the Monastery of SCV, analysed by Ref. [40], was also considered for comparison.

In Figure 12.4(b), the alumina and iron oxides were normalized to the silica content to characterize the composition of the silica source, irrespective of any added flux, and plotted against each other. Again, the two gourds with lower quantities of alumina (SCV-V115 and SCV-V210) are close to each other.

Four of the SCV samples (SCV-V079, SCV-V082, SCV-V352 and SCV-V365) appear to be very similar, presenting a positive correlation between the alumina and iron oxides normalized to the silica contents. The closeness of the SJT gourd to the four gourds from the SCV set suggests that the same source of silica was employed to produce these glasses. The SCV-V177 gourd is outside this correlation, indicating that it was made using a different source of silica. Finally, gourds SCV-

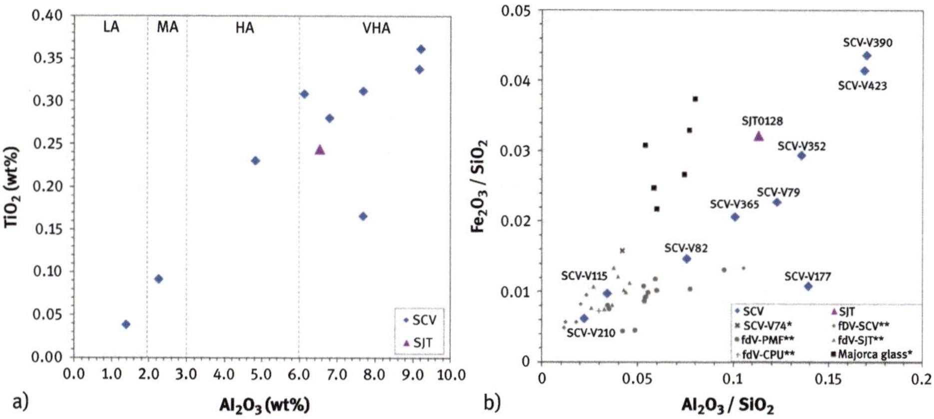

Figure 12.4: (a) Binary plot of Al_2O_3 versus TiO_2 with marked zones for the low alumina (LA), medium alumina (MA), high alumina (HA) and very high alumina (VHA) contents based on Ref. [40], and (b) alumina versus iron oxide, both normalized to the silica content and compared with another gourd from the SCV assemblage (SCV-V74*, reported in Ref. [40]), with results from contemporaneous *façon-de-Venise* objects found in Portugal and reported above, and production remains from a Majorcan furnace (Majorca glass***, reported in Ref. [81]). Results were obtained by μ-PIXE analysis.

V390 and SCV-V423 were probably made from the same silica source, one very high in impurities.

The gourds with high and very high alumina values are unrelated to any of the other objects. Based on their contents of oxides, the glasses from Majorca (in Figure 12.4(b), indicated as Majorca glass) are unrelated to any of the analysed gourds, and this demonstrates that the silica from which they were made came from a different source. Most of the analysed *façon-de-Venise* fragments from the Portuguese archaeological sites of Beja (PMF) and the Monastery of SJT (indicated as fdV-PMF** and fdV-SJT** in Figure 12.4), as well as the fragment of the small gourd SCV-V74 with *millefiori* decoration from the Monastery of SCV (shown in Figure 12.4(b)), do not show any similarity to the other gourds in our study, so their production, too, involved a different source of silica.

Looking at the gourds with low levels of alumina allows us to find some similarities. SCV-V210 is closely related to the three *façon-de-Venise* fragments from the SCV assemblage that have been identified as genuine Venetian imports. SCV-V115 also seems to be similar to the *façon-de-Venise* fragments with low alumina contents.

Zr and Hf have been used in the distinction of glass produced in Venice as seen above [19, 67, 73]. However, the ratio of Zr/Hf changes very little in zircon that have a granitic origin, resulting in similar (sometimes even coincident) ratios of these elements for glass made with different silica sources [4]. For this reason, the relation of Zr versus La was observed. The area marked in the binary chart for Venetian glass was based on the Zr values for this glass reported in the literature. Analysing these values (Figure 12.5), one can infer that SCV-V115 was made outside Venice. However, the silica source employed in the manufacture of this object was carefully chosen, as can be attested by the low levels of impurities in its composition. In contrast, the same figure shows that SCV-V210 is strongly compatible with Venetian glass. Because bottles of this shape are, so far, unknown among Venetian glasses, we can offer two hypotheses: (1) this object was ordered from a Venetian glass workshop in a form specified by the customer, or (2) it was made elsewhere, employing recycled Venetian glass. The second hypothesis seems more plausible, considering what was said above about the presence of lead and tin oxides in the composition of this gourd.

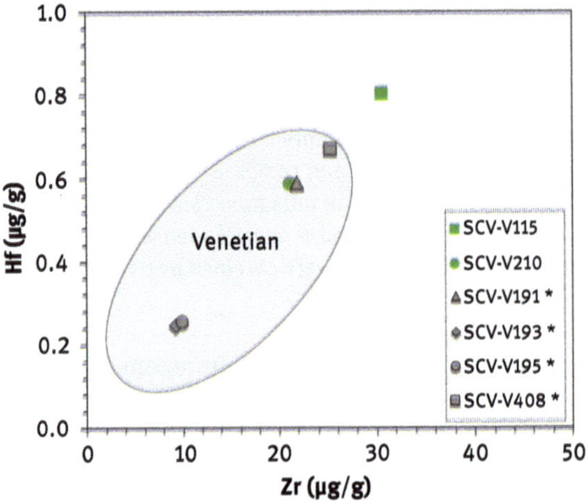

Figure 12.5: Binary plot of Zr versus La in µg/g, determined by LA-ICP–MS, for the SCV gourds. The area marked as "Venetian" refers to the Zr values reported in literature for genuine Venetian glass [73].

Analysing the oxides that come from the flux source (Na_2O, K_2O, MgO, P_2O_5 and CaO), these were further studied and examined in Ref. [68], where it was possible to reach the conclusion that for the gourds under study, two different manufacturing techniques could be identified, one with purified ashes and the other with unpurified ashes (SCV-V115, SCV-V210, SCV-V352 and SCV-V365 relate with the use of purified ashes and the remaining gourds relate with the employment of unpurified ashes in the batch). This reinforces the idea that gourds were produced in more than one location by glassmakers who employed different ways of preparing the raw materials introduced into the batch. However, this distinction could also be explained by the use of imported or recycled Venetian glass.

What is immediately striking about the gourd-shaped bottles found in Portugal is the palette used in making these objects. There are some instances in which the coloration is considered intentional, and this is the case for cobalt blue. The cobalt in the glass matrix is considered to have been added deliberately, as was also true for the purple glass, rich in manganese oxide. Considering the intense coloration of most of the recovered vessels, we can suggest that the glassmakers purposely used glass that they knew would result in bright, colourful objects. The most common colour among the sampled objects (in 5 of the 10 gourds) is green, and this is also reflected among the total number of gourds found to date (67 out of 133) [70].

The presence of iron is generally considered to be an impurity typically attributed to the silica source. Looking at the green tints in the SCV set, we can debate the intentionality of the presence of iron. It is important to clarify what is meant here by intentional: (a) iron can be added on purpose to obtain the green colour, (b) glassmakers purposely used sands rich in iron to obtain green glass and (c) sands rich in iron were used on purpose, to which manganese was added, not to decolorize but to enhance the green colour in the glass. The concentration of MnO in the glass seems to suggest that this component was introduced into the batch deliberately, pointing to the hypothesis described in (c). It is then proposed that manganese was not added to discolour the batch; instead, it may have been introduced as recycled cullet. The other suggestion is that, because MnO_2 (the form in which manganese is added to the batch) has an oxidizing effect, it was probably introduced into the batch to counteract the strong blue coloration provided by the reduced iron. In this way, manganese would oxidize the iron to a green colour.

12.3.2 Shadows and contours: grisaille painting

Many wrote about stained glass, but one must witness its splendour to understand the meaning of the words. Stories were told through a game of light and shadow, with coloured and colourless glass decorated with painted motifs. The glass-painter played a significant role to tell these stories. Stained glass painting was, from the beginning of the craft, one of the processes which experienced many changes and developments. The grisaille painting was the first painted decoration ever to be used on stained glass. The production of this paint involves the mixing of metal oxides, normally iron and/or copper, with ground lead glass. This mixture is applied on the glass panel with an agent such as gum Arabic and fired. It is used to depict on coloured and clear glass figures motives and other details as a thick opaque line, known as *grisaille à contourner*, or as a thin line, known as *grisaille à modeler* [83, 84]. Examples of sixteenth and seventeenth-century stained glass reveal the high-quality compositions that could be obtained using only the grisaille (Figure 12.6).

The treatise *De coloribus et artibus romanorum*, written by Eraclius between the tenth and the thirteenth centuries, indicates the mixture of a lead-rich glass with iron filings. Brass, a Cu–Zn alloy, could also be added to the base glass. At the present moment, there is knowledge of only two stained glass collections that present grisaille painting with a significant amount of zinc, one of them is the sixteenth-century stained-glass collection located at Batalha Monastery and the other belongs

(a) (b)

Figure 12.6: The grisaille painting followed the development of the painting on stained glass, allowing the creation of very detailed compositions with the use of grisaille. A portrait, dated to the sixteenth century (Portugal) (a), and Zabulon, dated to the seventeenth century (Low Countries) (b), are two examples which can be found in the collections from Convento de Cristo, Tomar, and Pena National Palace, Sintra, respectively. © Joana Delgado, © Luis Pavão, Parques de Sintra, Monte da Lua S.A. Convento de Cristo, Tomar, and Pena National Palace, Sintra, respectively. © Joana Delgado, © Luis Pavão, Parques de Sintra, Monte da Lua S.A.

to the stained glasses from the Salvatorkirche in Munique and dated to the sixteenth century [85, 86]. The twelfth-century treatise *De diversibus artibus* by Theophilus prescribes a mixture of a lead-rich glass with copper fillings. To apply the grisaille, wine, urine, or gum arabic were used as binders. The paint is heated at temperatures between 600 and 750 °C, allowing the attachment of the paint layer to the base glass, with a complete fusion of lead and the decomposition of some of the oxides, with the releasing of the metals which remain in its solid state dispersed in the glassy matrix (Figure 12.7) [87, 88]. This melting process results in a proper fusion of the paint, creating a diffuse interface between the grisaille and the base glass [89, 90]. The result is an opaque layer with an uneven surface. The quality of the grisaille will, thus, depend on the thickness of the paint, the correct amount of binder and the firing temperature [83].

12.3.2.1 Historic grisailles from Europe – morphology and compositions

This investigation focuses on the study of historical grisaille samples, with different provenances and production dates, aiming the creation of a spatiotemporal map of the grisaille compositions. This work constitutes the first steps for the creation of a database of grisailles compositions and morphology from which we

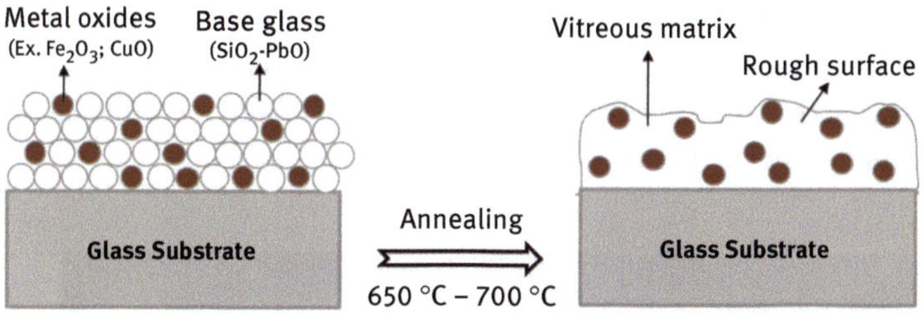

Figure 12.7: Schematic representation of a cross-section of grisaille before and after firing. (c) Andreia Machado after Olivier Schalm

will be able to study if and how the composition of grisaille changed over the centuries and countries, due to the materials available and the different artists that produced it.

Twenty-three samples from Low Countries, Switzerland, France and Portugal, dated between the thirteenth and twentieth centuries, were analysed. These were obtained from the private collection of Joost Caen (Belgium), Monastery of Batalha (Portugal), and Convent of Christ (Portugal). Some examples are presented in Figure 12.8.

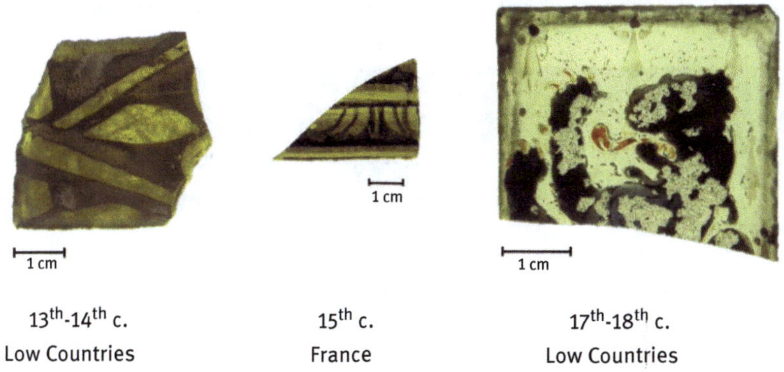

Figure 12.8: Examples of the analysed samples. Pictures by Márcia Vilarigues.

The morphological analysis of the grisailles was carried out in a cross-section in order to allow the observation and study of the layers. With this analysis, it was intended to see the thickness of the layers, the amount of pigment grains present in it, the interface between the paint and the glass and still perceive its structure. Figure 12.8 presents some examples of grisailles analysed.

The morphological analysis of the samples showed different thicknesses in the grisailles, going from around 10 to 70 µm. The interface of the paint with the glass is generally well defined, with some exceptions found in the seventeenth and nineteenth centuries. All grisailles have metallic grains in the composition, with diverse sizes and shapes, indicating the irregular milling of the metallic compounds and resulting in a heterogeneous layer. The only exception found was a sample of the Low Countries, dated to the sixteenth century (Figure 12.9), differing not only by its thickness but also by its distinct structure.

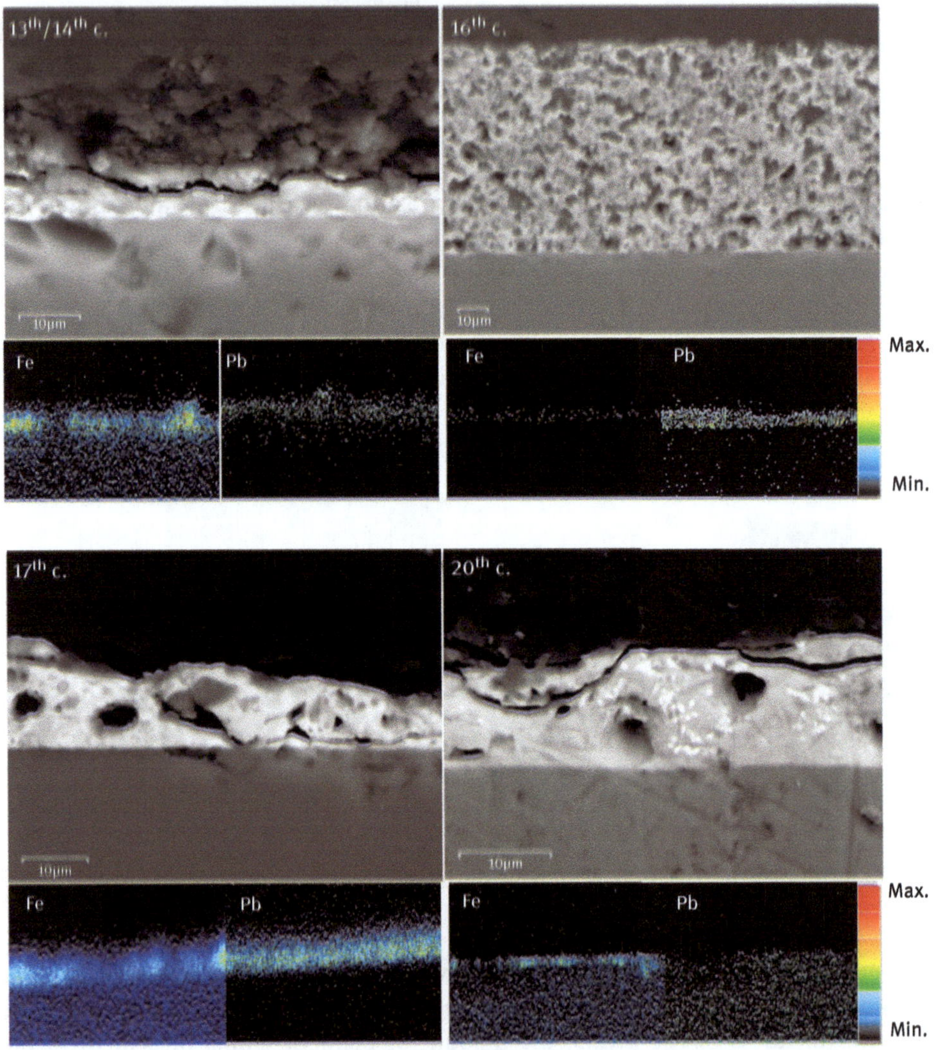

Figure 12.9: SEM–BSE images and elemental X-ray distribution maps of grisailles layers in historical samples.

The grisailles, as mentioned above, are constituted by a flux and a pigment, these corresponding to lead glass and iron or copper filings, elements that proved to be the main constituents through the X-ray maps shown in Figure 12.9.

Through the elementary analysis performed on the grisailles by μ-PIXE, the composition of the grisailles was determined and it was possible to follow differences in materials and technology according to region and period (see Table 12.2). The obtained results show also the presence of the main components of the glass substrate, although the PIXE quantification analysis was performed only on the grisaille region. This may result from the heterogeneities of the grisailles and consequent different penetration depths of the ion beam or to the diffusion (or lixiviation) of these elements to the grisaille due the corrosion of the glass.

The results obtained from the analysis show that in the Low Countries and France, iron is the main colouring agent used for the production of the grisaille. Iron was found as solo colourant until the fourteenth century, after which small amounts of copper were introduced. All samples dated from after this century have small amounts of copper, the iron being present in percentages ranging from about 85 % to 95 % of the colouring agents. Samples dated from the end of the seventeenth century start to show an increase in the percentage of copper to about 20 %, and it is therefore possible to conclude that the percentage of copper used in these countries has increased over time. In Portugal, iron concentrations are close to those found in the Low Countries, ranging from 85 % to 99 %. Samples from Switzerland have a very different composition, with a higher amount of copper than iron, with copper being present in percentages ranging from 25 % to 70 %, making this the colorant of choice for the production of these pictorial layers, probably due to an easier access to copper in these regions.

These analyses allowed us to learn that for the samples from the Low Countries, France and Portugal, iron was used as the preferential colorant agent, while in Switzerland, a lower percentage of iron was used with copper being an element of growing predilection.

The proportion of flux and colorant agents also varied between regions and time periods. It was ascertained that the percentage of flux varied between 40 % and 70 %, with the fifteenth and seventeenth centuries' grisailles having the highest percentage. Switzerland was the region that used the highest percentage of flux, in a percentage of around 80 %, followed by Portugal (*c.f.* Table 12.2). The remaining countries have significantly less flux, with this material going from 25 % to 50 %.

12.4 Conclusions

Over the last years, archaeometric studies and the constant evolving of the analytical techniques and the possibility of treatments one can apply to the results (e.g. chemometrics), together with the studies of manufacturing techniques, reproduction of recipes and so on, can bring valuable information on the transfer of technology between different people and locations.

Table 12.2: Composition of the major elements of the grisaille for the analysed samples.

Provenance	Date	Samples	SiO_2	P_2O_5	Cl	K_2O	CaO	TiO_2	MnO	Fe_2O_3	CuO	ZnO	BaO	PbO
P.B./France	Thirteenth/ Fourteenth	JC012-A1	44.4	0.86	0.78	10.84	21.44	0.23	0.52	6.48	0.22	0.08	0.15	14.02
Low	1480	JC013a-A1	45.8	2.89	0.89	9.31	20.26	0.23	0.97	9.18	1.26	0.24	0.26	8.73
Countries	1520	JC014-A1	32.7	1.85	2.49	5.63	23.57	0.24	0.52	17.28	1.05	0.27	0.13	14.25
	1550	JC015a-A1	17.6	<LD	0.30	0.59	2.56	0.06	0.06	42.52	4.41	2.58	<LD	29.29
	1560	JC016-A1	47.6	2.38	1.71	12.58	9.04	0.27	0.19	13.66	1.52	0.78	<LD	10.29
	1625	JC017-A1	49.3	2.50	0.74	6.76	17.56	0.43	0.47	17.61	0.12	0.09	0.10	4.26
		JC017-A2a	19.9	<LD	0.53	1.51	1.40	0.11	0.08	45.59	3.11	1.37	<LD	26.40
		JC017-A3-gris. inter	31.9	<LD	0.54	4.57	2.34	0.08	0.07	43.78	4.77	0.88	<LD	11.11
		JC017-A3-gris. Ext	59.9	<LD	0.75	12.90	1.77	0.07	0.04	0.72	8.04	0.06	0.03	15.66
	Seventeenth	JC018-A1	42.5	1.34	1.65	7.69	17.62	0.35	0.65	13.33	2.67	0.51	0.28	11.44
	1650	JC019-A1	43.1	<LD	0.39	2.29	10.20	0.15	0.32	28.95	1.97	<LD	0.07	12.50
		JC019-A2	33.6	<LD	0.47	2.49	4.53	0.06	0.31	39.12	3.71	<LD	<LD	15.76
		JC020-A2a	28.0	<LD	0.28	2.64	2.08	0.13	0.07	27.68	5.44	1.72	<LD	31.95
Switzerland	Seventeenth	JC021-A1	38.5	<LD	2.15	6.32	18.79	0.36	0.75	8.94	7.98	<LD	0.40	15.84
		JC021-A2	39.1	1.46	2.00	8.02	25.49	0.39	0.92	4.12	3.03	<LD	0.49	14.95
	Seventeenth	JC022-A1a	31.2	2.61	2.72	4.82	24.89	0.20	0.48	10.50	3.53	0.08	0.58	18.43
	Seventeenth	JC023-A1	52.2	1.88	0.66	5.77	24.90	0.29	0.65	1.92	2.32	0.02	0.56	8.80
		JC023-A2	40.5	1.17	0.54	4.11	16.31	0.18	0.34	3.79	7.81	<LD	0.28	24.91
Belgium	Nineteenth	JC024-A1	36.8	1.03	0.83	4.01	10.98	0.20	2.81	9.66	5.76	0.32	0.37	27.20
	1900	JC025-A1a	61.8		0.64	4.52	6.16	0.44	3.07	8.43	1.13	0.88	0.14	12.77
Portugal	1510/1518	T01	31.9		0.75	1.60	5.07	0.10	0.13	31.44	5.25	0.88	0.07	22.80
	Sixteenth	J18 a9 13	41.9		0.47	1.41	4.27	0.06	7.10	11.86	0.07	0.48	0.22	32.16
Certified*		CMOG C	34.30	0.10	0.10	3.00	5.00	0.800		0.30	1.20	0.040	12.00	37.50
Measured			36.5	<L.D	0.32	2.79	4.74	0.87		0.32	1.15	0.03	11.92	39.74

*Ref. [24].
Results are in weight percent of oxides and determined by means of μ-PIXE.

The growing knowledge on this field has also been useful to identify knowledge gaps as was the case of the glass unearthed in Portugal. Until a few years ago, little was known about the glass that circulated in the national territory and this information is needed in order to keep putting the puzzle together about the glass circulation routes in Europe and beyond.

The investigation about *façon-de-Venise* glass discovered in Portugal allowed for the identification of four genuine Venetian fragments, but what was striking was the impossibility (due to the glass composition) of attributing the majority of the fragments to known *façon-de-Venise* production centres in Europe. It proposed the existence of two more unknown *façon-de-Venise* production centres that could have been located in Portugal.

Concerning the gourd-shaped vessels, because this is a rare shape to find outside Portugal, these are considered good candidates to have been made in national territory. This assumption suggests that skilled glassmakers were working in Portugal in the early modern period. This calls into question the commonly accepted notion that glass made in Portugal at that time was of poor quality, crafted by less capable artisans. On the contrary, the results of our study reinforce the impression that local production should have attained some significance, at least from the sixteenth century onwards. This seems to be implied in an existing royal charter from King Dom Sebastian I (1554–1578), dated 1563, which appears to have forbidden the importation of Venetian glass because local production was judged to be of comparable quality. Once more, the composition allied to the shape and production methodology prevented us from finding comparable objects or fragments from any of the other known coeval European glass production centre.

Analytical investigation proved to be effective also on grisailles. With this work, it was possible to verify that most of the samples have a heterogeneous paint layer, with some granularity and differences of thickness. It was also proven that the preferred colourant agent is iron, being the countries of central and western Europe the ones with the highest proportion of this material. Countries from east Europe presented a higher content of copper than the others, being this material used with iron in a 1:1 or 1:2 ratio of iron/copper.

The proportion of flux in grisailles over the course of time is around 40–50 %, with the exception of the fifteenth and seventeenth centuries, where flux made up about 80 % of the grisaille on average. The country that had the highest concentration of flux in grisailles was Switzerland, around 80 %.

The results obtained in the presented case studies demonstrate the effectiveness of the used methodology. Combining stylistic, chemical and technological approaches, it has been possible to trace the provenience of the glass arriving in Portugal during the seventeenth century and to make hypothesis on local production, as well as to discuss and comprehend the technological differences along the centuries and on different locations for the production of the grisailles paintings.

References

[1] Henderson J. The science and the archaeology of materials. In: An investigation on inorganic materials. London and New York: Routledge, 2000.

[2] Artioli G. Scientific methods and cultural heritage. New York: Oxford University Press; 2010:278–81.

[3] Wedepohl KH, Simon K, Kronz A. Data on 61 chemical elements for the characterization of three major glass compositions in late antiquity and the middle ages. Archaeometry. 2011;53:81–102.

[4] Brems D, Degryse P. Trace element analysis in provenancing Roman glass-making. Archaeometry. 2014;56:116–36. DOI: https://doi.org/10.1111/arcm.12063.

[5] Rehren T, Freestone I. Ancient glass: from kaleidoscope to crystal ball. J Archaeol Sci. 2015;56:233–41. DOI: https://doi.org/10.1016/j.jas.2015.02.021.

[6] Šmit Ž, Janssens K, Bulskad E, Wagnerd B, Kose M, Lazar I. Trace element fingerprinting of façon-de-venise glass. Beam interactions with materials and atoms. Elsevier. 2005:94–9. DOI: https://doi.org/10.1016/j.nimb.2005.06.182.

[7] Cagno S, Mendera M, Jeffries T, Janssens K. Raw materials for medieval to post-medieval tuscan glassmaking: new insight from LA-ICP-MS analyses. J Archaeol Sci. 2010;37:3030–6.

[8] Shelby JE. Introduction to glass science and technology. 2nd ed. Cambridge: The Royal Society of Chemistry, 2005.

[9] Navarro JMF. El Vidrio, 3rd ed.Madrid: Consejo Superior de Investigaciones Científicas, Sociedad Española de Cerámica y Vidrio, Series Textos Universitários (Consejo Superior de Investigaciones Científicas), 6, 2003.

[10] Shortland A, Schachner L, Freestone I, Tite M. Natron as a flux in the early vitreous materials industry: sources, beginnings and reasons for decline. J Archaeol Sci. 2006;33:521–30.

[11] Verità M, Zecchin S. 2009. Thousand years of Venetian glass: the evolution of chemical composition from the origins to the eighteenth century. In: Janssens K, Degryse P, Cosyns P, Caen J, Van't Dack L editors. Annales du 17e Congrès de l'Association Internationale pour l'Histoire Du Verre (AIHV). Antwerp, Antwerp: Aspeditions, 602–13. 4th to 8th Sep 2006.

[12] De Juan Ares J, Schibille N. La Hispania antigua y medieval a través del vidrio: la aportación de la arqueometría. Boletin de La Sociedad Espanola de Ceramica Y Vidrio. 2017;56:195–204. DOI: https://doi.org/10.1016/j.bsecv.2017.04.001.

[13] Verità M. Venetian soda glass. In: Janssens K, editor. Modern methods for analysing archaeological and historical glass. vol. I. Chichester: Wiley, 2013:515–36.

[14] Silvestri A, Marcante A. The glass of Nogara (Verona): a "window" on production technology of mid-Medieval times in Northern Italy. J Archaeol Sci. 2011;38:2509–22. DOI: https://doi.org/10.1016/j.jas.2011.03.014.

[15] Barkoudah Y, Henderson J. Plant ashes from Syria and the manufacture of ancient glass: ethnographic and scientific aspects. J Glass Stud. 2006;48:297–321.

[16] Tite MS, Shortland A, Maniatis Y, Kavoussanaki D, Harris SA. The composition of the soda-rich and mixed alkali plant ashes used in the production of glass. J Archaeol Sci. 2006;33:1284–92.

[17] Wedepohl KH, Simon K. The chemical composition of medieval wood ash glass from Central Europe. Chemie der Erde e Geochemistry. 2010;70:89–97.

[18] Cílová Z, Woitsch J. Potash e a key raw material of glass batch for Bohemian glasses from 14the17th centuries? J Archaeol Sci. 2012;39:371–80. DOI: https://doi.org/10.1016/j.jas.2011.09.023.

[19] De Raedt I, Janssens K, Veekman J, Vincze J, Vekemans B, Jeffries TE. Trace analysis for distinguishing between Venetian and façon-de-Venise glass vessels of the sixteenth and seventeenth century. J Anal At Spectrosc. 2001;16:1012–17.

[20] De Raedt I, Jansenns KHA, Veeckman J. Compositional distinctions between sixteenth century 'façon-de-Venise' and Venetian glass vessels excavated in Antwerp, Belgium. J Anal At Spectrom. 1999;14:493–8. Royal Society of Chemistry, London.
[21] Šmit Z, Jansenns K, Schalm O, Kos M. Spread of façon-de-Venise glassmaking through central and western Europe. Nucl Instrum Meth Phys Res Sec B. 2004;213:717–22.
[22] Moretti C. Glossario del vetro veneziano. Dal Trecento al Novecento. Venezia: Marsilio, 2001.
[23] Ashtor E, Cevidalli G. Levantine alkali ashes and European industries. J Eur Econ Hist. 1983;12:475–522.
[24] Brill RH. Chemical analyses of early glasses, v. 2, The tables. Corning, New York: The Corning Museum of Glass, 1999.
[25] McCray PW. Glassmaking in renaissance Italy: the innovation of Venetian Cristallo. JOM-e, 1998:14–19. DOI: https://doi.org/10.1007/s11837-998-0024-0.
[26] Kunicki-Goldfinger JJ, Kierzek J, Kasprzak A, Malozewska-Bucko B. Analyses of 18th- century central European colourless glass vessels. In: Annales du 15e Congrès de l'Association Internationale pour l'Histoire Du Verre (AIHV). Corning: Nottingham, 2003:224–9. 15th to 20th Oct 2001.
[27] Kunicki-Goldfinger J, Kierzek J, Dzierz'anowski P, Kasprzak AJ, 2005. Central European crystal glass of the first half of the eighteenth century. In: Annales du 16e Congrès de l'Association Internationale pour l'Histoire du Verre (London 2003), London: AIHV, 258–62.
[28] Dungworth D, Brain C. 2013. Seventeenth- and Eighteenth-century English lead glass. In: Janssens KHA editor. Modern methods for analysing archaeological and historical glass. Chichester: John Wiley & Sons, 573–82.
[29] Pulido Valente F, Coutinho I, Medici T, Brain C, Vilarigues M. A group of early English lead crystal glass goblets found in Lisbon. J Glass Stud. 2016;58:211–25.
[30] Dungworth D, Brain C. Late seventeenth century crystal glass: an analytical investigation. J Glass Stud. 2009;51:111–37. The Corning Museum of Glass, Corning NY.
[31] Kurkjian CR, Prindle WR. Perspectives on the history of glass composition. J Am Ceram Soc. 1998;81:795–813. Westerville, Ohio.
[32] Moretti C, Hreglich S. Raw materials, recipes and procedures used for glass making. In: Janssens K, editor. Modern methods for analysing archaeological and historical glass. vol. I. Wiley: Chichester; 2013.
[33] Freestone I. Pliny on Roman glassmaking. In: Martinón-Torres M, Rehren T, editors. Archaeology, history and science: integrating approaches to ancient materials. California: Left Coast Press, 2008:77–100.
[34] Brems D, Degryse P, Hasendoncks F, Gimeno D, Silvestri A, Vassilieva E, et al. Western Mediterranean sand deposits as a raw material for Roman glass production. J Archaeol Sci. 2012;39:2897–907.
[35] Moretti C, Hreglich S. 2007. I vetri opachi. Sintesi delle tecniche usate dall'antichità all'Ottocento. In Ferrari D, VISSER TRAVAGLI AM editors. Il Vetro nell'Alto Adriatico, Atti delle IX Giornate Nazionali di Studio (Ferrara 2003), Comitato Nazionale Italiano AIHV, Università degli Studi di Ferrara, Soprintendenza per i Beni Archeologici dell'Emilia Romagna: Imola. 167–76.
[36] Tite MS, Pradell T, Shortland AJ. Discovery, production and use of tin-based opacifiers in glasses, enamels and glazes from the late iron age onwards: a reassessment. Archaeometry. 2008;50:67–84.
[37] Maltoni S, Silvestri A. A Mosaic of colors: investigating production technologies of Roman glass tesserae from Northeastern Italy. Minerals. 2018;8:255–86. DOI: https://doi.org/10.3390/min8060255.
[38] Fiorentino S, Vandini M, Chinni T, Caccia M, Martini M, Galli A. Colourants and opacifiers of mosaic glass tesserae from Khirbet al-Mafjar (Jericho, Palestine): addressing technological issues by a multi-analytical approach and evaluating the potentialities of thermoluminescence

and optically stimulated luminescence dating. Archaeolog Anthropolog Sci. 2017;1–23. DOI: https://doi.org/10.1007/s12520-017-0555-9.
[39] Verità M, Zecchin S. Scientific investigation of a Venetian polychrome goblet of the sixteenth century. J Glass Stud. 2008;50:105–15.
[40] Lima A, Medici T, Pires De Matos A, Verità M. Chemical analysis of seventeenth century Millefiori glasses excavated in the Monastery of Sta. Clara-a-Velha, Portugal: comparison with Venetian and *Façon-de-Venise* Production. J Archaeol Sci. 2012;39:1238–48.
[41] Neri E, Jackson M, O'Hea M. Analyses of glass tesserae from Kilise Tepe: new insights into an early Byzantine production technology. J Archaeol Sci Rep. 2017;11:600–12. DOI: https://doi.org/10.1016/j.jasrep.2016.12.036.
[42] Moretti C, Gratuze B. 2000. I vetri rossi al rame, confronto di analisi e ricette, Annales de l'Association Internationale pour l'Histoire du Verre. vol. 14. Venice and Milano: Association internationale pour l'histoire du verre. 227–32.
[43] Maltoni S, Silvestri A, Molin G. Opaque red glass tesserae from Roman and Early-Byzantine sites of north-eastern Italy: new light on production technologies. In: Wolf S, de Pury-Gysel A, editors. Annales du 20e Congrès de l'Association Internationale pour l'Histoire du Verre (Fribourg/Romont 7-11 septembre 2015). Romont: AIHV - Association Internationale pour l'Histoire du Verre, 2017: 280–7.
[44] Silvestri A, Nestola F, Peruzzo L. Multi-methodological characterisation of calcium phosphate in late-Antique glass mosaic tesserae. Microchem J. 2016;124:811–18. DOI: https://doi.org/10.1016/j.microc.2015.10.026.
[45] Silvestri A, Tonietto S, Molin GM, Guerriero P. The palaeo-Christian glass mosaic of St Prosdocimus (Padova, Italy): archaeometric characterisation of tesserae with antimony- or phosphorus-based opacifiers. J Archaeol Sci. 2012;39:2177–90.
[46] Machado C, Machado A, Palomar T, Vilarigues M. Grisaille in historical written sources. J Glass Stud. in press.
[47] Merrifield M. Original treatises dating from the XIIth to XVIIth centuries, on the arts of painting, in oil, miniature, mosaic, and on glass; of gilding, dyeing, and the preparation of colours and artificial gems, volume II. London: John Murray, 1849.
[48] Félibien A. Des principes de l' Architecture, de la Sculpture, de la Peinture, et des Autres Arts qui en dépendent. Paris: J.-B. Coignard; 1676.
[49] Santos Â, Vilarigues M. Sanguine paint: production, characterization, and adhesion to the glass substrate. Stud Conserv. 2018. DOI: https://doi.org/10.1080/00393630.2018.1482708.
[50] Cagno S, Brondi Badano M, Mathis F, Strivay D, Janssens K. Study of medieval glass fragments from Savona (Italy) and their relation with the glass produced in Altare. J Archaeol Sci. 2012;39:2191–7.
[51] Sarah G, Gratuze B, Barrandon J-N. Application of laser ablation inductively coupled plasma mass spectrometry (LA-ICP-MS) for the investigation of ancient silver coins. J Anal At Spectrom. 2007;22:1163–7.
[52] Chataigner C, Gratuze B. New data on the exploitation of obsidian in the southern Caucasus (Armenia, Georgia) and eastern Turkey, Part 1: source characterization. Archaeometry. 2014a;56:25–47.
[53] Chataigner C, Gratuze B. New data on the exploitation of obsidian in the southern Caucasus (Armenia, Georgia) and eastern Turkey, part 2: obsidian procurement from the upper Palaeolithic to the late bronze age. Archaeometry. 2014b;56:48–69.
[54] Gratuze B. Obsidian Characterization by Laser Ablation ICP-MS and its Application to Prehistoric Trade in the Mediterranean and the Near East: sources and Distribution of Obsidian within the Aegean and Anatolia. J Archaeol Sci. 1999;26:869–81.

[55] Gratuze B. Glass characterization using laser ablation inductively coupled plasma mass spectrometry methods. In: Janssens K, editor. Modern methods for analysing archaeological and historical glass. vol. I. Wiley: Chichester, 2013.
[56] Šmit Ž. Ion-beam analysis methods. In: Janssens K, editor. Modern methods for analysing archaeological and historical glass. vol. I. Chichester: Wiley; 2013: 156–83.
[57] Johansson SAE, Campbell JL, Malmqvist KG. Particle Induced X-Ray emission spectrometry (PIXE). New York, USA: John Wiley & Sons, Inc; 1995.
[58] Wang Y, Nastasi M. Handbook of Modern Ion Beam Materials Analysis. 2nd ed. Warrendale, USA: Materials Research Society, 2009.
[59] Ryan CG, Jamieson DN, Churms JV, Pilcher CL. A new method for on-line true-elemental imaging using PIXE and the proton microprobe. Nucl Instr Meth B. 1995;104:157–65.
[60] Campbell JL, Boyd NI, Grassi N, Bonnick P, Maxwell JA. The Guelph PIXE software package IV. Nucl Instr Meth B. 2010;268:3356.
[61] Uzonyi I, Szabó G. PIXEKLM-TPI – a software package for quantitative elemental imaging with nuclear microprobe. Nucl Instr Meth B. 2005;231:156.
[62] Grime G. 2018. Oxford Microbeams Ltd. http://wwwmicrobeams.co.uk/download.html. Accessed:5 Sep 2018.
[63] The CSIRO-GEMOC Nuclear Microprobe. http:/nmp.csiro.au/csiro-gemoc.html. Accessed: 5 Sep 2018.
[64] Oxford Microbeams LTD, http://www.microbeams.co.uk/. Accessed: 5 Sep 2018.
[65] Breese MBH, Jamieson DN, King PJC. Materials analysis using a nuclear microprobe. New York, USA: John Wiley and Sons, Inc., 1996.
[66] Janssens K. Electron microscopy. In: Janssens K, editor. Modern methods for analysing archaeological and historical glass. vol. I. Wiley: Chichester, 2013.
[67] Coutinho I, Medici T, Alves LC, Gratuze B, Vilarigues M. Provenance studies on *façon- de-Venise* glass excavated in Portugal. J Archaeol Sci Rep. 2016;7:437–48.
[68] Coutinho I, Medici T, Wiley R, Alves LC, Gratuze B, Vilarigues M. The Gourd-Shaped Vessel: A Portuguese Product? J Glass Stud. 2017;59:215–34.
[69] Coutinho I, 2016. New Insights Into seventeenth and eighteenth Century Glass From Portugal: Study and Preservation. PhD Thesis, Universidade Nova de Lisboa.
[70] Medici T. 2014. Vidros da Terra. O vidro tardomedieval e moderno em Portugal (séculos XIV-XVII). O contributo da arqueologia. PhD thesis, Faculdade de Letras da Universidade de Coimbra. Coimbra: Universidade de Coimbra.
[71] Ferreira M. Espólio Vítreo Proveniente da Estação Arqueológica do Mosteiro de Sta. Clara-a-Velha de Coimbra: resultados Preliminares. Revista Portuguesa De Arqueologia. 2004;7:541–83.
[72] Castro AS, Sebastian L. A componente de desenho cerâmico na intervenção arqueológica no Mosteiro de S. João de Tarouca, 1998–2001. Património/Estudos. 2002;2:33–42.
[73] De Raedt I, Janssens K, Veekman J. On the distinction between sixteenth and seventeenth century Venetian and façon-de-Venise glass. In: Veeckman J, editor. Majolica and Glass: from Italy to Antwerp and Beyond. the transfer of technology in the 16th - early seventeenth Century. Antwerp: Stad Antwerpen; 2002: 95–121.
[74] Custódio J. A Real Fábrica de Vidros de Coina (1719-1747) e o vidro em Portugal nos séculos XVII e XVIII. Lisboa: Instituto Português do Património Arquitectónico, 2002.
[75] Frothingham Wilson A. Hispanic glass with examples in the collection of the Hispanic society of America. New York: U.S.A. Hispanic Society of America; 1941.
[76] Valente V. O vidro em Portugal. Porto: Portucalense Editora; 1950.
[77] Medici T, Lopes FM, Lima AM, Larsson MA, Pires de Matos A. 2009. Glass Bottles and Jugs from the Monastery of Sta. Clara-a-Velha, Coimbra, Portugal. In: Janssens K, Degryse P, Cosyns P,

Caen J, Van't dack L, editors. Annales de l'Association Internationale pour l'Histoire du Verre, University Press Antwerp. vol. 17. Antwerp. 2006:391–400.
[78] Serrão VR, Sainte-Fare-Garnot P-N. Rouge et or: trésors du Portugal baroque. Portugal: Gabinete das Relações Internacionais, 2001: 140–1.
[79] Capellà Galmés MÀ. Ars vitraria: mallorca (1300–1700). Palma: Edicions UIB, 2015.
[80] Gudiol Ricart JM. 1936. Els vidres catalans, Monumen- ta Cataloniae vol. 3. Barcelona: Ed. Alpha.
[81] Capellà Galmés MÀ, Albero Santacreu D. El horno de vidrio del siglo XVII de Sa Gerreria (Palma, Mallorca): contextualización histórica y análisis preliminar de los materiales. Boletin de la Sociedad Española de Cerámica y Vidrio. 2015;54:142–52.
[82] Ulitzka S. Analysen von historischen Gläsern – licht im Dunkel der Geschichte? In: Theuerkauff-Liederwald A, editor. Venezianisches Glas der Kunstsammlungen der Veste Coburg. Die Sammlung Herzog Alfreds von Sachsen – coburg und Gotha (1844-1900). Venedig á la façon-de-Venise, Spanien, Mitteleuropa. Lingen: Luca verlag, 1994: 40–53.
[83] Caen J. The production of stained glass in the county of Flanders and the duchy of Brabant from the XVth to the XVIIIth centuries: materials and techniques. Antwerp: Brepols; 2009.
[84] Schalm O, Van der Linden V, Frederickx P, Luyten S, Van der Snickt G, Caen J, et al. Enamels in stained glass windows: preparation, chemical composition, microstructure and causes of deterioration. Spectrochim Acta Part B. 2009;64:812–20. DOI: https://doi.org/10.1016/j.sab.2009.06.005.
[85] Marschner H. 1996. Analyses de pigments de grisaille sur des vitraux Munichois de l'église du Sain-Saveur, réalisés vers 1500. In: C. royale des monuments sites et fouilles de la R. Wallonne editor. Grisaille, Jaune d'argent, Sanguine, Email et Peinture à Froid, techniques et conservation. Forum pour la Conservation et la Restauration des Vitraux, Liège, 19-22 Juin 1996, Liège: Commission royale des monuments, sites et fouilles, 53–9.
[86] Vilarigues M. 2008. Estudo do Efeito da Adição de Iões Metálicos na Corrosão de Vidros Potássicos. PhD Thesis, Universidade NOVA de Lisboa.
[87] Verità M. 1996. Composition, structure et mecanisme de detérioration des grisailles. In: C. royale des monuments sites et fouilles de la R. Wallonne editor. Grisaille, Jaune d'argent, Sanguine, Email et Peinture à Froid, techniques et conservation. Forum pour la Conservation et la Restauration des Vitraux, Liège, 19-22 Juin 1996, vol. 3 . 61–8. Liège: Commission royale des monuments, sites et fouilles.
[88] Vilarigues M, Da Silva RC. Ion beam and Infrared analysis of medieval stained glass. Appl Phys A Mater Sci Process. 2004;79:373–8. DOI: https://doi.org/10.1007/s00339-004-2538-9.
[89] Carmona N, Villegas MA, Navarro JMF. Study of glasses with grisailles from historic stained glass windows of the cathedral of León (Spain). Appl Surf Sci. 2006;252:5936–45. DOI: https://doi.org/10.1016/j.apsusc.2005.08.023.
[90] Machado A, Wolf S, Alves LC, Katona-Serneels I, Serneels V, Trümpler S, et al. Swiss stained-glass panels: an analytical study. Microsc Microanal. 2017;23:878–90. DOI: https://doi.org/10.1017/S1431927617000629.

Maria F. Guerra

13 Physicochemical approaches to gold and silver work, an overview: Searching for technologies, tracing routes, attempting to preserve

Abstract: Gold alloys and silver alloys have always been widely employed in the production of significant objects. With high reflectivity, precious metals are perceived as both materials and colours, and can be skilfully combined to produce metallic polychrome effects. Because their structure and composition contain information on their manufacture, use, disclaim and degradation, items in gold and in silver enclose major information on the technologies employed by past societies and on exchange networks. This information can be acquired using appropriate analytical protocols, established according to the nature of the query and the characteristics of the objects.

By using physicochemical techniques, it is possible to identify the technologies, materials and tools used by the artisan and, in particular cases, to situate the sources of raw materials and the workshops producing the objects, as well as to follow the trade routes. The aim of this work is to outline major achievements in the study of goldwork and silverwork based on the different physicochemical methods that are available, and to refer the analytical difficulties that have to be faced when studying objects made from precious metals. Based on several examples, three topics are addressed. The first concerns the major role of the techniques of exam when describing shaping, decorating, assembling and finishing; the second considers the search for metallic polychrome effects in some cultural areas; and the third discusses the challenging question of fingerprinting. A fourth section is dedicated to a short reflection on the difficulties related to the identification of the atmospheric corrosion mechanisms of precious metals.

Keywords: precious metal, gold, goldwork, silver, silverwork, alloy, jewellery, colour, polychrome effect, corrosion, provenancing, fingerprinting

13.1 Introduction

Expressions of social (political and economic) and religious power, prestigious objects are manufactured in specific materials. Among them metals, in particular gold and silver, because they have properties that result in perceived specific visual

---

This article has previously been published in the journal *Physical Sciences Reviews*. Please cite as: Guerra, M. F. Physicochemical approaches to gold and silver work, an overview: searching for technologies, tracing routes, attempting to preserve. *Physical Sciences Reviews* [Online] **2019**, 4. DOI: 10.1515/psr-2018-0015.

https://doi.org/10.1515/9783110457537-013

aspects [1] that can be enriched by illumination [2]. Transmission, refraction, absorption, scattering and reflection of light in gold and silver surfaces modify the appearance of the shapes. Perceived as both materials and colours [3], precious metals have a lustrous appearance [4] and are thus among the most appropriate for the production of significant objects.

More than their mechanical properties (ductility, malleability, strength, etc.) that make precious metals and particularly gold to be easily shaped by plastic deformation, their major asset is based on their optical properties, because metals have high reflectivity. In the past, silver was commonly alloyed with copper, which results in a whitish shining material (for example in Sumerian the same word is employed to designate silver, white, and bright, [5]), as silver is whitish over almost all the light spectrum [6]. Gold, silver and copper were alloyed to produce gold alloys with colours that go from yellow to red, green and white, depending on the amounts of alloying elements [7], and a reflectivity that increases with growing silver amounts [8]. It is also possible to employ thin gold foils to obtain polychrome effects, because the reflectivity of gold at the high extreme of the light spectrum makes it look yellow, whilst a gold leaf of 100 nm thick transmits a bluish-green light [9, 10].

It is consequently possible to obtain multiple colour effects in jewellery and ware with gold and silver alloys, even when these metals are employed alone (with no other materials, but with possible addition of metallic elements such as wires and granules), or when they are applied as thin layers on a substrate (platting, gilding, silvering, etc.). Precious metals can also be combined with more or less precious and colourful materials, such as gems and enamels, providing to the final object colour, lightness, volume, transparency, etc. Therefore, the use by the goldsmith and the silversmith of a wide-range of materials and techniques finely enhances the object by changing its visual perception. The artisan can also entrust it with a "role" or a "value" when certain colours are used ("colour-code"), when certain elements are added, or when certain motifs are applied by chasing, engraving, punching, etc. at the surface of a precise shape. Coins and liturgical cups are common examples of the roles assumed by objects with precise shapes, made from precious metals.

The technologies and materials employed in an object construction depend on the artisan's skill, on the workshop traditions, and on chronological and territorial knowledge diffusion at a specific time, as well as on access to materials [11]. There is indeed great reliance on access to sources of raw materials, and thus to opened trade routes and to the savoir-faire and workshop practices that voyaged along those routes and that had over time influenced creation. Nonetheless, the technologies and materials employed in a certain period are also related to fashion (indirectly connected to exchange). By the end of the nineteenth century, archaeological jewellery was produced employing the finest techniques of the art of working gold [12, 13]. In the early twentieth century, Art Nouveau creators, who reinvented antique jewellery, produced colourful items constructed with many materials that were combined with gold and silver [14], sometimes enhanced by patination [15].

Goldwork and silverwork enclose major information on a society, because their structure and composition contain information on their manufacture, use, disclaim and degradation. This information can be acquired by the detailed study of the many aspects of the objects, which includes their description (dimension, morphology, iconography, typology, etc.) and the data collected by employing physicochemical methods. With an appropriate analytical protocol, it is possible to finely describe the construction details of an object (including identification of the technologies and tools used by the goldsmith and the silversmith). It is also possible to identify the materials employed, and occasionally to situate the origin of the raw materials and of the items production, and follow their trade routes [16–21]. In the sections below are outlined major achievements in the study of goldwork and silverwork, based on the many physicochemical methods that are offered today. Because it has recently been observed an increase of the atmospheric corrosion of precious metals, a short section on this question was also included above.

13.2 Considerations on analytical protocols

A large number of physicochemical techniques, often developed in other areas of science and for other research goals, have been applied to the study of cultural heritage. Based on these techniques, efficient analytical protocols can be established after a good definition of the query and after considering the specificities of the objects under study. In our case, among the specificities of gold and silver objects are those that exclude sampling: their small size, their scarcity, their value, and their mounting by joining multiple parts. Consequently, and contrary to many other cases, the study of items made from precious metals is widely based on exam and non-destructive analysis.

The observation of the morphology and structure of an item, at several depths and scales, by using different lights and radiations, makes available information on quite a few technological steps related to mounting, decorating and finishing [11, 14, 22–24]. It is important to register detailed macrophotography images of those technological steps. Despite the difficulties caused by the high reflectivity of precious metals [25], attempts have been made to replace 2D by 3D imaging [26, 27].

The accurate description of an object's construction is one of the main phases of study, because understanding an object is essential to establish an analytical protocol. In fact, in a general way, the analytical techniques can be divided into three types – elemental, isotopic and structural – each of them providing specific information on the object. For example, if elemental analysis can be applied in evidencing wide-ranging questions, isotopic analysis typically gives information on the origin of the metals and on the circulation of metals and items. Because requiring a sample, isotopic analysis is occasionally applied to the analysis of silver alloys [28–30], but barely to the analysis of gold alloys [31].

Therefore, the development of an analytical protocol has to consider the choice of the most appropriate physicochemical techniques to solve one question, taking into account the object manufacture, and after considering the analytical limitations. After selection of the type of information that has to be searched – elemental, isotopic or structural – it is necessary to choose the most appropriate techniques that can provide this information, considering the nature of the query, the type of object under study, and the technical characteristics essential to achieve the objective. The technical performance parameters that play an important role when studying gold and silver work are typically: spatial resolution, depth of analysis, limits of detection, and portability.

Spatial resolution is a quite important parameter when analysing small objects consisting on several joined small parts. To give an example, Figure 13.1 shows a detail of an earring decorated with several elements in gold under the stereomicroscope and under the scanning electron microscope (SEM) combined with energy dispersive spectroscopy (EDS). The EDS analysis of the earring provides the elemental composition of the base-sheets and of the added parts (in this case granules), as well as the composition of the filler used to obtain the joints (Table 13.1). The filler metal, of lower melting point than the parts to be soldered [32, 33], contains, in the case of the earring, higher Ag and Cu contents than the base-sheets and the granules. The use of particle induced X-Ray emission (PIXE) with an incident micro beam of protons [12, 34, 35] or of X-ray fluorescence spectrometry (XRF) with polycapillary focusing optics [36] would have also been a possible analytical solution in this case. EDS, µPIXE and µXRF can be used to construct elemental maps, which are very useful in several situations, for example when estimating the composition of solders employed to join the object's parts [37, 38]. However, SEM-EDS has the unique advantage of providing together with elemental analysis high-resolution imaging of

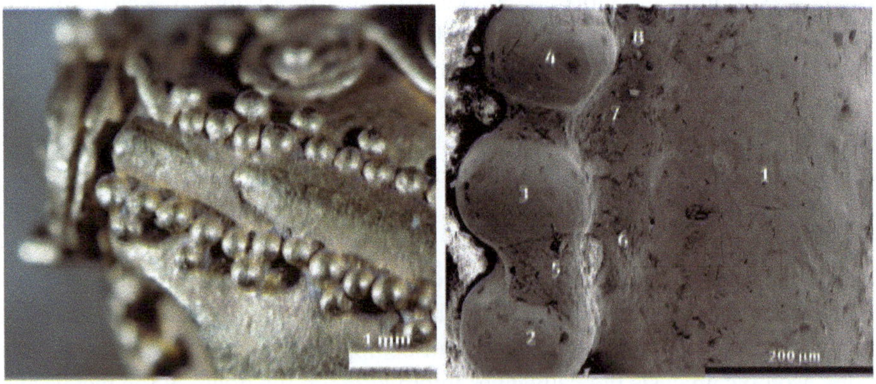

Figure 13.1: Details of an earring decorated with granulation under the stereomicroscope and under the SEM (FEI Philips XL 30 CP operated at 20 kV). The image obtained with the SE detector shows three soldered granules to the base-sheet.
(Images by M.F. Guerra).

Table 13.1: EDS data obtained for each region of interest in the earring decorated with granulation of Figure 13.1 (FEI Philips XL 30 CP with an integrated Link Isis 300 EDS system operated at 20 kV; analysis in spot mode with 300 s acquisition time; calculation with ZAF matrix correction model).

Region of analysis	Au wt%	Ag wt%	Cu wt%
1	95	3	1
2	96	3	1
3	95	4	2
4	95	4	1
5	87	8	4
6	85	12	3
7	85	11	4
8	83	13	4

the surfaces, which is essential when studying objects made from gold and silver. By using an equipment with a field emission gun (FEG), the images obtained can be improved for particular studies [19].

Depth of analysis has to be considered when either the surface or the bulk offer incomplete information on the object, for example when the surface was naturally or artificially modified. In this case, it is necessary to obtain information from both the first layers and the bulk by using specific techniques, or by combining techniques of surface and bulk analysis based on their effective penetration depths [39–42]. For example, in the case of gold alloys, the effective penetration depth of EDS is inferior to 1 µm, and that of XRF is between 8 and 60 µm [38]. In addition to elemental analysis, structural analysis (for example XRD, X-ray diffraction) can be employed in this case to identify the compounds present at the object surface and SEM to provide their morphology. These compounds give information either on the chemical surface treatments artificially applied, for example by patination [43], or on the natural corrosion products that grow on the surfaces [19, 44–46].

When searching for the origin and trade routes of silver and gold, the analytical techniques selected have to be either isotopic or elemental. In general, structural analysis provides no information on this point. Elemental analysis has to be carried out with good detection limits, because trace elements that may link the object to a source of metal have to be determined. In this case, PIXE can be used for the study of silverwork [47, 48], but more hardly of goldwork [49, 50]. A powerful technique in the area of provenancing, with high spatial resolution and providing elemental maps, is XRF with synchrotron radiation (SR), either with energy-dispersive [51] or with wavelength-dispersive [19, 52] spectroscopy. Another technique frequently used for provenancing that combines low detection limits with high spatial resolution, is LA-ICP-MS [53–56]. With the disadvantage of collecting small samples by laser ablation [57] and thus not appropriate when elemental maps are necessary or when the objects are made by joining multiple small parts, it has the advantage of when combined with a Multi-Collector (LA-MC-ICP-MS) to also provide the isotopic ratios [30, 31]. The small sampling might be a gain when sampling and analysis are carried out in separated phases [58].

At last, portability is widely searched when analysing silver and gold items. On one hand, the institutions where the objects are conserved or exhibited increasingly impose not to move them for analysis, and on another hand, as above-mentioned, the objects can barely be sampled. Therefore, the use of portable equipment has been growing for the study of precious metals, even if only a few techniques are available in portable configuration. Equipment that can easily be moved for *in-situ* analysis are mobile and handheld XRF, whose application in the study of gold and silver objects has been widely increasing [59–63]. Portable equipment has, however, lower levels of performance than large facilities hosted in laboratories. It would have been impossible to carry out the study of the earring shown in Figure 13.1 using handheld XRF. A spot beam of several millimetres (typically 3 mm) would have only provided an average composition of all the alloys employed in the earring construction, instead of the elemental composition of each of the joined parts and the solders.

13.3 Searching for technologies

Thermomechanical processes transform gold and silver into objects. The creative process of making silverwork and goldwork includes the use of sophisticated making techniques of shaping, decorating, assembling and finishing. These techniques that require strong skill, follow alloying and melting, used to obtain the base alloys to be worked.

The initial question regards whether the object is made by shaping one piece and how that piece was shaped. Otherwise, it has to be defined how the several pieces that constitute the object were shaped (casting, hammering, rolling, etc.) and how they were assembled (mechanical or thermal joining). Composite objects that are more complex require more consideration, because their shaping and mounting may employ several techniques. Much of the information on the creative process of silverwork and goldwork can be searched by using techniques of exam. Figure 13.2 shows the mechanical mounting of a comb in gold, horn and enamel, made by René Lalique in 1901–1902, under a stereomicroscope (white light) and using a portable digital X-ray radiography system. The head of the two gold screws are visible on the reverse of the comb, but the entire screws are only visible under X-rays. In spite of the high attenuation of X-rays in gold and silver alloys, X-radiography provides information on the invisible parts of the objects [65–67] and on their structure, which, in addition to manufacturing details, may reveal previous restorations and corrosion processes [68–70].

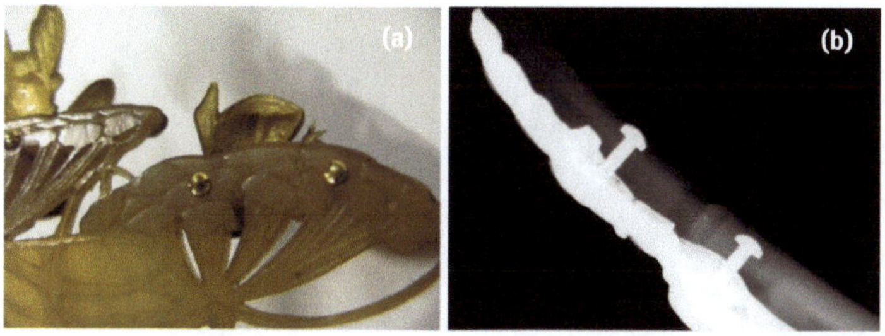

Figure 13.2: Details of comb "Bumblebees and umbels" in gold, horn and enamel, created by René Lalique in 1901–1902 [Calouste Gulbenkian Museum, Acc. n. 1160, [64]. (a) The head of the gold screws used in the mounting; and (b) the X-radiography showing the mounting system with screws (Smart 225 Andrex with Fuji and Dürr storage image plates and a Dürr digitizer; operating conditions: voltage tension 130 kV with 3 mA current, 90 s acquisition time, and 93 cm focus to detector distance). (photography by M.F. Guerra and radiography by project on Lalique's jewellery in Calouste Gulbenkian Museum headed by M.F.Guerra).

When the mounting technology is identified, observation under the stereomicroscope is frequently sufficient to verify whether the suggested techniques are employed. In spite of their many types and heights (3–25 cm), the traditional anthropomorphic and zoomorphic metallic figurines [61] produced for ritual offerings by the Incas are frequently made using standardised technologies, and are thus a good example of this situation. These figurines, made of gold, silver or spondylus shell, were used in the Capacocha ceremony [71], because several of them, often with their miniature garments in textile, feather, etc., were found with the sacrificed children in mountaintop shrines [72]. Traditional figurines are hollow, made by joining several hammered parts [73–75]. A female hollow figurine is regularly made by soldering five gold or silver sheets, as shown in Figure 13.3. Under the stereomicroscope, the seams corresponding to the expected assemblage by hard soldering can be searched. It is also possible to define under the stereomicroscope how the anatomical motifs were obtained (Figure 13.4), usually by using in repoussé and chasing.

The resolution limitations and the highly reflective gold and silver surfaces under white light make, however, difficult to observe very small details under a stereomicroscope. It is in certain cases more convenient to use SEM, because with an incident electron beam it is possible to obtain high-resolution images and, for metals, to overcome the problem of surface reflectivity. As already mentioned, another advantage of SEM is that it can be combined with EDS to provide elemental information with high spatial resolution. To understand the soldering technology employed in the construction of the ritual Incan figurines, it is necessary to observe and analyse the joints, as shown in Figure 13.5 for a male gold figurine studied using SEM-EDS. In addition to the surface morphology, the composition of the parts to be soldered and the solders could be determined, showing that the solder alloys contain high Cu contents to lower the melting point.

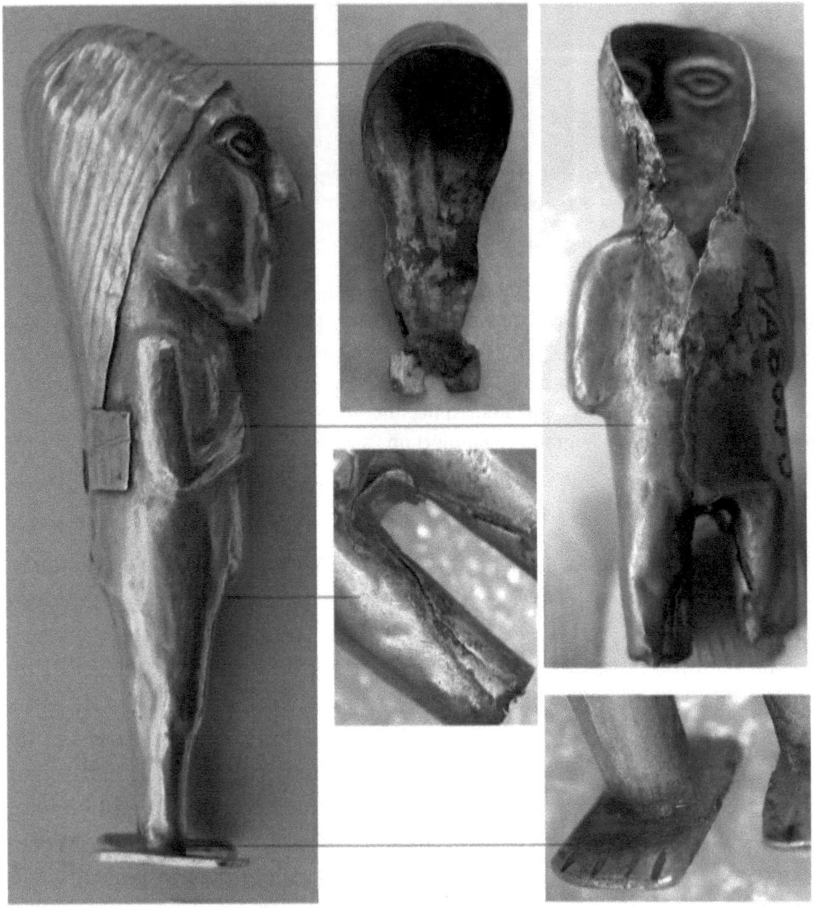

Figure 13.3: Details under the stereomicroscope of the mounting of an hollow female figurine produced in Incan times for ritual offerings by joining one element that reproduces the body, legs and head to the other elements that represent the hair, the crotch and the feet. (photography by M.F. Guerra).

13.4 Silver and gold as colours

Silver whitish and gold yellowish, reddish, greenish and whitish alloys can be combined to produce polychrome objects. From the analytical point of view, the search for the use of metallic polychrome effects is one of the easiest questions to be solved, when are excluded objects with surface treatments. In fact, it is sufficient to determine the elemental composition of the alloys employed in the production of the metallic parts, or the nature of the applied compounds in the case of, for example, niello inlays [76].

The combination of silver whitish and gold yellowish alloys produce bicolour objects, a very widespread practice for example in the Andean area, such as in the

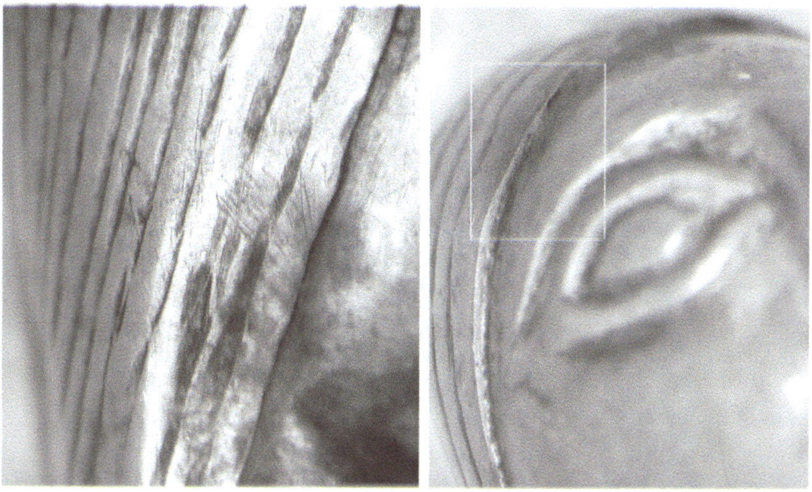

Figure 13.4: Under the stereomicroscope it can be seen for the female figurine of Figure 13.3 the solder in the joints and the tool marks that show the use of chasing to represent the hair. (photography by M.F. Guerra).

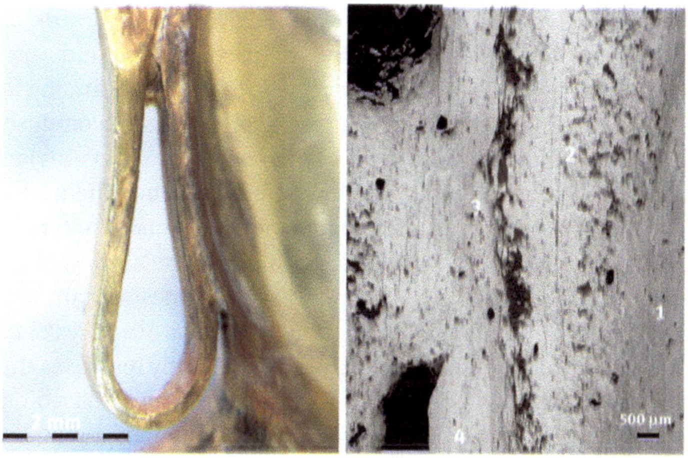

Figure 13.5: Details under the stereomicroscope and the SEM of one Incan male figurine in gold. The rectangular in section wire bent to form the ear was hard soldered to the head. EDS data show that the solder contains high Cu contents: (1) head sheet 1 wt% Cu, 28 wt% Ag, 71 wt% Au; (2) solder 6 wt% Cu, 14 wt% Ag, 80 wt% Au; (3) solder 5 wt% Cu, 20 wt% Ag, 75 wt% Au; (4) ear wire 2 wt% Cu, 23 wt% Ag, 75 wt% Au. (images by M.F. Guerra).

Vicús' workshops. Many bicolour metallic objects are attributed to that pre-Columbian culture that developed around 400 BC in the Piura area, north of Peru, in particular remarkable nose rings of several types [77]. The bicolour part of one of those nose rings is shown in Figure 13.6, with a detail of the mounting by hard-

Figure 13.6: Bicolour area of an unprovenanced nose ring attributed to the Vicús culture with a detail under the stereomicroscope of the region where the whitish and yellowish wires are assembled using hard-soldering.
(photographies by M.F. Guerra).

soldering the whitish and the yellowish parts. The gold alloys employed are quite common, containing about 10 wt% Ag. The only analytical difficulty in this case is the analysis of the smallest parts, which means the wires and joints. They are, however, in this case large enough to be analysed *in-situ*, using a mobile XRF equipment with a spot beam of 1 mm.

The Andean area also gives one of the major examples of the skilful work of combining the alloying metals to obtain gold alloys of particular colours and reflectivity. The symbolic Incan figurines above mentioned (Figure 13.3) are made using standardised technologies and by employing artificial alloys that accomplish precise requirements [61]. Their analytical study was carried out *in-situ*, in Peruvian and European institutions [78], because their size allows employing handheld XRF with a 3 mm spot beam. It was shown that the alloys employed in the manufacture of the silver figurines are of very high quality (the copper content in the alloy is often below 1 wt%) and those employed in the production of the gold ones contain low copper contents and high silver amounts. Silver frequently reaches 30–50 wt% in the gold alloys, which keeps the yellow colour of gold, but significantly increases the reflectivity of the surface. The searched effect could be an interpretation based on visual perception, connecting the object and the shining sun, as there is a relation between the sun and gold [79] and between the sun and shiny objects [80].

The skilful combination of gold with the alloying metals to obtain sophisticated polychrome effects can be found in Ancient Egypt, where the art of the goldsmith includes the combination of alloys of different colours and shades to produce composite objects. One example comes from tomb 72 excavated in Cemetery A at Harageh [81], an intact burial of a young girl, dating to the mid-late 12th Dynasty (*c.* 1875–1795 BC), where were found five fish pendant amulets [82] today in the collection of the National Museums Scotland. The analysis using XRF, with a spot beam of about 1 mm, of the body and tail of one of the fishes (Acc. n. A1914.1081) has shown that it was produced with gold alloys of different colours. A yellowish alloy (containing 17 wt% Ag) was used for the body, and a whitish one (containing 56 wt% Ag) was used for the tail [83].

The use of gold alloys only to produce jewellery with different colours is a traditional practice in the Egyptian workshops, well represented by the more or less elaborate beads in the shape of cowrie shells, which in the Middle Kingdom (*c.* 2050–1710 BC) were regularly strung in girdles found in female burials [82]. Beads of this type were also found in burials from other periods [38]. The simplest ones are those dated to the First Intermediate Period (*c.* 2181–2055 BC). Pierced at each end, they were made by embossing gold sheets, where the shell's aperture was represented by a groove. In the collection of the Petrie Museum of Egyptian Archaeology (University College London) are one string of twelve gold cowrie shell beads (UC18092) enhanced with small whitish ring beads (Figure 13.7), found in the undisturbed tomb 7923 at Qau [84], p. 45, pl. 46], and a group of four specimens (UC42397). Using a portable XRF spectrometer with a spot beam of 1 mm (the same used in [62]), the cowrie shell beads could be analysed *in-situ*. The small ring beads over and on each side of the cowrie shell beads, were also analysed, but data obtained represent average compositions.

Figure 13.7: One of the yellowish beads in the shape of cowrie shells threaded with small whitish gold ring beads from string UC18092 found in undisturbed tomb 7923 at Qau, under the stereomicroscope.
(image by M.F. Guerra).

Data plotted in the diagram of Figure 13.8 for the two groups of cowrie shell beads show that despite the use of gold alloys only, two different polychrome effects were searched. String UC18092 is a metallic bicolour item, as it contains elements in two colours of gold: the yellow-green cowrie shell beads are strung with whitish ring beads. The four cowrie shell beads UC42397 were made to produce another colour effect, a kind of colour gradient. In fact, their alloys have similar Ag/Au ratios, but increasing Cu contents, which produces gold alloys in shades of yellow, yellowish and reddish.

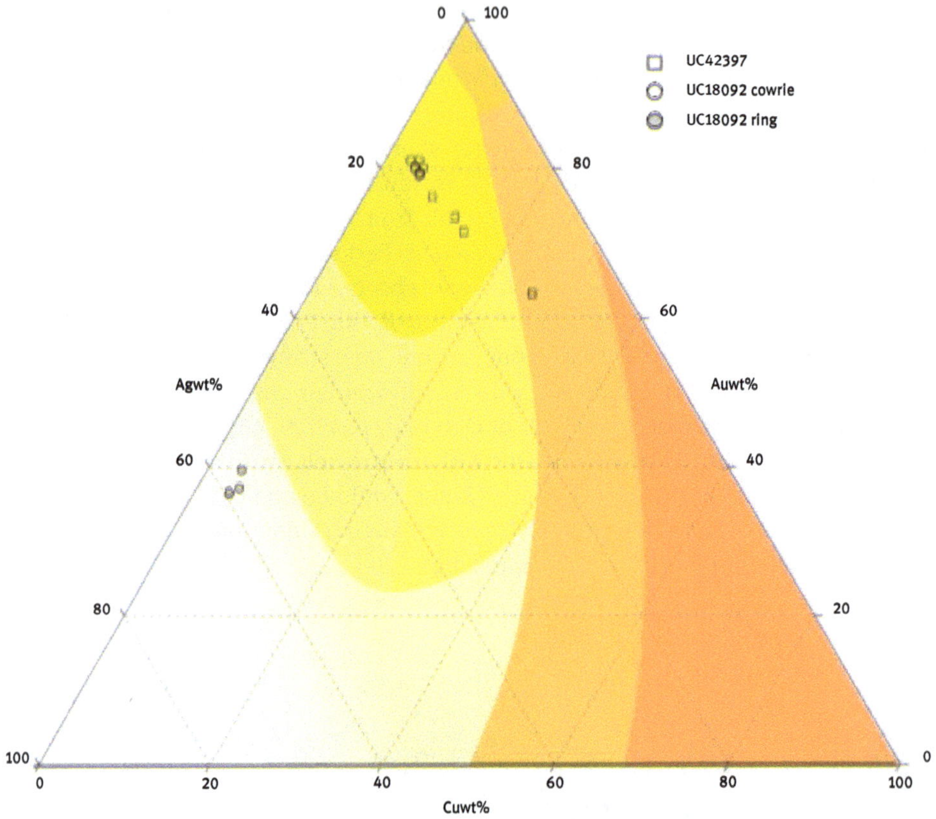

Figure 13.8: Ternary diagram representing the Au, Ag and Cu contents (in wt%) obtained by XRF for the gold cowrie shell beads and ring beads UC18092 and UC42397.

13.5 Tracing trade routes and marking the dates

The movements of people have contributed to the spread and transmission of ideas, beliefs, savoir-faire, technologies, and of goods and materials. Extended distance pathways resulted in trade routes and exchange networks that much influenced past economy, but which are challenging to reconstruct, in particular when documentary sources and historical evidence are missing. Attempts to trace trade routes are consequently carried out based on finds of objects.

Among the goods that travelled along the pathways, were rare materials and items that needed specific skill for their manufacture. Consequently, prestigious objects made from gold and silver contain information that should contribute to the definition of networks, when the exploited sources of metals or the workshops that produced the objects can be localised, or when a change of supply can be identified. Nonetheless, it is difficult to relate an object to a metal or to a workshop. Gold and silver processing is thermomechanical, resulting in loss of information between the first and the last steps of production. Indeed, gold and silver objects are usually made from alloys produced by mixing different metals, in general previously refined. These processes cause a loss of information on the original metals. Besides the alloy production, other actions contribute to the loss of information on the initial metals. After use, the object is disclaimed and usually reused (gold and silver are too expensive to be thrown away), which means melted (possibly refined) to produce other objects. During periods of weak economy, precious metals are constantly mixed and recycled.

It is, however, possible for certain items – in particular coinages, because their issue is controlled by an authority and information is often cited in documentary sources – and for specific periods, situations, and areas of influence, to obtain through the analysis of goldwork and silverwork evidence on exchange networks. The information obtained very much depends on the analytical technique(s) that is(are) used. It is of course very difficult to search for hints on the metal provenance when the object cannot be moved for analysis, or when it cannot be sampled. As referred above, provenancing involves the use of either isotopic analysis or elemental analysis with good detection limits.

Even though, in the particular case of gold, it is possible using exam to determine whether its origin is alluvial. In fact, the presence of PGE inclusions (naturally occurring platinum group elements) at the object's surface is an indicator of this type of gold deposits. However, their absence simply indicates the use of a different gold, which can also be alluvial, but from a geological region deprived of PGE deposits. It is interesting to refer that the few published work on the analysis of Egyptian gold jewellery often refers the presence of variable quantities of PGE inclusions in the items, indicating the exploitation of alluvial sources [38, 85, 86]. Figure 13.9 shows one PGE inclusion at the surface of a hollow Egyptian penannular

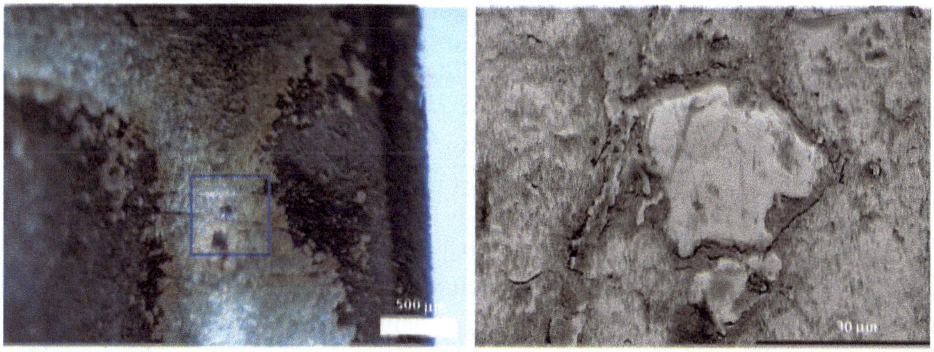

Figure 13.9: PGE inclusion at the surface of an hollow Egyptian penannular earring under the stereomicroscope and under the SEM. EDS analysis shows that the base sheet contains high silver contents (55 wt% Au, 42 wt% Ag and 3 wt% Cu) and that the PGE inclusion is a rutheniridosmine (24 wt% Ru, 32 wt% Os and 44 wt% Rh).
(images by M.F. Guerra).

earring made from an alloy containing high Ag contents. Almost invisible under the stereomicroscope using grazing white light, the inclusion clearly appears under the SEM. The EDS analysis has shown that it is, as expected, a rutheniridosmine.

The use of techniques that determine a few minor and trace elements might not provide information on the localisation of the sources of metals, but might allow identifying a change of supplies that occurs at a certain period, and provide a *terminus post quem* for certain productions. It is representative of this situation the Mapuche female ritual and daily use silver jewellery. Because of migration and of local political problems, the use codes and the technologies employed in the jewellery manufacture were regrettably nearly lost.

Said to be connected to the women's social status, the earrings, brooches, finger rings, etc. in silver significantly flourished in the nineteenth century in Chile and Argentina, in particular in Araucania and Patagonia. The ritual items, among them earring (of several types) and brooches (tupos), may reach very large dimensions [87, 88]. Like the tupos, the earrings have served as support for codified messages, and the ritual ones could have been transmitted through several generations. The daily use earrings were presumably constructed in different phases, related to the several life stages of women. Symbolic motifs and elements in silver were added to the earrings to represent life events like puberty, pregnancy, etc.

The exam using optical microscopy, SEM, and X-radiography of ritual and daily use earrings of different types revealed that the first ones were made by skilled silversmiths contrary to many of the second ones [89, 90]. μPIXE analysis of the different parts of the earrings [90] has shown that the silver alloys are of very variable quality, with silver contents within the range of 60–90 wt%. Nevertheless, the most interesting result is related to the alloying elements. In addition to Cu, the alloys of several pieces also contained Zn and Ni. The amounts of those three elements in the silver alloys used in the production of the earrings are shown in Figure 13.10, together with those found for a few ceremonial tupos and daily use finger rings analysed using the same protocol.

The intensification during the eighteenth century of the commercial relations in Araucania and Patagonia, resulted in the use of silver coins to produce silver jewellery either (as shown in Figure 13.11) by directly joining them (mechanically or thermally) to base plates and other supports [87, 88], or by reuse, which includes reshape and melting as described and photographed by Claude Joseph [91, 92]. From 1871 onwards in the Chilean mints and from 1896 onwards in the Argentinean mints, a new whitish alloy, the cupronickel, replaced silver [93]. The recycling of coins and jewellery certainly integrated the new alloy, justifying the presence of these elements in the composition of the jewellery produced after introduction of the cupronickel in the area. The presence of Zn and Ni in the alloys is thus a chronological indicator, because these elements point out a *terminus post quem* for the jewellery production. Interestingly, the objects dedicated to ceremonies are those that, in general, do not contain Zn and Ni, which could indicate a former production and thus reinforcing the suggestion of transmission through several generations.

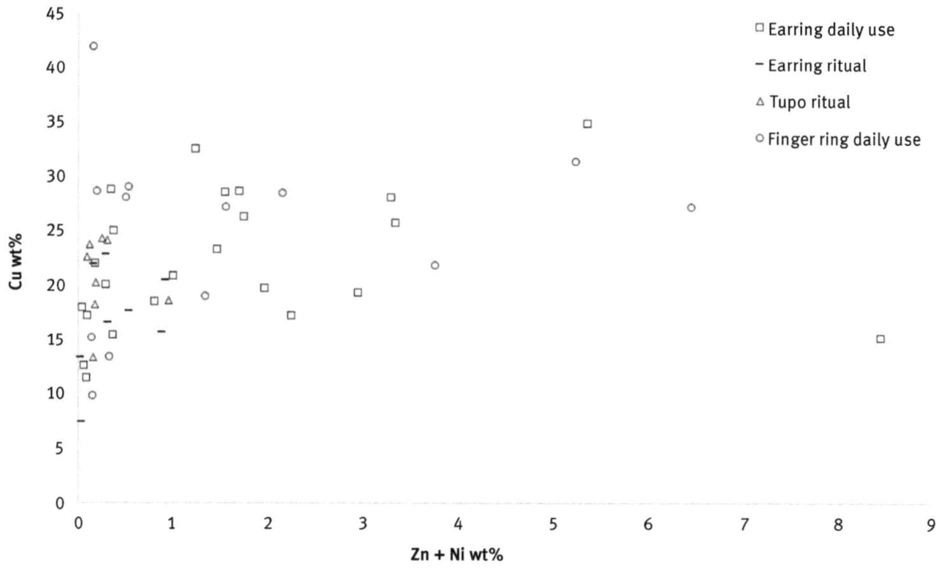

Figure 13.10: Diagram representing the Cu, Zn and Ni contents (in wt%) in the silver alloys obtained by μPIXE for the ritual and daily use silver Mapuche feminine jewellery.

Figure 13.11: Mapuche finger ring consisting of a silver strip to which was soldered one 10 cents Argentinean silver coin, struck by the Buenos Aires mint in 1881–1883.
(images by M. F. Guerra).

Several silver supplies could have been used for the manufacture of the (suggested) oldest objects, but constant recycling of items to produce new jewellery makes more difficult to search for the origin of the silver. Despite the determination by PIXE of several trace elements in the silver alloys, such as Au, As, Sn and Sb, which can be used for silver fingerprinting, it is difficult to relate the Mapuche silver to South American sources. Silver from one of the main Andean sources, the mines at

Potosi, in Bolivia, is characterised by high Au and low In contents [20, 94]. If high Au contents are, in the case of silver alloys, easy to determine using PIXE, this technique has inadequate detection limits for the determination of low In contents. It should be emphasized, however, that gold and silver from several sources situated in the American continent could be characterized using precise trace elements, determined using nuclear reactions, such as Pt in Colombian gold, Pd in Brazilian gold, In in Bolivian silver, and Sn and Sb in Mexican silver [20, 94].

Fingerprinting gold and silver could be successfully achieved in other areas of influence, even in the case of complex societies, during difficult periods, and when long-distance trade networks were used. By using ICP-MS, SR-XRF, and nuclear reaction analysis, it was possible to address questions related to, for example, the Islamic expansion [56, 95], the Byzantine Empire [96, 97], the Roman Empire [30], the Asiatic nomadic confederations [51], etc. Based on the analysis of appropriate items it is sometimes even possible to search for the evolution of the gold and silver supplies used in the manufacture of coins and jewellery during long periods [53, 98, 99]

13.6 Attempting to preserve

Because atmospheric corrosion threats several collections, the aim of this section is to offer a brief introduction to the challenging analytical questions associated to the study of the corrosion mechanisms of objects made from alloys of precious metals. In indoor conditions, gold alloys and silver alloys corrode mainly in the presence of sulphur (sulphidation), through the formation of sulphide salts [17, 100, 101]. The grown corrosion products result in multihued spots and stains at the surface of the tarnished objects, modifying their aspect, as shown in Figure 13.12, and thus their perception, leading to their restoration.

Figure 13.12: Evolution of the corroded surface colour of a gold Iron Age torc in a showcase at the National Museum of Archaeology (Lisbon, Portugal) between 2009 (on the left) and 2012 (on the right). (Photography M. F. Guerra).

When not properly carried out, restoration interventions are operations that might have serious consequences in the objects preservation as well as on the information related to their production, like tool marks, that might also be removed. Not to cause damage, neither to the object nor to the information contained in it, only the corrosion layer should be removed. Its efficient removal depends on the understanding of the corrosion processes, which requires the previous full description of the involved corrosion mechanisms. To achieve this goal, both the corrosion layer and the substrate have to be characterised and the environmental conditions described.

As stated, jewellery, ware, coins and other objects are made not from pure metals but from alloys. The role of the alloying elements in the corrosion processes has to be searched, at different stages of corrosion, because the corrosion mechanisms of an alloy and of the pure alloying elements are not the same. Using XRD, ultraviolet visible (UV-vis) spectrophotometry, contact angle goniometry, ellipsometry and SEM-EDS, recent studies on the corrosion mechanisms of pure Cu, pure Ag, and sterling silver (an alloy containing 7.5 wt% Cu and 92.5 wt% Ag, widely used in the production of silverwork), submitted to artificial sulphidation at the same environmental conditions, demonstrated that the corroded layers grow in different structures. Contrary to those of pure Ag and Cu [102], the corroded layers of sterling silver grow in layer-by-layer structures [103].

The colours attained at different stages of corrosion by the corroded surfaces of sterling silver are related not only to the thickness of the corrosion layer, but also to the formed layer-by-layer corrosion structure, where are contained corrosion products which composition is variable over time. In fact, at the early stages of corrosion of sterling silver, Cu compounds prevail over the Ag compounds, but the latter become dominant for longer periods [102, 103]. There are no publications focusing on the use of the same type of described protocol to determine the corrosion mechanisms of gold alloys submitted to sulphidation, but as these alloys consist of a mixture of silver, copper and gold, comparable corrosion processes are expected.

The layer-by-layer corrosion mechanism observed for sulphidated silver-copper alloys makes the removal of the corrosion layer grown at the surface of objects made from precious metals difficult to carry out without the analytical identification of the corrosion products whose composition, as referred, depend on the stage of corrosion. Another difficulty comes from the corrosion layer thickness that for sterling silver and alloying elements only reaches a few hundreds of nanometres [102]. Despite the availability of successful removal techniques based on mechanical, chemical, electrochemical, and laser ablation cleaning methods [104–108], the thinness of the corrosion layers, grown at the surface of precious metals submitted to sulphidation, limits their effectiveness. Difficulties encountered by several researchers when using such cleaning techniques [104, 106, 109] might be explained by the thinness of the corrosion layers.

Therefore, cleaning operations are risky when both the composition of the corrosion products and the thickness of the corroded layer are unknown. The removal of the corrosion layer is far more difficult to achieve when the alloys are patinated, a decoration technique often employed to create polychrome effects in jewellery and ware [14, 110], that consists of a controlled corrosion process in specific areas to form the motif. Corroded patinated silver surfaces become dull and present spots and stains in shades of grey, caused by a two-phase corrosion mechanism that consists on the dissolution of the artificial patina followed by the natural formation of Ag and Cu sulphur compounds [103] [15]. The artificially grown compounds that constitute the original patina and the naturally grown compounds that constitute the corroded layer are chemically similar and can hardly be distinguished before cleaning.

13.7 Conclusion

Used in the production of prestigious objects, precious metals are perceived as both materials and colours. Their skilful combination can either entrust the object with a role or a value, or enhance it with polychrome effects. From the elaboration of the alloy until the finished object, the production of jewellery, coins, ware, etc. in gold and in silver includes several steps that require specific expertise. Information on those different steps can be acquired using physicochemical methods, which are of exam or analysis. According to the query, the latter are elemental, isotopic, or structural. In some cases, the objects – as well as the raw materials – circulated in exchange networks, sometimes to be used far from the workshops where they were produced. Information on their circulation, use, disclaim and degradation can also be searched using precise analytical protocols.

If, in general, the techniques employed in shaping, decorating, assembling and finishing an object can be detected with rather common physicochemical methods, provenancing gold and silver was only successfully achieved for some particular cases. In fact, it remains difficult to carry out provenancing, because physicochemical techniques with particular technical characteristics are necessary, and because gold and silver are reused by melting. Re-melting operations involve the mixing of metals and alloys from different origins with consequent loss of information.

In spite of the large number of physicochemical methods nowadays available for the construction of analytical protocols, the analysis of objects in gold and in silver has certain constraints. In addition to non-destructivity, which makes exam a major part of those protocols, when studying gold and silver work, spatial resolution, depth of analysis and limits of detection are technical characteristics that play an important role. In addition, *in-situ* analysis using portable equipment has been widely increasing, because the objects do not have to be moved for analysis. Many questions on gold and silver work were solved using *in-situ* analysis; however, the information obtained on the objects is, in general, more limited, because not covering queries on exchange and circulation.

Even if provenancing remains a difficult analytical question, the atmospheric corrosion mechanisms of gold and silver alloys provides presently the most challenging analytical queries. Made from alloys, the objects are submitted to corrosion processes where the alloying elements play important roles. The corrosion mechanisms of silver alloys (Ag-Cu) were recently studied showing that the very thin (a few hundreds of nanometres) corrosion layer contains compounds that change with time. The corrosion mechanisms of ternary gold alloys, certainly of the same type, are today not yet fully identified.

References

[1] Todd JT, Norman JF. The visual perception of metal. J Vis. 2018;18:1–17.

[2] Matsumoto T, Fukudak K, Uchikawa K. Appearance of gold, silver and copper colors of glossy object surface. Int J Affective Eng. 2016;15:239–47.

[3] Okazawa G, Koida K, Komatsu H. Categorical properties of the color term "GOLD". J Vis. 2011;11:1–19.

[4] Fleming RW. Visual perception of materials and their properties. Vision Res. 2014;94:62–75.

[5] Warburton DA. Ancient color categories. In: Luo R, editor. Encyclopedia of color science and technology. New York: Springer-Verlag, 2014. DOI: 10.1007/978-3-642-27851-8_75-12.

[6] Roberts EF, Clarke KM. The colour characteristics of gold alloys. Gold Bull. 1979;12:9–19.

[7] Rapson WS. The metallurgy of the coloured carat gold alloys. Gold Bull. 1990;23:125–33.

[8] Shiraishi T, Tilley RJD. An estimation of the reflectivity of gold- and platinum-group metals alloyed with copper. J Mater Sci. 2014;49:3461–68.

[9] Saeger KE, Rodies J. The colour of gold and its alloys. The mechanism of variation in optical properties. Gold Bull. 1977;10:10–14.

[10] Choudhury AKR. 2014. Principles of colour appearance and measurement. Object appearance, colour perception and instrumental measurement. Cambridge: Woodhead Publishing Limited, Elsevier. ISBN 9780857092298.

[11] Guerra MF. An overview on the ancient goldsmith's skill and the circulation of gold in the past: the role of X-ray based techniques. X-ray Spectrom. 2008;37:317–27.

[12] Guerra MF. Examen et analyse élémentaire de bijoux étrusques de la collection Campana. In: Metzger C, Gaultier F, editors. Les bijoux de la collection Campana: de l'antique au pastiche. Paris: École du Louvre, 2007:145–77. ISBN10 2904187200.

[13] Guerra MF. Etruscan gold jewellery pastiches of the Campana's collection revealed by scientific analysis. In: Cavallini M, Gigante GE, editors. De Re Metallica: dalla produzione antica alla copia moderna. Rome: Studia Archaeologica 150, L'Erma Di Bretschneider, 2006:103–28. ISBN 9788882654009.

[14] Tissot I, Manso M, Guerra MF. Unveiling the art of René Lalique with XRF and Raman spectroscopy. Technological innovation in jewellery production. J Cult Heritage. 2018;33: 83–9.

[15] Tissot I, Monteiro OC, Barreiros MA, Guerra MF. Atmospheric corrosion of patinated silverwork: a conservation challenge. Corrosão e Protecção de Materiais. 2016;35:15–19.

[16] Guerra MF, Calligaro T. The analysis of gold: manufacture technologies and provenance of the metal. Meas Sci Technol. 2003;14:1527–37.

[17] Guerra MF, Tissot I. Bronze Age and Iron Age gold torcs and earrings from the Iberian Atlantic façade: a non-invasive multi-analytical approach to the characterization of the alloys and the corrosion. X-ray Spectrom. 2016;45:5–13.
[18] Guerra MF. Role of radiation physics in the study and authentication of ancient gold work. Radiat Phys Chem. 2014;95:356–61.
[19] Tissot I, Troalen LG, Manso M, Ponting M, Radtke M, Reinholze U, et al. A multi-analytical approach to gold in Ancient Egypt: studies on provenance and corrosion. Spectrochim Acta B. 2015;108:75–82.
[20] Guerra MF. The circulation of South American precious metals in Brazil in the end of the seventeenth century. J Archaeol Sci. 2004;31:1225–36.
[21] Guerra MF. La circulation des objets en or. In: Dillmann P, Bellot-Gurlet L, editor(s). Circulation et provenance des matériaux dans les sociétés anciennes. La contribution des méthodes archéométriques. Paris: éditions des Archives Contemporaines, 2014:161–73.ISBN 9782813001634
[22] Mongiatti A, Suleman F, Meeks N. Beauty and belief: the endangered tradition of Omani silver jewellery. Br MusTech Res Bull. 5;2011:1–14. ISBN 9781904982678.
[23] Guerra MF. Bijoux étrusques en or: une recherche de faux et pastiches par des méthodes physico-chimiques. In: Cattelain P, Bozet N, Di Stazio GV, editors. La parure de Cro-Magnon à Clovis. Il n'y a pas d'Âge(s) pour se faire beau. Treignes: Ed. CEDARC, 2012;99–107. ISBN 2871490708.
[24] Ashkenazi D, Gitler H, Stern A, Tal O. Metallurgical investigation on fourth century BCE silver jewellery of two hoards from Samaria. Sci Rep. 2017;7:40659.
[25] MacDonald L, Guerra MF, Pillay R, Hess M, Quirke S, Robson S, et al. Practice-based comparison of imaging methods for visualization of toolmarks on an Egyptian scarab. In: Elmoataz A, Lezoray O, Nouboud F, Mammass D, editors. Image and signal processing. Lecture Notes in Computer Science, vol. 8509. Switzerland: Springer, 2014:239–46. DOI: 10.1007/978-3-319-07998-1_27.
[26] Esquès C, Guerra MF, Plé E, Stutz F. Techniques de décoration sans apport de matière par mesure optique sans contact de l'état de surface: première approche aux bijoux mérovingiens. ArcheoSciences. 2008;32:71–81.
[27] Regert M, Guerra MF. 2015. Physico-chimie des matériaux archéologiques et culturels. Paris: Éditions des archives contemporaines, Collection Sciences archéologiques. ISBN 9782813001924.
[28] Gale NH, Stos-Gale ZA. Cycladic lead and silver metallurgy. Annu Br Sch Athens. 1981;76: 169–224.
[29] Stos-Gale NH, Gale ZA. Metal provenancing using isotopes and the Oxford archaeological lead isotope database (OXALID). Archaeol Anthropol Sci. 2009;1:195–213.
[30] Ponting M, Evans JA, Pashley V. Fingerprinting of Roman mints using laser-ablation MC-ICP-MS lead isotope analysis. Archaeometry. 2003;45:591–7.
[31] Standish C, Dhuime B, Chapman R, Coath C, Hawkesworth C, Pike A. Solution and laser ablation MC-ICP-MS lead isotope analysis of gold. J Anal At Spectrom. 2013;28:217–25.
[32] Maryon H. Archæology and metallurgy. I. Welding and soldering. Man. 1941;41-85:118–24.
[33] Humpston DM, Jacobson G. Gold in metal joining. In: Corti C, Holliday R, editors. Gold: science and applications. New York: CRC Press, Taylor and Francis Group, 2010:161–90. ISBN 9781420065237.
[34] Šimić K, Zamboni I, Fazinić S, Mudronja D, Sović L, Gouasmia S, et al. Comparative analysis of textile metal threads from liturgical vestments and folk costumes in Croatia. Nucl Instrum Methods Phys Res Sect B: Beam Interact Mater Atoms. 2018;417:115–20.

[35] Ager FJ, Moreno-Suárez AI, Scrivano S, Ortega-Feliu I, Gómez-Tubío B, Respaldiza MA. Silver surface enrichment in ancient coins studied by micro-PIXE. Nucl Instrum Methods Phys Res B. 2013;306:241–44.
[36] Scrivano S, Ruberto C, Gómez-Tubío B, Mazzinghi A, Ortega-Feliu I, Ager FJ, et al. In-situ non-destructive analysis of Etruscan gold jewels with the micro-XRF transportable spectrometer from CAN. J Archaeol Sci: Rep. 2017;16:185–93.
[37] Lemasson Q, Moignard B, Pacheco C, Pichon L, Guerra MF. Fast mapping of gold jewellery from ancient Egypt with PIXE: searching for hard-solders and PGE inclusions. Talanta. 2015;143:279–86.
[38] Troalen LG, Guerra MF, Tate J. Goldwork in Ancient Egypt. Workshop practices at Qurneh in the 2nd Intermediate Period. J Archaeol Sci. 2014;50:219–26.
[39] Demortier G. Targeting ion beam analysis techniques for gold artefacts. ArcheoSciences. 2009;33:29–38.
[40] Martinón-Torres M, Valcárcel Rojas R, Sáenz Samper J, Guerra MF. Metallic encounters in Cuba: the technology, exchange and meaning of metals before and after Columbus. J Anthropol Archaeol. 2012;31:439–54.
[41] Demortier G, Ruvalcaba-Sil JL. Quantitative ion beam analysis of complex gold-based artefacts. Nucl Instrum Methods Phys Res B. 2005;239:1–15.
[42] Fourdrin C, Pagès Camagna S, Pacheco C, Radepont M, Lemasson Q, Moignarda B, et al. Characterization of gold leaves on Greek terracotta figurines: A PIXE-RBS study. Microchem J. 2016;126:446–53.
[43] La Niece S, Craddock P, editors. Metal plating and patination. Cultural, technical and historical developments. Oxford: Butterworth-Heinemann, Elsevier, 1993. ISBN 978-0-7506-1611-9.
[44] Caridia F, Torrisi L, Cutroneo M, Barreca F, Gentile C, Serafino T, et al. XPS and XRF depth patina profiles of ancient silver coins. Appl Surf Sci. 2013;272:82–87.
[45] Reiff F, Bartels M, Gastel M, Ortner HM. Investigation of contemporary gilded forgeries of ancient coins. Fresenius' J Anal Chem. 2001;371:1146–53.
[46] Gusmano G, Montanari R, Kaciulis S, Montesperelli G, Denk R. "Gold corrosion": red stains on a gold Austrian Ducat. Appl Phys A. 2004;79:205–11.
[47] Guerra MF, Abollivier P. Monetary alloys in Iron Age Armorica (Finistère, France): the singular case of the Osismi tribe. Nucl Instrum Methods Phys Res B. 2016;377:1–11.
[48] Rodrigues M, Schreiner M, Melcher M, Guerra MF, Salomon J, Radtke M, et al. Analysis of the Hoard of Beçin using X-ray-based techniques. X-Ray Spectrom. 2012;41:416–24.
[49] Guerra MF. Fingerprinting ancient gold with proton beams of different energy. Nucl Instrum Methods Phys Res B. 2004;226:185–98.
[50] Nixon S, Rehren T, Guerra MF. New light on the early Islamic West African gold trade: coin moulds from Tadmekka, Mali. Antiquity. 2011;85:1353–68.
[51] Radtke M, Reiche I, Reinholz U, Riesemeier H, Guerra MF. Beyond the great wall: gold of the Silk Roads and the first Empire of the steppes. Anal Chem. 2013;85:1650–56.
[52] Radtke M, Buzanich G, Reinholz U, Riesemeier H, Scharf O, Guerra MF. Double Dispersive X-Ray Fluorescence (D^2XRF) Bbased on an energy dispersive pnCCD detector for the detection of platinum in gold. Microchem J. 2016;125:56–61.
[53] Guerra MF. Trace elements fingerprinting using accelerators and ICP-MS: circulation of gold from the sixth century BC to the twelfth century AD. In: Griecken RV, Janssen K, editor(s). Cultural heritage conservation and environmental impact assessment by non-destructive testing and micro-analysis. London, Balkema: Taylor and Francis group, 2005:223–44.ISBN 9789058096814.
[54] Watling RJ, Herbert HK, Delev D, Abell ID. Gold fingerprinting by laser ablation-inductively coupled plasma-mass spectrometry. Spectrochim Acta Part B: Atom Spectrosc. 1994;49:205–19.

[55] Guerra MF, Sarthre CO, Gondonneau A, Barrandon JN. Precious metals and provenance enquiries using LA-ICP-MS. J Archaeol Sci. 1999;26:1101–10.

[56] Gondonneau A, Guerra MF. The gold from Ghana and the Muslim expansion. In: Young S, Pollard M, Budd P, Ixer R, editors. Metals in Antiquity, BAR international series 792. Oxford: Archaeopress, 1999:262–70. ISBN 1841710083.

[57] Wagner B, Syta O, Sawicki M. A moderate microsampling in laser ablation inductively coupled plasma mass spectrometry analysis of cultural heritage objects: a review. In: Targowski P, Walczak M, Pouli P, editors. Lasers in the conservation of artworks. Toruń: Proceedings of LACONA XI, NCU Press, 2017:155–78. DOI: 10.12775/3875-4.11.

[58] Glaus R, Koch J, GüNther D. Portable Laser Ablation sampling device for elemental fingerprinting of objects outside the laboratory with Laser Ablation Inductively Coupled Plasma Mass Spectrometry. Anal Chem. 2012;84:5358–64.

[59] Melcher M, Schreiner M, Bühler B, Pülz AM, Muss U. Investigation of ancient gold objects from. Artemision at Ephesus using portable µ-XRF. ArcheoSciences. 2009;33:169–75.

[60] Guerra MF, Rehren T. In-situ exam and analysis of the gold jewellery from the Phoenician tomb of Kition (Cyprus). ArcheoSciences. 2009;33:151–59.

[61] Guerra MF, Fischer M, Radtke M, Reinholz U. Inca figurines from the Ethnologisches Museum in Berlin: study of some typical and atypical productions. Bull l'Institut Français d'Etudes Andines. 2017;46:221–52.

[62] Adrimi-Sismani V, Guerra MF, Walter P. La tombe mycénienne de Kazanaki (Volos) et le mythe de la Toison d'or. ArcheoSciences. 2009;33:135–41.

[63] Hauptmann A, Klein S. Golden artifacts from the Royal Tombs of Ur, Mesopotamia. Metalla. 22;2016:75–146. ISSN 0947-6229.

[64] Gomes Ferreira MT. 1997. Lalique. Jóias, Lisbon: Museu Calouste Gulbenkian. ISBN 9789728128272.

[65] Mongiatti A, Meeks N, Simpson J. A gold four-horse model chariot from the Oxus Treasure: a fine illustration of Achaemenid goldwork. Br MusTech Res Bull. 4;2010:27–38. ISBN 1904982557.

[66] Perea A, Gutiérrez-Neira PC, Climent-Font A, Fernández-Esquivel F, Rovira-Llorens S, Ruvalcaba-Sil JL, et al. Pre-hispanic goldwork technology. The Quimbaya Treasure, Colombia. J Archaeol Sci. 2013;40:2326–34.

[67] Guerra MF, Demortier G, Vitobello ML, Bobomulloev S, Bagault S, Borel T, et al. Analytical study of the manufacturing techniques of Kushan gold jewellery (National Museum of Antiquities of Tajikistan). ArcheoSciences. 2009;33:177–85.

[68] Barham E. Controlled lifting and X-radiography of gold threads from ancient archaeological textiles. In: O'Connor S, Brook M, editors. X-Radiography of textiles, dress and related objects. Oxford, Butterworth-Heinemann: Elsevier, 2007:302–6. ISBN 9380931921.

[69] Morigi MP, Casali F. Radiography and Computed tomography for works of art. In: Russo P, editor. Handbook of X-ray Imaging: physics and Technology. New York: CRC Press, Taylor & Francis group, 2018:1259–74. ISBN 978-1-4987-4152-1.

[70] Lang J, Middleton A, editors. Radiography of cultural material. Oxford: Butterworth and Heinemann series, 1997. ISBN 0750663472.

[71] Benson E. Why sacrifice? In: Benson EP, Cook AG, editors. Ritual sacrifice in ancient Peru. Austin: University of Texas, 2001:1–20. ISBN 978-0-292-70894-5.

[72] Ceruti C. Human bodies as objects of dedication at Inca mountain shrines (north-western Argentina). World Archaeol. 2010;36:103–22.

[73] La Niece S, Meeks N. Diversity of goldsmithing traditions in the Americas and the Old World. In: McEwan C, editor. Pre-Columbian gold. Technology, style and iconography. London: British Museum Press, 2000:220–39. ISBN 0714125342.

[74] Schuler-Schömig IV. Werke indianischer Goldschmiedekunst. Berlin: Staatliche Museen, Preußischer Kulturbesitz, Bilderhefte der Staatlichen Museen Berlin 17. 1981. ISBN 3786160686.
[75] Rowe JH. Inca. In: Boon E, editor. Andean art at Dumbarton Oaks, vol. 1. Washington D.C.: Dumbarton Oaks Research Library and Collection, 1996:301–19. ISBN 978-0-88402-225-1.
[76] Oddy WA, Bimson M, La Niece S. The composition of niello decoration on gold, silver and bronze in the antique and mediaeval periods. Stud Conserv. 1983;28:29–35.
[77] Vetter L, Guerra MF. Las narigueras: simbolismo y tecnología. Bulletin de l'Institut Français d'Etudes Andines, submitted 2018.
[78] Guerra MF, Nuñez-Regueiro P, eds. La orfebrería en los Andes en la época inca (siglos XV-XVI), Lima. Bulletin de l'Institut Français d'Études Andines 2017;Tome 46. DOI: 10.4000/bifea.8120.
[79] Berthelot J. The extraction of precious metals at the time of the Inka. In: Murray JV, Wachtel N, Revel J, editors. Anthropological history of Andean polities. Cambridge: Cambridge University Press, 2009:69–88. DOI: 10.1017/CBO9780511753091.010.
[80] Sauders NJ. Stealers of light, traders in brilliance: Amerindian metaphysics in the mirror of conquest. Anthropol Aesthetics. 1998;33:225–52.
[81] Engelbach R, Gunn B. Harageh, vol. 28. London: British School of Archaeology in Egypt, 1923.
[82] Grajetzki W. 2014. Tomb treasures of the Late Middle Kingdom: the archaeology of female burials. Philadelphia: University of Pennsylvania Press. ISBN 9780812245677.
[83] Troalen LG, Tissot I, Maitland M, Guerra MF. Jewellery of a young Egyptian girl: middle Kingdom goldwork from Haraga tomb 72. Hist Metal. 2015;49:75–86.
[84] Brunton G. Qau and Badari, vol. 1. London: British School of Archaeology in Egypt, 1927:29.
[85] Guerra MF, Pagès S. On the way to the New Kingdom. Analytical study of Queen Ahhotep's gold jewellery (17th Dynasty of Egypt). J Cult Heritage 2018. in press
[86] Troalen LG, Guerra MF. Gold from the tomb of Scribe Beri: a comparative analytical approach to the New Kingdom gold grave goods from Riqqa (Egypt). Appl Phys A. 2016;122:210–22.
[87] Bennewitz RM. 1992. Platería mapuche. Santiago de Chile: Kactus/ Sipimex. ISBN 9567013100.
[88] Bennewitz RM. 1997. Plateros en la frontera y la platería Araucana: en el proceso caratulado "Salteo al Cacique Huenul" (1856-1960). Temuco-Chile: Ediciones Universidad de La Frontera. ISBN 956-236-070-9.
[89] Núñez-Regueiro P, Guerra MF. Los aros de plata de Patagonia Septentrional: aportes de la colección Henry de la Vaulx (1896) sobre forma, tecnología y metalurgia. Chungara Revista de Antropología Chilena. 2016;48:331–45.
[90] Guerra MF, Nuñez-Regueiro P. Indigenous silver jewellery of Northern Patagonia (nineteenth century): analytical approach to composite objects. X-Ray Spectrom. 2012;41:342–49.
[91] Joseph HC. La Platería araucana, Anales de la Universidad de Chile, Año VI, 2da serie. Santiago: Establecimientos Gráficos Ballcells & Co, 1928.
[92] Joseph HC. Los adornos araucanos de Lanalhue. Rev Univ. 1930;15:512–18.
[93] Auer E. Das Nickel im Münzwesen. Von der Antike bis zum EURO. In: Auer E, Müller S, Slotta R, editors. 250 Jahre Nickel: vom Nickel zum Euro. Nickel als Münzmetall. Bochum: Selbstverlag des Deutschen Bergbau-Museums, 2001:111–202. ISBN 3921533813.
[94] Guerra MF. The mines of Potosi: a silver Eldorado for the European economy. In: Demortier G, Adriaens A, editors. Ion beam study of art and archaeological objects. Luxembourg: Office for Official Publications of the European Communities, 2000:88–94. ISBN 92-828-7652-2.
[95] Gondonneau A, Guerra MF. The circulation of precious metals in the Arab Empire: the case of the Near and the Middle East. Archaeometry. 2002;44:573–99.
[96] Meyers P. Elemental compositions of the Sion treasure and other Byzantine silver objects. In: Boyd SA, Mango MM, editors. Ecclesiastical silver plate in sixth century Byzantium.

Washington DC: Dumbarton Oaks Research Library and Collection, 1992:169–90. ISBN 088402203x.

[97] Meyers P. Production, distribution, and control of silver: information provided by elemental composition of ancient silver objects. In: van Zelst L, editor. Patterns and Process: a Festschrift in Honor of Dr Edward V. Sayre. Washington DC: Smithsonian Center for Materials Research and Education, 2003:271–88.

[98] Guerra MF. The circulation of gold in the Portuguese area from the 5th to the eighteenth century. In: Perea A, Montero I, Garcia-Vuelta O, editors. Ancient gold technology: america and Europe. Madrid: Anejos de AespA XXXII, CSIC, 2005b:423–31. ISBN 9788400082932.

[99] Guerra MF, Calligaro T, Perea A. The treasure of Guarrazar: tracing the gold supplies in the Visigothic Iberian Peninsula. Archaeometry. 2007;49:53–74.

[100] Ingo GM, Angelini E, Riccucci C, Caro TD, Mezzi A, Faraldi F, et al. Indoor environmental corrosion of Ag-based alloys in the EgyptianMuseum (Cairo, Egypt). Appl Surf Sci. 2015;326:222–35.

[101] Faraldi F, Angelini E, Mezzi A, Riccucci C, Caro TD, Di Carlo G. Surface studies of environmental reactive species during exhibition or storage of ancient Ag-based artefacts. Surf Interface Anal. 2014;46:796–800.

[102] Tissot I, Monteiro OC, Barreiros MA, Correia J, Guerra MF. The influence of the constituent elements on the corrosion mechanisms of silver alloys in sulphide environments: the case of sterling silver. RSC Adv. 2017;7:28564–72.

[103] Tissot I, Monteiro OC, Barreiros MA, Corregidor V, Correia J, Guerra MF. Corrosion of silver alloys in sulphide environments: a multianalytical approach for surface characterization. RSC Adv. 2016;6:51856–63.

[104] Palomar T, Oujja M, Llorente I, Ramírez Barat B, Cañamares MV, Cano E, et al. Evaluation of laser cleaning for the restoration of tarnished silver artifacts. Appl Surf Sci. 2016;387:118–27.

[105] Novakovic J, Vassiliou P, Georgiza E. Electrochemical cleaning of artificially tarnished silver. Int J Electrochem Sci. 2013;8:7223–32.

[106] Palomar T, Ramírez Barat B, García E, Cano E. A comparative study of cleaning methods for tarnished silver. J Cult Heritage. 2016;17:20–26.

[107] Siano S, Salimbeni R. Advances in laser cleaning of artwork and objects of historical interest: the optimized pulse duration approach. Acc Chem Res. 2010;43:739–50.

[108] Hacke AM, Carr CM, Brown A, Howell D. Investigation into the nature of metal threads in a Renaissance tapestry and the cleaning of tarnished silver by UV/Ozone (UVO) treatment. J Mater Sci. 2003;38:3307–14.

[109] Lee JM, Yu JE, Koh YS. Experimental study on the effect of wavelength in the laser cleaning of silver threads. J Cult Heritage. 2003;4:157–61.

[110] Rudoe J. Oxidized silver in the nineteenth century: the documentary evidence. In: La Niece S, Craddok PT, editors. Metal plating and patination: cultural, technical, and historical developments. London: Butterworth- Heinemann, 1993:161–70. ISBN 978-0-7506-1611-9.

Giacomo Eramo and Annarosa Mangone

14 Archaeometry of ceramic materials

Abstract: Ceramics are among the most studied findings, one of the best markers for providing technological and functional information in archaeological contexts. Their chemical-mineralogical characterization allows to answer a large number of historical-archaeological questions about classification, provenance, production technologies, trade routes, economic exchange, etc. The best methodological approach not only integrates morphological-stylistic studies to the archaeometric ones, but also includes a synergic instrumental strategy aimed both to take advantage of each different analytical technique to the best of its potentiality and to over step the problems connected to the preciousness and uniqueness of the objects. As far as the mineralogical and petrographical composition of preindustrial ceramics is concerned, its determination is crucial to answer provenance and technological issues like raw materials procurement and the production processes in this respect, equivalent firing temperature, redox atmosphere during firing are important factors that help in understanding the relevant mineralogical and micro-structural transformations. In this paper, we illustrate how an integrated approach of analytical techniques, tested on different classes of ceramics – pottery with spathic calcite, Apulian red figure pottery and technical ceramics – can provide answers to archaeological questions.

Keywords: pottery fabric, pertrography, ICPMS, XRPD, multivariate statistics

14.1 Introduction

14.1.1 Methodology of ceramic analysis

The issue of sharing a common ground to choose the most appropriate analytical methods to achieve results useful in interpreting cultural behaviour was (and still is) a key point since the first analyses of archaeological materials with methods used in earth sciences, chemistry, physics, biology, etc. (i. e. archaeometry). This should be kept in mind in all case studies, from the formulation of archaeological questions to address to archaeometry, to the sampling up to the interpretation of the data.

As concern the archaeometric study of ceramics, the best methodological approach not only integrates morphological-stylistic studies to the archaeometric ones, but also includes a synergic instrumental strategy aimed both to take advantage of each different analytical technique to the best of its potentiality and to overstep the

This article has previously been published in the journal *Physical Sciences Reviews*. Please cite as: Mangone, A., Eramo, G. Archaeometry of ceramic materials. *Physical Sciences Reviews* [Online] **2019**, 4. DOI: 10.1515/psr-2018-0014.

https://doi.org/10.1515/9783110457537-014

problems connected to the preciousness and uniqueness of the objects [1, 2]. An organic archaeometric approach can provide diversified information on the body and the coatings (composition, microstructure, microstructural profile at various depths, different phases, inclusions and segregations), to identify the procedures of production of the artefacts and to spot their dissimilarities and origin [3–7]. This is vital in order to draw a complete map of the production centres of the objects and of their transfer, and to answering the question concerning the provenance of raw materials. Despite the fact that this premise is widely accepted, the above-mentioned plan of action is scarcely adopted in archaeometric investigations. In detail, the recognition of the geological sources of the ceramic raw materials provides data concerning the artefacts' provenance [8, 9] and the trade and economic routes. However, to trace the used clay location, it is compulsory to consider the possible clays mixture with specific minero-chemical features, the deliberate addition of different tempers (grog, sand, etc.), and thus the unavoidable modifications of the final paste minero-petrographic composition due to all the phases of the manufacturing process (firing, painting, glazing, etc.). To determine the provenance of the raw materials, the average elemental composition of the ceramic bodies must be obtained, even though this can display the same value for different mixes because, even for the simplest manufacture, using basic untreated clay, pelitic sedimentary deposits are involved. For instance, a deposit richer in coarse silt-sand than in clay and one much richer in clay than in coarse silt-sand could exhibit analogous average chemical compositions, because they are characterized by the same minerals in the same ratio; notwithstanding, their fabric would be totally different and their final paste would have different technological features. Accessory minerals – whose presence does not condition the average chemical composition – can suggest different clayey sedimentary basins [10]. Only a concurrence of all the results obtained with different analytical techniques could guarantee the assessment of the raw materials provenance. Furthermore, the mineralogical composition of the artefacts pastes is also necessary to answer technological questions, like the pottery manufacturing processes or modifications in the production techniques. For example, important information concerning these transformations are the maximum firing temperatures, the firing duration and the redox atmosphere of the kiln: neo-formation phases are highly informative because they are related to heating temperatures and – due to their low quantity and tiny size, often below 1 μm – can be identified only with peculiar analytical techniques. As well as the ceramic bodies', also the coatings' characterization by a multi-technique approach is supported by the same motivations.

14.1.2 Analytical methods commonly employed in ceramic investigation

1. Ceramic paste

The most used elemental analysis techniques for ceramic bodies are neutron activation analysis (NAA), X-ray fluorescence spectroscopy (XRF) and inductively coupled plasma mass spectrometry (ICPMS). The latter, while being the most popular due to its excellent detection limit for all the elements and to the instrumentation availability [11–14] needs the decomposition of a small part of the object into its elements, which can be done with the help of a hydrofluoric, nitric and hydrochloric acids mixture, in order to measure the abundance of a number of elements by mass spectrometry [15]. It can be associated to various instrumental facilities, of which laser ablation (LA) seems to be the most successful [16–18]. LA-ICPMS constitutes an important alternative to solids sampling: even though less sensitive than wet ICPMS, it has remarkable advantages, such as its practical implementation and its speed of analysis, besides it does not require sample preparation and it is almost non-destructive (only a few nanograms of material are ablated with each laser pulse). The last years these attractive features have interested many researchers including archaeologists, art historians and conservators. When employing LA-ICPMS, various multiple samples (lines or points) must be analysed and appropriately averaged on the same item, because the ceramic fabric is heterogeneous (mixture of clay, mineral inclusions and fillers), and this can greatly affect the chemical composition data. The most employed techniques for the technological investigation of pottery are: polarized-light optical microscopy (POM), scanning electron microscopy (SEM) with energy dispersive spectrometry (EDS) and X-ray powder diffraction (XRPD). POM gives knowledge about the mineralogy, the structure and the texture of the ceramic body. Together with bulk chemical methods, POM and XRPD were the earliest methods to analyse archaeological ceramics [19]. Although they are sometimes used separately, they provide complementary data of *fabric* that cannot substitute each other. Petrographic characterization provides the most accurate "map" of the ceramic body as "territory," since it allows to identify the components and their reciprocal distribution (i. e. fabric). Whilst two ceramics with the same chemical composition may have sensible mineralogical and petrographical differences (e. g. different mineral assemblages and/or texture), two ceramics with the same fabric are chemically more akin to each other. This inner variability of petrographic groups with respect to chemical groups should be taken into account when an archaeometric study is envisaged, unless bulk chemical determining the concentrations of dozens of elements. As well as the bulk mineralogical composition (XRPD) lacks information about the shape, the structural arrangement of phases, the presence of polycrystalline and/or polymineralic grains – grain size may be inferred from spectrum features [20] – the bulk elemental composition lacks information about the way elements are combined in order to form the phases occurring in the ceramic body [21].

The petrographic analysis of ceramic body, thus represents a key analytical approach that can be preliminarily carried out to orient further bulk and/or in situ analyses [22]. The preparation of this section out of small ceramic fragments allows to a trained petrographer in few minutes to have first insight of the petrofacies, texture and alteration state. Although the optical resolution limits (about 0.3 μm) and the masking effect of the clay matrix do not allow to observe grains smaller than 5 μm, the potentialities of optical microscopy are nowadays improved by the modern technological developments in computing and photonics.

Scanning electron microscopy represents a common analytical extension of optical microscopy – often on the same thin sections – to identify the particles of the matrix, the reaction domains and the microstructures [23–28], using the higher magnifications (>500,000×) and the greater depth of field available compared to light microscopes (<2,000×). Moreover, several detectors may be available to expand analytical capability of SEM (e. g. EDS, EMPA, CL, EBSD, etc.) to obtain point chemical data and enhance microstructures documentation.

Consequently, optical and electron microscopes have different areas of application and they complement each other. The specialized skills required to operate electron microscopes compared to those needed for light microscopes do not necessarily mean lower amount of information that can be obtained by an experienced microscopist. If the very high costs to buy and maintain electron microscopes are considered, the capacity of cheaper light microscopes integrated with video/photo cameras and accessories to provide effective data should not be overlooked.

2. Ceramic coatings

In regard to the analysis of coatings (gloss, glazes and engobes) information on the chemical composition can be provided by NAA, XRF, LA-ICPMS, on the structure and thickness by POM and SEM-EDS and on pigments by spectroscopic techniques, such as FT-IR, Raman, XPS. Analyses can be performed on the surface of sample in the case of a single coating layer, whereas in case of overlapped coatings, analyses should be carried out on cross section.

Since the chemical method of analysis have been thoroughly illustrated and discussed in the previous chapters, it will be given here a brief description of POM technique and information it can provide for the study of ceramic materials.

14.1.2.1 Petrography on thin section

The petrographical analysis of ceramics has more than a century of history and originated from the necessity to understand the specific compositional and microstructural features of an artificial material that signed the life and cultural evolution of humans from the Mesolithic to current times. Since the end of nineteenth century, ceramic petrography developed in the field of archaeology [29] and industry [30, 31]. Petrographical methods applied to ceramics analysis were essential to identify the mineral constituents and/or to understand the changes which take place during firing. Regardless of the type of ceramics involved, the scientist is fundamentally interested in the chemical (mineralogical) and physical changes occurring during firing and the way they affect the behaviour of the ceramic body in usage. Since no essential changes in chemical composition may occur during firing, bulk chemical analysis is frequently useless in determining the mineralogical phases and the microstructures that constitutes the body after firing. In one way or another, microstructure is one of the fundamental characteristics of ceramics able to affect the physical and chemical properties and this is why petrography became more and more important. In principle, ceramics may be considered as an artificial analogue of metapelites, but the temperature, pressure and kinetics involved are significantly different. This is considered as the inverse of the weathering of rocks that yielded clayey sediments by a man-made shortcut compared to what diagenesis and/or metamorphism would have made [32].

Since in most of the preindustrial ceramics, the use of natural clayey raw materials and the partial mineralogical transformation of them made recognizable *petrofacies* [33, 34], there was methodological and vocabulary transfer from sedimentary petrology and soil micromorphology. The former contributed to the study of the non-plastic inclusions (NPIs) and the latter to a better characterization of the matrix and voids [35]. However, these are not sufficient to describe and interpret the "human side" of ceramics, which is indeed *artificial*, consisting of technological choices and modes of use. The mechanical (i. e. kneading, forming), compositional/textural (i. e. levigation, tempering, clay mixing) and thermal (i. e. firing, use) processing of the clayey raw materials yield typical man-made microstructures that escape from the framework of sedimentary petrology and soil micromorphology, and requires specific descriptive strategies. The combination of all these information constitutes a network that involve mineralogy, petrology, chemistry, physics, archaeology and anthropology. Other than the ability to observe and record this complexity, the ceramic petrographer must also act as an interpreter [36]. She/he must act as a *node* in the knowledge network to aim at a true interpretation of the artefacts, as a repository of technological, use-wear and post-depositional information.

The term used to identify and express the combination of *natural* (i. e. original mineralogical composition, etc.) and *artificial* features of the ceramic body is *fabric*. This should be preferred to the locution *petrographic group* because it highlights the "human nature" of ceramics (*fabrica* = product of art → fabric). However, there are several interpretations of ceramic fabric. For example, Peacock [37], Maggetti [38] and Capelli [39] support an interpretation that only considers the mineralogical and petrographical composition of the ceramic body, whereas the writers share with Whitbread [35], Orton [40], Santacreu [41] and several others an interpretation that includes also the textural and structural features. The *petrofacies* – i. e. the original mineralogical and petrographical content of the raw material(s) – conveyed by the ceramic body is a deliberate selection of a raw material(s) and then it is intended even so an artificial trait, as clay levigation, tempering and mixing are. Differently to what proposed by Stoltman [42], the term *paste* is intended here as the final result of the compositional/textural processing, not the *body* [43], as well as *clay* is intended as the staring clayey raw material, named *paste* by the same author. Bearing in mind that the petrofacies identified in the ceramic body can be or not an unmodified portion of a clayey source, it represents in any case, at least in part, a distinct clay (sand) composition zone [42] that can be related to a depositional environment or, more generally, to a sedimentary basin. Aware of that, Capelli [39] proposed a description protocol of pottery fabric that put in evidence the geological fingerprint of the fabric (i. e. *gruppi minero-petrografici*) and that was applied to thousands of pottery thin sections of different age and provenance in the Mediterranean area. Santacreu [41] uses the term *petrofabric* for fabric and *petrogroup* to intend the petrofacies [34]. If an unprocessed clay was used as paste, it brings important features of depositional environment in which formed, providing both technological information and insights on paleoenvironment existing at time the ceramics were made [44, 45].

Several scholars [35, 38, 46–49] agree in distinguish three main components in the ceramic body: a non-plastic (or aplastic) fraction; a matrix (or groundmass) fraction; and voids (Figure 14.1). The thresholds to separate the non-plastic domain from the matrix one is matter of debate. Since the coarse and fine fractions of the fabric are identified in ceramic petrography according to their plastic (or not) behaviour when mixed with water, the separation between the NPIs and the matrix is a grain-size threshold that is placed differently by several authors: 10 μm [35, 49], 15 μm [38], 20 μm [40], 60 μm [46], and 63 μm [42, 50]. Unfortunately, plasticity is not strictly related to grain size of particles, but is also influenced by the amount of clay minerals, solid/water ratio, presence of organic matter, etc. This implies that an unambiguous limit between NPIs and matrix is not available.

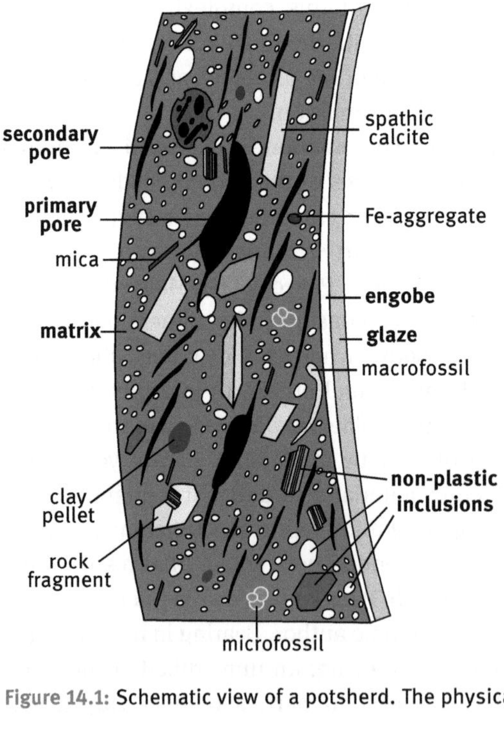

Figure 14.1: Schematic view of a potsherd. The physical components of a coated ceramic fabric.

However, this technological motivation interferes with the analytical methods used to characterize the fabric. One of the problems associated to the study of ceramics in thin section resides in its matrix-rich nature. A recent paper by Aprile [51] shows the limits of POM in qualitative and quantitative analysis of matrix-rich samples as traditional ceramics are. Certainly, on the one hand, a lower threshold (e. g. 10 μm) risks to be a theoretical than practical choice if a petrographical microscope is considered, because of the masking effect of matrix in thin section. On the other hand, a higher threshold (e. g. 63 μm) is more practical and accurate from an analytical point of view, but distorts the plastic/non-plastic sense of the separation related to the hydrated state of the paste. The use scanning electron microscopy (SEM) may be resolutive to improve the accuracy of determination of silt grade inclusions [51, 52]. Furthermore, to work on the surface of the thin section, not only reduces the masking effect of the matrix on the smaller inclusions and pores, but also allow to better estimate the amount of porosity and its morphological and dimensional features (see below).

NPIs are the most reliable evidence of the original petrofacies, since they are the coarsest fraction of the ceramic body, less subject to mineralogical transformation during firing and use. The use of qualitative and quantitative methods applied in sedimentary petrology is a solid basis to record the compositional and textural features of the NPIs.

The maximum size of the NPIs determines whether or not the thin section is representative of the fabric. Especially for quantitative analyses, the ratio between the maximum grain area and the thin section area should be below 0.05 [44, 52]. Taking into account an average thin section area of 3 cm^2 the maximum diameter of NPIs to obtain accurate data should not exceed 4 mm.

The term *temper* is used for wittingly added NPI non-plastic inclusions [35, 42, 53] and thus implies a technological meaning (Figure 14.2). The interpretation of a given fraction of NPIs as temper must not be separated from its relative amount in the fabric. For example, if few fragments of pottery (i. e. grog) occur as NPIs in a new pottery fabric, they should be considered as accidental, otherwise it should be inferred that the potter was aware of the physical mechanical changes on the firing and use of the ceramic body, caused by such few quantities of inclusions. However, if technological reasons for an intentional addition of few fragments of grog can be excluded, a possible accidental presence or immaterial cultural reasons cannot. A correct interpretation of the fabric should take into account the *etic* and *emic* aspects of the artefacts [54, 55] to avoid spurious speculations.

The matrix is generally not considered in the optical petrography as the NPIs are. Although its particle is too small to be accurately described in their optical properties, a series of important features can be documented. Firstly, the main compositional characteristics, relevant for technological analysis of the ceramic, can be identified. The presence or not of microcrystalline calcite (i. e. micrite) dispersed in the clay matrix, the abundance of iron oxi-hydroxides, or micas, may orient the research of

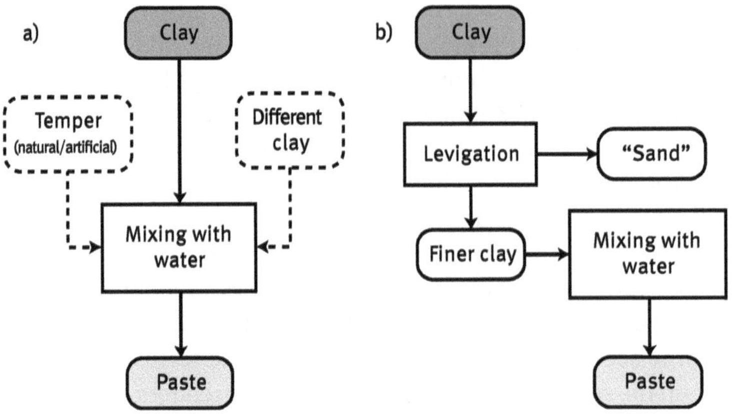

Figure 14.2: Modes of paste preparation. (a) Three possible cases are considered: the use of unprocessed clay; the addition of temper; the mixing of two (or more) clays. The levigation of clay is reported on (b).

the clay source. The occurrence of clay pellets in the matrix should be also recorded, since it may help to identify the depositional environment (or facies) of the clay.

An important feature to describe is the oxidation structure of the body through the colour/birefringence pattern, essentially provided by the matrix phases (i. e. aggregate polarization).

Preindustrial ceramics were generally fired using biomass as fuel (e. g. wood, straw, dung, etc.). Such fuel use implies an array of thermal/redox effects on the paste that were more easily recorded by the matrix, due to its finer grain size and the presence of free iron phases (i. e. iron oxi-hydroxides). An attempt to document these feature was made by Rye [53] and Orton et al. [40] using graphical schemes and by Santacreu [41] with acronym indicating an oxidation pattern (e. g. "O-R-o: oxidised external margin, reduced core and thinly oxidized internal margin"). The presence of carbonaceous matter (i. e. partially burned organic matter) dispersed in the matrix may mask birefringence and appear optically inactive [35]. The persistence of carbonaceous matter in the paste demonstrates non-oxidizing atmosphere episodes during firing. In Figure 14.3, a visual scheme to classify such structures, very important for the technological characterization is proposed. The acronyms refer to the sequence of reduced (R) or oxidized (O) domains of the fabric from the core outward. The combination of reduced and oxidized domains may assume different arrangements. As an example, when a reduced domain is followed by two symmetrical oxidized domains the acronym is "RO" (RO2 = double RO sequence). Asymmetric zoning is indicated adding "I" (internal) or "E" (external) at the end of the acronym (ORI = oxidized and then reduced on the inner surface). The "marbled" (M) structure is characterized by patched reduced and oxidized domains.

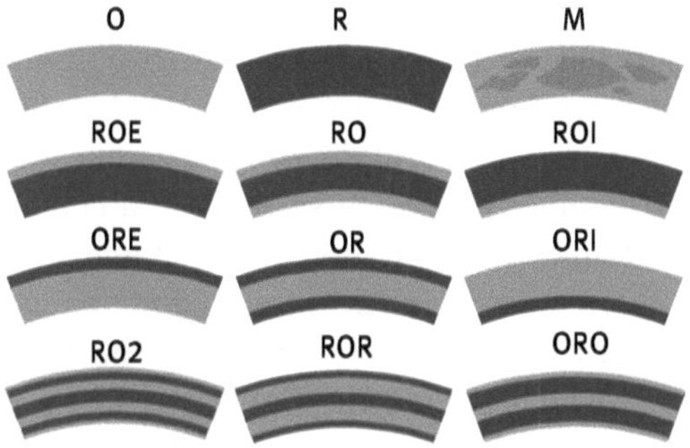

Figure 14.3: Oxidation structures of the fabric. Abbreviations key: O = oxidized domains; R = reduced domains; M = marbled structure; E = external; I = internal. From left to right, acronyms report the oxidation sequence from the core to the surfaces. See text for details.

Argillaceous inclusions [56] are special components of the body that deserve particular attention. They may be considered as part of the NPIs or matrix according to their solidity and origin. Other than the very accurate classification proposed by Whitbread [56] to distinguish natural from artificial origin of argillaceous inclusions, a further class of artificial inclusions should be considered: *chamotte*. Despite the quite common use of this term as synonym of *grog* [57], it has specific features and technological meaning that should not be overlooked by a ceramic petrographer. In Figure 14.4, two flow charts that explain the technological differences underling these two terms are shown. Although grog and chamotte are both thermally altered clays, in the first case a recycled existing ceramic is meant; whereas, in the second case, a calcinated clay deliberately produced to be crushed and added to (generally the same) clay for paste preparation is intended [58]. Chamotte is a term used in modern ceramics (its origin was probably in Germany

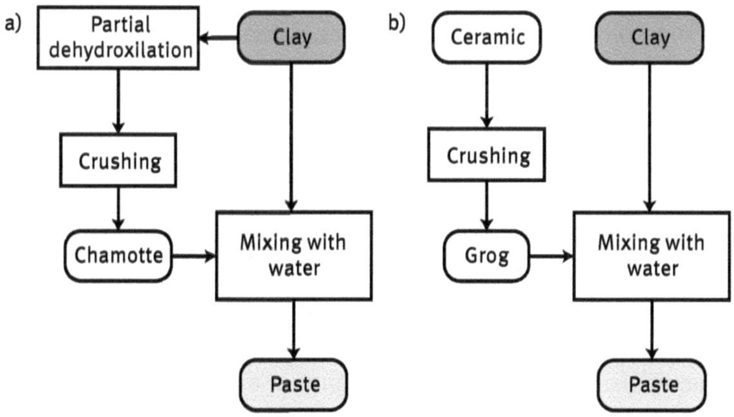

Figure 14.4: Technological differences in preparation: (a) chamotte and (b) grog-tempered ceramics.

in eighteenth century: *Schamott* = fired clay) that was erroneously applied to ancient ceramic production. Chamotte and grog are not only different in their origin but also in the way they interact with the matrix of the ceramic body (see below). Although are still defined in various ways in ceramic glossaries, sometimes as synonyms, some others not, it is advised to clearly distinguish them in archaeometry and archaeology. The former was obtained by calcination of clay at temperatures lower than those reached by the ceramic body during firing, so chamotte tend to better sinter with the matrix. Moreover, the calcinated clay is usually the same to which is added to obtain the paste and reduce shrinkage. The latter, being a recycled ceramic, was originally fired in a temperature range comparable to that of the ceramic body in which was added as temper, so grog generally tends to be partially sintered with the matrix, unless firing temperatures higher than those reached to fire the grog were attained. It is worth to note that the risk to confuse chamotte with grog arises only in modern ceramics, whereas in prehistory and antiquity only grog was attested.

Unless frequent elongated NPIs occur in the fabric, voids are the best indicators of microstructures [35, 59]. The description and classification of voids take into account their size and shape, as well as the orientation in relation to the outer surfaces of the artefacts and to the grain boundary of sand-sized NPIs. The kneading and preparation of paste for forming incorporate air, which becomes primary porosity. Successively, the loss of plasticity and adsorbed water during drying and early stages of firing originate the secondary pores that closes at higher degrees of sintering. Differences between primary and secondary pores are generally marked by differences in size and shape, other than reciprocal relations. Low primary porosity indicates good paste preparation, while scarce presence of secondary porosity and rounded shape of residual pores point to higher degree of sintering. In Figure 14.1, elongated fissure-like pores cut the primary. Such more or less oriented porosity with the outer surfaces of the ceramic body improves the tenacity [27, 60] and the thermal shock resistance of ceramic. Shrinkage porosity, which forms during drying and firing, is more sensitive than NPIs to the iso-orientation produced by forming and finishing actions on the plastic paste. As a rule, compression stress results in particles being aligned and pores flattened perpendicularly to the stress direction, whereas shear stress results in alignment of particles and pores stretched along the direction of shear [59]. Experimental works on microstructures produced by different forming techniques show the application of image analysis on POM photomicrographs to measure the disorientation of NPIs and voids in thin section. Ceramics formed with coiling show higher disorientation compared to those formed with pinching or wheel throwing [59]. As is the case of NPIs, the clay matrix causes some masking of the smaller pores.

If finishing(s) are applied on the surface of the ceramic, they should be described as further ceramic materials, with their own NPIs, matrix and porosity. It is also important to annotate the stratigraphic relation(s) existing between the ceramic body and layer(s) of finishing (Figure 14.1).

In order to already convey petrofacies information in the name of the fabric, it is suggested to use as prefix the abbreviation(s) of the more abundant NPI(s), followed by a suffix that includes the abbreviation(s) of the characteristic components of the matrix. As an example, if a given fabric has NPIs dominated by quartz (Q) and grog (G) and a matrix relatively rich in iron oxi-hydroxides (i) and carbonaceous matter (o), the resulting acronym will be: QG_io. However, the suffix could be omitted if acronym is too long and/or the other fabrics eventually distinguished among the studied potsherds are very similar. Such terminology focus on the basic composition of the fabric, to highlight the similarity (compatibility) between the matrix and NPIs, as well as the details useful to identify sedimentary facies. Since the scale of archaeological problems is not necessarily the same of that of the geological maps, the answer to the provenance question should be found in the field, at the human scale. It is not excluded the possibility to group the fabrics according their geological fingerprint [39, 41].

Scanning electron microscopy (SEM), together with polarized optical microscopy (POM) is typically used for ceramic petrography purposes. POM provides some advantages with respect to SEM. Firstly, for specimens that show structures in the scale size of 0.1–10 mm, optical microscopy is better suited, since the field of view is quite large at low magnification. Secondly, it may provide a wider array of discriminant features about mineralogical phases of interest (e. g. identify polymorphs, distinguish crystals in monomineralic rock fragments, etc.). Conversely, SEM analyses are preferred to probe the internal structure of optically opaque phases, to achieve higher spatial resolutions, to avoid the masking effect of transmitted light, to have a 3-D impression of the specimen surface, to determine chemical-physical parameters (e. g. chemical composition), which cannot be measured by electromagnetic radiation.

Point-counter and visual estimation charts have been usually used in ceramic petrography to perform quantitative investigations via thin section that is to provide modal estimation and identification of mineralogical phases as well as semi-quantification of their morphometry and texture [34, 35, 51, 61, 62]. Since these practices may be subjective and time-consuming, digital image analysis techniques were proposed to automatize the work. In this way a set of parameters (like grain-size and shape descriptors) can be easily measured by using multisource digital images (e. g. POM, SEM) of the interested sample as acquired by thin sections [44, 51, 52, 63, 64].

14.1.2.2 X-ray diffraction

X-ray powder diffraction (XRPD) is a typical method to investigate the fine grained mineral fraction of the ceramic body, difficult to study under the petrographic microscope. The indirect mineralogical information deriving from the powdered ceramic samples provides rapid identification of crystalline phases, commonly constituted by relic phases (i. e. raw materials), firing/use phases (e. g. anorthite, maghemite, mullite, etc.) and post-depositional phases (e. g. calcite, gypsum, smectite, etc.). Other than crystalline phases identification, XRPD data are used to quantify them with different strategies (e. g. Rietveld, RIR, etc.) and, even though difficult to perform on a polycrystalline mixture, to determine crystal sizes and structure [22]. The use of synchrotron radiation (SR) as X-ray source make available higher beam energy and diffraction resolution, allowing more precise qualitative and quantitative mineralogical characterization of the ceramic body (e. g. [65–67]).

As discussed before, the use of a single analytical method is not resolutive and the XRPD is not an exception. The bulk mineralogical composition provided by XRPD does not distinguish the origin of a given phase (e. g. original calcite or post-depositional mineralization) and a combined use with optic microscopy and/or SEM/EDS is advised.

XRPD typically provides information about the mineral assemblage of the ceramic body. Thermally induced transformations in mineralogy and microstructure are useful to estimate the firing temperature ranges of the archaeological ceramics. It must be kept in mind that the temperatures estimated are fictive, since they are the result of a combination of temperature-time-atmosphere-composition, and that the maximum temperature range of firing is not an indicator in itself of the degree of sintering of the ceramic body [32]. Moreover, ceramics are metastable polyphasic systems characterized by microchemical domains [24, 27, 68] that never attain thermodynamic equilibrium. Such compositional/textural differences yield difference of sintering in the same ceramic body, especially if short firing durations occurred. If coatings applied on the surface are considered, the difference in sintering is even more clear (e. g. red figure pottery, glazed pottery, etc.). More generally, the same sintering degree of the ceramic body can be obtained by firing either with longer time/lower temperature combination or with shorter time/higher temperature [32, 38, 69]. Experimental firing evidences show how temperatures measured within the same object [70] or in the firing structure [71–74] may vary in a range of few hundreds of Celsius degrees. For this reason, to talk about estimated punctual firing temperatures is meaningless. Given this variability of recorded temperature within the same ceramic artefact and its magnitude as a result of a T/t evolution during firing, it is more appropriate to consider a temperature range, inferred by different methods (e. g. XRPD, FTIR, microstructure evolution by SEM, etc.) that provide *equivalent firing temperatures* [69], that is a set of temperatures corresponding to their equivalent obtained in standardized laboratory conditions (i. e. soaking time = 1 h in a controlled atmosphere). Even if this definition is related to an experimental firing under known conditions able to replicate the characteristics of the archaeological ceramic body, the concept involved stays for itself.

If the sintering degree of a given ceramic body may be obtained by different T/t curves, its variability by changing composition and texture of the components is even higher. As a rule, a first difference can be observed between carbonate-rich and carbonate-poor matrices [24, 32, 68, 75, 76]. Experimental evidences show that the presence of fine calcite dispersed in the clay matrix yields an equivalent sintering degree to that of the same clay without calcite at temperature even lower of 200 °C [77]. If carbonate-poor clays are considered, a difference exists between the kaolinitic-rich and the illite-rich clays. Even though kaolinite yields a partially dehydroxylated metastable phase, and consequent densification, at lower temperature than illite, the sintering degree may be significantly influenced by the amount of quartz present in the clay (e. g. [78–80]. Smectite-rich clays are not suitable for ceramic production because are responsible for strong shrinkage, however they sinter at relatively lower temperatures compared to kaolinitic clays [81, 82]. Such temperature differences to densificate between the various clay pastes have played a role in the technological evolution of ceramics and pyrotechnology [68]. The determination of firing temperatures is thus important for at least two archaeological aspects: to understand (i) the pyrotechnology and (ii) the ceramic technology. The estimation of both temperature range and maximum temperatures reached for a series of potsherds can give information about the level of control over the firing process (i. e. standardization) and type of pyrotechnology (i. e. firing structure, combustible, etc.) involved.

It should not be overlooked that the thermal exposition of ceramics could have been of the same intensity of that occurred during firing, as is the case of technical ceramics used for glassmaking [28, 58] or metallurgy [83].

14.2 Case studies

The case studies reported hereafter exemplify the role of POM to identify the temper in the ceramic body, as well as the finishing coatings as well as the combined use of SEM on thin section and bulk mineralogical (XRPD) and chemical (ICPMS and XRF) analyses to solve archaeological problems about provenance, technology and function. The proposed selection took into account ceramics from Late Neolithic to Antiquity and Modern times, in order to provide a compositional, techno-functional and age array of artefacts.

14.2.1 Pottery with spathic calcite

The presence of carbonates (mainly calcite) in the ceramic body may have different forms, which implies different type of reactivity with the clay matrix. In Mediterranean area, several examples of calcite tempered pottery are reported [84–86]. The presence of calcite in the paste may have very different effects on sintering degree of the ceramic body [24, 27, 55, 87]. If microcrystalline calcite dispersed in the clay matrix has a positive effect on sintering new phases formation [68], the presence of coarse calcite inclusions may be potentially destructive if they are not stabilized with prevalent reducing atmosphere during firing, fast heating rates and/or short soaking time in the range between 600 °C and 800 °C [55, 86].

Spathic calcite deliberately added to the clay paste in Late Neolithic and Bronze age pottery from Apulia (southern Italy), represents an interesting case study [85]. Most of the Apulian territory is characterized by the presence of Cretaceous limestones belonging to the Apulian microplate [88–90] and of transgressive and regressive calcarenite deposits of Neogene and Quaternary age [91, 92]. Such calcareous substrate is interested by karst phenomena that yielded residual clays (i. e. *terra rossa*) and calcite mineralizations in speleothems and veins. Although residual clays and continental calcite are products of the same process, the natural presence of spathic calcite in terra rossa is not so common. In karst environment, surface water flow is scarce, so the erosion/deposition mechanism associated. The natural presence of spathic calcite is generally associated to faulted limestone zones, where the karst forms are more concentrated, and where the occurrence of clastic calcite is higher [93]. Away from the fractured zones, sand-sized clasts of limestone and/or spathic calcite are less frequent.

In Figure 14.5, a coarse ware from Masseria Fragennaro (Laterza) shows the presence of angular spathic calcite added as temper [85]. The coarse size of clasts and the prevalent reducing conditions during firing (RO, see above) stabilized calcite in the EFT range of 800–900 °C. The presence of secondary porosity (i. e. shrinkage) accounts for a medium-low sintering of the body.

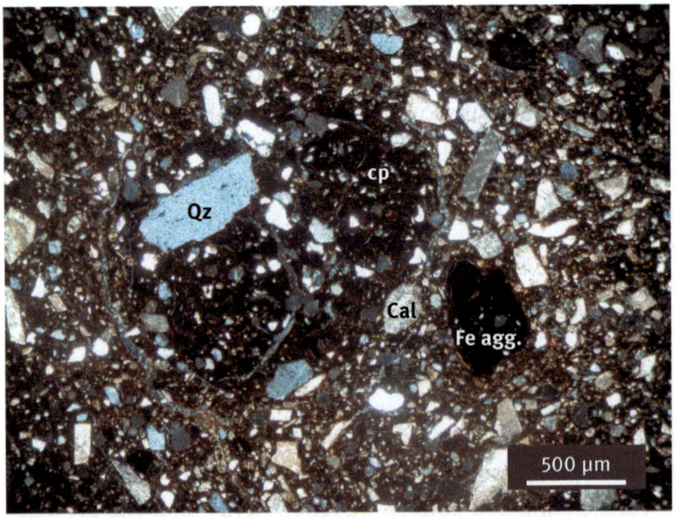

Figure 14.5: Crushed spathic calcite (Cal) added as temper in Late Neolithic pottery from Masseria Fragennaro (Laterza). The clay pellet (cp) in the centre of the photomicrograph does not contain spathic calcite, thus demonstrating its intentional addition to the clay (Qz = quartz; Fe agg. = iron aggregate). Figure modified from Muntoni [85].

14.2.2 Apulian red figure pottery: provenance and production technology assessment from ceramic body analyses

Apulian red figure pottery (3rd quarter of the fifth-end of the fourth century BCE) is a local variant of Attic red figure pottery [94–96]. It is based on the figural depictions in red colour – saved from the ceramic paste – on a black glossy background painted on the ceramic body. Its noteworthy features being a very high quality both in manufacturing and in drawing, it became, throughout the fourth century, inside the figured pottery produced in *Magna Grecia*, the most numerous, widespread and marketed handicrafts – within and outside the Apulia. The production started in Taranto and spread in the other Apulian sites thanks to a temporary transfer of skilled workers and artists from Taranto. These sites became soon branch centres of production outside the main polis, added to the activity of the workshops of the Magna Grecia colony.

Many questions are still unresolved for this production, especially during the phase of increasing production (fourth century BCE), in which vessels of very different both size and formal/iconographic quality are found together [94–96]. Among them, the identification of workshops and production technology (equal to the Attic? Diversified depending on the final use of the item, e. g. for daily use, for a burial, etc.) and/or on the customer target?

A major contribution to the solution of these questions can come from the results of the archaeometric studies that can shed light on the production technology employed for the manufacturing of this ceramic class on the basis of objective data; furthermore, they can recognize dissimilarities between objects, which can lead to a differentiation of production processes and workshops.

Studied samples come from sites among the most relevant in Apulia – Monte Sannace (Gioia del Colle), Altamura, Conversano, Botromagno (Gravina), Taranto, Egnazia, Canosa and Arpi.

They were analysed by a multi-technique approach – optical and scanning electron microscopy, X-ray diffraction and atomic spectroscopy – that allowed a complete physical-chemical, mineralogical and morphological characterization of finds.

14.2.2.1 Provenance imports and local productions from compositional analysis

The multivariate statistical analyses of compositional data of ceramic bodies [97–101] clearly distinguish samples coming from the three different areas in in which the ancient Apulia was divided (*Peucetia*, *Messapia* and *Daunia*) (Figure 14.6), allowing us to hypothesize a polycentric production.

The same figure shows a great compositional diversity of the T38 sample from the Apulia ones. It was discovered in Taranto, but considered an Attic find, on the basis of archaeological studies. The import hypothesis is confirmed by the greater compositional

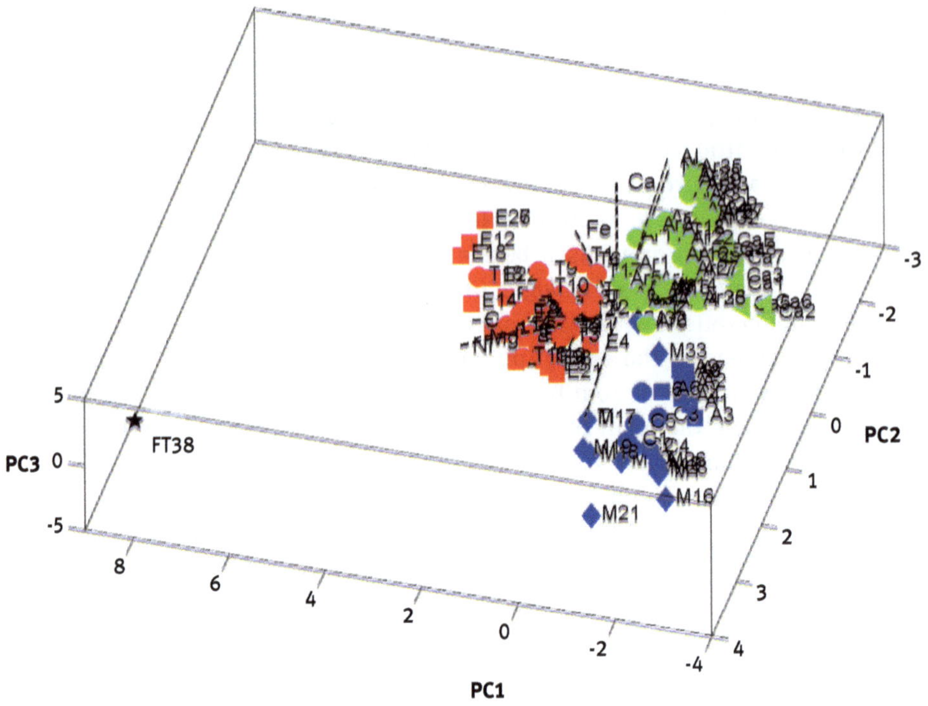

Figure 14.6: Scores and loadings diagram for the first three principal components related to the examined objects. Three clusters, gathering finds from Peucetia (blue glyphs), Messapia (red glyphs) and Daunia (green glyphs), respectively, and an outlier (T38), are displayed. The accounted variance is 67% of the total variance.

similarity of its ceramic body to that of coeval Greek red figure samples (Figure 14.7) compared to the Apulian ones.

As concern samples from Taranto, stylistically datable to the fourth century BCE, since we could analyse a fragment definitely made in Taranto – a probable firing tester [104] – and two vitrified clay samples both coming from a pottery workshop located in Taranto [105], it is possible to prove a local production for them (Figure 14.8).

In Figure 14.9, the differences of texture and microstructure between the Attic (T38) and the Apulian (T64A) pottery fabric are observed. The carbonate-rich matrix of Apulian red figures pottery is more sintered than the Attic one.

On the base of results of XRPD and SEM-EDX analyses, revealing the presence of newly formed pyroxenes and gehlenite [101], we can affirm that all vases were fired betwwen 900 and 1,000 °C. Such high temperatures were kept for a shorter time for cluster B samples with respect to cluster A ones, as indicated by the lower sintering degree and the presence of calcite clusts in their ceramic bodies.

However the low amount of pyroxenes and the scarce sintering, revealed by SEM-EDX analysis for FT38 sample suggests an insufficient firing time [101].

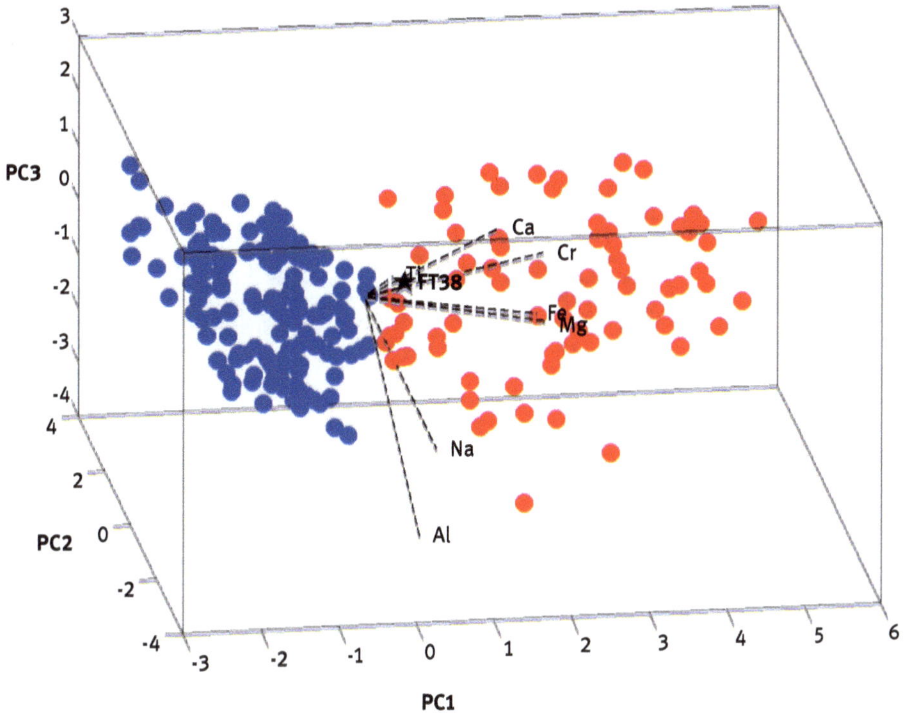

Figure 14.7: Scores and loadings diagram for the first three principal components related to Apulian (blue circles) and Attic (red circles) red figure pottery. The accounted variance is 78% of the total variance [97–103].

In particular, the diversity of T38 compared to the Apulian samples is due to the lower quantity of calcium oxide, the greater quantity of micas (mostly muscovite) and to the presence of clino-pyroxenes in its ceramic body.

14.2.2.2 Production technology: tradition and innovation from morphological and compositional analyses

For each Apulian sites, the results of multivariate statistical treatment of compositional data of the ceramic bodies of the fourth century samples highlights the separation of samples into two clusters. On the base on morpho-mineralogical analysis it is possible to justify this separation and attribute it to the use of two different production technologies (Figure 14.10 and Figure 14.11). Certain vases were produced using the "classic" Attic technology and others with a different technique, never reported before in the literature and not highlighted in Attic items, which involved the application of a red engobe layer on the clay paste, before the black gloss painting, in order to imitate the red/black contrast of Attic pottery, otherwise unavailable with the local carbonate-rich clay.

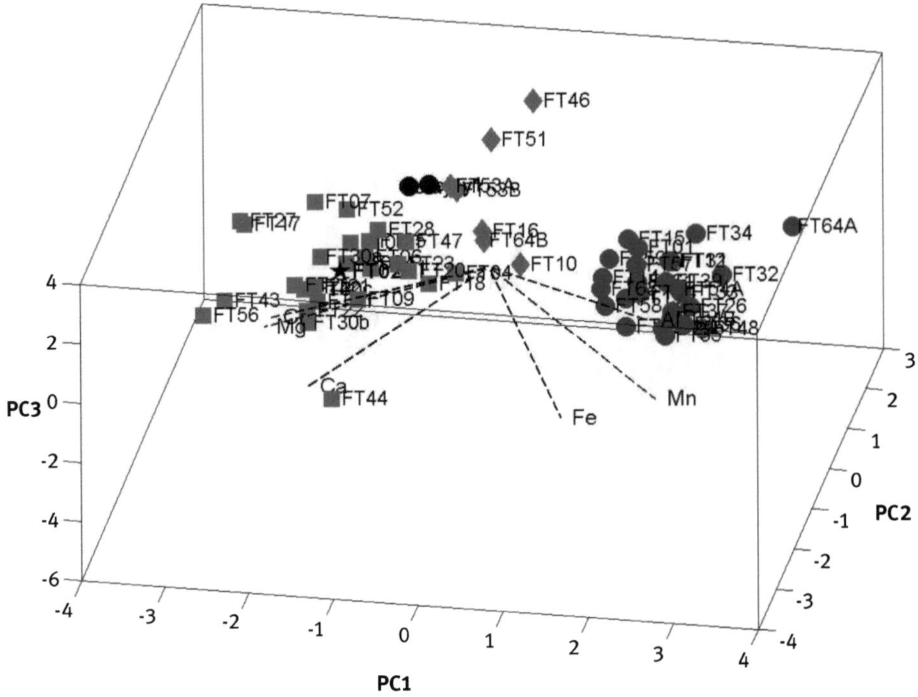

Figure 14.8: Scores and loadings diagram for the first three principal components related to finds from Taranto. The accounted variance is 78% of the total variance. Glyphs: red squares and diamonds: finds stylistically datable to the fourth century BCE, black star: probable firing tester [104], black points: vitrified clay samples, both coming from a pottery workshop located in Taranto [105].

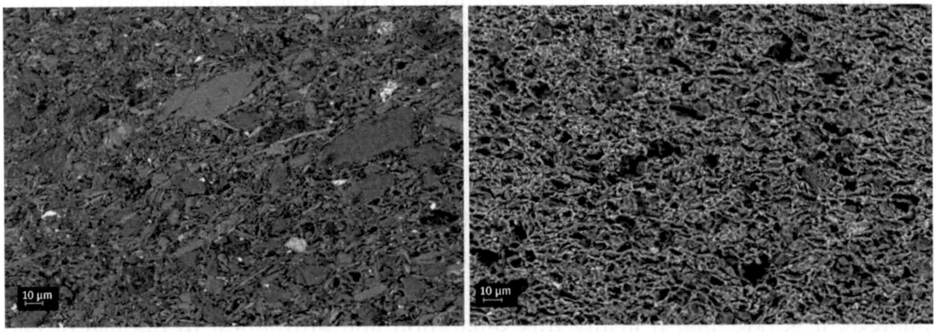

Figure 14.9: Scanning electron microscopy-back scattered electrons (SEM-BSE) photomicrographs of ceramic bodies of samples T38 (on the left) and T64A (on the right).

Though the ceramic body and the engobe layer share an analogous assemblage of minerals, they are present in different ratios: the body is richer in calcite, quartz and feldspars, whereas the engobe shows a higher abundance of micas (biotites and

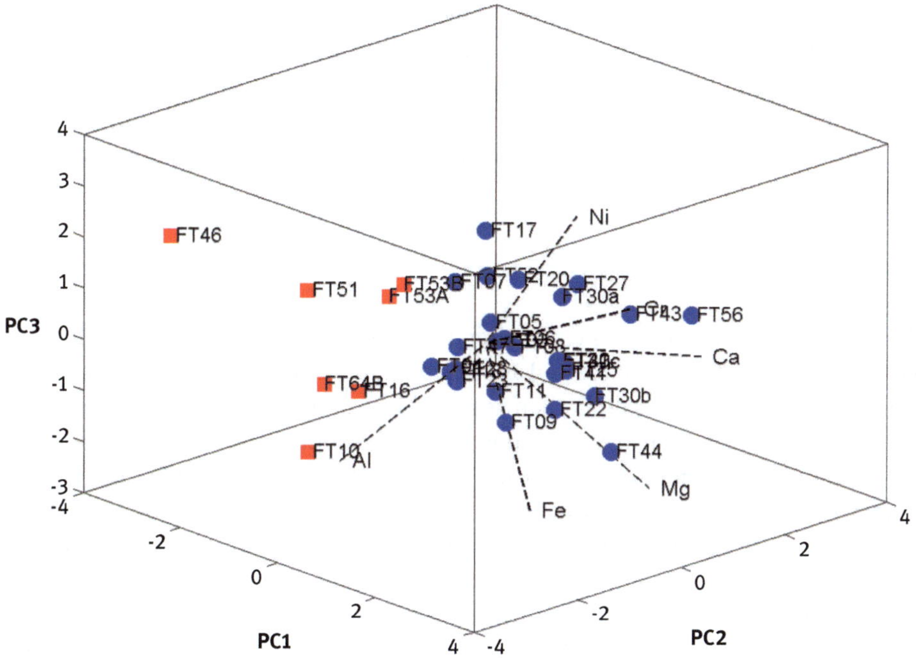

Figure 14.10: Scores and loadings diagram for the first three principal components related to samples from Taranto, stylistically datable to the fourth century BCE. Two clusters are evident. The blue circles are related to samples with engobe layer on the ceramic body. The accounted variance is 81% of the total variance.

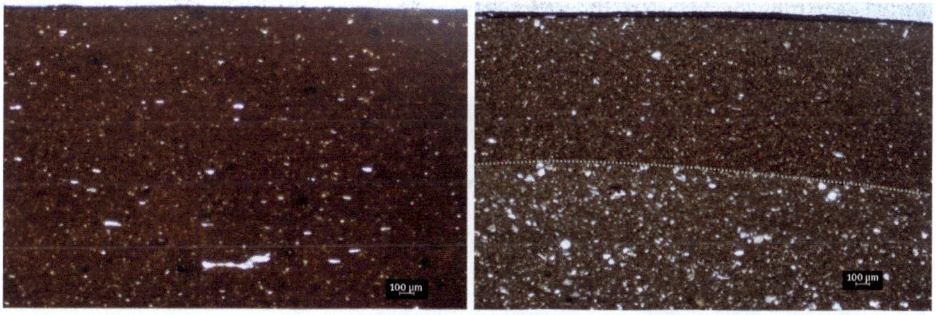

Figure 14.11: Optical microscopy images highlighting fabric differences between shards T10 (left) and T41 (right). In T10, the black gloss was directly applied on the ceramic body, whereas a Fe-rich engobe was applied on the carbonate-rich body of T41, before the black gloss.

muscovites), clayey minerals and Fe-oxides. As a consequence, being richer in clay than in silt, the engobe appears more compact than the body. These results lead to the hypothesis of the use of the same raw materials for both, but refined in the case of the engobe.

The finding of fragments with the same technological characteristics in all the Apulian sites allow us to hypothesize that the use of this technology was not an isolated expedient, limited to a single area of the region, but a peculiar characteristic of the Apulian pottery, representing a distinctive attribute of Late Apulian manufacture, since we are aware of similar experimental evidences from other geographical areas.

14.2.3 Technical ceramics for glassmaking

The locution *technical ceramics* comprehend a series of artefacts used as tools in other production cycles. They are classified as *transformation ceramics* by Rice [106] and may include structural ceramics other than vessels. The examples here reported are refractory bricks sampled from the protoindustrial glassworks of Roches (Moutier, CH), where the use of different recipes for the ceramic body was justified by the different thermal exposition and chemical alteration of bricks [107]. In this case study is clear the difference between grog and chamotte (see above). In Figure 14.12, chamotte inclusions are easily recognizable after their higher optical density, angular shape and very fine texture. The characteristics of crushed crucibles recycled as grog in the refractory are visible in Figure 14.13. In the first case, the refractory bricks used to line the melting chamber of the furnace were tempered with chamotte to minimize shrinkage and glass corrosion. In the second case, the refractory panels used outside the furnace to cover the glory holes were tempered with recycled bricks and/or crucibles since they were less exposed to heat and wearing. The shrinkage fissures around the larger chamotte fragments highlight the difference in clay content between them. Grog can be identified for its grey tones compare to the yellowish original clay, the higher degree of thermal

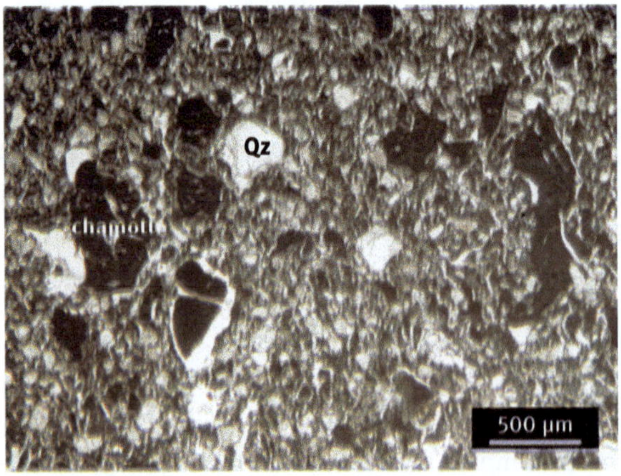

Figure 14.12: Chamotte fragments in a refractory brick of the glassworks of Roches (Moutier, CH). Abbreviation: Qz = quartz.

Figure 14.13: Grog fragments (i. e. recycled glass crucibles) in a refractory panel to close a glory hole of the melting furnace of the glassworks of Roches (Moutier, CH).

alteration of quartz grains and the presence of glass fragments, originally stick to the surface of the crucibles. In both cases the grain size distribution is seriate as consequence of crushing. As concern refractoriness and resistance to chemical alteration, the use of grog introduces more fluxes into the ceramic body and locally lowering the eutectic with consequent formation of glass and beginning of ceramic corrosion [28, 58]. Whilst the sum of K_2O, CaO and Fe_2O_3 concentrations is below one weight percent in refractory bricks, it ranges between three and five weight percent in refractory panels. Such different paste recipe and function can be also inferred by mineral assemblages determined by XRPD. Refractory components show equivalent maximum temperatures between 1,200 °C (mullite in) and 1,500 °C (co-occurrence of tridymite/ cristobalite), though bricks show quartz grains almost completely substituted by cristobalite and tridymite.

References

[1] Fabbri B. Lo studio dei materiali ceramici nei beni culturali. La La Chimica e l'Industria. 2002;84:43–8.

[2] Brunetti B, Sgamellotti A. Lo studio delle ceramiche storico-artistiche. La La Chimica e l'Industria. 2002;84:49–54.

[3] Riccardi P, Nodari L, Gualtieri S, De Simone D, Fabbri B, Russo U. Firing techniques of black slipped pottery from Nepal (twelveth–third century BC): the role of Mössbauer spectroscopy. J Cultural Heritage. 2008;9:261–8.

[4] Riccardi P, Nodari L, Fabbri B, Gualtieri S, Russo U. Contribution for a mineralogical thermometer to be applied to low fired and/or non-carbonate ceramics. BAR Internatl Ser. 2007;1691:13–18.

[5] Maritan L, Mazzoli C, Nodari L, Russo U. Second Iron Age grey pottery from Este (Northeastern Italy): study of provenance and technology. Appl Clay Sci. 2005;29:31–44.

[6] Laviano R, Muntoni IM. Provenance and technology of Apulian Neolithic pottery. Geol Soc London Spec Publ. 2006;257:49–62.

[7] Amato F, Gualtieri S, Ricciardi P. Potenzialità dell'archeometria nello studio del materiale ceramico. In Lega A, editor. Ceramica: materia e tecnica, vol. 20. Faenza: Phase srl ed, 2006:57–82.

[8] Weigand PC, Harbottle G, Sayre EV. Turquoise sources and source analysis: Mesoamerica and the southwestern U.S.A. In: Earle TK, Ericson JE, editor(s). Exchange systems in prehistory. New York: Academic Press, 1977:15–34.

[9] Descantes C, Speakman RJ, Glascock MD. Compositional studies of Caribbean ceramics: an introduction to instrumental neutron activation analysis. J Caribbean Archaeol. 2008;Special Publication 2:1–14.

[10] Peacock DP. The heavy mineral analysis of pottery: a preliminary report. Archaeometry. 1967;10:97–100.

[11] Glascock M. Characterization of archaeological ceramics at MURR by neutron activation analysis and multivariate statistics. In: Neff H, editor(s). Chemical characterization of Ceramic pastes in archaeology. Madison (WI): Prehistory Press, 1992:11–26.

[12] Glascock M, Neff H. Neutron activation analysis and provenance research in archaeology. Meas Sci Technol. 2003;14:1516–26.

[13] Vaggelli G, Cossio R. μ-XRF analysis of glasses: a non-destructive utility for cultural heritage applications. Analyst. 2012;137:662–7.

[14] Stoner WD, Glascock MD. The forest or the trees? Behavioral and methodological considerations for geochemical characterization of heavily-tempered Ceramic pastes using NAA and LA-ICP-MS. J Archaeol Sci. 2012;39:2668–83.

[15] Hein A, Tsolakidou A, Iliopoulos I, Mommsen H, Buxeda i Garrig´os J, Montanac G, et al. Standardisation of elemental analytical techniques applied to provenance studies of archaeological ceramics: an inter laboratory calibration study. Analyst. 2002;127:542–53.

[16] Gratuze B, Blet-Lemarquard M, Barrandon JN. Mass spectrometry with laser sampling: a new tool to characterize archaeological materials. J Radioanal Nucl Chem. 2001;247:645–56.

[17] James WD, Dahlin ES, Carlson DL. Chemical compositional studies of archaeological artifacts: comparison of LA-ICP-MS to INAA measurements. J Radioanal Nucl Chem. 2005;263:697–702.

[18] Resano M, Garc´ıa-Ruiz E, Vanhaecke F. Laser ablation-inductively coupled plasma mass spectrometry in archaeometric research. Mass Spectrom Rev. 2010;29:55–78.

[19] Tite MS. Methods in physical examination in archaeology. London and New York: Seminar Press, 1972.

[20] Dinnebier RE, Billinge SJ. Powder diffraction: theory and practice. Cambridge: Royal Society of Chemistry, 2008.

[21] Maggetti M. Mineralogical and petrographical methods for the study of ancient pottery. In: Burragato F, Grubessi O, Lazzarini L, editor(s). Proceedings of the 1st European workshop on archaeological ceramics. Dipartmento Scienze della Terra, Università degli Studi di Roma 'La Sapienza', 1994:23–35.

[22] Artioli G, Angelini I. Scientific methods and cultural heritage: an introduction to the application of materials science to archaeometry and conservation science. Oxford: Oxford University Press, 2010.

[23] Tite MS, Maniatis Y. Examination of ancient pottery using the scanning electron microscope. Nature. 1975;257:122.

[24] Riccardi MP, Messiga B, Duminuco P. An approach to the dynamics of clay firing. Appl Clay Sci. 1999;15:393–409.

[25] Rathossi C, Pontikes Y. Effect of firing temperature and atmosphere on ceramics made of NW Peloponnese clay sediments. Part I: reaction paths, crystalline phases, microstructure and colour. J Eur Ceramic Soc. 2010;30:1841–51.
[26] Hein A, Müller NS, Day PM, Kilikoglou V. Thermal conductivity of archaeological ceramics: the effect of inclusions, porosity and firing temperature. Thermochim Acta. 2008;480:35–42.
[27] Allegretta I, Pinto D, Eramo G. Effects of grain size on the reactivity of limestone temper in a kaolinitic clay. Appl Clay Sci. 2016;126:223–34.
[28] Eramo G. The glass-melting crucibles of Derriere Sairoche (1699–1714 AD, Ct. Bern, Switzerland): a petrological approach. J Archaeol Sci. 2006a;33:440–52.
[29] Richardson C. Petrology of the Galapagos Islands. In Lawrence John Chubb, Geology of Galapagos, Cocos, and Easter Islands. Bernice B Bishop Museum-Bull. 1933;110:45–67.
[30] Hussak E. Feldspatporzellan. Sprechsaal. 1889;22:153.
[31] Behrens H. Ueber das Porcellan und einige verwandte Entglasungsproducte. Ann Phys. 1874;226:386–99.
[32] Heimann RB, Maggetti M. Ancient and historical ceramics. Stuttgart: Schweizerbart Sche Vlgsb, 2014.
[33] Ingersoll RV. Actualistic sandstone petrofacies: discriminating modern and ancient source rocks. Geology. 1990;18:733–6.
[34] Heidke JM, Miksa EJ. Correspondence and discriminant analyses of sand and sand temper compositions, Tonto Basin, Arizona. Archaeometry. 2000;42:273–99.
[35] Whitbread IK. Greek transport amphorae: a petrological and archaeological study Fitch Laboratory Occasional Paper, vol. 4. London: British School of Athens, 1995.
[36] Peck AB. The rôle of the ceramic petrographer. A brief review of the development of the ceramic petrography. Bull Am ceramic Soc. 1927;6:297–305.
[37] Peacock DP. The scientific analysis of ancient ceramics: a review. World Archaeol. 1970;1:375–89.
[38] Maggetti M. Phase analysis and its significance for technology and origin. In: Olin JS, Franklin AD, editors. Archaeological ceramic. Washington, DC: Smithsonian Institution Press, 1982: 121–33.
[39] Capelli C, Mannoni T. Proposte per una scheda descrittiva delle sezioni sottili e per una classificazione minero-petrografica delle ceramiche. Archeol Medievale. 1996;23:689–97.
[40] Orton C, Hughes M. Pottery in archaeology. Cambridge: Cambridge University Press, 2013.
[41] Santacreu DA. Materiality, techniques and society in pottery production: the technological study of archaeological ceramics through paste analysis. Berlin, Boston: Walter de Gruyter GmbH & Co KG, 2014.
[42] Stoltman JB. The role of petrography in the study of archaeological ceramics. In: Goldberg P, Holliday VT, Ferring CR, editor(s). Earth sciences and archaeology. Boston, MA: Springer, 2001:297–326.
[43] Heidke JM, Miksa EJ, Wallace HD. A petrographic approach to sand-tempered pottery provenance studies: examples from two Hohokam local systems. In: Glowacki DM, Neff H, editor(s). Ceramic production and circulation in the greater Southwest: source determination by INAA and complementary mineralogical investigations, vol. 44. Los Angeles: Cotsen Institute of Archaeology, University of California, 2002:152–78.
[44] Eramo G, Giannossa LC, Rocco A, Mangone A, Graziano SF, Laviano R. Oil lamps from the catacombs of Canosa (Apulia, fourth to sixth centuries AD): technological features and typological imitation. Archaeometry. 2014;56:375–91.
[45] D'Ercole G, Eramo G, Garcea EA, Muntoni IM, Smith JR. Raw material and technological changes in ceramic productions at Sai Island, northern Sudan from the 7th to the 3rd millennium BCE. Archaeometry. 2015;57:597–616.

[46] Bishop RL. Aspects of compositional modeling. In: Fry RE, editor. Models and methods in regional exchange, Washington, DC: Society for American Archaeology. 1980:47–66. SAA Papers No. 1.
[47] Shepard AO. Ceramics for the archaeologist. Washington, DC: Carnegie Institution of Washington, 1985.
[48] Tite MS. Ceramic production, provenance and use – a review. Archaeometry. 2008;50:216–31.
[49] Quinn PS. Ceramic Petrography: the interpretation of archaeological pottery and related artefacts in thin section. Oxford: Archaeopress, 2013.
[50] Levi ST. Dal coccio al vasaio. Manifattura, tecnologia e classificazione della ceramica. Bologna: Zanichelli, 2010:1–275.
[51] Aprile A, Castellano G, Eramo G. Classification of mineral inclusions in ancient ceramics: comparing different modal analysis strategies. Archaeol Anthropol Sci. 2019;11:2557–67.
[52] Aprile A, Castellano G, Eramo G. Combining image analysis and modular neural networks for classification of mineral inclusions and pores in archaeological potsherds. J Archaeol Sci. 2014;50:262–72.
[53] Rye OS. Pottery technology: principles and reconstruction. Washington, DC: Taraxacum, 1981.
[54] Arnold DE. Ethnomineralogy of Ticul, Yucatan potters: etics and emics. Am Antiq. 1971;36:20–40.
[55] Rye OS. Keeping your temper under control: materials and the manufacture of Papuan pottery. Archaeol Physl Anthropol Oceania. 1976;11:106–37.
[56] Whitbread IK. The characterisation of argillaceous inclusions in ceramic thin sections. Archaeometry. 1986;28:79–88.
[57] Cuomo Di Caprio N, Vaughan SJ. An experimental study in distinguishing grog (chamotte) from argillaceous inclusions in ceramic thin sections. Archaeomaterials. 1993;7:21–40.
[58] Eramo G. The glass-melting furnace and the crucibles of Südel (1723–1741, Switzerland): provenance of the raw materials and new evidence of high thermal performances. J Cultural Heritage. 2006b;7:286–300.
[59] Thér R. Identification of pottery-forming techniques using quantitative analysis of the orientation of inclusions and voids in thin sections. Archaeometry. 2016;58:222–38.
[60] Tite MS, Kilikoglou V, Vekinis G. Strength, toughness and thermal shock resistance of ancient ceramics, and their influence on technological choice. Archaeometry. 2001;43:301–24.
[61] Schubert P. Petrographic modal analysis – a necessary complement to chemical analysis of ceramic Coarse Ware. Archeometry. 1986;28:163–78.
[62] Eramo G. Pre-industrial glassmaking in the Swiss Jura: the refractory earth for the glassworks of Derrière Sairoche (ct. Bern, 1699–1714). Geol Soc London Spec Publ. 2006c;257:187–99.
[63] Reedy C, Anderson J, Reedy T, Liu Y. Image analysis in quantitative particle studies of archaeological ceramic thin sections. Adv Archaeol Prac. 2014;2:252–68.
[64] Dal Sasso G, Maritan L, Salvatori S, Mazzoli C, Artioli G. Discriminating pottery production by image analysis: a case study of Mesolithic and Neolithic pottery from Al Khiday (Khartoum, Sudan). J Archaeol Sci. 2014;46:125–43.
[65] Tang CC, MacLean EJ, Roberts MA, Clarke DT, Pantos E, Prag AJ. The study of Attic black gloss sherds using synchrotron X-ray diffraction. J Archaeol Sci. 2001;28:1015–24.
[66] Pantos E, Tang CC, MacLean EJ, Roberts MA, Murphy BM, Collins SP, et al. Applications of synchrotron radiation to archaeological ceramics. BAR Int Ser. 2002;1011:377–84.
[67] Barone G, Bartoli L, Belfiore CM, Crupi V, Longo F, Majolino D, et al. Comparison between TOF-ND and XRD quantitative phase analysis of ancient potteries. J Anal At Spectrom. 2011;26:1060–7.
[68] Heimann RB. Assessing the technology of ancient pottery: the use of ceramic phase diagrams. Archeomaterials. 1989;3:123–48.

[69] Tite MS. Firing temperature determinations how and why? Kungliga Vitterhets Historie Och Antikvitets Akademien Konferenser. 1995;34:37–42.
[70] Maggetti M, Neururer C, Ramseyer D. Temperature evolution inside a pot during experimental surface (bonfire) firing. Appl Clay Sci. 2011;53:500–8.
[71] Gosselain OP. Bonfire of the enquiries. Pottery firing temperatures in archaeology: what for? J Archaeol Sci. 1992;19:243–59.
[72] Livingstone Smith A. Bonfire II: the return of pottery firing temperatures. J Archaeol Sci. 2001;28:991–1003.
[73] Wolf S. Estimation of the production parameters of very large medieval bricks from St Urban Switzerland. Archaeometry. 2002;44:37–65.
[74] Thér R. Identification of pottery firing structures using the thermal characteristics of firing. Archaeometry. 2014;56:78–99.
[75] Letsch J, Noll W. Phase formation in several ceramic sub-systems at 600–1000 °C as a function of oxygen fugacity. Ceram Forum Int/Ber DKG. 1983;60:259–67.
[76] Duminuco P, Messiga B, Riccardi MP. Firing process of natural clays Some Microtextures and Related Phase Compositions. Thermochim Acta. 1998;321:185–90.
[77] Duminuco P, Riccardi MP, Messiga B, Setti M. Modificazioni tessiturali e mineralogiche come indicatori della dinamica del processo di cottura di manufatti ceramici. Ceramurgia. 1996;26:281–8.
[78] Cruciani G. Processi di idratazione e disidratazione: fillosilicati. Plinius. 1999;21: 237–52.
[79] Aras A. The change of phase composition in kaolinite-and illite-rich clay-based ceramic bodies. Appl Clay Sci. 2004;24:257–69.
[80] Fernandez R, Martirena F, Scrivener KL. The origin of the pozzolanic activity of calcined clay minerals: A comparison between kaolinite, illite and montmorillonite. Cement Concr Res. 2011;41:113–22.
[81] Abdrakhimov DV, Abdrakhimova ES, Abdrakhimov VZ. Sintering quality of clay materials. Glass Ceram. 1999;56:190–3.
[82] Velde B, Druc IC. Archaeological ceramic materials: origin and utilization. Berlin: Springer Science and Business Media, 2012.
[83] Bayley J, Rehren T. Towards a functional and typological classification of crucibles. London: Archetype/British Museum, 2007.
[84] Levi S, Cazzella A, Moscoloni M, Fratini F, Pecchioni E, Conticelli S, et al. Analisi archeometrica della ceramica dell'età del bronzo di Coppa Nevigata (FG): alcune implicazioni archeologiche. Scienze dell'Antichità. 1995;8:101–60.
[85] Muntoni IM, Eramo G, Laviano R. Production of Mid-late Neolithic "Serra d'Alto" ware in the bradanic trough (south eastern Italy), In: Vessel: inside and outside, Proceedings of the conference Emac'07, 9th European meeting on ancient ceramics. 2009:53–62.
[86] Fabbri B, Gualtieri S, Shoval S. The presence of calcite in archeological ceramics. J Eur ceramic Soc. 2014;34:1899–911.
[87] Picon M. Quelques observations complémentaires sur les altérations de composition des céramiques au cours du temps: cas de quelques alcalins et alcalino-terreux. ArchéoSciences, revue d'Archéométrie. 1991;15:117–22.
[88] D'Argenio B. Le piattaforme carbonatiche periadriatiche. Una rassegna di problemi nel quadro geodinamico mesozoico dell'area mediterranea. Società Geologica Italiana, Memorie. 1974;13:137–59.
[89] Spalluto L, Pieri P, Ricchetti G. Le facies carbonatiche di piattaforma interna del Promontorio del Gargano: implicazioni paleoambientali e correlazioni con la coeva successione delle Murge (Italia meridionale, Puglia). Bollettino-Societa Geologica Italiana. 2005;124:675.

[90] Morsilli M, Hairabian A, Borgomano J, Nardon S, Adams E, Gartner GB. The Apulia carbonate platform – Gargano promontory, Italy (Upper Jurassic–Eocene). Am Assoc Pet Geol Bull. 2017;101:523–31.
[91] Tropeano M, Sabato L. Response of Plio-Pleistocene mixed bioclastic-lithoclastic temperate-water carbonate systems to forced regressions: the Calcarenite di Gravina Formation, Puglia, SE Italy. Geol Soc London Spec Publ. 2000;172:217–43.
[92] Tropeano M, Spalluto L, Moretti M, Pieri P, Sabato L. Depositi carbonatici infrapleistocenici di tipo foramol in sistemi di scarpata (Salento–italia meridionale). Il Quaternario. 2004;17: 537–46.
[93] Bögli A. Karst hydrology and physical speleology. Berlin: Springer Science and Business Media, 2012.
[94] Trendall AD. Red figure vases of south Italy and Sicily: a handbook. London: Thames-Hudson, 1989.
[95] Trendall AD, Cambitoglou A. The red-figured vases of Apulia; early and middle Apulian I. Oxford: Oxford University Press, 1989.
[96] Trendall AD, Cambitoglou A. The red-figured vases of Apulia; late Apulian II. Oxford: Oxford University Press, 1989.
[97] Mangone A, Giannossa LC, Ciancio A, Laviano R, Traini A. Technological features of apulian red figured pottery. J Archaeol Sci. 2008;35:1533–41.
[98] Mangone A, Caggiani MC, Giannossa LC, Eramo G, Redavid V, Laviano R. Diversified production of red figured pottery in Apulia (Southern Italy) in the late period. J Cult Her. 2013;14:82–8.
[99] Giannossa LC, Laviano R, Mastrorocco F, Giannelli G, Muntoni IM, Mangone A. A pottery jigsaw puzzle: distinguish true and false pieces in two Apulian red figured vases by a poli-technique action plan. Appl Phys A-Mater Sci Process. 2016;122. DOI: 10.1007/s00339-016-9615-8.
[100] Bitetto A, Mangone A, Mininni RM, Giannossa LC. A nonlinear principal component analysis to study archeometric data. J Chemom. 2016;30:405–15.
[101] Giannossa LC, Mininni RM, Laviano R, Mastrorocco F, Caggiani MC, Mangone A. An archaeometric approach to gain knowledge on technology and provenance of Apulian red figured pottery from Taranto. Archaeol Anthropol Sci. 2016;9:1125–35.
[102] Jones RE. Greek and Cypriot pottery: a review of scientific studies. Athens: British School of Athens, 1986.
[103] http://helios.unive.it/-termo/DataBank/Attica/AiginaandMegaride/Athens.htm.
[104] Castoldi M. I vasi a figure rosse del periodo protoapulo e apulo antico: Taranto e le officine ceramiche. In: Sena Chiesa G, Slavazzi F, editors. Ceramiche attiche e magnogreche, collezione Banca Intesa. Milano: Electa, 2006:178–81.
[105] Dell'Aglio A. La forma della città: aree e strutture di produzione artigianale. In: Taranto e il Mediterraneo, Proceedings of the 41st convegno di studi sulla Magna Grecia, 12–16 october 2001. Taranto: Istituto per la storia e l'archeologia della Magna Grecia, 2002:171–93.
[106] Rice PM. Pottery analysis: a sourcebook. Chicago: University of Chicago Press, 2015.
[107] Gonda C, Évéquoz E, Eramo G. Découverte d'une verrerie du XIX e siècle: rebeuvelier (Jura, CH). In: Bélet-Gonda C, Mazimann JP, Richard A, Schifferdecker F, editors. Actes des Premières Journées Archéologiques Frontalières de l'Arc Jurassien Delle (F) – Boncourt (CH) 21–22 octobre 2005., Besançon-Porrentruy: Presses Universitaires de Franche-Comté, Office de la culture et Société jurassienne d'Émulation; 2007.

Irina Crina Anca Sandu, Antono Candeias, Klaas Jan van den Berg, Erika Gohde Sandbakken, Eva Storevik Tveit and Henk van Keulen

15 Multi technique and multiscale approaches to the study of ancient and modern art objects on wooden and canvas support

Abstract: The present chapter will illustrate few examples of multi-technique and multi-scale approaches for objects belonging to past or modern periods in the history of art chronology. Thus, four studies on four different topics are brought to the attention: multi disciplinary and multi-techniques characterization of gilded altar-pieces and sculptures in Portugal between 1500 and 1800 (3 years research project); an interdisciplinary approach to easel paintings from fifteenth and sixteenth century in Portugal (3 years research project); characterization of ketone resins varnishes (2 years research project) and of salts developed on canvas paintings from the Munch Museum collection (past investigation to be continued with an ongong 3 years research project). These studies show how important is to make use of different analytical tools when investigating a specific material, artistic technique or degradation phenomenon. The objects studied in the above mentioned research projects were created on wooden or canvas supports and they present paint, polichromy or gilded layers, and varnish coatings, therefore their structure and composition is complex and requires a complex, interdisciplinary approach. In many cases to the surface materials from a polychrome or painted object, degradation products are present and their identification as well the understanding of their mechanism of formation is possible only we have a complete overview of the materials present in the layered structure of the object.

Keywords: multiscale multitechnqiue investigation, ancient and modern art objects, polychrome and gilding layer

15.1 Introduction

Scientific approaches in cultural heritage studies are in most of the cases focused on case studies, meaning individual or series of art-objects with specific questions around their materials composition, artistic techniques, authenticity and conservative issues, including the elucidation of degradation and deterioration mechanisms and effects of restoration treatments on their long-term preservation.

This article has previously been published in the journal *Physical Sciences Reviews*. Please cite as: Sandu, I. C. A., Candeias, A., van den Berg, K. J., Sandbakken, E., Tveit, E. van Keulen, H. Multi technique and multi-scale approaches to the study of ancient and modern art objects on wooden and canvas support. *Physical Sciences Reviews* [Online] **2019**, 4. DOI: 10.1515/psr-2018-0016.

https://doi.org/10.1515/9783110457537-015

Objects and/or material-based scientific research are thus required to make use of different analytical tecniques in a multiscale and as much as possible noninvasive, nondestructive approach. Neverthless, micro-destructive techniques are still widely used and useful for giving in depth information on multilayered structure of an artifact or sample, execution techniques, conservation condition, past treatments and interventions, aging and degradation/deterioration phenomena etc.

There are cases when conventional technique, noninvasive or microdistructive (optical microscopy [OM], FTIR or Raman spectroscopy, SEM-EDS, XRF spectroscopy, XRD, GC-MS/PyGC-MS, DTMS) need to be complemented by more specific and sophisticated techniques (synchrotron based-techniques; surface analyses such as SIMS, XPS, AFM; SEM-FIB or TEM; SERS imaging and antibody assays; MALDI-TOF-MS, nanoLC-MS etc.) able to amplify our observation and analytical capability up to nano and atomic scale. One sample can thus unveil many misteries hidden in the sequence of layers and materials, imaged or analyzed from a macro until a nano-scale level. One sample might not always be representative of the entire object, but if the objects are previously screened with noninvasive imaging or spectroscopic techniques (such as macroXRF or multi and hyperspectral imaging) it is easier to decide on the relevant sampling areas in order to extract the maximum amount of data from the less amount/ number of samples.

The present chapter will illustrate few examples of such multitechnique and multiscale approaches for objects belonging to past or modern periods in the history of art chronology. Thus 4 studies on four different topics are brought to the attention: multidisciplinary and multitechnique characterization of gilded altarpieces and sculpures in Portugal between 1500 and 1800 (3 years research project); an interdisciplinary approach to easel paintings from fifteenth and sixteenth century in Portugal (3 years research project); characterization of ketone resins varnishes (2 years research project) and of salts developed on canvas paintings from the Munch Museum collection (past investigation to be continued with an ongoing 3 years research project). These studies show how important is to combine different analytical tools when investigating a specific material, artistic technique or degradation phenomenon.

The objects studied in the above mentioned research projects were created on wooden or canvas supports and they present paint, polichromy or gilded layers, and varnish coatings, therefore their structure and composition is complex and requires a complex, interdisciplinary approach. In many cases to the surface materials from a polychrome or painted object, degradation products are present and their identification as well the understanding of their mechanism of formation is possible only we have a complete overview of the materials present in the layered structure of the object.

Even when a specific type of materials is characterized (such as synthetic varnishes, surface efflorescences or gilded layers made of precious metal alloys), the study has to take into account the whole underlaying structure and its behavior in relationship with it. To this respect, interfaces between layers and different materials represent an important source of information on the behavior of specific materials in complex structures. The discontinuities in these interfaces can reveal characteristic markers and parameters for the study of degradation and deterioration phenomena.

15.2 Materials, techniques and imitation of gilding in polychrome decorations from Portugal

Irina Sandu

Between 2012 and 2015 an inter- and multidisciplinary research project was developed in Portugal under the title "*Gilt-Teller: an interdisciplinary multi-scale study of gilding techniques and materials in Portugal, 1500–1800*" (http://www.gilt-teller.uevora.pt) and funded by the Portuguese Foundation for Science and Technology (FCT) [1].

It was the first time that "talha dourada" (gilded carved wood) decorations all over the Portuguese territory, from churches and museums, retables and other wooden and gilded decorations, were systematically investigated from the material and technical points of view. Due to the limited amount of time versus the huge amount of altarpieces and other woodcarving typologies of artefacts (ecclesiastic furniture, sculptures etc.) found all over Portugal, the project focused on some geographic areas (Algarve, Portalegre, Santarém, Viseu, Evora, Guarda, Lisbon and surroundings, Miranda de Douro, Aveiro, Coimbra, and Funchal in Madeira island) and on three typologies of objects: *main, lateral and collateral altarpieces* from churches and chapels, *gilded choir stalls* and *polychrome sculptures* with gilded decoration ("estofado"/"esgrafito" techniques) [2].

The criteria of selection of the case studies were: aesthetical, historical and cultural relevance of the case studies in close relationship with: presence or not – better if not, of past interventions; presence of documents, either of legal contracting or the creation moment and authorship; prediction of future treatments – sampling campaign to be organized, where possible, before the restoration intervention or during work campaign. Examples of the three typologies of artefacts chosen are given in the Figure 15.1.

In most of the cases, sampling was performed in the front part of the altarpieces (known as "predela" in Portuguese), from columns and decorative motives presenting also polychromy. Besides the gilding layers, the sampling areas considered also flesh tones and garments from figurative motives or sculptures integrated into the altarpiece structure (including the religious scenes from the tabernacle's door), imitation of precious stone (such as "marmoreado" technique) and zoomorphic or

Talha dourada - main altarpiece and side walls
Vale de Figueira Church

Main altarpiece of
Miranda do Douro Cathedral

Polychrome gilded
sculpture of St. Joseph
Museum of Aveiro

Figure 15.1: Examples for the main typologies of gilded artifacts involved into study.

vegetal decorations. The sampling criteria took also into account the lack of previous interventions of restoration (where possible) and possible presence of historical documents attesting the data and authorship of the carving and gilding work [2–5].

Around 604 samples were collected from 84 objects with gilding decorations (67 gilded altarpieces, sculptures, wooden choir stalls, and two pulpits in Portugal and 16 altarpieces, polychrome sculptures, icons and other decorations with gilding layers from abroad: Spain, Czech Republic, Romania, Belgium, Germany and Italia-Bulgaria).

An integrated protocol of investigation of gilded composites [2], following different steps aimed to describe a multi-scale (from macro- to nano-scale) and multi-technique approach: sampling documented by digital photography in situ; registration of the samples in a database according the chosen ACRONYM of the sampled altarpiece/object and documentation of each sample under stereomicroscope (SM); basic stratigraphic analysis on cross-sections using OM under visible and fluorescent light; surface characterization of selected samples using AFM and eventually SEM-EDS for elemental mapping; micro-structural and phase-characterization using micro-computerized Tomography (microCT) and X-ray Diffractometry (XRD); characterization of organic and inorganic phases on micro-fragments using microFTIR, FTIR, SEM-EDS, immuno-enzymatic assays (ELISA), MALDI-TOF-MS, Py-GC/MS and LC-MS; other analyses on cross-sections (staining tests, microRaman, microFTIR imaging, XRF, MALDI-TOF-MS) for identification and mapping of organic and inorganic materials.

Thus the main questions to be answered by the integrated protocol were:
- which are the sequence and characteristics of the layers in the stratigraphy of gilded and polychrome samples from the sampled objects;
- is it possible to correlate the structure and characteristics of the inner architecture of the polychrome samples to the surface characterization/assessment;
- what kind of ground and paint materials and if mixtures are present (for mineral/inorganic phases);
- what kind of leaf/gilding material or layer, if it is original or later applied;
- what is the distribution of proteinaceous materials in the paint layers (binders mainly);
- if a tempera technique was used for polychromy what kind ("tempera grassa" or just egg tempera or casein tempera) and in which samples;
- are there any other organic components as coating materials (varnishes, glazes);
- which are the original layers/materials and which the later added?

Within the research activities a special attention was paid to the creation and application of a noninvasive multi-scale and multi-technique methodology for imaging and analyzing surfaces and interfaces in gilded artifacts, mainly using AFM, microCT, SEM-EDX and OM [6–9].

The stereomicroscopy (SM), the optical microscopy (OM, Vis & UV) and the scanning electron microscopy (SEM) allowed to image the different gilded areas and interfaces between layers from macro to micro scale, both in visible (Vis) and UV light (OM) and as grayscale images taken under bse (SEM). The combined use of these techniques on surfaces give a whole range of parameters (color, fluorescence answer, surface morphology, granulometry, roughness profile, patterns of degradation/deterioration such as cracks, lacunae, fissures, discontinuity in the varnish layers etc.) to be measured and eventually evaluated before, during and after aging ing of materials.

The Atomic Force Microscopy (AFM) in particular is a surface technique used to perform 3D imaging of the nano/micro areas of the sample while measuring the roughness values (RMS - Root Mean Square). Advantages of the use of AFM consists of measurements executed in nondestructive (non contact) scanning mode and directly on sample, not being limited to conductive sample such as the traditional SEM [8, 9]. Topography and frequency modulation scans allowed to retrieve the elastic modules of the surface material. The elastic modulus was then resolved into the phase diagram. Phase diagram allowed qualitative analysis of the surface. This method is close, but yet not accurate, to the EDS analysis. The AFM allows statistical studies of the sample surface contamination, topological structure and material distribution. The phase diagram is an unique mark to that material and also to the material phase. AFM data interpretation had to be done in correlation with the observation of the sample and the results of other analytical techniques.

The use of AFM on few samples taken from altarpieces and sculptures demonstrated how important is to analyze and compare features of a given surface at submicron scale as additional investigation to other surface techniques. Thus, the understanding of certain technical aspects was possible, such as: application technique and finishing of the metal leaf (matte or burnished gold); conservation state – presence of dirt, protection layer, losses and lack of adherence, biological contamination.

The AFM results showed that roughness value changes are related to mechanical damage and/or to the chemical etching or abrasion of the surface. For example, in the case of the gold leaf, it was possible to understand the application procedure and the composition of the gold (e. g. pure or mixed with other elements) just comparing the roughness profiles for different treatments. In other words it was possible to retrieve surface information about the sample and also subsurface information. For a sample taken from the eighteenth century St Francis sculpture from St Domingos church in Lisbon the following values were retrieved: the phase angle for the Au leaf to be ~130° and the phase angle for the bole and the ink ~58° (Figure 15.2).

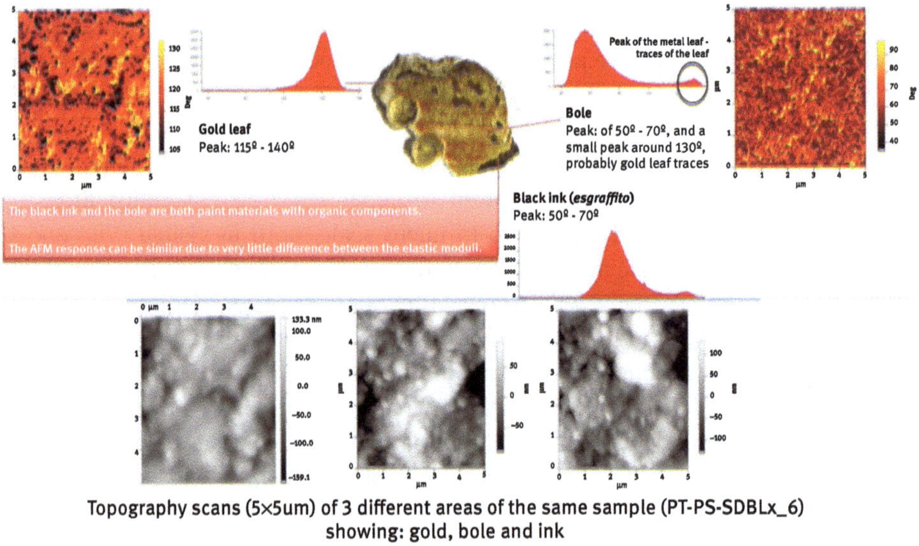

Figure 15.2: AFM results on a sample taken from a "estofado" punched decoration (PT-PS-SDBLx_6).

Using the μCT and SEM techniques it was possible to study and visualize in a dynamic stratigraphy (video) the microstructure/architecture of the gilded/polychrome samples: presence and shape of the pores in the ground, granulometry and discontinuities between layers [6, 7]. The conservation state of the polychrome gilded layers (cracks, transversal fissures, chromatic alterations, etc) can also be evaluated on micro-fragments and on cross-sections obtained from samples. Correlating the XRD results from ground layers with the μCT data it was possible to characterize and distinguish between layers of "gesso grosso" and "gesso fino" (Figure 15.3).

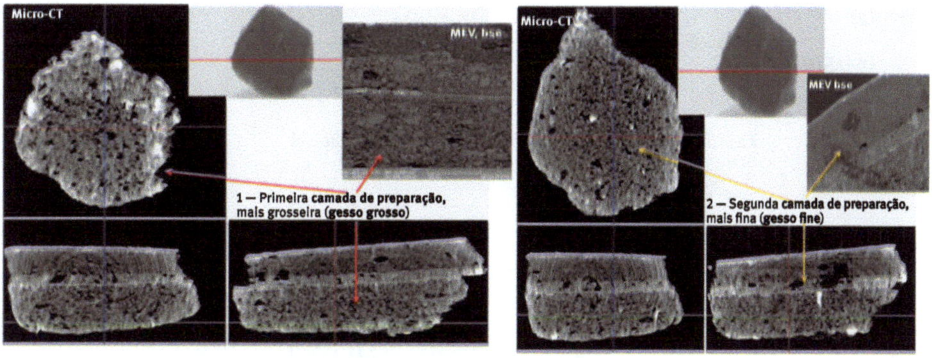

Figure 15.3: microCT investigation of a sample from the main altarpiece of Bucelas church, eighteenth century.

The shape/size/orientation of pores (spherical or irregular) or mineral grains, and other physical parameters can be also correlated to the degree of craftsmanship and the nature of the applied materials. The simple analysis of the X-ray radiographs can provide very useful information concerning the distinct elements composing the samples, including the wooden support. Unfortunately, in the reconstructed slices, of combined wood/polychromy composition, the wood became "transparent" due the extreme contrast of phases (Ca and metal rich layers, such as Au, Cu, Fe, Pb, vs. organic materials). On other hand, the separated tomographic study of the wood supports constitutes a powerful tool for the recognition of the species and for the characterization of conservation pathologies and eventually effects of restoration interventions.

Another innovative trend proposed by the project was the development and testing of a methodology for identifying and mapping organic binders (mainly proteinaceous) directly on cross-sections, using OM, staining tests (visible and fluorescent), antibody-assays, MALDI-TOF-MS and nanoLC-ESI-Q-TOF mass spectrometry [10–12]. Thus, several samples from sculptures and altarpieces were analyzed using complementary protocols for proteinaceous binders characterization (ELISA, Sypro Ruby, MALDI-TOF-MS and nanoLC-ESI-Q-TOF-MS) [10]. This was the first testing of four analytical methods used together for simultaneous identification and mapping of proteinaceous binders in real samples (fragments and also cross-sections, Figure 15.4).

From the point of view of stratigraphic components, the analyzed altarpieces generally display common characteristics to water gilding composites, with exception of interventions using mordant gilding technique or imitations of gold leaf with silver leaf or "purpurins".

The stratigraphic structure of the gilding layers is quite similar: thick ground with "gesso grosso" and "gesso matte (fino)" differentiation, yellow/reddish layers of bole, metallic alloy leaf, varnish or particulate deposits or residues of the wax-resin treatments or varnishes [3–5].

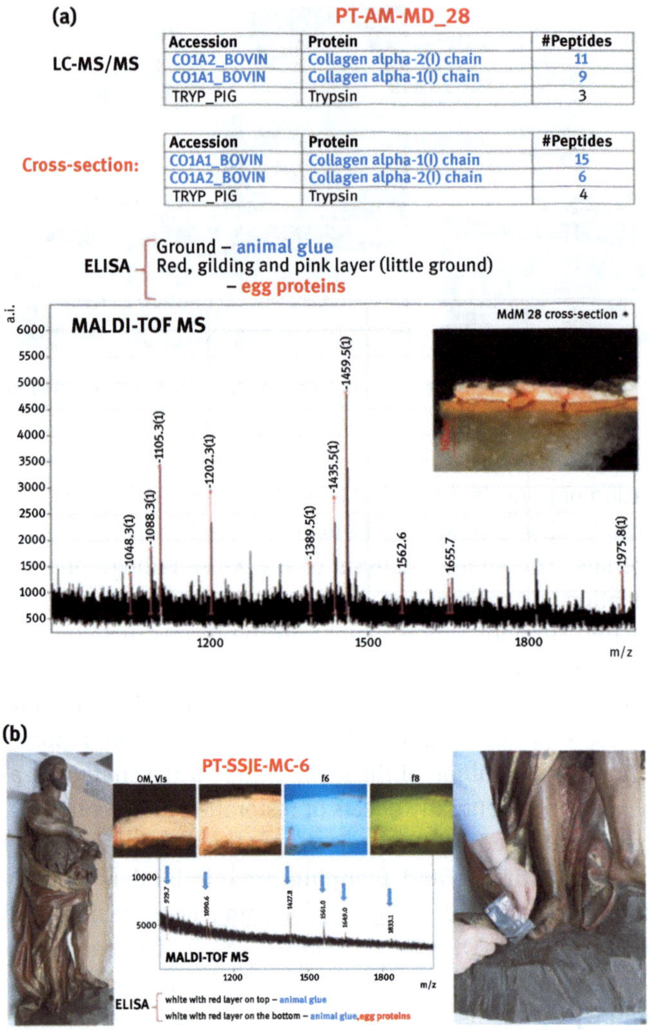

Figure 15.4: Identification of proteinaceous binders on samples/cross-sections from: (a) Miranda de Douro Main altarpiece (PT-AM-MD_24) and (b) a sculpture by Machado de Castro (PT-SSJE-MC_6).

The overall composition of ground and bole layers is based on mineral materials: gypsum and anhydrite for gesso grounds with inclusions of silicates and quartz; in few cases calcite was identified in mixture with the gesso; clay minerals (kaolinite, illite, muscovite) and (Fe, Ti)-oxides (hematite and goethite) in the bole. Celestite was also detected (Sr identified by SEM-EDS) in grounds from different altarpieces and this points toward a Portuguese provenance of inert materials used in the ground layers. In two cases (main altarpiece from Bucelas and Mourning sculpture from Czech Republic) fossils remnants (coccoliths) were found in the ground layers of the upper composite

(Figure 15.5). For the first case, the higher concentration of Ca in the upper layer of gesso would possibly indicate a mixture of Ca-based materials (calcite and Ca sulphate). The bole layers were of yellow or red/orange color and sometimes two different layers of different color (yellow and red) could be observed.

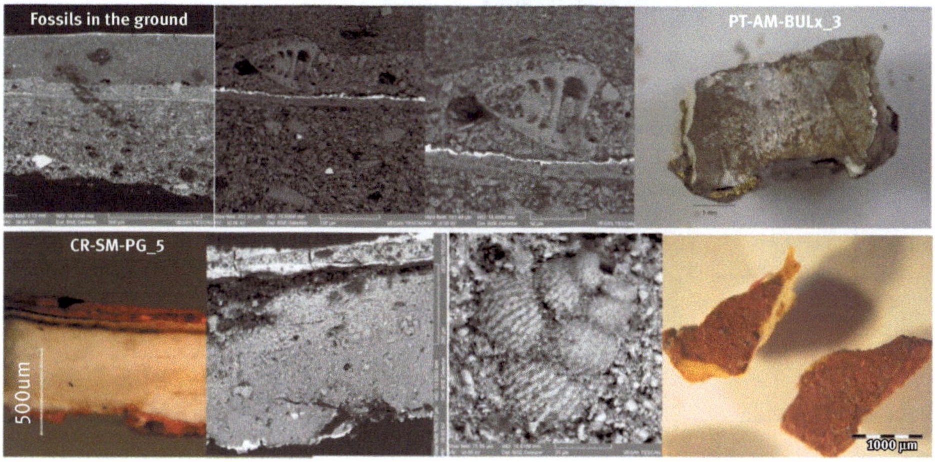

Figure 15.5: Coccoliths in the ground of two samples.

As far as the composition of the leaf is concerned, the major part of elemental SEM-EDS analyses detected the presence of a leaf made of ternary or binary alloys: Au/Ag/Cu or Au/Cu/Ag; Au /Ag. The purity of the leaf is very variable, but in most of the cases a high quality (between 22 and 23.75 carat) can be detected (just few cases of lower % of Au were encountered, such as for the lateral altarpiece in the Tancos church, PT-AL-Ta and for the gilded carved wood of the chapel of Albertas, MNAA: leaf of 20–21 carat). The thickness of the leaf is an important issue to be discussed but the tools used to measure it (OM and SEM) did not make possible an accurate evaluation of it. Nevertheless, in some comparative studies, such as the one between the gilded carved wood in the Church of Albertas (MNAA) and the Vale de Figueira (VF) altarpiece, the thickness of the leaf is slightly variable (5 µm for MNAA and 10 µm for VF, respectively) and this can point on a different quality of the gilding layer application between the church in Lisbon and the rural one or better skills to laminate the gold leaf (in this respect economic issues can be considered as reason for the use of a thicker leaf for the second case against the thinner leaf in the first case). The leaf is made of an alloy (Au/Ag/Cu) of 23 carat for VF samples and of 20÷22 carats for MNAA samples, confirming data already reported in the literature about "gilding" materials on the territory of Portugal (Figure 15.6). Usually the average value obtained with measurements made on cross-sections by OM is around 1–3 µm but this can be due to the folding of the leaf leading to errors in the evaluation of its thickness.

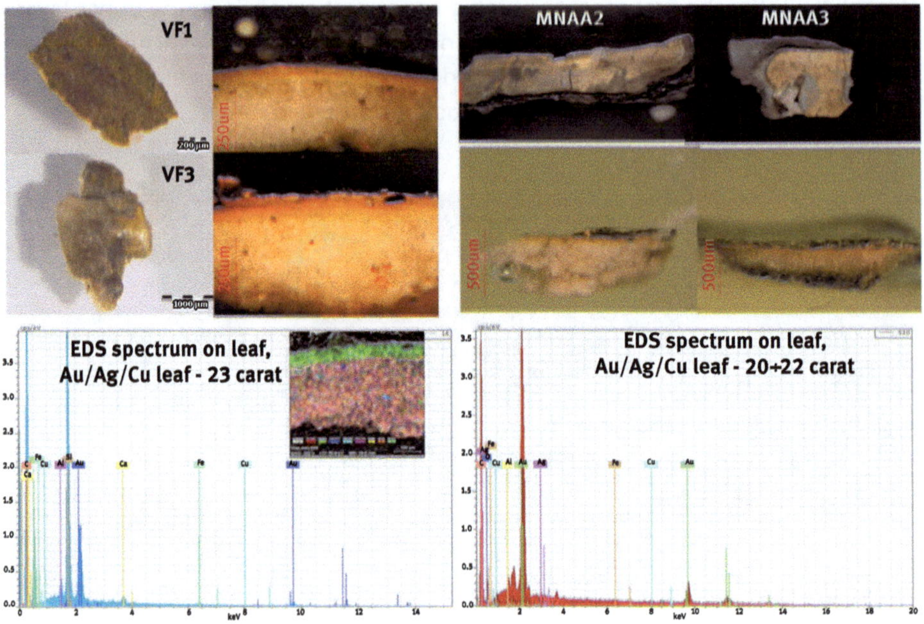

Figure 15.6: SEM-EDS compositional analyses on cross-sections of samples from MNAA and VF altarpieces for the identification of the leaf.

The use of highly gold-content alloys (more than 87% Au) and of alloys made of Au/Ag/Cu was already reported from an extensive and comparative SEM-EDX research performed on several Portuguese polychrome reliquaries from Alcobaça, Açores, Coimbra, Aveiro, Tibães (sixteenth century) and in other studies on retables from Coimbra (Cathedral's main altarpiece, dated from 1502 to 1685) and Porto, dated from seventeenth-eighteenth centuries [13, 14]. In these cases, 22–23 carat gold was reported (applied as water or mordant based techniques, the stratigraphy being characterized by the presence of ground, red bole layer and gold leaf). Our results confirm the extensive use of such alloys during seventeenth and eighteenth centuries up to nineteenth century in Portugal.

As substitute of the gold leaf the presence of Cu/Zn alloys (regilding areas) was identified in several cases. Silver (Ag) leaf covered with shellac varnish of yellow-orange fluorescence was used also in some cases as an imitation of the gold leaf and as regilding (silvering) layers over original ones (Figure 15.7).

Polychromy of different color (white for flesh tones, blue or green for garments, red for decorations alternating gold leaf and pigments or dyes/lake) can be found on Tabernacle's doors, flesh tones or garments of biblical figures and sculptures, on columns and imitation of precious stones (marble-like decorations).

The main pigments identified in garments, fleshes and decorative elements (fruits, birds, vegetal motives) were: lead white (cerussite /hydrocerussite) sometimes mixed

(a)

(b)

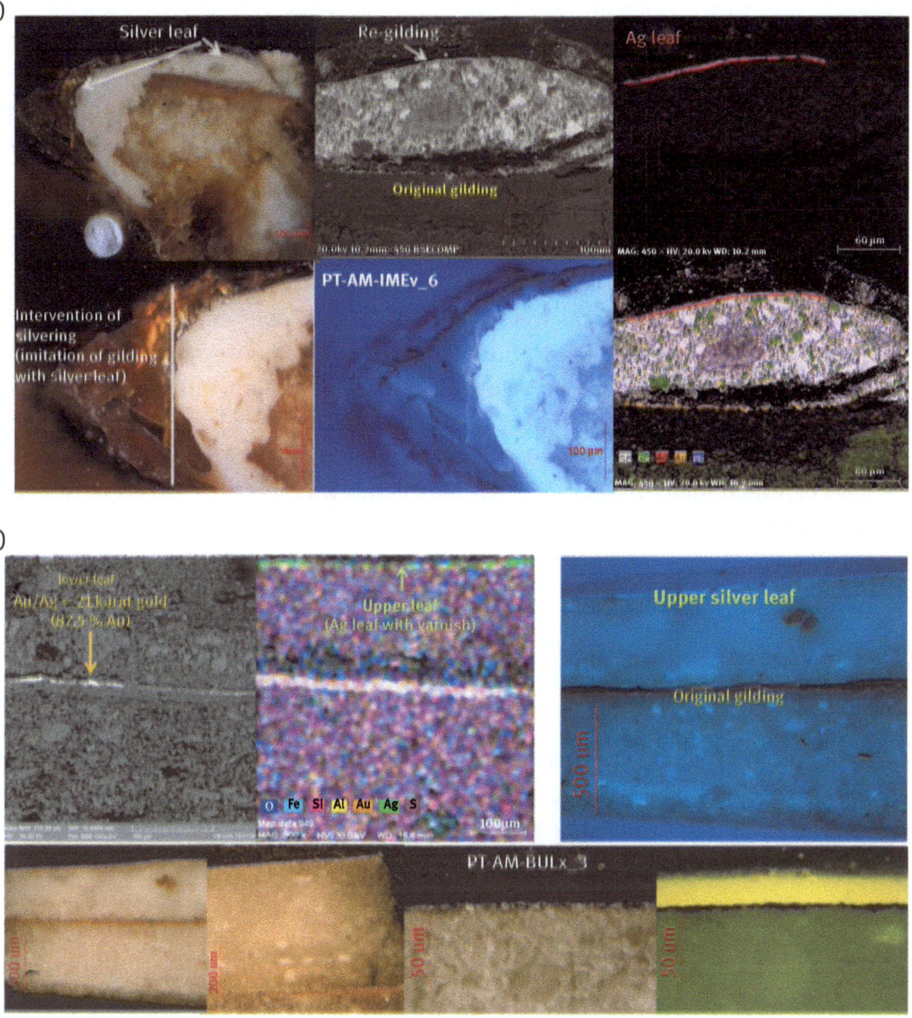

Figure 15.7: Two phases of gilding identified in two samples, the later being made with Ag leaf and varnish.

with calcite, anglesite and in few cases barite (the later could indicate a posterior intervention of polychromy), red and yellow ochres, cinnabar and red lead (minium) grains inside the white layers; goethite for yellow layers; azurite, copper resinate and sodalite for blue and green tones; carbon black etc.

More recent pigments, indicating repainting interventions, are: barite and chromium green found in the altarpieces of St Domingos in Benfica, Baraçal, Pinhel, Tancos, Prussian blue as substitute for lapislazuli in the "marmoreado" decoration of the columns in the case of St Francis altarpiece in Viseu. For the altarpiece of Cidadela Palace in Cascais, residues of an waxy protective material (with approximately 50 µm thickness and with no specific fluorescence answer to the protein staining with Sypro Ruby) with fissures and cracks along its thickness were observed in cross-section and the SEM mapping in this layer detected the presence of Zn, probably a zinc oxide contamination from a posterior intervention on the altarpiece. Polychrome layers belonging to more recent overpainting (containing barite and zinc white) were also found in one lateral altarpiece in Santárem.

The organic materials mainly identified and mapped in the stratigraphy are the animal glues in the ground and bole layers, wax and shellac as coating layers, egg tempera for "estofado"/"esgrafito" decorations. Interventions with wax-resin as surface consolidation material were detected in few cases such as the gilding in the Church of Albertas from the National Museum of Ancient Art. Oil was identified in few cases, in the silicate layers below the leaf (Tancos main altarpiece) and in the polychromy of green and red areas ("regraxos") in the two altarpieces from Evora. For these later cases (altarpieces from churches of Misericórdia and Cartuxa in Evora) cochineal lake of Mexican provenance and wax were identified in red areas (Figure 15.8).

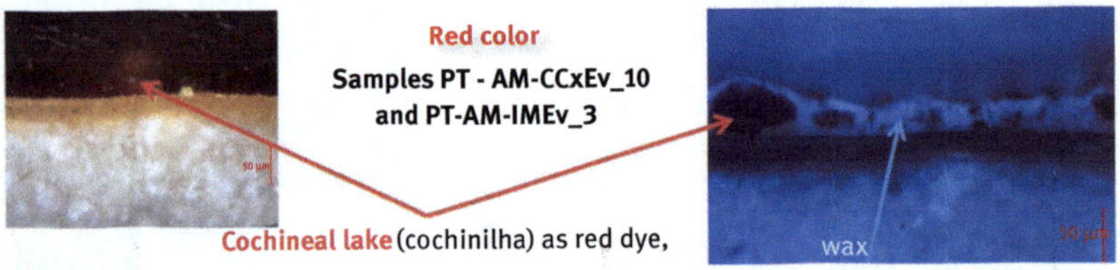

Tempo de retenção DAD (min.)	LC/DAD data (nm)	LC/MS data (m/z)	Identificação	Observações
14.55	238, 269, 490	491, 447, 357	Ácido carminico	Cochinilha. Como não foram detectados os compostos minoritários das cochinilhas polaca e arménia, poderá tratar-se de cochinilha mexicana.

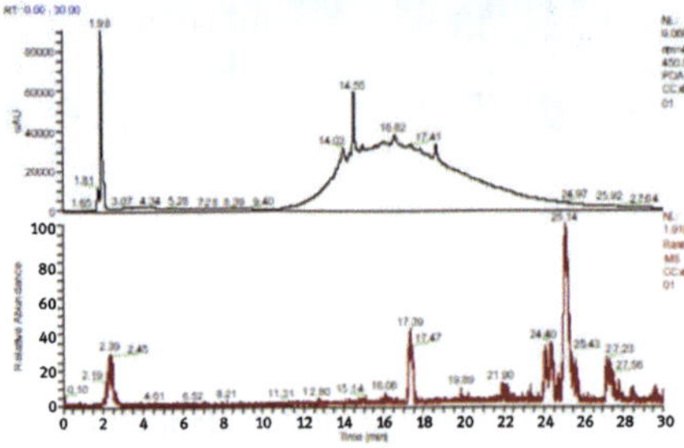

Figure 15.8: Cochineal lake and wax identified in two samples from eighteenth century altarpieces in Evora.

The absence of organic materials as protective or aesthetic coatings over the leaf indicates that cleaning or other restoration interventions were performed. In some cases, such as the three altarpieces in Santarem area, these interventions were documented while for others (e. g. Bucelas altarpiece) the visual assessment of conservation state of the altarpiece decoration and the interview with the parish of the church confirmed them.

15.3 Panel paintings from fifteenth and sixteenth centuries in Portugal

Antonio Candeias

Since 2009, the Portuguese Laboratory for Conservation and Restoraton (Laboratorio Jose de Figueiredo) of the General Directorate for Cultural Heritage established strategic partnerships with different universities, in special with HERCULES Lab from Evora University, to develop integrated studies on easel paintings of the Portuguese arts golden period: the fifteenth and sixteenth centuries [15–22]. Thus, two major projects were funded by the Portuguese FCT: the project *ON-FINARTS – On the Flemish Importance for the National Arts in the sixteenth century* (PTDC/EAT-HAT/115692/2009) and the project *The invisible preparation layer and its influence on Portuguese painting of the XV and XVI centuries: an issue to be solved* (PTDC/EAT-HAT/100868/2008).

At the beginning of the sixteenth century, the styles and techniques of painting in vogue in Portugal were strongly influenced by the new artistic movements from Flanders, conveyed through the intense commercial relations established between the two countries. The materials were essentially imported from Bruges and Antwerp, while Portuguese and Flemish masters alternated between the two countries, introducing Flemish style and technique in Portugal [20]. The study of the work of these painters is a fundamental aspect for the understanding of the profound artistic changes that took place in Portugal during the reigns of D. Manuel and D. João III. The perception of the contours of these changes required that what is specifically Flemish was kept in the activity of these masters, while working in Portugal, and what is due to the Portuguese way of painting, and to what extent was its contribution to the subsequent transformation of the national artistic milieu.

The integrative approach of the ONFINARTS project combined existing knowledge of historical documental sources with the results of a wide and comprehensive set of imaging and surface analysis and material characterization techniques [16, 17, 23].The analytical and surface characterization of the works of these masters are particularly important in view of the scarce existing documental information on the Flemish Primitives, to which must be added the systematic absence of signed or dated pieces, leading to the attribution of a work to a painter and his dating usually based on stylistic issues. Thus, the (re)definition of the corpus of works by Luso-Flemish masters and their workshops, as well as the relations between these workshops and other art production centers in Portugal and in Europe, was a major objective of this project together with the later project that allowed to develop and apply a methodology for the thorough analysis and semi-quantification of the components of painting preparations from the fifteenth and sixteenth centuries.

The study of the panel paintings started in most cases by the visual examination of the paintings and screening by imaging techniques, which are non-destructive and noninvasive in nature, and consist on the differential response of the painting materials and supports when subjected to radiations with different energy, from infrared to X-rays. These encompass standard light photography, U.V. fluorescence photography (UVF), infrared reflectography (IRR) and X-ray radiography. Figure 15.9 depicts the use of these techniques in two paintings from the Funchal altarpiece. UVF photography was particularly relevant to study discontinuities in the paintings surfaces including repaints, cleaned areas, dirt, layers of varnish and some pigments while IRR allowed the visualization of the underdrawing which is characteristic of each Master and Workshop and a fundamental piece of information for the art historians. On the other hand, radiography takes advantage of the penetrating power of X-rays, revealing the internal structure of the paintings including assemblies and information about its state of conservation (eg insect galleries), study of the polychrome layers with identification of areas denser to X-rays, effects of previous treatments and discovery of dissimulated details or underpaintings.

Figure 15.9: Details of the images obtained by the different imaging techniques on two paintings of the Funchal altarpiece (top) Last Supper (bottom) Saint Gregory's Mass: (a) visible photography, (b) UV fluorescence photography (c) IR reflectography and d) X-ray radiography.

After identification of the wood of the different boards of the panels it was possible in the case of oak wood, and whenever relevant, to date the supports by dendrochronology.

In addition to the area examinations (imaging techniques), noninvasive spot analyzes were performed in different areas of the paintings and aimed at a first approximation to the material composition of the paintings (Figure 15.10). In the methodology adopted in the study of fifteenth and sixteenth centruy Portuguese paintings the techniques for *in situ* analysis were:

- energy dispersion X-ray fluorescence spectrometry (EDXRF) that allowed elemental analysis of chromatic layers;
- the colorimetry / visible absorption spectroscopy that allowed to obtain color parameters and the reflection / absorption spectra of the paintings areas in the visible region;
- Fourier Transform Infrared Spectroscopy (FTIR) in total reflection mode, which allowed to obtain information about superficial organic layers.
- portable Raman spectrometry (p-Raman), which allowed the identification of some pigments of the chromatic layers.

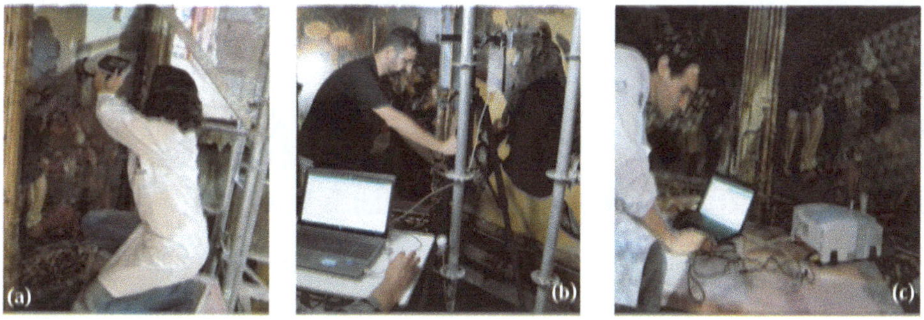

Figure 15.10: Snapshots of in-situ analyses performed during the Funchal altarpiece study. (a) colorimetry, (b) EDRXF, (c) p-FTIR.

Regarding EDXRF, it is necessary to take into account that, given the penetrating power of X radiation, the elemental composition obtained is not only the surface layer but also the underlying layers [24, 25].

With the aim to develop a detailed material and technical study of the paintings, microsampling was considered taking into account the objectives of the study, the minimum sample size for analysis purposes and the fact that the samples should only be removed if the type of information that was intended was only possible by study of micro-samples. The rationalization and definition of the sampling strategy was defined by the combined interpretation of the information obtained by the area examinations and point analyzes with that of the on-site examinations. This micro-sampling was always carried out under the supervision of conservators. Figure 15.11 shows the micro-sampling record for one of the paintings and the details of a cross-section, where the sample size can be observed.

Figure 15.11: (a) Registration of micro-sampling of one painting; (b) indication of the sampling points; (c) cross-section of the microsample under optical microscopy.

In general, the micro-samples collected, using a sampling pin or stylus, were divided into two subsamples. One of the subsamples was embeded in epoxy resin in transversal position and later polished to allow the stratigraphic analysis of the various layers by OM (with visible light and UV) including number and thickness of the chromatic layers, optical properties, morphology and color of the particles, and the subsequent analysis by different techniques of microanalysis, namely:

- scanning electron microscopy coupled to energy dispersion X-ray spectrometry (SEM-EDS) for microstructural analysis and semi-quantitative, punctual and bi-dimensional elemental analysis which allowed the identification of pigments and fillers and their distribution in the various layers
- Raman microspectrometry for confirmation of the identification of inorganic phases (pigments and fillers).

Figure 15.12 shows the results by OM, SEM-EDS and micro-Raman of green layers in one painting by Flemish-portuguese painter Frei Carlos. The combined analysis of the data allowed the identification of the unusual pigment brochantite which was found to be characteristic of the palette of this Master and has not been found in other painters active in Portugal during the same period [26, 27].

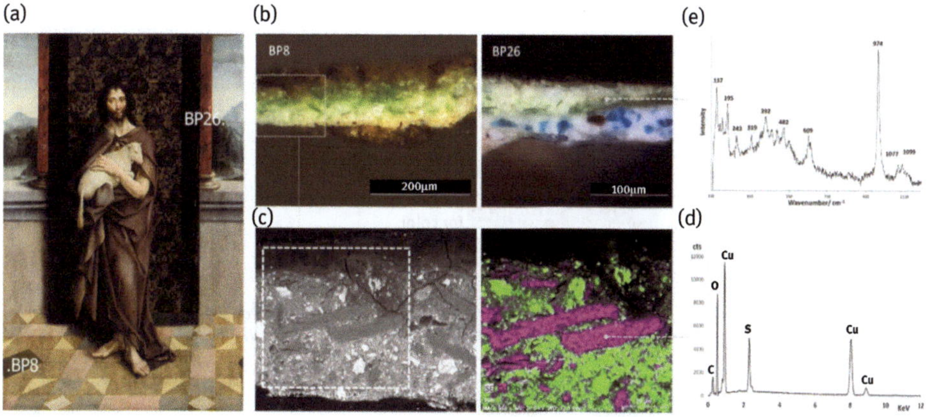

Figure 15.12: (a) Frei Carlos painting "Good Shepherd" by light photography showing the sampling areas for BP8 and BP26; (b) cross-sections under optical microscope for both samples; (c) SEM micrograph of sample BP8 and EDS combined map; (d) EDS analysis of a green particle; (e) Raman spectrum of a green particle showing presence of brochantite. Adapted from [26].

The results obtained from the mounted cross-sections by these microanalytical techniques were complemented by those obtained by other techniques of microanalysis and high resolution chemical analysis in the various layers of the unassembled sub-sample, after separation of the several layers, namely:

- micro-FTIR, which allowed qualitative analysis of organic compounds (binders, varnishes) and some inorganic compounds (including some pigments and fillers),
- micro-X-ray diffraction to confirm the identification of crystalline compounds (pigments and fillers),
- gas chromatography with mass spectrometry preceded by pyrolysis to confirm the identification of varnishes and binders,
- liquid chromatography with mass spectrometry for the identification of dyes from the layers of varnish and lake pigments in chromatic layers.

The multianalytical methodology used in the study of these Masters allowed a deep material and technical characterization of the paintings and painting materials of the different workshops active in that period and allowed a new insight on the way these artists worked and collaborated.

As an example, Figure 15.13 depicts the results by Py-GC/MS that allowed to distinguish, among the works initially attributed to Flemish-portuguese painter Frei Carlos, the work of a follower. The Py-GC/MS showed the used of a different binder from the ones used by Frei Carlos, while the other techniques showed the use of different pigments and also technique of colour modelling. These were fundamental to revise the paintings attributions [16].

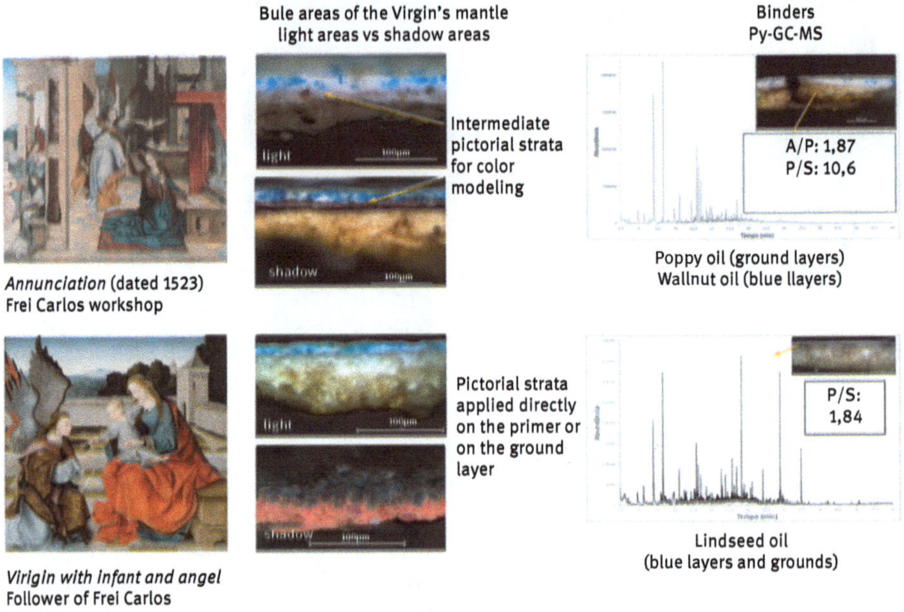

Figure 15.13: (a) Cross-sections of blue areas by optical microscopy from two paintings initially attributed to Frei Carlos workshop; (b) Py-GC/MS results. Adapted from [16].

Particularly significant in the characterization of the technical options of the different workshops were the preparation (ground) layers of the paintings [18, 27–40].

In a first approach to the study of preparation layers, a methodology was adopted combining elemental analysis by SEM-EDS with micro-FTIR and micro-XRD analysis (Figures 15.14, 15.15). By SEM-EDS it was possible to obtain the elemental composition and morphology of the particles and their distribution, differentiating the various constituents, such as calcium sulphates, calcium carbonates, silicates, etc. The X-ray micro-diffraction, although it did not have the spatial resolution of SEM-EDS, allowed to identify the types of crystalline compounds and thus to distinguish the constituents in gypsum, anhydrite, calcite, quartz, complementing the analysis obtained by the first one.

The micro-FTIR analysis allowed to obtain molecular information on the binders and to confirm the information of the previous techniques on the inorganic

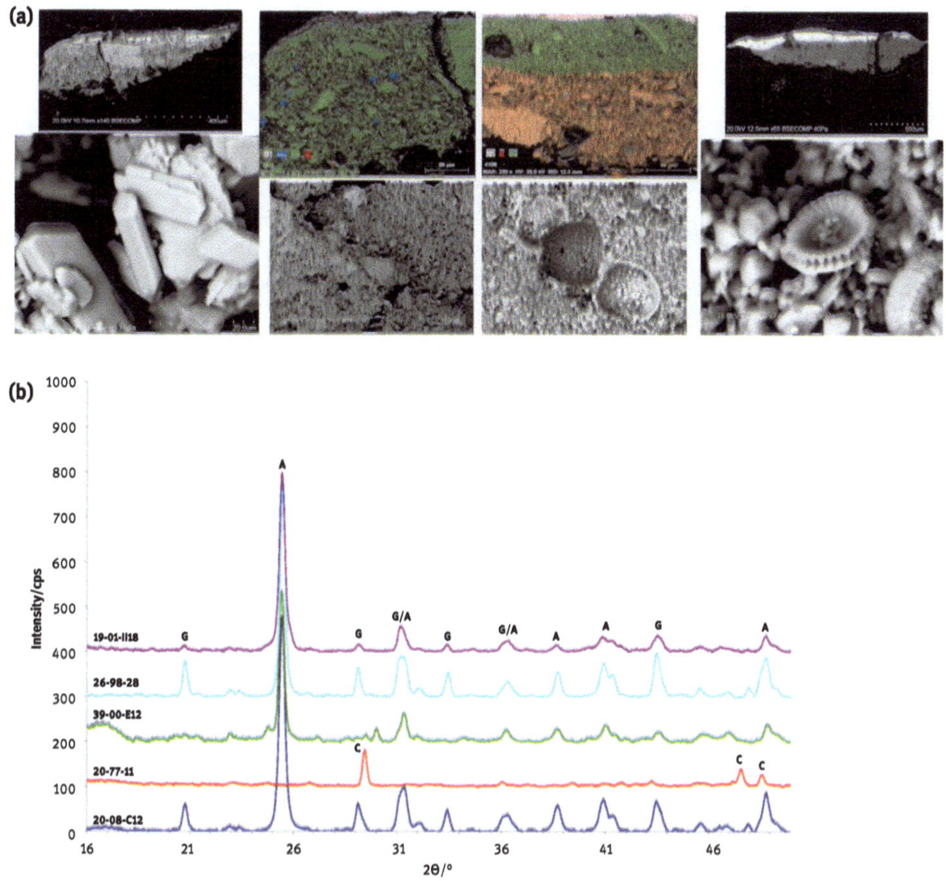

Figure 15.14: (a) Micromorphology and elemental composition of ground layers by SEM-EDS – left: gypsum ground (Ca, S) and right: chalk ground (Mg, Ca, Fe); (b) Micro-X-ray diffraction patterns of different paintings analysed. gypsum (G), anhydrite (A), and calcite (C); Adapted from [29].

components. However, from an analytical point of view, micro-FTIR has limitations in terms of the distinction between gypsum and mixed anhydrite because the hydroxyl bands of the hydrated compound hide its low intensity or possibly nonexistence in the anhydrous compound. Throughout this investigation [22, 41], it was found that, in addition to the hydroxyl bands, a deviation in the S-O stretching and bending bands of standards and original samples appeared to be a strong indication of the presence of anhydrite and thus be linked to the degree of hydration of the material. To further investigate this issue that is not mentioned in the scientific literature on artistic goods, it was decided to investigate the relevance of these bands in relation to the degree of hydration of the material based on investigations on the structure and behavior of water molecules in calcium sulfate. Reconstitutions of formulations with calcium sulfate with varying degrees of hydration, were agglutinated in a solution of

Figure 15.15: (a) Secondary electrons image of a laboratory test sample showing the sizing layer; (b) secondary electrons image of the sizing layer of a sixteenth century Triptych.

animal glue. Their analysis by micro-FTIR and XRD and their comparison with the analytical results of the original preparations allowed to highlight in micro-FTIR spectra the importance of δS-O bands in determining the degree of hydration of this material. It has been found that the deviation of the S-O bending bands from the original samples to wave numbers greater than or equal to 673 cm^{-1} should be taken as indicative of a high proportion of anhydrite in the preparations irrespective of the intensity of the hydroxyl bands. The results obtained thus contributed to overcome some of the analytical limitations of μ-FTIR in the differentiation between mixtures of calcium sulfate with various degrees of hydration, which is particularly useful in routine analyzes when it is difficult to access XRD.

It was thus possible to identify and contextualize the paintings preparation layers of the different paintings and to study their specificities and technique of production. Accelerated aging tests of preparation layers test samples applied with materials of regional origin, contributed to a new insight at the relationship between the artists, their criteria and the material resources used in the creation of works, trying to define the existence of identities between workshops.

Relating the type of substrates to the preparations of the paintings studied, it was verified that both the Lisbon workshop and the Évora workshop mainly use calcium sulphate preparations on oak wood support, preferably imported from the Baltic region, and the workshops in Coimbra and Viseu, when using this preparation, they do it mainly on chestnut wood support, a local resource. The analysis of the relationship between the dimension of the boards and the thickness of the preparations highlighted the increase of the thickness of the preparations with the increase of the area to be painted in the case of the paintings with chestnut support, with the inverse relation in the case of the paintings on oak support. The comparison between the types of support and the types of preparation allowed to verify that both the use of *gesso matte* (thinner calcium sulphate, more gypsum than anhydrite) and that of *gesso grosso* (coarse calcium sulphate, more anhydrite than gypsum) in the paintings of this period is variable, corresponding to the selection of this material also to a workshop tradition. The use of *gesso grosso*, typically occurs in the Lisbon workshop, from the end of the fifteenth century to the end of the sixteenth century, is also verifiable in the Coimbra workshop, being applied on boards of different types of support, encompassing variable dimensions. The use of *gesso matte* is mostly found in the Viseu and Coimbra workshops, and rarely in the Lisbon workshop, and is usually applied on smaller supports, both chestnut and oak. Considering the above aspects, the use of calcium sulphate, both on oak pannels (predominantly from the center to the south of the country) and on chestnut supports (mainly from the center to the north of the country), can be considered as a typical national formulation. Thus, the use of calcium carbonate on chestnut support can also be considered as a national, rather than a regional specificity of the Viseu workshop.

The study showed also that a glue sizing appears in most of the paintings and workshops studied. The stability of the animal glue, besides an insulating function, offers adhesion between the movements of the support and the preparation. The layers of glue are of thin thickness, almost invisible to the optical microscope, characterized by a transparent and bright film. The morphological characterization of this layer was carried out by SEM (Figure 15), directly in samples without previous manipulation and inclined to about 45° [34, 37]. Further analysis by μ-FTIR of the layers observed by SEM-EDS showed the absorption bands characteristic of the animal glue. This protein material is readily identified by μ-FTIR taking into account the position of its absorption bands in the infrared. Several bands were identified: 2930 and 2875 cm^{-1} (symmetrical and asymmetric elongation bands CH2), 1650 cm^{-1} (amide I, vibration C=O), 1540 cm^{-1} (amide II, CN and CH vibrations) and 1450 cm^{-1} (amide III, CH2 deformation). The proposed methodology contributed to the characterization of the sizing, bringing a new challenge for future research in technical art studies.

15.4 Characterization of ketone resin varnishes on modern canvas paintings

Irina Sandu, Eva Tveit, Klaas Jan van den Berg, Henk van Keulen

Among the synthetic resins adopted in the field of conservation as coating or varnish materials, cyclohexanone resins (ketone resins) are still in use nowadays both in private and institutional practice [42–45] They are oligomeric compounds of 6–10 monomeric units of cyclic structure and an average molecular weight of 600–800 (low molecular weight resins) [45], products of a condensation reaction of cyclohexanones and methyl cyclohexanone (AW2, MS2) or of cyclohexanones only (Ketone Resin N®/ Laropal K80) [42]. The photosensitive carbonyl groups in cyclohexanones are prone to oxidation [46]. Therefore, since the first production of AW2 by Farben (1921) and after the Second World War by BASF, the carbonyl functional groups were reduced to hydroxyls (MS2A, MS2B) and thus more stable materials were obtained [47–51].

A recent survey on the paintings collection of Munch Museum showed that a number of 23 canvases from the period 1882–1940 are reported as being varnished with ketone resins varnishes between the 1950s and 80s. The documentation on the conservation treatment reports and files in the archives of the Museum showed that 15 paintings have been varnished with AW2 between the years 1957 and 1970 [52].

Multi-technique analytical investigation using FTIR-ATR, microFTIR, Py-GC/MS and DTMS was carried out in collaboration with laboratories from Los Angeles (GCI) and RCE (Amsterdam). From the 15 paintings reported to be varnished with AW2, only for 7 paintings the presence of a ketone resin was confirmed (Figure 15.16).

Figure 15.16: 4 of the 15 paintings reported as being varnished with AW2 in 1957 (Photo credits: Munch Museet).

The Archlab access provided through the Iperion platform, funded by the European Commission, H2020-INFRAIA-2014–2015 and pursued between May and September 2017 allowed to extend the survey to other collections and institutions in Europe [52]. The Access visits to three European institutions – OPD (Italy), CATS (Denmark) and RCE (Netherlands) also made possible to gather small quantities of ketone resins in beads and as varnishes which together with the ones already present in the collection of the Conservation Department at Munch Museum constitute now a database of reference materials for this project and also for further studies.

It is possible to identify chemical differences in the ATR-FTIR spectra between the reduced ketone resins (e. g. MS2A) and two unreduced ketone resins (AW2 among them). AW2 is the first synthesized - cyclohexanone- methyl-cyclohexanone condensation compound, Ketonharz (also known as Ketone resin N) is a poly-condensation product derived from cyclohexanone alone, highly susceptible to light-induced oxidation. In the MS2A the carbonyl groups of the cyclohexanone resins were reduced to hydroxyl groups (the chemical reduction product is made of a complex mixture of methylation and condensation products formed by reactions involving methyl cyclohexanone, methanol and their derivatives). Therefore, -C=O band absorption is present in Ketonharz and AW2 spectra but is lacking in the MS2A. Infrared spectra of AW2 and MS2 show a band at around 1380 cm^{-1} (1374–1375 cm^{-1} in Figure 15.17) where methyl groups absorb, whereas this band is much diminished in intensity in the spectra of Ketone Resin N and Laropal K 80 [53]. Laropal K 80, currently available from BASF, is identical to Ketone Resin N according to its manufacturer; the only difference appears to be that the latter product was manufactured in a discontinuous process, while Laropal K 80 appears to be produced in a continuous one [48]. De la Rie and Shedrinsky showed that Ketone Resin N® and Laropal®K80 have identical characteristics when analyzed spectroscopically [48]. Gas chromatography of the fraction of these resins also yields an identical fingerprint of the two resins.

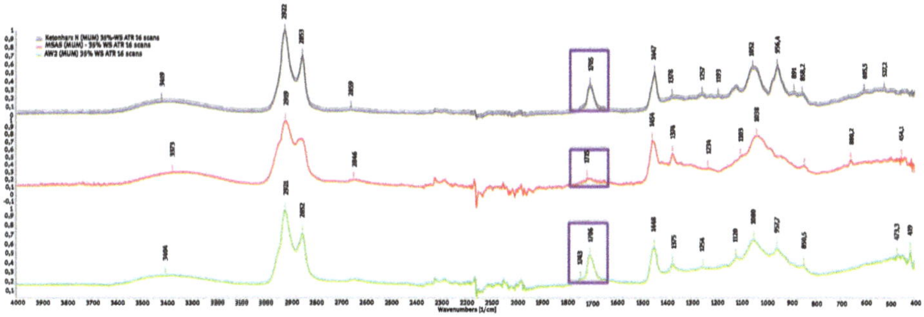

Figure 15.17: MID-ATR FTIR spectra of 3 types of ketone resins: AW2, Ketonharz N and MS2A.

Laropal K80 is still in use, for example, in Italian and Danish laboratories (OPD and SMK) as final varnish covering a coating of natural resin (Dammar or Mastic) [54, 55].

When comparing natural resins as Dammar and Mastic with ketone resins there are some similitudes in the spectral features (Figure 15.18). The regions were the absorption peaks have similar values are: CH stretching (3100–2860 cm^{-1}) and CO (carbonyl) bands (around 1700–1730 cm^{-1} and 1440–1470 cm^{-1}) [56]. It is difficult to discriminate if mixtures of vegetal (Dammar, Mastic) and ketone resins are present, as the peaks would overlap and create slightly different values than the absorption for the pure material. The interval specific for ketone resins containing -CO groups is 1700–1720 cm^{-1}, while the natural resins have the interval between 1695 and 1715 cm^{-1} (if oxidation products are present this band can be larger). Bands between 1200 and 1300 cm^{-1} (C-O) can also help in identified a natural vegetal resin if it is not mixed with other materials [57].

If wax is added to the ketone resin some new features are observed, such as: a shoulder at 1736–1741 cm^{-1} for -COO bond (in 1800–1700 region), around 1171–1184 cm^{-1} (esters groups) and also in 700–600 cm^{-1} (C-C str) region (Figure 15.18a).

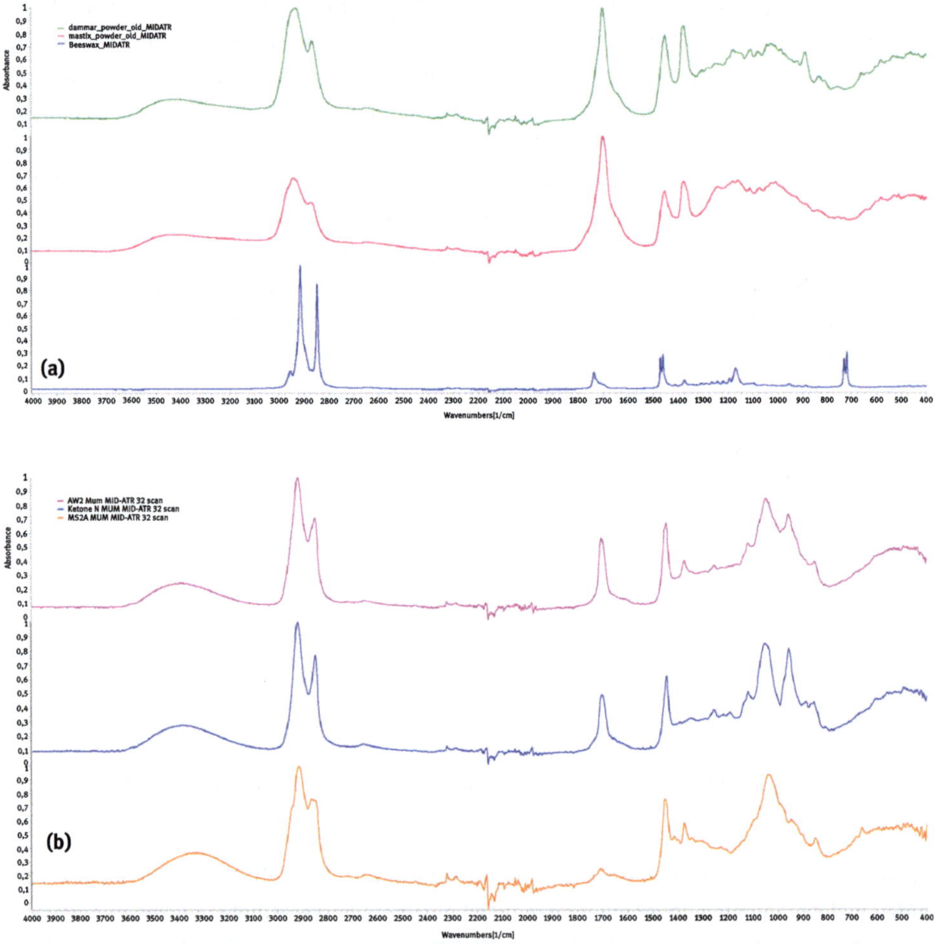

Figure 15.18: MID-ATR-FTIR spectra of 2 natural resins and wax (a) compared with ketone resins: AW2, Ketonharz N and MS2A (b).

Figure 15.19 shows examples of FTIR spectra of varnish samples taken from paintings at Munch Museum, where ketone resins are identified together with other components from paint or varnish layers (e. g. starch in sample MM.M513_A1; drying oil, egg and lead white in sample MM.M927_A4P1).

FTIR spectroscopy is an useful tool to screen the presence of a resin (natural or synthetic, or both) as varnish material, but will never be conclusive regarding the identification of the exact type in an unknown real sample. Additional method of analysis, such as Py-GC/MS and DTMS can be used for discriminating, each of them with specific advantages and disadvantages.

Among the chromatographic techniques, Pyrolysis Gas-Chromatography Mass Spectrometry (Py-GC/MS) is one of the analytical tools for accurate characterization

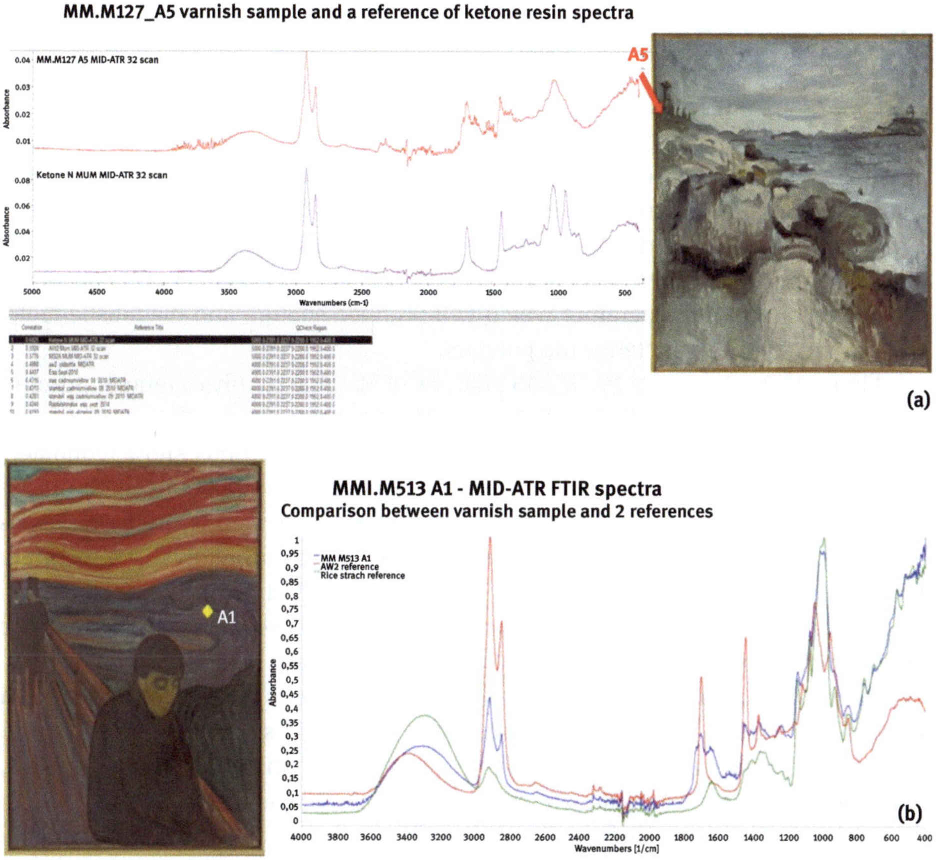

Figure 15.19: MID-ATR-FTIR spectra of several varnish samples: (a) Ketone N/AW2 identified in MM. M127 varnish; (b) AW2 identified in mixture with starch and proteinaceous material in MM.M513 painting; c) MS2A probably identified in MM.M927b-A2A3 sample and drying oil in mixture with lead white and egg in sample A4P1.

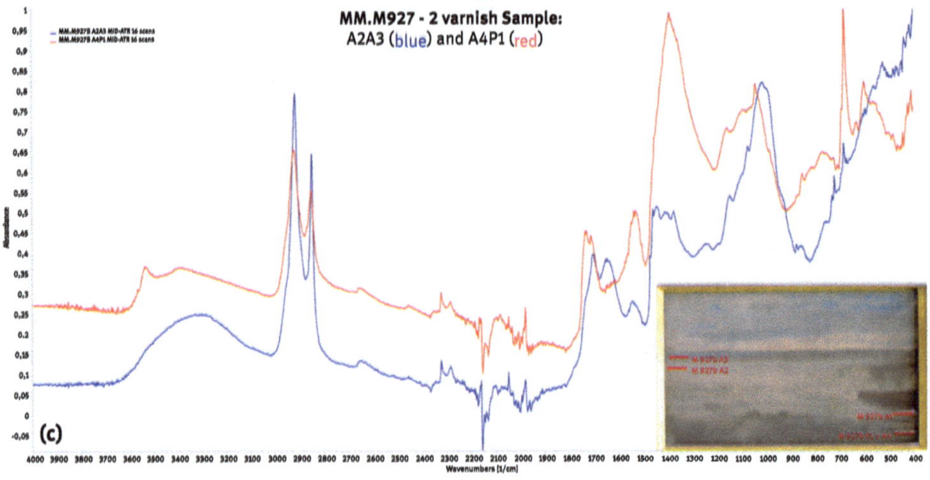

Figure 15.19: (continued).

of ketone resins and other synthetic polymers [57]. Pyrolysis is defined as thermochemical decomposition of organic material at elevated temperatures in the absence of oxygen; large polymeric molecules are converted into smaller ones, suitable for analysis and identification with GC/MS. Every polymer has its own optimum pyrolysis temperature, a temperature which will produce enough and diagnostic molecular fragments to characterize the polymer.

Figure 15.20 shows the Py-GC/MS analysis of MS2A at pyrolysis temperatures of 500 °C, 600 °C and 700 °C. Small fragments of resin are transferred to a steel pyrolysis cup then placed in the pyrolyzer. The chromatograms show monomers (0–7 min), dimers (10–15 min) and trimers (17 min and more) of the ketone resin. Noticeable is a reduction of the trimers and dimers of the resin at higher pyrolysis temperature resulting in abundant monomers. An optimum pyrolysis temperature for the MS2A resin is around 650 °C. The analysis will show a balance between typical, diagnostic monomer fractions without losing the information from the dimer fraction of the polymer.

As real samples from paintings varnishes often are a mixture of natural and synthetic materials, the standard analytical approach is applying a Py-GC/MS method dedicated for natural materials. It might not be optimal for synthetic resins, but good enough for an unambiguous identification of those resins.

Pyrolysis-Gas Chromatography/Mass Spectrometry using tetramethylammonium hydroxide for thermally-assisted hydrolysis and methylation (THM-Py-GC/MS) is presently the best analytical technique for characterizing (natural) organic materials in samples obtained from works of art [57]. Samples are ground with 20 µl of TMAH (5% in methanol; diluted 25% TMAH in methanol, Sigma Aldrich 334901) in a small glass mortar, 5 µl of the suspension is transferred to a steel pyrolysis cup

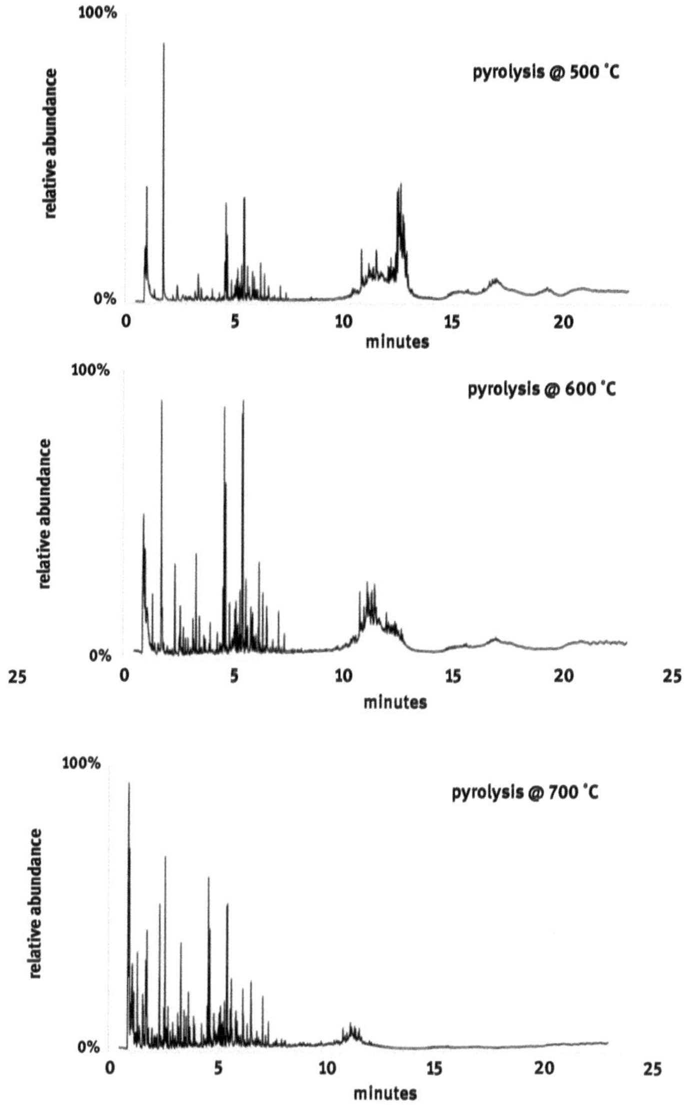

Figure 15.20: Py-GC/MS analysis of MS2A resin at different pyrolysis temperatures.

then placed in the pyrolyzer. The pyrolizer unit used was a Frontier Lab 3030D pyrolizer on a Thermo Scientific Focus GC/ISQ mass spectrometer. The pyrolizer is used in the program temperature mode, and after the sample introduction the pyrolyzer is heated from 360 °C to 660 °C at 500 °C/min, total time of 1 min. Both pyrolyzer interface and GC inlet were at 290 °C. The split flow was set at 1:30. Column: Supelco SLB5 MS (20M × 0.18mm × 0.18μm); helium at 0.9 ml/min. GC oven program: 35 °C (1 min), 60 °C /min to 110 °C, 14 °C/min to 240 °C, 5 °C/min to

315 °C (2 min). MS interface 250 °C; ion source 220 °C; scan range 29 to 600 amu at 7scans per sec.

Figure 15.21 shows the chromatograms of the ketone resins Laropal K80, MS2A, AW2 and Ketone N 200. The THM-Py-GC/MS runs show that, although the pyrolysis temperature is not optimal, a THM-Py-GC/MS method for natural materials can be used to characterize synthetic resins. All the resins show abundant compound peaks in the dimer section of the GC/MS run (around a retention time of 10 min). The mass spectra of the ketone resin dimers, those of the monomers, and the retention times of the compounds, except those of Laropal K80 and Ketone N 200, are specific enough for an unambiguous identification of the synthetics resins, also in the presence of natural materials and other synthetics. Laropal K80 and Ketone N 200 are both based on cyclohexanone and will produce more or less the same result when analysed with Py-GC/MS or with THM-Py-GC/MS.

DTMS (Direct Temperature resolved Mass Spectrometric analysis) is a very versatile technique used to chemically identify varnishes in minute paint samples because it gives information about the chemical composition of a broad range of materials including some indication of the degree of oxidation and polymerisation. The temperature resolution allows the separation into mobile "free" fractions and chemically bound fractions, which require a higher temperature [28–60].

Figure 15.22 shows comparative spectra generated with DTMS of three ketone resins (two from the Collection of Munch Museum and one from the Conservation laboratory of paintings at OPD in Florence).

DTMS is able to easily distinguish between synthetic and natural resins (such as gum dammar and mastic – Figure 15.23). The distinction with different types of ketone resins is straightforward since these resins show clusters with mass increments of 90–120 Da. Distinction between those resins, such as AW2 and MS2A or Ketone N and Laropal K80 is less straightforward, but is done by comparing the mass distribution in e. g. the dimer clusters, with e. g. m/z/ 192 and 204 as the main peaks for Laropal, 202 and 220 for MS2A and 206 and 218 for AW2 (Figure 15.22).

The DTMS analysis made on some samples of varnish from these paintings corroborated the presence of AW2 as hypothesized initially according the survey on the collection and preliminary screening with ATR-FTIR spectroscopy (e. g. in sample MM.M127 (1888) where the presence of the ketone resin was indicated by FTIR). Py-GC/MS also confirmed the preliminary screening of a ketone resins in several samples (e. g. Laropal was identified in sample MM.M927b-A1 (1882)). The painting was varnished in 1957, and Keton N arrived on market in 1960 and its production changed to Laropal in 1979, therefore the varnish is likely AW2.

Figure 15.23 shows that the DTMS spectrum for MM.M628 (1882–1883) sample points toward the presence of Keton N, but could also be attributed to a mixture of this resin with AW2/MS2A. In fact, m/z 125, present in MM.M628, is not found in Keton N but is found in AW2/MS2A. This painting, as M927b was also varnished in 1957,

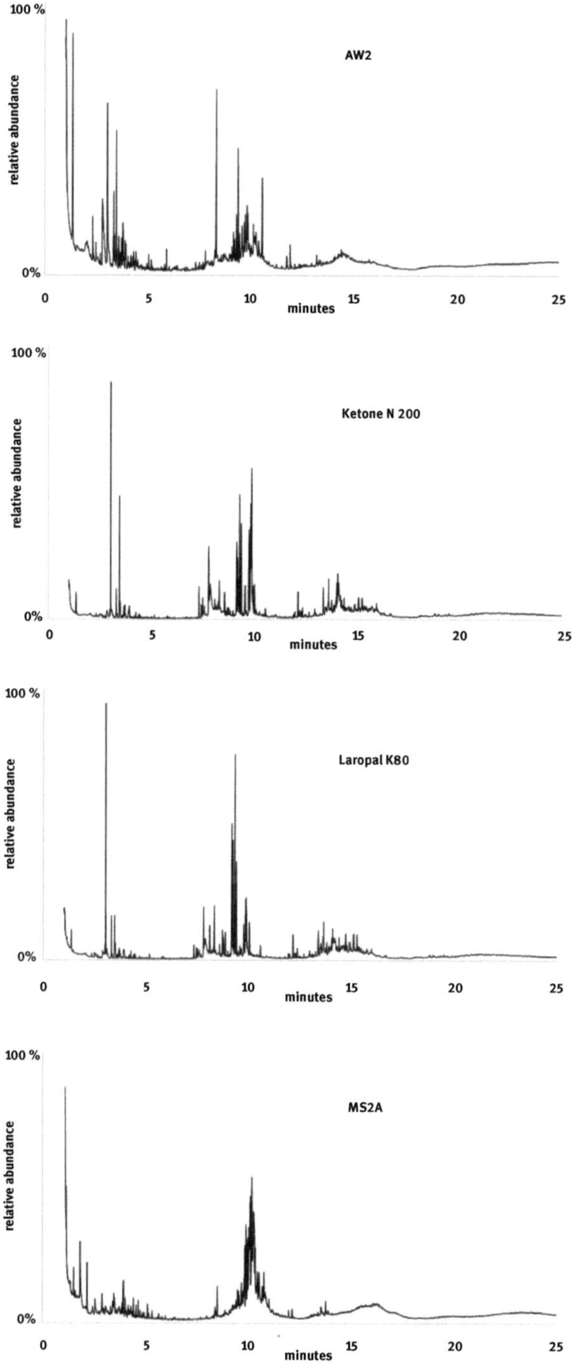

Figure 15.21: THM-Py-GC/MS chromatograms of 4 ketone resins: Laropal K80, MS2A, AW2 and Ketone N.

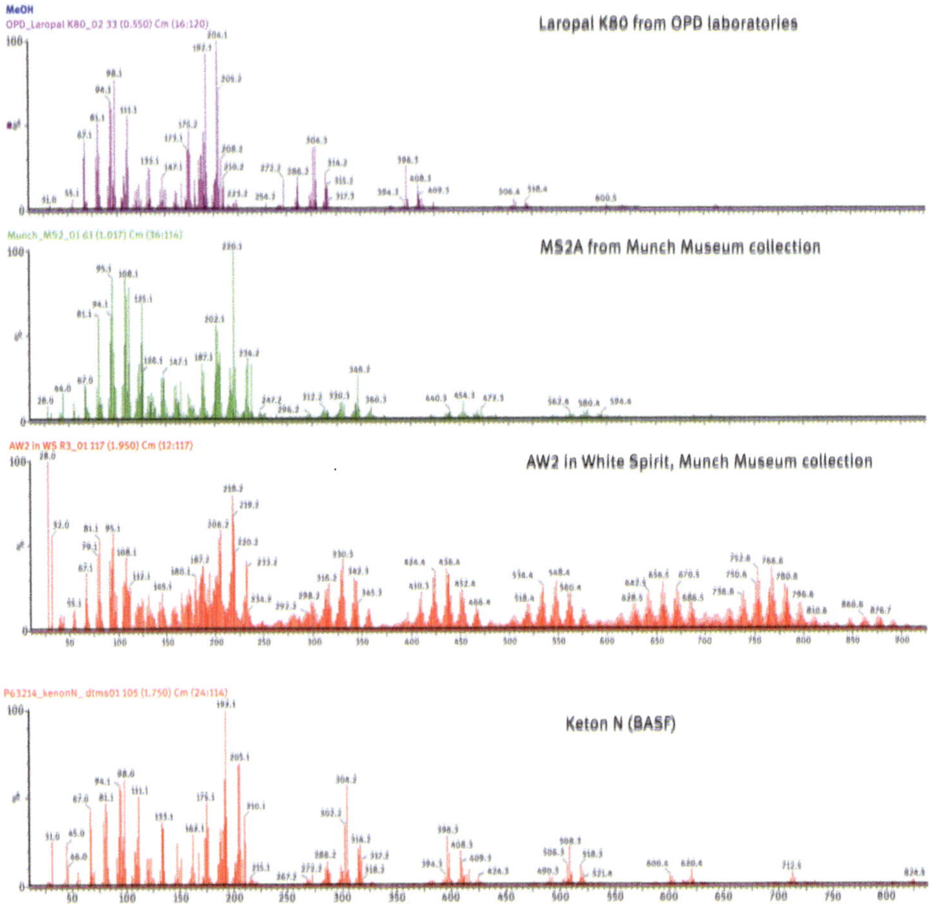

Figure 15.22: DTMS mass spectra for Laropal K80, MS2A, AW2 and Ketone N.

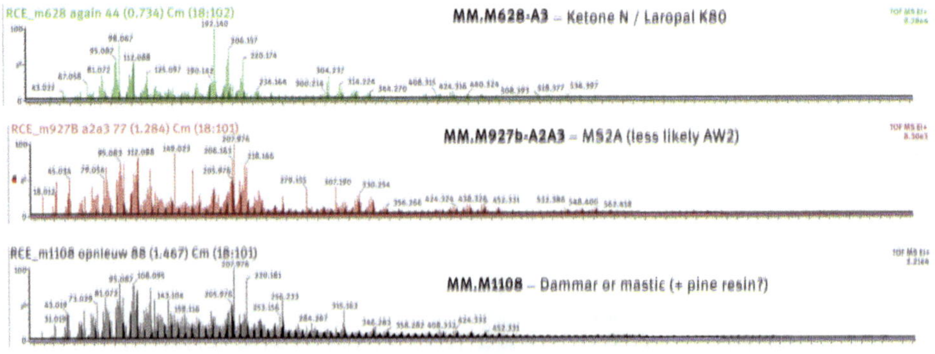

Figure 15.23: Examples of DTMS results on 3 samples from paintings.

before Laropal entered the market. We speculate that here a Keton N varnish layer was applied on top of a layer of AW2.

DTMS spectrum of MM.M927b varnish sample matches MS2A closest, but AW2 is also similar, as the resins are apparently too similar chemically to be distinguished reliably with this technique.

For MM.M1108 (1883) sample the spectrum shows the presence of Dammar (mastic) and pine resin: if a ketone is present it would be AW2/MS2A, but due to the presence of the natural resins this is less clear. To elucidate this we would need GC-MS and a larger sample.

In conclusion, while DTMS of pure synthetic resin varnish samples allows for clear identification of these varnishes, this may be less straightforward in real aged samples from paintings, especially if samples are small and may contain more than one resin.

In this study, much of the evidence from analysis with FTIR, DTMS and/or GCMS methods could be corroborated with each other and supported with evidence from conservation records. However, when samples are small and conservation records are incomplete, this remains a challenge.

15.5 Materials characterization and sulphate crystals formation on Edvard Munch's aula sketches

Erika Gohde Sandbakken

The Munch Museum beholds 131 Aula sketches painted by Edvard Munch between 1909 and 1916. Of his earliest sketches (1909 and 1910), 15 of them, all on cotton fabrics, have proved to have extensive areas with white surface crystals of Magnesium, Zinc and Zinc-magnesium sulphate. The mechanism of formation of the crystals on these sketches is still unclear. These salts are not very often found on canvas paintings, but are more common on porous building structures and fresco paintings. Seven of the 15 sketches have been studied using a multianalytical and multiscale approach. Different analytical techniques, such as X-ray fluorescence (XRF), scanning electron microscope (SEM-EDX), X-ray Diffraction (XRD), Fourier Transform Infrared (FTIR) and Raman spectroscopy (Raman), Gas Chromatography Mass Spectrometry (GC-MS), and Direct Temperature resolved Mass Spectrometry (DTMS) have been used. In addition, light microscopy, Hirox 3D microscopy, a UV-radiation source, simple lab tests, thorough ocular examinations, different archival material, such as diary notes and letters have supplemented the research.

15.5.1 The general condition of the sketches today

Some of Munch's Aula sketches were painted and stored up to four years in his outdoor studio (Figure 15.24) [61]. Outdoor storage and in general poor storage during

Munch's lifetime suggests that the sketches have been exposed to extreme unstable climates including sub-zero temperatures.

Figure 15.24: Munch standing next to the "Researchers" in his outdoor studio. The studio had wooden walls and a canopy of 1.5 m. Skrubben, 1910–1911. The red square indicates what is left of the motif today. (Photo: A.F. Johansen).

All layers in the sketches appear matte and dry, whether they are crayons or painted media. Gloss meter measurements from several sketches revealed values of 0.6–0.1 Gloss Units (GU), were 70–100 is high gloss and below 10 low gloss (measurements made with a SADT GTS 60N advice). These points to porous paint films with low binding media content. In a porous paint film the pigment grain will have less protection as they can lay open on the surface, as opposed to a well-bound oil paint where the pigment grains will be embedded in oil and create a smooth surface [62]. Porous materials applied to thin cotton fabrics would rapidly deteriorate in an outdoor environment and become more porous. This would in particular affect water-based paint media which is known to be more prone to mechanical and physical damage than changes related to chemical activity between pigment and binder [63]. The deteriorated materials are however, not only a consequence of unstable environments and storage, but also of Munch's choice of materials and painting techniques as he expressed that he wanted to achieve matte fresco-like surfaces [64].

15.5.2 Analytical approach

Hitachi S-3600N SEM with a BSE and Thermo Noran system Six EDS detectors, FEI ESEM Quanta Fei 200, coupled with micro analyser EDS and a Jeol JSM-84 EDX Oxford Instruments model 6506 X-ray scanning electron energy dispersive X-ray microscopes were used for elemental analyses and maps (4 samples of crystals and 2 paint samples were analysed). A Philips-FEI XL-30 FSEM coupled to an EDAX energy dispersive analyser was used for 3 paint samples. Four paint samples and 2 samples of crayon were analysed using a 6890N Network GC System (Agilent Technologies) coupled with a 5973 MS detector with quadruple analyser (4) [65]. Nine samples of crystals were analysed with a X'pert Pro Pan analytical X-ray diffractometer. A handheld Thermo Scientific Niton XL3 XRF (pXRF) was used to detect elements in the bare canvas areas of which results were compared to those from new cotton canvases. EIS P/N 9910 XRF-generator XRF was used to detect pigments and canvas areas in 2 sketches. Two paint samples were analysed with a Direct temperature resolved mass spectrometer (DTMS) at 16 eV electron ionisation on a JEOL – SX102-102 4 sector instrument. A Hirox KH-8700 3D Digital Camera was used for direct imaging (Hirox-europe.com), mounted on a MOPAS stand (wwwjaap-enterprise.com). Two paint samples were analysed with an imaging FTIR type Bruker Hyperion 3000 microscope equipped with focal-plane-array (FPA) detection using an attenuated total reflection (ATR) crystal.

15.5.3 The original materials and techniques of the sketches

Most likely the same type of thin cotton fabric was used for all of the 15 sketches according to visual similarity and thread counts ($20 \times 22\,cm^2$). None of the 15 sketches have ground layers or any visible remnants of preparatory layers. All of them have smaller or larger areas of bear canvas as a result of not having been covered with any coloured layer. Consequently the canvases thus form parts of the motifs, either as background or as detail elements.

Among the 15, 10 sketches were executed mainly with crayons and three with diluted paint. The crayons are forming lines and zigzag patterns with large areas of bare canvas (Figure 15.25a & b). The liquid paint has to some degree soaked into the bare canvases. Munch gave the impression that he sought after absorbent materials, as he also had expressed that he wanted the paint to soak into the supports [66].

Sampling from a blue and a yellow layer of crayon (Figure 15.25a) proved to be so small in amount and was thereby soaked up with cotton swabs wetted with a mixture of acetone, methanol and hexane. The GC-MS analyses proved the two samples to have a similar quantitative composition. The complex mixture of lipid molecules was attributed to ruminant fats and the acid profile the presence of lanolin [67, 68] (Figure 15.26). Lanolin is a yellow waxy substance secreted of the

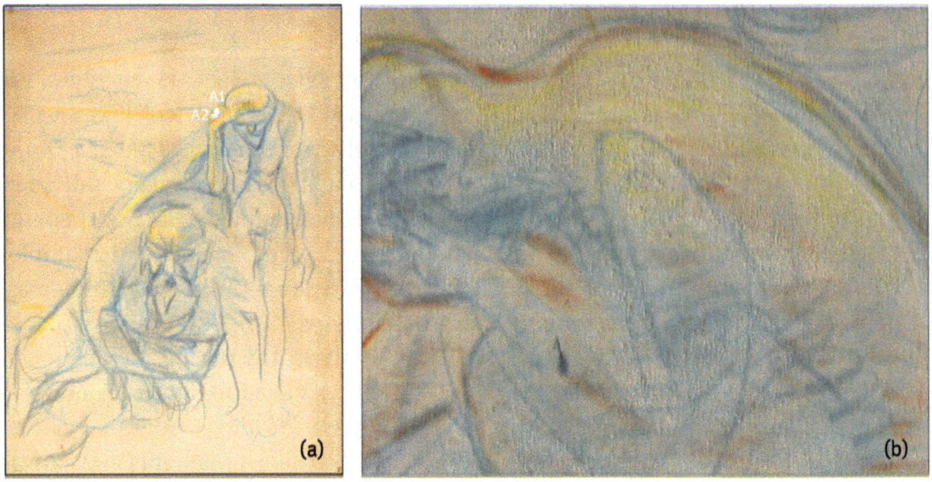

Figure 15.25: (a) "Old men in sunlight", sampling areas of the crayons are indicated upper right. (b) Detail from "People in sunshine (I)". Zig zag patterns of Prussian blue and lines of chrome yellow and vermillion crayons. Large areas of the sketch have extensive crystal distribution, particularly in the lighter chrome yellow area.

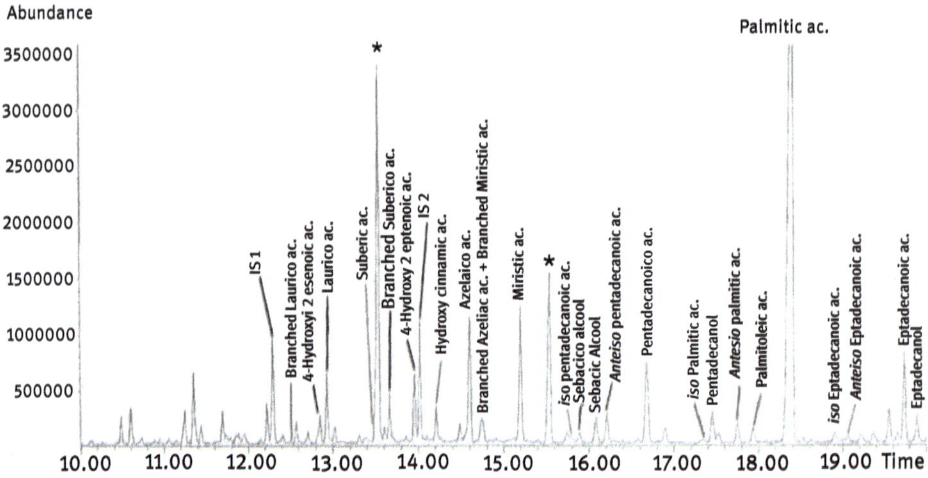

Figure 15.26: GC-MS chromatogram of the acidic fraction of a sample of crayon from "Old Men in Sunlight" (Table 15.1) indicating Lanolin (Time range 10–20 min. IS1: injection internal standard hexadecane; IS2: derivatization internal standard tridecanoic acid. *= phthalate contamination [Graph: 61].

sebaceous glands from wool bearing animals, like sheep (wikipedia). In addition a vegetable oil was present. The crayon technique was thus identified and labeled as a wax crayon [69].

Many of the 13 sketches have additional small strokes of what resembles thicker oil paint. This is indicated by visible brush marks and the, so-called, halo-effect (oil stains) seen in the canvas areas surrounding these layers. This has been connected to other artists using unprepared canvases [70]. The halo effect is seen encircling lines of crayons as well in some of Munch's sketches; these have in literature been described to deteriorated crayons or crayons exposed to heat [71].

The 2 sketches "*The Researchers* (I)" (Figure 15.27) and "*Nude Figures* (II)" have more covering painted layers, as opposed to the ones mainly executed in crayons that have more resemblance of drawings. Friends of Munch wrote that he bought jars of pigments and that he mixed his own paint with the help from his assistant [72]. In 1915 Ludvig Ravensberg (1871–1958) (Munch's cousin) wrote that the monumental three-part sketch of "*The Researchers*" was painted on a common fabric with glue-paints [73].

Figure 15.27: Mid part and right end part of the 3-part sketch seen in Figure 15.24, "*The Researchers* (I)". The red areas are documented as being affected by white crystals.

In 3 samples taken from the vast green grass area large and uneven pigment grains was seen under the light microscope (Figure 15.28a). In another red paint sample air bubbles and swirl-like paint can be seen in a SEM-BSEimage (Figure 15.28c) [74]. These samples support that Munch used self-made paint.

One of the green samples was divided into four. One micro-fragment was studied in DTMS and revealed a proteinaceous media, but not animal glue [74]. Another

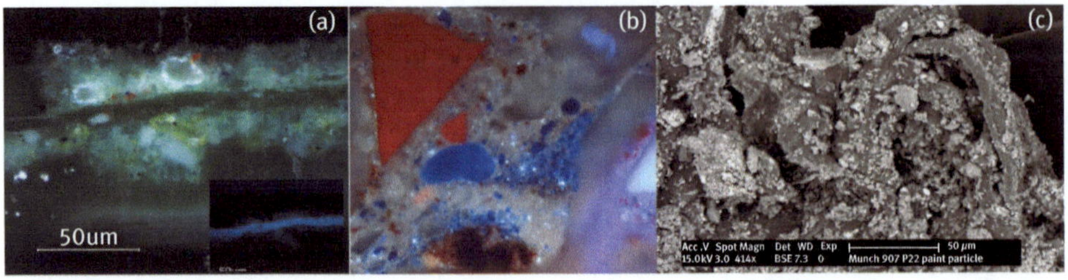

Figure 15.28: (a) SEM-BSE image of the green paint sample, large pigments grains in a porous paint film, a fluorescent glue layer is seen between the two green layers (Photo credits: J.J. Boon) (b) UV-fluorescent detail-image from a purple paint sample (light microscope). Elemental analyses and mapping with SEM-EDX indicated use of an organic red (C, O), vermillion (Hg, S) a red lake pigment (Al, Na, S, K) and ultramarine (Na, Al, Si, O, S) to create the purple colour (Photo: Erika Gohde Sandbakken). (c) SEM-BSE image showing a red paint sample with swirl-like paint (Photo credits: J.J. Boon).

fragment was prepared into a cross-section and studied in SEM and FTIR. The ATR-FTIR amide vibrations (1644 and 1550 cm^{-1}) also indicated proteins. Transmission FTIR measurements were compared to a database for casein and strengthened the identification of casein [74].

In 1910 the painter Erik Werenskiold (1855–1938) recommended Munch of getting casein elsewhere than in Oslo, because it was of bad quality and tended to crack heavily [75]. In a large area of the sky, a thick semi translucent heavily cracked layer visually resembled characteristics of casein as casein paint is well known to crack heavily under unstable environments (Figure 15.29a). A sample from this area was analysed with GC-MS. The results point to values of a palmitic and stearic acid ratio that ascribe to the presence of linseed oil with traces of pinaceae resin [76].

Other three samples (one from the vast area of the green grass, one from a thin media rich pink layer, Figure 15.29b - and the third from a dry darker pink brush stroke) were taken for their different visual appearance. A proteinaceous material was identified as animal glue (hydroxyproline (Hyp) was identified as the specific

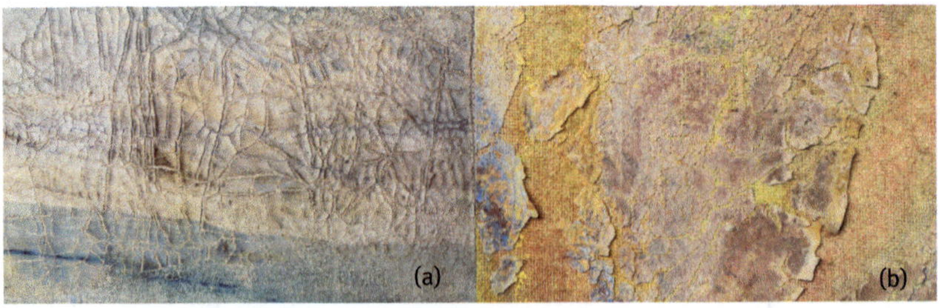

Figure 15.29: Two details of paint layers from "*The Researchers* (I)": (a) A heavily cracked layer analysed as linseed oil. (b) Thin unstable medium rich pink layer, analysed as animal glue, pinaceae resin and linseed oil, suggesting the use of an emulsion paint (Photo credits: Erika Gohde Sandbakken).

marker by GC-MS in all three samples) [76]. Principal Components Analysis (PCA) confirmed the results. In addition, these three samples also contained a presence of glycerolypidic material identified as drying oil. The latter results point to the use of a possible glue in oil emulsion paint. Animal glue would make a good emulsion, compared to using vegetable glue [77]. It would have made more easily workable paint, and despite the oil content, have the advantage of drying relatively fast [77].

XRF-measurements have pointed toward the presence of certain pigments, and in some samples pigments were further identified with FTIR and SEM-EDX [74, 78, 79] (Table 15.1). It is difficult to detect trends in the appearance of the crystals, as they seem to grow on all types of areas, also on the bare cotton fabrics. In several areas however, the crystals follow some of the drawn lines of crayons or the painted layers, with nearly clear borders where there are no crystals on the adjacent bare canvas areas (Figure 15.32b). It has been observed that some colours are more affected than others: Emerald green and the lighter of the two chrome yellows and the lighter of the two Prussian blues.

Particularly the lighter hue of the chrome yellow have a more intense amount of crystals present, than the darker hues (Figure 15.32b). Chrome yellow contains lead sulphate and chromate. Traditionally, more sulphate, often potassium sulphate, sulphuric acid or barium sulphate was added to the lighter hues [80]. The XRF analyses from "*People in Sunshine* (I)" does not show S or Ba, but in addition to Cr, contain Ca, Pb and Zn (some of the elements could belong to the support). After the industrial revolution zinc replaced previous fillings like magnesium and potassium sulphates in pastels [81]. It is difficult, however, to know what could have been used as a lightening agent in the lighter chrome yellow in Munch's sketches, particularly as Mg and S can be difficult to detect with XRF.

The Emerald green crayon used in the sketch "*People in sunshine* (I)" is severely affected compared to the lines of ultramarine and Prussian blues, which only have minor areas affected.

These observations point to the fact that the different qualities of the pigments to some degree affect, or at least exacerbate the crystal formation. Pigment mixtures containing chalk and silicates [82], and hygroscopic pigments such as green earth, ochre, smalt have in other studies been connected with white surface phenomena called *blanching* [83].

15.5.4 Characterization of crystals and discussion around their formation process

Preliminary tests of crystal material proved solubility in water and only partly in ethanol. Examination under a UV-radiation source (Rescolux UV 365 LED set. Willard) showed the crystals to have no characteristic fluorescence.

Under the light microscope the crystal form resembled a bacteroid particle form. Bacteroid particle forms belong to crystal forms too small to be more specifically

Table 15.1: Results obtained from several analytical methods on 7 sketches (they were selected as representative when it comes to materials, distribution of crystals and visual looks).

Title	Dimensions in cm	Suggested pigments	Method	Media	Method	Canvases (XRF)**
The Researchers(I)*	270 × 738	Red Earth, Zinc oxide, Zinc Potassium Chromate, ultra-marine, vermillion, organic red	SEM-EDX	Casein, linseed oil, oil in animal glue (emulsion)	FTIR, DTMS GC-MS GC-MS	Zn,S,Ca,Fe,Cl Zn,Ca,S,K,Fe,Cl
The Researchers (II)	132 × 383	–	–	Crayons, Drying oil	Visual GC-MS	S,Mg,Cl,Ca,K (Fibres: SEM-EDX) S8 (Fibres: XRD)
People in Sunshine (I)	134 × 232	Chrome yellow, vermillion, Prussian blue Emerald green	XRF	Crayons, oil paint?	Visual	K,S,Ca,Cl,Fe,Zn,Cr S, Cl,Ca,K,Fe,Zn,Cr S, Cl, Ca, K, Fe K, S, Zn, Ca, Fe, Cr
Geography	141 × 72	–	–	Protein based paint, oil paint?	FTIR Visual	
*Astronomy**	141 × 77	Cadmium yellow, chrome yellow, Prussian blue, Vermillion, Lead white	XRF	Diluted paint	Visual	Zn,Fe,S,Cl,K,Ca Zn, Fe,S,Cl, K,Ca
Nude figures (II)	62 × 138	Chrome yellow, vermillion, red pigment 83:1, $CaCO_3$	XRF, FTIR, Raman	Protein based paint, Drying oil	FTIR GC-MS	
Old men in Sun light	203 × 133	Ultramarine, Prussian blue, vermillion, chrome yellow	–	Lanolin and a vegetable oil (Crayon)	GC-MS	

* The sketch was lined (1948–1966) with a Henkel A 22ED (PVA), rye flour and arabin (starch-based glue). Often zinc or lead white was added to the lining glue mixture. ** The results are not indicating areas with or without crystals present as there were no indication of significant differences between the elements found.

identified under the light microscope. They however have a similar round capsular-shaped form [84].

The nine samples of crystals analyzed by XRD revealed their hydration states, with four, six and seven water molecules of hydration. Zinc ($ZnSO_4 \cdot 6H2O$ and $ZnSO_4 \cdot 7H2O$) and zinc-magnesium sulphates ($ZnMgSO_4 \cdot 4H2O$) were ascribed (identified as Boylite, Hexahydrite and Epsomite) [78].

Four samples of efflorescent material that were analyzed with SEM-EDX proved to be magnesium sulphates (Figure 15.30). Zinc-magnesium sulphate was also identified, but not zinc sulphate [85]. SEM-EDX does not allow recognizing their hydration states, but the magnesium sulphate which exists naturally on earth is either, anhydrase, mono, hexa- or heptahydrite [86].

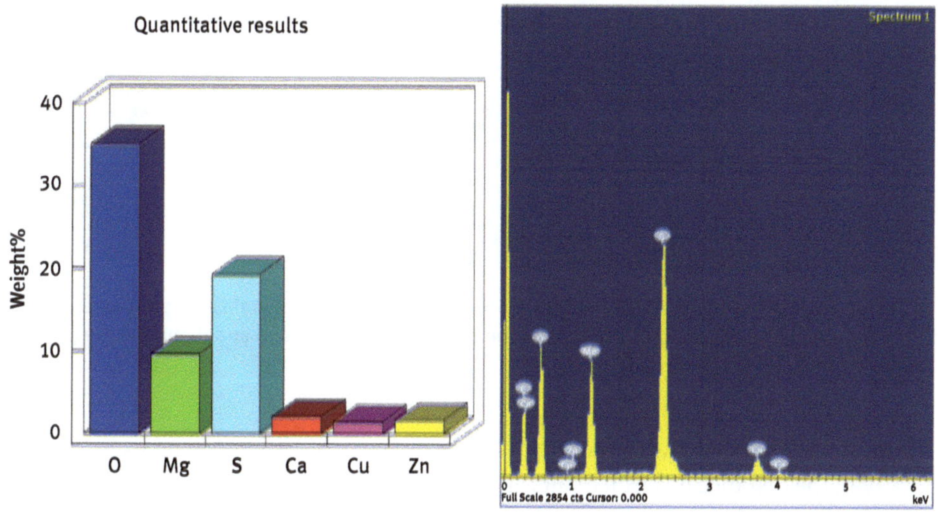

Figure 15.30: SEM-EDX quantitative graph and spectrum of a crystal sample showing the presence of magnesium sulphate (Graph and spectrum: Hartmuth Kutzke).

These crystals are hygroscopic. Hexahydrate hydrates at 41–51% RH, but in its mono hydrated form, kieserite, already at 21% [87]. The physical migration of the crystals to the surface is worrying with respect to the disruption of the crayons and paint layers. When these salts hydrate, their physical size and weight increases. This is also recognised as one of the detrimental factors in porous building structures [88]. In Hirox microscope photos it becomes evident how the crystals have created disruption and stress in paint layers from Munch's sketches as well (Figure 15.31).

Crystals are in a large extent also found on the bare canvas areas. These crystals are known to exhibit the ability to spread as they influence nucleation and crystal-propagation in adjacent areas [89]. It was therefore first thought that the crystals could

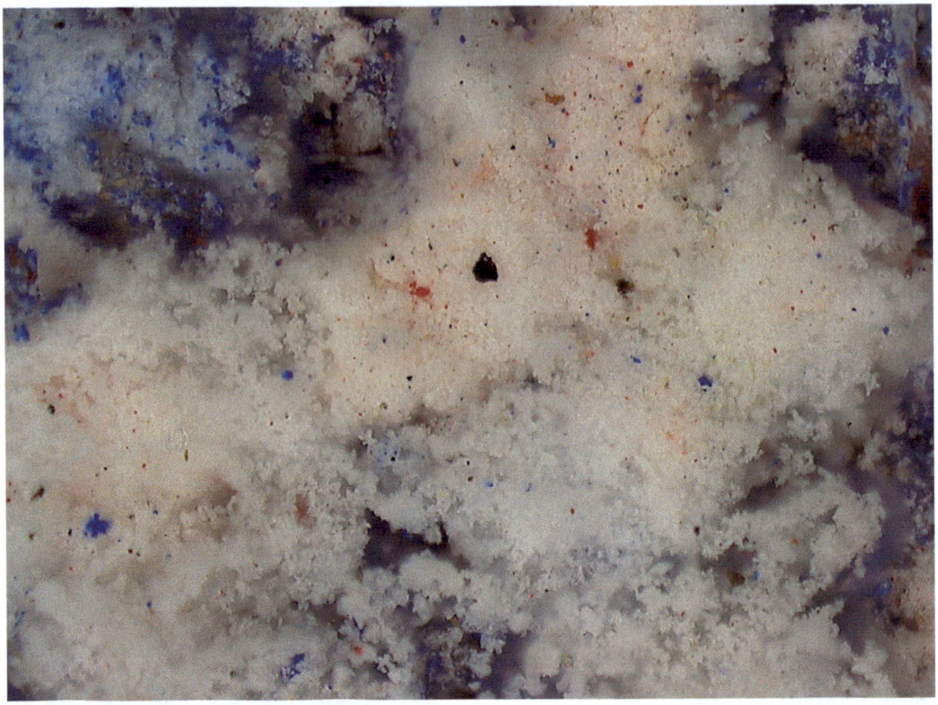

Figure 15.31: Hirox microscope photo from "*Nude Figures* (II)". Pigment grains are seen on top of the crystals (particularly upper right and left) as the crystals have protruded and disrupted the paint (Photo credits: Emilien Leonhardt).

have spread from areas with paint and layers of crayons. During visual examination, however, the distance of surface salts and coloured layers could in some areas be up to 40 cm and does not indicate this. In several areas of the supports the crystals seem randomly spread, but in other areas they seem to follow the weave of the canvas, as they appear in either weft or warp threads (Figure 15.32a).

Portable XRF-analyses were executed on the bare canvas; both in areas where crystals were present and absent (Table 15.1). The results were compared to pXRF-analyses made on new cotton canvases that had gone through different treatment stages, as well as surgical cotton wool (Mediq Norway AS). The analyses suggested a consistent presence of sulphur in relatively larger amounts in Munch's sketches as opposed to the new fabrics. In the new cottons several of the three readings of each of the fabrics, revealed that sulphur was within the limit of detection (LOD) and could suggest that there is an absence or very small amounts of sulphur present. Sulphur was however detected in the cotton wool and on the new fabric that was not bleached, but exposed to anti-crease and anti-shrinking treatments.

Two fibres obtained from the reverse of one sketch, "*The researchers* (II)", were analyzed in SEM-EDX and K, Ca, Mg and a consistent presence of S was detected

Figure 15.32: (a) Close up photo from the cotton canvas ("*The Researchers* (II)") showing a white crystal distribution probably related with preparation before weaving of the fabric; (b) Crystals following the lighter of the chrome yellow zig zag line on "*People in sunshine* (I)", the adjacent lines are darker chrome yellow (Photos: Erika Gohde Sandbakken).

[30]. These elements coincide with the pXRF measurements of the canvas surfaces (Table 15.1). XRD analyses of the two fibres could provide more specific information of the sulphur identified as it matched with *elemental sulphur*, S8 (Table 15.1).

Elemental sulphur is today widely used in agriculture as a pH-control or to lower the pH in fertilizers or alkaline soil [90]. Cotton plants with a sulphur deficiency have shown to have slower growth rate and lower crops levels than cotton receiving high rates of sulphur from fertilizers [91].

Earlier theories about the origin of the consistent presence of sulphur were rejected with the XRD findings of elemental sulphur. A theory on the source of sulphur from external pollution in the environment (from industries generating sulphur) was thus rejected. So was also the lack of Al (SEM-EDX) as a possible indication of canvas treatments with alum, aluminium potassium sulphate ($KAl(SO_4)_2 \cdot 12H2O$). Alum-sulphate has been used in the art paper production [92] and alum-water is recorded within the common use for cotton tents (with the intent to make them water resistant) [93].

Most probably, the elemental sulphur could correspond to molten sulphur (Colombini & Cantisani, 2015b). Molten sulphur-treated textiles were patented in 1929, but already in use since the 1890s with granular sulphur applied with a solvent vehicle [94].The molten sulphur treatment made the textiles stiff so they could be formed into odd shapes. Sulphur treated strips of cotton textiles were also used as undelays for wall papers to prevent damp and mould [94]. This could provide an answer as to why only small areas on four of Munch's canvases have spear mould occurrence (Cladosporium sp.). There are other factors however, that does not match well with a molten sulphur treatment: Munch's cotton supports are not stiff and measured pH levels of cotton fibres from two sketches proved to contain surprisingly low acidity (pH 6.98–7.1: ISO standard and 3071 method) [78].

15.6 Conclusions

The formation of the sulphate salts does not seem to have a correlation with any types of analyzed paint media, which have confirmed the use of wax crayons, linseed oil, animal glue-oil emulsion and suggested casein. Ravensberg largely seem to have been right about Munch using glue for some of his paint for the "*Researchers* (I)". It is also evident however, that Munch experimented with different combinations of binding media and different visual expressions using the same binding media.

The crystals seem to be more dependent on the paint's ability to degrade and to form porous structures. In that sense the materials of the sketches are interacting with the sulphate mineral salts as they contribute and ease and exacerbate the migration onto the surfaces. Strokes of more well-bound oil paint has so far withstood protrusion of crystals.

Aspects like air pollution, RH, light and temperature are important factors, particularly in the early drying phase of a painting [95]. It is believed that it was in the early years of these sketches that the process of the formation of salts occurred and that they have crystallised, migrated and become visible on these surfaces decades later. As the crystals in most areas seem only to be loosely attached or are easily smeared into the fabrics, it is likely that the migration has taken place after the 1960s when five of these sketches were lined. It is believed that a lining treatment would have disturbed the crystals, which is not evident on the sketches. On the other hand the lining treatment would have involved another cycle of high humidity.

Some questions have been answered in this study, but others, such as where the detected zinc comes from and how some of the possible pigment content is exacerbating the efflorescence are pending. Munch's 15 sketches are over one hundred years old. During more than half of these years they have undergone alternating cycles of humid, freezing and dry environments which have deteriorated the paint materials and at the same time been favourable for crystal formation. But were Munch bought his cotton fabrics and how he or the manufacture might have treated them is unknown and might include many numbers of possibilities. The detection of elemental sulphur is however a fact, more sample analyses are needed to study the implications of this finding. The chemical and physical reactions which have taken place continue to be the subject of ongoing research.

Acknowledgements: The Portuguese FCT is acknowledged for the funding provided to the research project "Gilt-Teller: an interdisciplinary multi-scale study of gilding techniques and materials in Portugal, 1500-1800", grant no. PTDC/EAT-EAT/116700/2010 (2012-2015). Important contributions to the project were given by colleagues working in different conservation and academic institutions in Portugal (Lisbon, Evora, Porto, Santarem), Czech Republic (Prague), Romania (Iasi), Germany (Hildesheim), Spain (Asturias) and Italy (Florence, Biella).

The Portuguese FCT is acknowledged for the funding provided to the research projects: *ON-FINARTS – On the Flemish Importance for the National Arts in the sixteenth century* (PTDC/EAT-HAT/115692/2009) and *The invisible preparation layer and its influence on Portuguese painting of the XV and XVI centuries: an issue to be solved* (PTDC/EAT-HAT/100868/2008).

The ketone varnish project was provided with Archlab Access within the Iperion CH platform, grant 654028, funded by the European Commission (H2020-INFRAIA-2014-2015).

Specialists from OPD, RCE, and SMK/CATS are also kindly acknowledged for their support during the Archlab access: Hilde Le Clerq - IPERION Archlab Access office, Monica Galeotti – OPD, Florence; Julian van den Berg & Melissa Daugherty - RCE, Amsterdam; René de la Rie - University of Amsterdam; Jan Dorscheid, Petria Nobel, Anne Krekeler – Rijksmuseum, Amsterdam; Margje Leeuwestein - Kröller-Müller Museum; Vera Blok, Laura Kolkena - Stedelijk Museum; Oda van Maanen - Van Gogh Museum; Kathrin Kirsch – freelancer; Jørgen Wadum, David Buti, Pauline Lehmann Banke, Troles Filtenborg – CATS-SMK; Dr Wim Genuit, Shell Global Solutions International B.V. Amsterdam. Charlotte Stahmann is acknowledged for all her precious work during the project and the involvement in the sampling campaign and the Archlab Access. Andrea Casini, Erasmus Plus internship at Munch Museum (March-September 2018) is also kindly acknowledged for the help in improving the images included in the chapter.

Acknowledgements for their contribution to the analytical campaigns are given to: Perla Columbini, Emma Cantasani, Marco Zanaboni, Francesca Modugni (GC-MS, SEM-EDS, XRD analyses), Jaap J. Boon (SEM-EDX, FTIR, DTMS analyses with assistance from Stefan Zumbühl for imaging FTIR and Karen Wyss with diamond cell FTIR), Ida Bronken (pXRF), Renie Birkeland Nielsen (XRD on crystal samples), Hartmuth Kutzke (SEM-EDX), Hans Jørgen Berg (in memoria) for SEM-EDX, Costanza Miliani (Molab - XRF).

References

[1] Sandu ICA. GILT-Teller: um estudo interdisciplinar multi-escala das técnicas e dos materiais de douramento em Portugal, 1500–1800. VIIII Jornadas de Arte e Ciência - Conservação e Restauro de Artes Decorativas de Aplicação Arquitectónica. Porto: CITAR-UCP, 201220122013a. 29September2012 2013a.

[2] Veiga R, Sandu ICA, Busani T, Costa PEREIRA M, Muralha S, Leal N. Proposal for an analytical methodology of study of gilded wood carving techniques and materials. ICOM CC meeting of working group on polychrome sculpture. Tomar: , 2013a 28–29thof May(poster presentation).

[3] Sandu ICA, Murta E, Ferreira S, Costa Pereira MF, Hrdlickova Kuckova S, Valbi V. A comparative multi-technique investigation on material identification of gilding layers and the conservation state of 7 Portuguese mannerist altarpieces. Int J Conserv Sci. 2015;6:439–54.

[4] Sandu ICA, Murta E, Veiga R, Muralha S, Costa Pereira M, Kuckova S. An innovative, interdisciplinary, and multi-technique study of gilding and painting techniques in the

decoration of the Main Altarpiece of Miranda do Douro Cathedral (XVII-XVIIIth centuries, Portugal). Microsc Res Technol. 2013c;76:733–43.

[5] Sandu ICA, Murta E, Neves ER, Pereira AM, Sandu AV, Kuckova S. A comparative, multi-scale and interdisciplinary study of gilding techniques and materials in two Portuguese Baroque “talha dourada” complexes. ECR – Estudos de conservação e restauro. Porto: , 2013b. DOI: 10.7559/ecr.4.3080 ISSN1647-2098no. 4.

[6] Costa Pereira MF, Maurício AM, Sandu ICA, Veiga R, Le Gac A, Leal N. Complementary imaging and characterization methodology of polychrome composites in gilded woodcarving using Micro-CT, SEM-EDX and OM. Microsc Microanal. 2015;21:148–9. DOI: 10.1017/S1431927614014445.Special issue of the SPMicros International Congress: «Microscopy in Research 2013».

[7] Veiga R, Sandu ICA, Leal N, Barata C, Busani T. Frequency and amplitude modulation atomic force microscopy (AFM) combined with H-SEM micro analysis as a novel approach for a multi-scale study of gilded samples. Book of abstracts of in art 2013 conference (First international conference on innovation in art research and technology). InArt 2013b, Evora Iasi: IJCSp.5

[8] Busani T, Paba F, Sandu ICA, Dias L. Atomic force microscopy (AFM), optical microscopy (OM) and scanning electron microscopy (SEM) in a complementary approach for the study of documented reconstructions of water gilded composites. Book of abstracts of Gilt-EnArt2015 (International conference on Gilding materials and techniques in European art). Èvora, Portugal; Iasi: IJCS, 2015;89 25th–27thof May(poster).

[9] Sandu ICA, Murta E, Ferreira S, Costa Pereira MF, Candeias A, Mirão J, Cuzman OA, Manganelli Del Fà R, Tiano P, editors. More than gold – an interdisciplinary, complementary study of gilding materials and techniques in baroque altarpieces from Portugal. Proceedings book of ESRARC 2014 (6th European symposium on religious art, restoration & conservation). Florence; Florence: , 2014;218 ISBN: [Amazon][WorldCat][Amazon][WorldCat][Amazon][WorldCat][Amazon][WorldCat]9788840443652(poster extended abstract).

[10] Hrdlickova Kuckova S, Schultz J, Veiga R, Murta E, Sandu ICA. Proteomics tools for the contemporary identification of proteinaceous binders in gilded samples. Special issue of GILT-EnArt (International conference on gilding materials and techniques in European art). Evora, Portugal; Iasi: IJCS, 201520152015;507–18. 25–27thof May2015 62015.

[11] Sandu ICA, Schafer S, Magrini D, Bracci S, Roque AC. Cross-section and staining-based techniques for investigating organic materials in polychrome works of art - a review. Microsc Microanal. 2012;18:860–75. DOI: 10.1017/S1431927612000554;.

[12] Kuckova S, Sandu ICA, Crhova M, Hynek R, Fogas I, Muralha VS. Comparison of MALDI-TOF and nanoLC-ESI-Q-TOF mass spectrometry on identification of proteins in cross-sections of medieval sculpture. MaSC meeting. Pisa, Italy: , 2013 3–7thof June(oral presentation).

[13] LE GAC A, Nogueira I, Seruya AI. The 1685 blue smalt coating of the Coimbra Old Cathedral main altarpiece under scrutiny. Microsc Microanal. 2015;21:158–9. DOI: 10.1017/S1431927614014494;. Special issue of the SPMicros international congress: «Microscopy in Research 20136».

[14] Le Gac A, Oliveira P, Dias Costa I, Dias Costa MJ. Materials for painting and gilding used in the Benedictine Community of Portugal from 1638 to 1822. Other times, other ways. In: Dubois H, Townsend JH, Nadolny J, Eyb-Green S, Neven S, editor(s). Making and Transforming Art: Technology and Interpretation, Proceedings of the ICOM-CC 5th International Symposium of the Art Technological Source Research-ATSR Working group, Brussels, Belgium, 22–23 November 2012. London: Archetype Publications Ldt. ISBN 2014:54–74.978-1-909492-16-5.

[15] Gomes S, Ferreira C, Nascimento G, Piorro L, Cardoso A, Candeias A. Identification and removal of non-original layers in the sixteenth century paintings of Funchal's Cathedral altarpiece. Color Res Appl. 2016;41:283–8.

[16] Valadas S. Flemish-Lusitan Early Masters - interacting Sciences, with History and Art. PhD Thesis in Chemistry, Universidade de Évora, 2016.

[17] Valadas S, Cardoso A, Mirão J, Dias C, Freire R, Candeias A. A pintura flamenga em Évora no Século XVI: novas perspetivas sobre a obra atribuída a Frei Carlos. In: Serrão V, Antunes V, editors. As preparações na Pintura Portuguesa Séculos XV e XVI, 2013:115–22. (Overprint).

[18] Campos Maia B, Calvo A, Candeias A, Frade J, Mirao J, Dias L. The altarpiece of the church of Freixo de Espada-à-Cinta: a study on its artistic materiality. In: Rogerio-Candelera M, Lazzari E M, Cano E, editors. Science and technology for the conservation of cultural heritage. Routledge – Taylor and Francis Group, 2013:197–200. DOI: 10.1201/b15577-46.

[19] Antunes VH, Anunciação O, Candeias A, Seruya AI. Data mining analysis over a database of materials of Portuguese paintings from fifteenth and sixteenth centuries. e-Preserv Sci. 2012;9:17–22.

[20] Caetano JO. «Un Siglo de Pintura Portuguesa. Artistas, obras, interrogantes». In: Primitivos. El siglo dorado de la pintura portuguesa. 1450–1550. Valladolid: Museu Nacional Collegio de San Gregorio, 2011:21–57.

[21] Antunes V, Candeias A, Oliveira MJ, Carvalho ML, Dias CB, Manhita A, et al. Uncover the mantle: rediscovering Gregório Lopes palette and technique with a study on the painting “Mater Misericordiae”. Appl Phys A: Mater Sci Process. 2016a;122:965.

[22] Melo H. O pintor Francisco João – Materiais e técnicas na pintura de cavalete em Évora na Segunda Metade do Século XVI. PhD Thesis in Conservation and Restoration, Universidade Católica Portuguesa, Porto, 2013.

[23] Candeias A, Carvalho G. Considerações sobre o estudo de Pintura - A Experiência do Laboratório José de Figueiredo. In: Serrão V, Antunes V, editors. As preparações na Pintura Portuguesa Séculos XV e XVI, 2013:1–6. (Overprint).

[24] Antunes V, Candeias A, Carvalho ML, Oliveira MJ, Manso M, Seruya AI, et al. Gregorio Lopes painting workshop: characterization by X-raybased techniques. Analysis by EDXRF, μ-XRD and SEM-EDS. J Instrum. 2014d;9:05006.

[25] Valadas S, Freire R, Cardoso A, Mirão J, Vandenabeele P, Caetano JO, et al. New insight on the underdrawing of 16th Flemish-Portuguese easel paintings by combined surface analysis and microanalytical techniques. Micron. 2016;85:15–25.

[26] Valadas S, Freire R, Cardoso A, Mirao J, Dias CB, Vandenabeele P, et al. On the use of the unusual green pigment brochantite in the sixteenth century Portuguese-Flemish paintings attributed to the Master Frei carlos Workshop. Microsc Microanal. 2015;21:518–25.

[27] Melo H, Cruz AJ, Candeias A, Mirão J, Cardoso A, Valadas S. Os estratos preparatórios das obras do pintor eborense Francisco João (ativo entre 1563 e 1595). In: Serrão V, Antunes V, editors. As preparações na Pintura Portuguesa Séculos XV e XVI, 2013:143–53. (Overprint).

[28] Anunciação O, Antunes V, Candeias A, Seruya AI. Prospeção de dados sobre uma base de dados de pinturas portuguesas dos séculos XV e XVI. In: Serrão V, Antunes V, editors. As preparações na Pintura Portuguesa Séculos XV e XVI, 2013:85–92. (Overprint).

[29] Antunes V, Serrão V, Oliveira MJ, Dias L, Candeias A, Mirão J, et al. Técnicas e materiais de preparação na pintura portuguesa dos séculos XV e XVI". In: Serrão V, Antunes V, editors. As preparações na Pintura Portuguesa Séculos XV e XVI, 2013:55–74. (Overprint).

[30] Antunes V, Candeias A, Oliveira MJ, Longelin S, Serrão V, Seruya A, et al. Characterization of gypsum and anhydrite ground layers in fifteenth and sixteenth centuries Portuguese paintings by Raman Spectroscopy and other techniques. J Raman Spectro. 2014a;45:1026–33.

[31] Mendes J, Cruz AJ, Carvalho ML, Candeias A, Mirão J. As preparações das pinturas de Nuno Gonçalves. In: Serrão V, Antunes V, editors. As preparações na Pintura Portuguesa Séculos XV e XVI, 2013:107–14. (Overprint).

[32] Coroado J, Antunes V, Serrão V, Oliveira MJ, Dias L, Candeias A. Presença de celestite em retábulos portugueses. In: Serrão V, Antunes V, editor(s). As preparações na Pintura Portuguesa Séculos XV e XVI. (Overprint) 2013:75–83.

[33] Antunes V, Candeias A, Oliveira MJ, Lorena M, Seruya AI, Carvalho ML. Calcium sulfate fillers and binders in Portuguese fifteenth and sixteenth centuries: ground layers from a family painting workshop - Study by multianalytical spectroscopic techniques. Microche J. 2016c;125:290–8.

[34] Antunes V, Oliveira MJ, Vargas H, Candeias A, Seruya A, Dias L. Characterization of glue sizing layers in portuguese wood paintings from the fifteenth and sixteenth centuries by SEM secondary electron images and µ-FTIR. Microsc Microanal. 2014b;20:66–71.

[35] Miguel C, Antunes V, Serrão V, Seruya AI, Coroado J, Candeias A. Estudo das formulações proteicas das camadas de preparação de um conjunto de retábulos portugueses dos Séc. XV-XVI por micro-FTIR: uma abordagem quimiométrica. In: Serrão V, Antunes V, editors. As preparações na Pintura Portuguesa Séculos XV e XVI, 2013:45–54. (Overprint).

[36] Antunes V, Candeias A, Coroado J, Serrão V, Cachão M, Carvalho ML. A multidisciplinary approach to the study of the brightening effects of white chalk ground layers in fifteenth and sixteenth century painting. Anal Methods. 2016b;8:4785–97. DOI: 10.1039/c6ay00435k.

[37] Oliveira MJ, Antunes V, Serrão V, Candeias A, Seruya AI, Mirão J. Caracterização material da camada preparatória de pintura portuguesa por micro-XRD e SEM-EDS. In: Serrão V, Antunes V, editor(s). As preparações na Pintura Portuguesa Séculos XV e XVI. (Overprint) 2013:39–44.

[38] Antunes V, Oliveira MJ, Vargas H, Serrão V, Candeias A, Carvalho ML. Characterization of glue sizing under calcium carbonate ground layers in Flemish and Luso-Flemish painting-analysis by SEM-EDS, µ-XRD and µ-Raman spectroscopy. Anal Methods. 2014c;6:710–7.

[39] Serrão V, Antunes V, Oliveira MJ, Dias L, Costa S, Piorro L. Considerações técnicas, materiais e artísticas sobre o retábulo da Misericórdia de Almada e os seus pintores (1589–1591). In: Serrão V, Antunes V, editor(s). As preparações na Pintura Portuguesa Séculos XV e XVI. (Overprint) 2013:259–72.

[40] Campos Maia B, Frade J, Calvo A, Candeias A, Mirão J, Cardoso A. Obras de Grão Vasco e da sua oficina. In: Serrão V, Antunes V, editor(s). As preparações na Pintura Portuguesa Séculos XV e XVI. (Overprint) 2013:123–32.

[41] Melo H, Cruz AJ, Candeias A, Mirao J, Cardoso A, Oliveira MJ. Problems of analysis by FTIR of calcium sulphate-based preparatory layers: the case of a group of sixteenth century Portuguese paintings. Archaeometry. 2014;565:513–26.

[42] De la Rie ER, Mcglinchey CW. New synthetic resins for picture varnishes. In: Mills JS, Smit P, editors. Cleaning, retouching, and coatings. London: International Institute for Conservation of Historic and Artistic Works, 1990:168–73.

[43] Feller RL, Stolow N, Jones EH. On picture varnishes and their solvents. Oberlin, Ohio: Intermuseum Conservation Association, 1959.

[44] Horie CV. Materials for conservation. Organic consolidants, adhesives and coatings. London: Butterworth-Heinemann, 1992.

[45] Borgioli L, Cremonesi P. Le resine sintetiche usate nel trattamento di opera policrome. Il Prato: Collana I Talenti, 2001:142–5.

[46] Thomson G. New picture varnishes, recent advances in conservation. In: Contributions to the IIC Rome conference. Rome, 1963.

[47] De la Rie ER. Stability and function of coatings used in conservation. In: Allen NS, Edge M, Horie CV, editors. Polymers in conservation: Proceedings of an international conference organized by Manchester Polytechnic and Manchester Museum, Manchester, 17–19 July 1991. Cambridge: Royal Society of Chemistry, 1991:62–81.

[48] De la Rie ER, Shedrinsky AM. The Chemistry of ketone resins and the synthesis of a derivative with increased stability and flexibility. Stud Conserv. 1989;34:9–19.

[49] Ciabach J. Plasticized and stabilized cyclohexanone resins. In: Resins: ancient and modern, preprints of the SSCR's 2nd resins conference. Edinburgh: Scottish Society for Conservation and Restoration, 1995:85–7.

[50] De la Rie ER. Polymer additives for synthetic low-molecular-weight varnishes. In: 10th Triennial Meeting, ICOM-CC. Washington D.C., Lawrence, Kan: Allen Press, August 1993:566–79. Preprints, Vol. 2.

[51] Lank H. Picture varnishes formulated with resin MS2A. In: Brommelle N, Smith P, editors. Conservation and restoration of pictorial art. London: Butterworths, 1976:148–9.

[52] Stahmann C, Tveit E, Sandu ICA, Van Den Berg KJ, Joosten I, Mazurek J. Ketone resins varnishes on canvas paintings from the collection of Munch Museum: an interdisciplinary transnational research project. Modern oil paints conference. Amsterdam; 2018a. 23–25th of May 2018.

[53] Mills JS, White R. The organic chemistry of museum objects. London: Butterworths, 1987:118–20.

[54] Bracco P, Sartiani O, Toso C. «Per te parte porto». Tecnica artistica, stato di conservazione, e intervento di restauro dei dipinti. In: M Ciatti, Paola Marini, editor. Andrea Mantegna. La Pala di San Zeno. Studio e conservazione. Firenze: Edizioni Edifir, 2009:125–74.

[55] M Ciatti, C Frossinini, editors. La Madonna di S. Giorgio alla Costa di S. Giorgio. Studi e Restauro. Firenze: Edifir-Edizioni Firenze, 2009:80.

[56] Adrover Gracia I. Applicazioni della spettrofotometria IR allo studio dei beni culturali. Padova, Il Prato: Collana I Talenti, 2001.

[57] Van Keulen H. The analysis and identification of transparent finishes using thermally assisted hydrolysis and methylation pyrolysis-gas chromatography-mass spectrometry. In: Dias MV, editor. Twelfth International symposium on wood and furniture conservation. Amsterdam: Stichting Ebenist, 2014:134–41.

[58] Van Der Doelen GA, Van Den Berg KJ, Boon JJ. Comparative chromatographic and mass spectrometric studies of triterpenoid varnishes: fresh material and aged samples from paintings. Stud Conserv. 1998;43:249–64.

[59] Van Den Berg KJ, Boon JJ, Pastorova I, Spetter LF. Mass spectrometric methodology for the analysis of highly oxidised diterpenoid acids in old master paintings. J Mass Spectrom. 2000;35:512–33.

[60] Boon J, Van Den Berg KJ, Pureveen J, Van Der Doelen G, Groen K, Van Och J, et al. Direct temperature resolved mass spectrometry (DTMS) as a tool to study painted art: a case study of "De Eendenfarnilie" by W. Maris, a nineteenth century Dutch impressionist painter. In: MM Ross, editor. Proceedings of the 43rd ASMS conference on mass spectrometry and allied topics, May 21-26 1995. Atlanta, Georgia, 1, 1995.

[61] Edvard Munch's diary entry 1938, N 47. https://www.emunch.no. Accessed: 18 Oct 2018.

[62] Burns T. The invention of pastel painting. London: Archetype Publication, 2007:3.

[63] Newmann R. Tempera and other nondrying-oil media. In: Dorge V, Hawlett FC, editors. Painted wood: history and conservation, proceedings of the Symposium in Williamsburg. Virginia: The Getty Conservation Institute, 1994:33–64.

[64] Ravensberg diary entry. LR 536; Munch Museum archives, 1910a.

[65] Colombini P, Cantisani E. Unpublished Scientific Report. Department of Chemistry and Industrial Chemistry, University of Pisa, 2015.

[66] Ravensberg diary entry. LR 537: Munch Museum archives, 1910b.

[67] Colombini MP, Modugno F, Sandbakken EG, Tveit ES, Zanaboni M. Chemical investigation of paint media in Edvard Munch's Aula sketches (1909-1916). In: Frøysaker T, Streeton N, Kutzke H, Hanssen-Bauer F, Topolova-Casadiego B, editors. Public paintings by Edvard Munch and his contemporaries – change and conservation challenges. London: Archetype Publication, 2015:197–204.

[68] https://en.wikipedia.org/wiki/Lanolin. Accessed: 18 Oct 2018.
[69] Colombini P, Cantisani E. University of Pisa. Private communication, 2015.
[70] Zöld K-J. What will happen to a Frankenthaler? An investigation into the oil stain effect on un-primed and un-sized cotton canvases. Thesis, Queens University, Kingston, 2002.
[71] Reid Z, Mcguinne N, Fields E. The yeats archive: a method of identifying wax crayon, in works of art on paper, books, documentation and photographs, techniques and conservation. IIC Baltimore Congress. 2002;176–9.
[72] Gierløff C. Edvard Munch selv. Oslo: Gyldendal Norsk Forlag, 1953:205 & 208.
[73] Ravensberg diary entry. LR 562: Munch Museum archives, 1910c.
[74] Sandbakken EG, Boon JJ. An analytical survey of painted areas in poor condition in Munch's first monumental sketch for the researchers. In: Frøysaker T, Streeton N, Kutzke H, Hanssen-Bauer F, Topolova-Casadiego B, editors. Public paintings by Edvard Munch and his contemporaries – change and conservation challenges. London: Archetype Publications, 2015:181–96.
[75] Werenskiold Erik in letter to Edvard Munch 1910, MM K 1239: https://www.emunch.no. Accessed: 16 Jul 2018.
[76] Zanaboni M, Modugno F, Columbini MP. Unpublished Scientific Report. Department of Chemistry and Industrial Chemistry, University of Pisa, 2013.
[77] Doerner M. The materials of the artist, and their use in painting. Rev ed. New York: Harvest, Hartcourt Inc., 1984.
[78] Sandbakken EG, Sink, Magnesium og sinkmagnesium sulfater i overflaten på 15 av Edvard Munchs Aula-skisser, Master thesis, University of Oslo, 2014:35.
[79] EU-ARTECH. Molab campaign to the Munch museum. unpublished results of XRF-analyses, 2009.
[80] Mactaggart P, Mactaggart A. A pigment microscopist's notebook. 7th ed. Somerset: Published by the authors, 1998.
[81] Shelley M. American pastels, of the late ninetheenth and early twentieth centuries: materials and technique. In: American pastels in the metropolitan museum of art. New York: Harry N. Abraham Inc., 1989:33–45.
[82] Epley B. Jan Both's Italian landscape. In: Massing A, editor. Hamilton Karr Institute Bulletin, number 3. Cambridge University, 2000:127–35.
[83] Groen K. Scanning electron-microscopy as and aid in the study of blanching. In: Mcclure I, editor. Hamilton Karr institute bulletin, number 1. Cambridge University, 1988:48–65.
[84] Easthaugh N, Walsh V, Siddal R. Pigment compendium: a dictionary and optical microscopy of historical pigments. Oxford: Butterworth-Heinemann, 2008.
[85] Kutzke H. Private communication. University of Oslo, 2014.
[86] Strang T, CCI. Private communication, 2014.
[87] Beck W. Treatment of efflorescent salts on twentieth-century frescos in the parliamentary precinet, Ottawa. In: The decorative, conservation and the applied arts, IIC vienna congress. 2012, 342–3 (extended poster).
[88] Mora P, Philippot P. The conservation of wall paintings. London: Butterworth-Heinemann, 1999:180.
[89] Derluyn H. Salt transportation and crystallization in porous limestone. Neutron-X-ray imaging and poromechanical modeling. PhD thesis, Leuven University, 2012: 56.
[90] Roig A, Cayuela ML, Sanchez-Monedro MA. The use of elemental sulphur as organic alternative to control pH during composting of olive mill wastes. Chemosphere. 2009;57:1099–105.
[91] Yin XH, Gwathmey CO, Main CL. Sulfur effects on cotton yield components. Better Crops. 2012: 96. Tennessee.

[92] Green S. An outline history of sizing methods with special references to practises at hayle mill. In: Fairbass S, editor. The institute of painting conservation, conference papers. Manchester, 1992:197–206.
[93] Zarchy H. Let's go camping: a guide to outdoor living. New York: Knopf, 1951.
[94] Hoffman W. Patent, serial number 75.437, New York, 1929.
[95] Van Den Berg JD. Analytical chemical studies on traditional linseed oil paints. Molart report 6, Amsterdam: AMOLF, 2002:39.

Maria J. Melo, Paula Nabais, Rita Araújo and Tatiana Vitorino

16 The conservation of medieval manuscript illuminations: A chemical perspective

Abstract: Illuminated manuscripts are the most abundant and well-preserved surviving medieval cultural artefacts. Created to contain sacred texts, their visual structure allows the reader to identify divisions and delight in their beautiful ornamentation and iconography. Western European manuscripts were written on parchment, which was the main writing support in the Middle Ages prior to the rise of paper production in the fourteenth–fifteenth centuries. Highly functional, durable animal skins were also used in medieval bookbinding (covers) and have been crucial in the preservation of the illuminations. These illuminations make wonderful use of form and colour. This chapter focuses on recent advances in the molecular characterization of these colours, used in medieval manuscripts produced by Western Europeans, and the challenges inherent in analysing materials that are intrinsically heterogeneous. We then discuss how molecular characterization may reveal conservation conditions and extend our analysis to parchment and iron gall inks. Finally, we address the challenges and possibilities for this flourishing field of research.

In short, we show in this chapter how analytical methods are used:

- To understand how medieval illumination colours were made and what palette defines them;
- To predict original colours and to understand why certain colours have remained in excellent condition, preserving both adherence and luminosity, while others have changed over the centuries;
- To assess deterioration (paints, parchment, writing inks) with the goal of determining what potential restoration measures could be taken, after careful consideration.

Keywords: manuscript illumination, medieval pigments & dyes, iron gall ink, parchment, molecular characterization

16.1 Medieval Illuminations materials and techniques

16.1.1 An interdisciplinary approach in the study of manuscript illuminations [1–3]

We start this chapter by introducing the scholars who pioneered a systematic interdisciplinary research on the colourants used in medieval manuscript collections at

This article has previously been published in the journal *Physical Sciences Reviews*. Please cite as: Melo, M., Nabais, P., Araújo, R., Vitorino, T. The conservation of medival manuscript illuminations: a chemical perspective *Physical Sciences Reviews* [Online] **2019**, 4. DOI: 10.1515/psr-2018-0017.

https://doi.org/10.1515/9783110457537-016

the molecular level, from a global, holistic perspective: Mark Clarke, Claude Coupry, Robert Fuchs, Doris Oltrogge, Mary Virginia Orna, Salvador Muñoz-Viñas and Eugene F. Farrell [4–10]. Historically, one of the primary obstacles to analyzing illuminated manuscripts has been the restriction on sampling [5]. Until recently, that meant it was only possible to identify a certain proportion of inorganic pigments using Raman microscopy and Energy dispersive X-ray fluorescence (EDXRF) [11] and obtaining vague indications of organic pigments, usually by UV-VIS spectroscopy, without any analysis of the binding media. To understand the strengths and limitations of the various physical methods used at the time to characterize paints in medieval illuminated manuscripts, please see Clarke's review [5].

On the one hand, over the past decade, advanced analytical techniques and new approaches have enabled these obstacles to be overcome. Surface Enhanced Raman Spectroscopy (SERS) combined with microspectrofluorimetry in the visible (VIS) has proved to be a powerful technique for in-depth characterization of dye-based paints [12–14]. Fiber Optic Reflectance Spectroscopy, FORS, in the VIS-NIR is a promising tool for both colourants and binder identification [15–18]. On the other, our research on medieval illuminations and interdisciplinary approach has proven that micro-sampling minimizes handling of the manuscript, and in situ techniques minimize the number of micro-samples collected, enabling quality information and a complete description of the paint to be obtained [3, 19–23]. Given the fragility of the manuscripts and the illuminations, this *damage to knowledge-obtained* ratio is considered highly favourable by conservators [24, 25]. Micro-sampling combined with in situ analysis allows us to understand the complex degradation mechanisms that cause the collapse of certain paints. This knowledge establishes the groundwork for the development of crucially needed sustainable conservation treatments.

16.1.2 A *modus operandi* that combines *in situ* analysis and micro-sampling

After examining the illuminations with the naked eye and under the microscope, we began to explore the molecular universe of the manuscripts. Our analysis combines elemental with spectroscopic techniques. Raman microscopy (microRaman), Fourier Transform Infrared spectroscopy (microFTIR), emission fluorescence spectroscopy (microspectrofluorimetry), and Fiber Optic Reflectance Spectroscopy in the UV–VIS region are powerful complementary spectroscopic techniques for the characterization of colourants in medieval manuscripts. Raman and infrared spectroscopy reveal a "molecular fingerprint": if a single compound is present, it is possible to unequivocally characterize it. Complexity arises because we are usually faced with aged, superimposed mixtures of compounds. Another layer of difficulty is added when a compound is not present in our reference database.

The creation of a database of colourants, binders, and colour paints was critical to our studies of illuminations, Figure 16.1 and Figure 16.2. These standards were

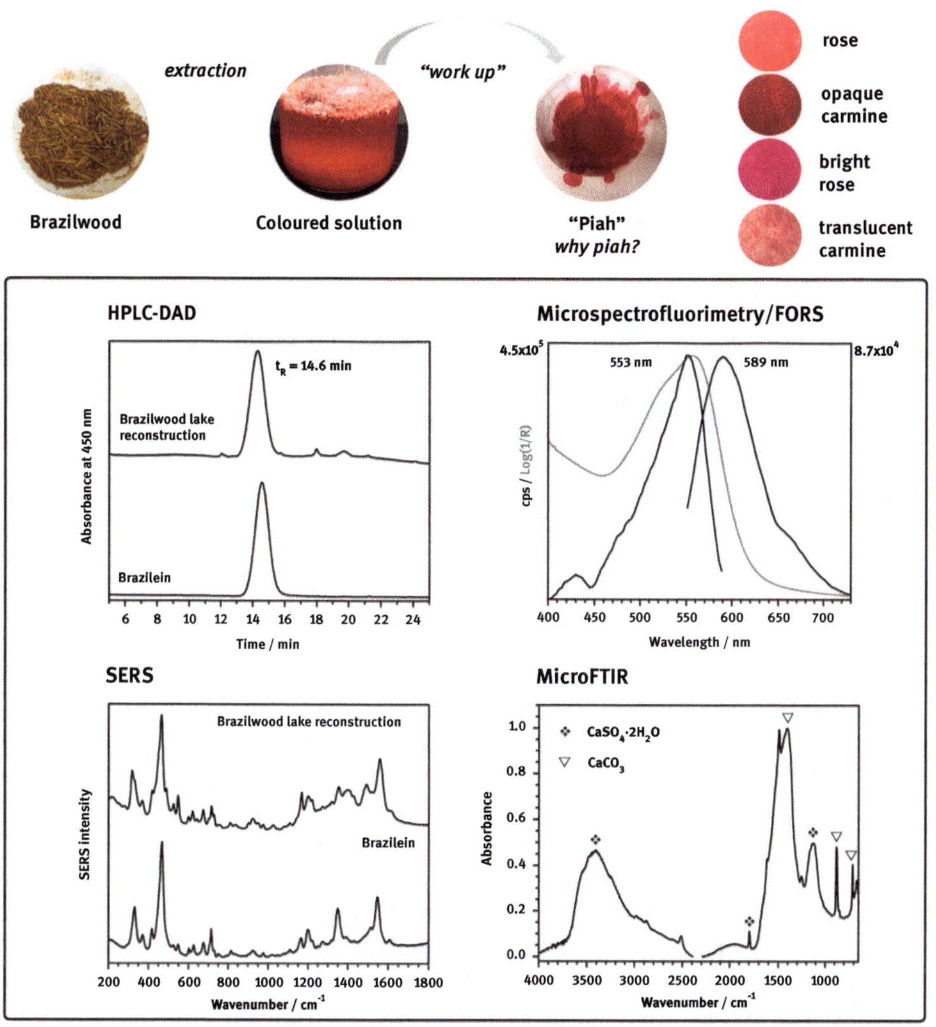

Figure 16.1: Production and characterization of brazilwood lake pigments following the four recipes to produce rose colours from brazilwood in *Livro de como se fazem as cores* [26–28]. The acid or basic extraction determines the hue of the pigment and the additives added during "work up" influence its opacity. Filtration over a chalk or gypsum stone (*piah feitah de gis o de pedra kri*) improves the process and introduces chalk or gypsum into the final formulation. HPLC-DAD confirms the presence of brazilein in the pigments produced. Microspectrofluorimetry and FORS are used for the in situ identification of the chromophore. SERS and microFTIR provide a molecular fingerprint of brazilein and the additives, respectively.

determined by our own analyses of medieval paints and were produced by following and adapting the procedures described in written sources, such as recipe books and medieval technical treatises.

Figure 16.2: The main colourants characterized in medieval manuscripts (eleventh to the fifteenth century).

During an on-site investigation, the first screening involves FORS and microEDXRF, the latter of which indicates the possible colourants and extenders present and allowed for the initial quantification of these elements. Moreover, the 70 μm offered by the microEDXRF enabled us to obtain data on the distribution of a certain paint colour throughout the manuscript, Figure 16.3.

MicroRaman and microspectrofluorimetry allow for high spatial resolution (1–5 μm spot), so that diverse paint components can be analyzed separately. In addition, paints were also analyzed by optical microscopy to understand how the final colour is built up, to detect possible degradation phenomena, and to sample the colour paints that will be subjected to a more detailed characterization. In the laboratory, a sample typically will first be analyzed using microFTIR to characterize binders and gain insight into the full paint formulation. If a dye is present, microspectrofluorimetry may be carried out first, as it requires no contact with the colour paint and employs low-intensity radiation. SERS will be utilized on at least one micro-sample to determine the molecular fingerprint of the dye. Depending on degradation phenomena or other challenging aspects, other analytical techniques may also be used.

16.2 The making of medieval pigments and paints: The quest for a reference database

In situ techniques, such as microfluorimetry and FORS, can only be used for an in-depth analysis if supported by a made-to-measure reference database. Likewise,

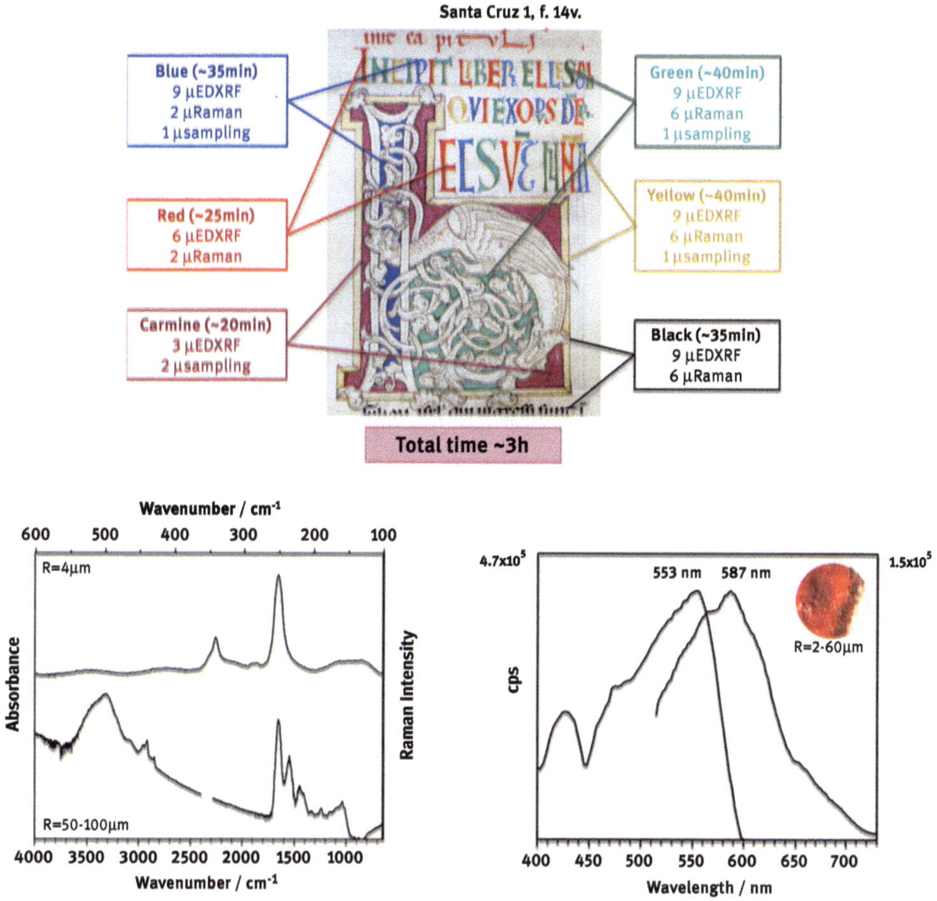

Figure 16.3: The carmine/pink colour used in this Romanesque initial was made admixing vermilion with lac dye in a proteinaceous binding medium. Vermilion was identified through **Raman microscopy** by its characteristic bands at 252 cm^{-1} and 282 cm^{-1}. **Infrared spectroscopy** (*bottom left*) clearly shows the proteinaceous binder fingerprint. **Microspectrofluorimetry** (*bottom right*) and surface-enhanced Raman spectroscopy identified lac dye by comparison with references. Spatial resolution indicated for each technique (R). Santa Cruz 20 manuscript, 12th c., in the collection of Biblioteca Pública Municipal do Porto.

when considering the inherent heterogeneity of the materials studied, the development of model references has been critically important for molecular fingerprint techniques such as Raman and infrared spectroscopy, Figure 16.1. They are also useful in assessing the strengths and limitations of the various physical methods used to characterize paints in medieval illuminated manuscripts.

For this reason, over the past 15 years, we have been studying and reconstructing medieval processes for making pigments and paints that were used to create medieval manuscript illuminations [26, 29, 30]. Our primary approach has involved the establishment of a dialogue between the medieval written sources and the multi-analytical molecular characterization of the original colours. Reproductions are made with as much historical accuracy as possible, using shortcuts offered by modern chemistry, and validated as a reference through comparison with the molecular analysis of the original colours found in the illuminations [Appendix 1 in 26]. The success of the validation depends on the depth of the molecular characterization, since the paints found in medieval illuminations are heterogeneous formulations which include pigments, binders, additives and varnishes. If from a molecular and morphology perspective, the reproduced colour paint matches the original, it is considered a reconstruction and is included in a database.

In this approach, the processes for making pigments and paints for manuscript illuminations, presented in the remarkable Portuguese Hebrew text, copied in the 15th c. [31, 32], have been reconstructed and the knowledge obtained has now been published [26, 29, 30]. Ten years of systematic experimentation, now available in the electronic appendix of [26], allowed us to conclude that the descriptions for making colours and paints are brief and accurate, and in many cases, they include additional information on the most critical steps, which can be crucial to the practitioner. Other important medieval technical texts have been recently edited and studied, adding to the number of written sources that can be used to reconstruct the medieval processes to make pigments and paints [33–36]. Colourants described therein are included in our reference material [26, 29, 30], Figure 16.1.

16.3 Medieval colours: Challenges to analysis and recent advances

16.3.1 The colourants in medieval manuscript illuminations: Towards a chronology

Our investigations of colour materials in Portuguese monastic manuscripts, French and Flemish books of hours have led us to argue that organic colourants are one of the most important materials for establishing a chronology, in particular the red hues [1, 19, 21, 28]: carmine,[1] pink, and purple.[2] Therefore, based upon studies that have characterized the red dyes used in medieval European manuscripts [14, 27, 28, 37], and also upon a critical analysis of medieval technical treatises, we find the following chronology for the use of red/carmine/purple colours;[3] lac dye red was used in

1 Carmine is used thorough the text as "hue", the attribute of colour perception denoted by carmine, and is not related with the molecule carminic acid.

2 Purple hues, the colourant typically absorbs between 500–550/600 nm.

3 Orcein purple is not considered in this study.

monastic scriptoria during the twelfth and thirteenth centuries [14, 38], while brazilwood-based carmine and pink colours were used from the fourteenth century on, both in monastic scriptoria and lay workshops [21, 27, 28, 37, 39].[4]

With respect to inorganic pigments, Figure 16.2, the chronology of the use of yellow is perhaps the most revealing [19]. The synthetic pigment known as mosaic gold may have been introduced in the thirteenth century. Variations in its production can provide important clues as to *when* and *where* a colour paint was prepared. Based on the results of studies of European manuscripts, we can draw the following conclusions about the use of inorganic yellows: orpiment was used in monastic scriptoria from the eleventh to thirteenth centuries; mosaic gold was used possibly from the end of the thirteenth century or beginning of the fourteenth century; and lead tin yellow began to be used in the fifteenth century.

16.3.2 Dyes: disclosing a paint's formulation

Dye characterization in works of art is a challenging endeavour [19, 27, 28, 37, 39–41]. We need to address the fundamental chromophore as well as the full "formulation" to get in-depth knowledge about the culture that produced these precious and beautiful colours - which are centuries or even millennia - old. High-performance liquid chromatography with a diode-array detector (UV-VIS) and electrospray ionisation-mass spectrometry (HPLC-DAD-MS) combines high separation power with an almost perfect dye characterization (only the binder will remain invisible), but it is very difficult to have a micro-sample from an illumination with enough material to be extracted for HPLC. So, we devised other approaches that combine high spatial resolution, the highest sensitivity (microfluorimetry) and a molecular fingerprint (SERS). We needed to combine at least two analytical techniques to achieve our goal (dye identification), but to know more about lake pigment formulations, we also needed to include FORS and microFTIR. Finally, it is critical to have a reliable reference database. UV-VIS spectra in absorbance or reflectance are molecular spectra, however, they do not have the fingerprinting capability of Infrared and Raman techniques. Nowadays, there are many publications using UV-VIS spectra to provide strong evidence on the presence of a main chromophore when sometimes only an indication is possible. For the authors, data representativeness and a clear identification of the chromophore is fundamental for an appropriate investigation of medieval dyes.

In Figure 16.3, we describe how lac dye is identified in a 12th c. initial. Microspectrofluorimetry and SERS allowed for the identification of lac dye mixed

4 Before the arrival of the Portuguese in Brazil, in 1500, redwood was imported to Europe from Ceylon (Sri Lanka). Its presence in European trade peaked in the fifteenth to sixteenth centuries. In medieval technical treatises this colour was referred to as "brazil" or a similar term.

with the vermilion. Infrared spectroscopy allowed the identification of a proteinaceous binder and the filler (gypsum) [38]. If a semi quantification of vermilion was considered necessary, references could have been prepared for microEDXRF (by using, e. g. the intensity ratio for Ca present in the filler and Hg).

16.3.3 Egyptian blue, an unusual pigment in medieval ages?

This is a case whereby advances in scientific analysis have challenged the prevalent romantic perspective that the production and usage of Egyptian blue ended with the collapse of the Roman Empire [42]. Making us rethink the sources and significance of the use of Egyptian blue in medieval Europe.

Egyptian blue is an extraordinary and inspiring material that dates to the 4th Dynasty in Egypt (around 2500 BC) [43–45]. It shows an exceptionally high luminescence quantum yield for an infrared emitter, ϕ_{Φ} = 10.5%, and a long luminescence lifetime, making it a perfect example for identification relying on luminescence techniques [44].

It is considered the first synthetic pigment to be used in antiquity;

> on tomb paintings in Egypt from around 2300 BC, and during the subsequent 3000 years, its use as a pigment and in the production of small objects spread throughout the Near East and Eastern Mediterranean and to the limits of the Roman Empire [46].

According to Marco Verità, "Egyptian blue is a multicomponent material whose colour is due to the presence of calcium-copper tetrasilicate crystals (cuproivaite: $CaCuSi_4O_{10}$) produced by firing a batch (mixture) of quartz, lime, copper and a small amount of a soda flux", and depending on the time and the geographic area where it was produced the matrix composition displays interesting variations [47].

Italian researchers were the first to publish irrefutable data on its continued use in medieval Italy; e. g. in Rome until at least the 11th c. AD [47–49], and Genoa and Treviso (Italy) during the 13th c [49].

Particularly interesting, is the first description of its use admixed with lapis-lazuli (by optical microscopy and SEM-EDS) in the wall paintings of the church of San Saba (Rome), dating to the first half of the eighth century AD [49]. This type of mixture was more recently reported to be found in, (i) four manuscripts associated with Canterbury scriptorium, from c. 1000 to c. 1100" [50]; and a Carolingian manuscript, the Godescalc Evangelistary dated 781–783 (BnF 1203), in a detail of peacock bodies [51].

16.3.4 Binding media: towards an in-depth characterization

Medieval binders, such as a protein or a polysaccharide, display a very clear infrared signature. The interpretation of these molecular spectra is made following a stepwise approach. In a first step, the main bands are identified and, simultaneously, their relative intensities and positions are verified. Exemplifying the second step for a proteins' spectra that are very well described in textbooks on infrared spectroscopy, Figure 16.3 and Figure 16.4, the protein structure is analysed, to check if there are other peaks present, e. g. peaks for oxalate compounds known to be final degradation products of proteins, oils and polysaccharides. This in-depth analysis provides a great deal of relevant information and is seldomly performed [52–58]. An example of a paint where the proteinaceous binder displays extensive degradation is the colour we named as *bottle green* [57].

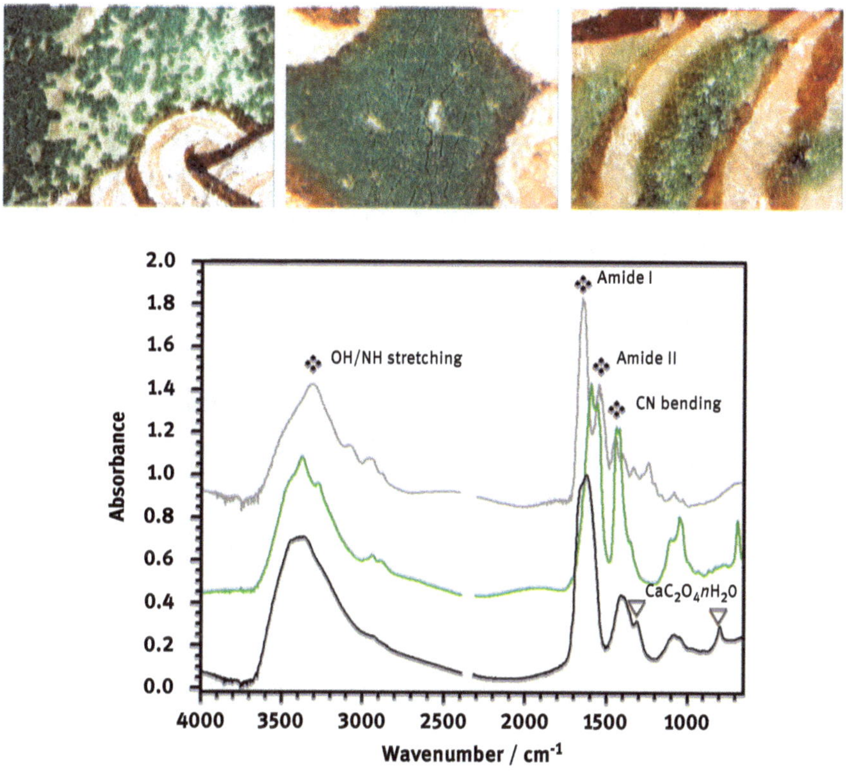

Figure 16.4: Details of bottle green paints in 12th c. Portuguese manuscripts in the collection of Biblioteca Pública Municipal do Porto. Infrared spectra of parchment glue (collagen based binder) and an original bottle green paint, where the binder degradation is visible through the presence of calcium oxalate at 1132 and 784 cm^{-1} (▽) as well as through the loss of structure in the two main amide bands (amide I and II,❖).

The original *bottle green* paint was a shiny colour, but at present, even though the brightness and the saturation of the deep green has been preserved, the proteinaceous binding media shows extensive chain scission and cross-linking, which is confirmed by the presence of calcium oxalate as will be discussed in more detail in the next section. The consequences of this molecular evolution are an extensive loss of paint cohesion and adhesion to the support. The green colour is physically detached, and thus lost. Given that the green occupies 20 % of the entire coloured area, this is a critical issue for the preservation of Romanesque Portuguese manuscripts.

Thus, infrared spectroscopy is an important technique not only for determining the binding media, but also its conservation condition. The rich information included in an infrared spectrum may further provide precious information on a protein structure as Barbara Brosky Doyle's 1975 work clearly discusses [58].

16.3.5 Polyphenol complexity in medieval iron gall inks disclosed by HPLC-DAD-HRMS

Iron gall inks pose a major conservation issue in heritage collections due to degradation. In medieval treatises, recipes for making these inks described the use of plant extracts such as *Quercus infectoria*, which were combined with iron salts [59–62]. This mixture resulted in an iron-polyphenol complex, to which a polysaccharide, such as gum arabic, was added [61, 62], Figure 16.5.

Figure 16.5: Main steps and ingredients in the production of iron gall ink: an extract solution from galls, to which iron sulphate is added, forming a black precipitate. Gum arabic is added to keep the precipitate in suspension.

The models used by the scientific community to describe the iron complexes assume that gallic or tannic acid were available in the gall extracts to bind iron [63]. To test if this is the case, reproductions of medieval Iberian inks were prepared to investigate the molecular structures of the chromophores in iron-gall inks [61]. Quantification of gallic acid both by HPLC-DAD and HPLC-ESI-MS show that it is a minor compound in most gall extracts prepared following Iberian recipes, Figure 16.6. These extracts are characterized by complex elution patterns, which include gallic acid as well as the derivatives of gallic acid in the form of polygalloyl esters of glucose, also called gallotannins (mono- to hepta-galloyl glucose structures were identified) [61].

Iron gall ink reproductions and a rigorous chemical analysis proved essential to show that we must shift from model compounds acid gallic-based to polygalloy ester-based.

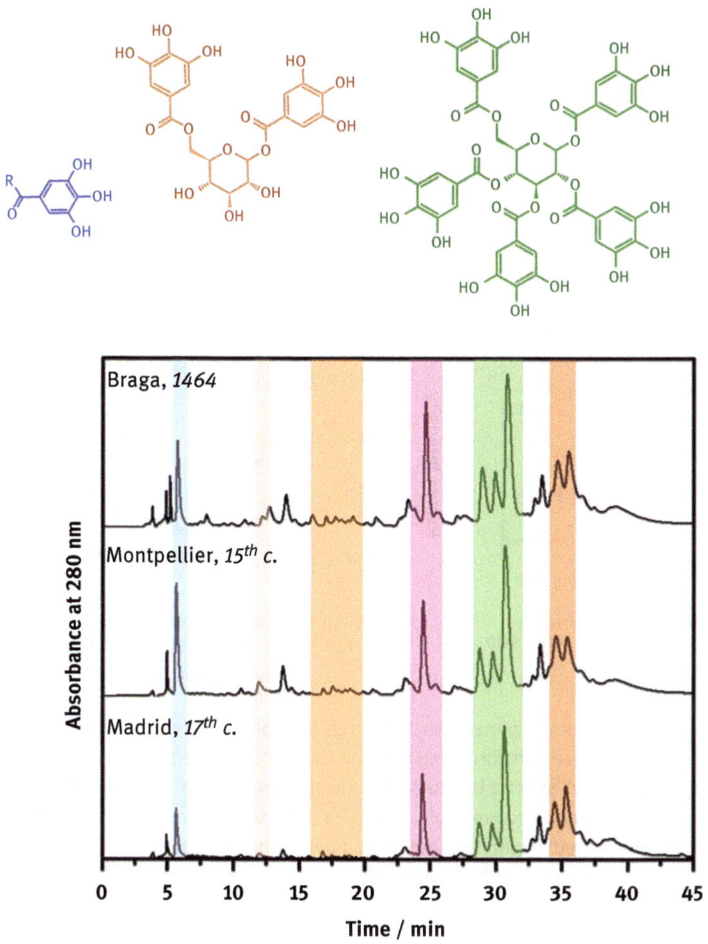

Figure 16.6: The characterization of historical inks proved that polygalloyl esters of glucose, such as digalloyl glucose (brown) or pentagalloyl glucose (green), are present in much higher quantity than gallic acid (blue). This characterization reveals that the paradigm must shift from model compounds acid gallic-based to polygalloy ester-based.

16.4 Medieval paints degradation at the molecular level

The primary finding of our studies on Books of Hours (14th-15th c.) and on three important medieval Portuguese Romanesque manuscript collection's illuminations (12th-13th c.) show that the most frequently observed deterioration pathologies fall into two categories [52], Figure 16.4, Figure 16.7 and Figure 16.8.

(i) Changes in colour owing to chemical transformation (*pigment based paint degradation*), as found in *minium* and lead white, which have darkened dramatically

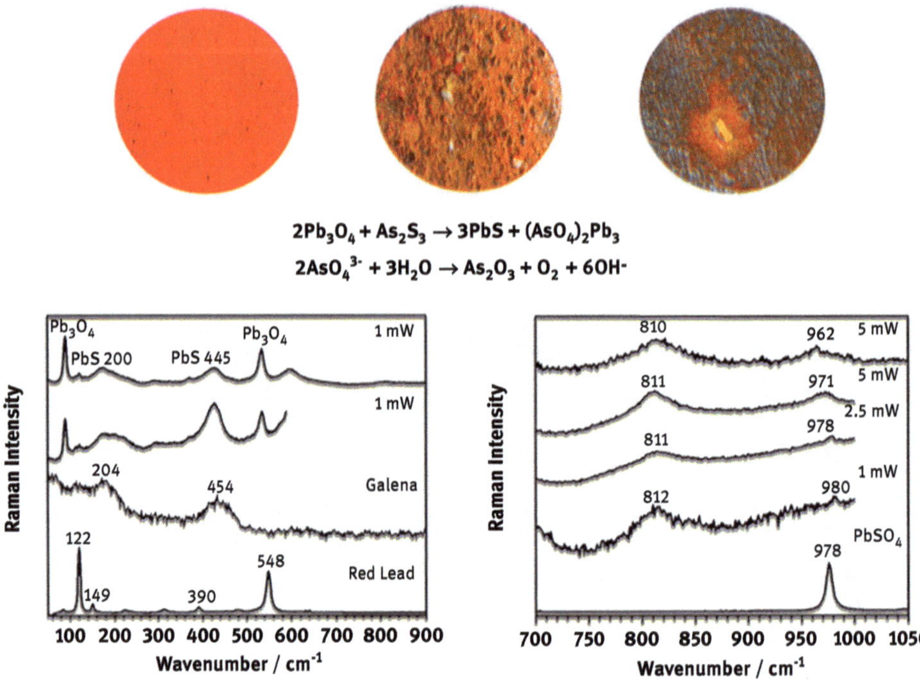

Figure 16.7: In Lorvão Beatus (1189), where orpiment (As_2S_3) has been applied as a strong yellow colour (never admixed with Pb_3O_4), the extensive transformation of red lead into black galena (PbS) was proved by Raman microscopy (acquiring "noisy" spectra in situ and better signals in micro-samples). Raman spectra were acquired, in situ, on the darkened areas of Lorvão Beatus (laser power set at 1 mW), where the main bands of red lead are seen together with two large broad bands that appear at ca. 200 and 445 cm^{-1}, corresponding to galena (spectra on the left). Additionally a 810 cm^{-1} band was observed and attributed to a lead arsenate species (spectra on the right). For more details please see [53].

[53, 54], Figure 16.7. The same holds true for silver-based colours, profusely applied in the Books of Hours [55], Figure 16.8. In both cases, we found black sulfide based compounds as final products, such as PbS and Ag_2S [53, 55].

A closer look at the mechanism for Ag_2S formation reveals that the equation used by the cultural heritage community to describe how silver tarnishes, falls short of a true understanding of the primary factor causing the severe and extensive darkening in medieval illuminated manuscripts [55, 64], Figure 16.8. This equation depicts the oxidation of silver instantly with a sulfide containing species as H_2S. Recent research has confidently proved that the first intermediate of silver oxidation is Ag_2O [55, 64]. First, oxygen is absorbed on the surface of silver. The biradicaloid nature of molecular oxygen is the driving-force for the first electron transfer from silver to oxygen, forming intermediates of the type O_2^{2-}. The complete transfer of 4 e$^-$ enables the formation of O^{2-} and consequently of Ag_2O. Oxygen is replaced

	E^o/V	ΔG^o/kJ/mol
$O_2(g) + 4Ag\ (s) \rightarrow 2Ag_2O\ (s)$	0.43	– 22.42
$Ag_2O\ (s) + H_2S\ (g) \rightarrow Ag_2S\ (s) + H_2O$ (aq	–	– 233.26
$Ag_2O\ (s) + COS\ (g) \rightarrow Ag_2S\ (s) + CO_2\ (g)$	–	– 256.94

Figure 16.8: Silver-based colours in 15th c. Books of Hours exhibit a complex stratigraphy that impacts conservation conditions. To understand how silver colours are formed, and how the several layers and components influence silver degradation, different sample preparation methods were designed to enable the analysis of the fragile and degraded stratigraphy by SEM–EDS. SEM studies were supported by a previous in situ analysis (optical microscopy, μ-EDXRF, μ-Raman and μ-FTIR) [55]. In the original silver acanthus leaf depicted in the figure silver is severely altered, the metal lustre is lost and the colour is perceived as black due to Ag_2S that was formed following the main steps depicted in the equations [64]. Originally silver was applied over a ground prepared with chalk and gypsum, within a contour draw with brazilwood lake pigment. The ground, in turn, is partially applied over the blue colour made with azurite and lapis lazuli.

by S^{2-}, when available, due to its very low solubility constant (K_{sp} $Ag_2O = 2.6 \times 10^{-8}$; K_{sp} $Ag_2S = 6.0 \times 10^{-50}$).

(ii) Loss of cohesion and adhesion of the pictorial layer to the support, exemplified with *bottle green* and lapis lazuli blue [52], Figure 16.4. Both phenomena led to extensive detachment of the original colour due to polymer degradation (*binder based paint degradation*). The presence of low quantities of binding media in the paint formulation might explain the lack of adhesion sometimes observed on lapis lazuli and lead white paints.

Bottle green degradation is not related to a low amount of binder, as we already pointed out. When analyzing for the presence of calcium oxalate in all the infrared spectra collected for medieval colours in three important Portuguese Romanesque manuscript collections (12th-13th c.), we concluded that it is frequently present in bottle green colour, but rare in others. More importantly, in bottle green paints a high amount of calcium oxalate strongly correlates with the collapse of amide I and II bands (C=O and O=C-N stretching at 1653 and 1550 cm^{-1}, respectively) in a single broad band, and to the loss of cohesion of the bottle green. Therefore, we posit that the copper pigment is promoting protein degradation through a radical mechanism that led to extensive chain scission and cross-linking, where calcium oxalate is one of the final products [52, 57], Figure 16.4. This hypothesis is in agreement with the findings of Nati Salvadó and her colleagues, proposing that the presence of calcium oxalate, as weddellite (dihydrate form) and/or whewellite (monohydrate) in Cataluña medieval paints, may be ascribed to binding media degradation [56].

In brief, oxalate has been proposed as a marker for binding media degradation in medieval art by Nati Salvadó's team [56], and it can be monitored straightforwardly through its characteristic infrared bands at 1132 and 784 cm^{-1} [57].

It is also important to note that when we compare the general state of conservation of the pictorial layers of the 12th–13th c. illuminations with the 15th c. Books of Hours in Portuguese collections, we conclude that the latter are generally in worse conditions [52]. We propose that this is related to their interesting history of possession, where the same book could be rebound several times throughout its lifetime, being rebound in a manner that followed the style of the time, and were poorly protected. Additionally, these small books - produced for private devotion for its owners - were much handled and read. On the other hand, Romanesque books maintained their original bindings to the extent possible and were only replaced once functionality was lost.

16.5 Assessing parchment degradation using micro and nano IR

If the paint's parchment support fails, the paint is lost. Thus, condition assessment of the support is critically important as proven in the EC project *Methods in the Microanalysis of Parchment* and *Improved Damage Assessment of Parchment* (IDAP) [65]. In IDAP, gelatinization was considered the last degradation phase that leads to irreversible collapse. When denaturation occurs, the native fibrillar collagen, organized in a right-handed triple helix of three polypeptide chains [66], unfolds into a random coil structure. This leads to the formation of water-soluble gelatin[5] [67]. Thus, early-warning systems are necessary to identify collagen in a "pre-gelatinized" state, triggering suitable treatments or conservation conditions.

Fortunately, as stated by Holl et al.,

> recent advances in the development of atomic force microscopy (AFM) combined with infrared spectroscopy offer the promise of direct chemical detection (molecular vibrations) with a spatial resolution of ~20–100 nm and thus the possibility of studying collagen, including its intrinsic heterogeneity, over important length scales [66].

In fact, the degradation of collagen was studied in parchments supports and book covers using *(i)* microIR and, *(ii)* nanoIR combined with AFM and nonlinear optical microscopy [68, 69]. These approaches, although still in their developmental phases, offer substantial promise for measuring collagen in a 'pre-gelatinized' state as the two case studies summarized below show.

---

5 In manufacturing, the creation of gelatin results in the partial cleavage of cross-links that stabilize the triple helix, accompanied by partial hydrolysis of the polypeptide chains. The original "structure is broken down to such an extent that "warm water-soluble collagen is formed"; i. e. gelatin", p48 in [67].

In 2016, the team lead by Latour and Robinet presented, in the authors' own words,

> the correlative imaging of collagen denaturation by nonlinear optical microscopy and nanoscale infrared spectroscopy to obtain morphological and chemical information at different length scales. Such multiscale correlated measurements are applied to the investigation of ancient parchments (. . .) [68]. The main issue is to characterize gelatinization, the ultimate and irreversible alteration corresponding to collagen denaturation to gelatin.

In 2018, the Kazarian group studied samples from original book covers dated between sixteenth and nineteenth century from the collection of the Mesrop Mashtots Institute of Ancient Manuscripts, in the Matenadaran, Yerevan, Armenia [69]. An FTIR spectroscopic imaging apparatus coupled with a single reflection diamond ATR accessory for the macro mode allowed the visualization of the gelatinization of collagen and the detection of metal carboxylates such as calcium stearate. The setup used provided an image size of ~600 x~ 550 µm with a spatial resolution of 15–20 µm, in the range of 3800–850 cm^{-1} [69].

16.6 Summary

To preserve manuscript illuminations for future generations, we must combine both human and natural sciences methodologies. This holistic approach provides an in-depth examination of the creative processes, material composition of medieval illuminations, techniques used, and deterioration processes.

Over the past seven years, breakthroughs supported by advanced analytical techniques have fostered a flourishing field of research, and the study of the materials and techniques used to produce medieval illuminations is reaching maturity. Yet, we have crucial gaps in our knowledge that hinder the development of informed conservation strategies, e. g. on binding medium and lake pigments. Binding media and parchment support are key to the permanence of colours, but their degradation, at the molecular level, is yet to be assessed. A key issue to overcome in the next few years, with a little help from paleoproteomics, that will allow confident sequencing of ancient proteins [70]. Breakthroughs in our knowledge of molecular structures of lake pigments are anticipated through new powder XRD technologies, solid state RMN and theoretical previsions [71]. Multi and hyperspectral mapping systematically applied to the study of important manuscript collections [72, 73] will bring a wealth of important data, but also the challenge of handling and sharing "big data".

16.7 Final thoughts

A word of caution concerning "fear" of microsampling that has led to the pervasive one-sided perspective that deems in situ techniques as "non-destructive" and sampling as "destructive" [25]. Now is the time for Jan Wouters' "*damage to knowledge-obtained* ratio" to bring us to an appropriate perspective [24]. This discussion is important because imaging techniques are increasingly accessible and employ high-energy radiation in the form of X-rays or very intense beams as lasers. Both can produce damage at a molecular level that will only be revealed over the medium-term. This is particularly true in works of art like illuminations, where no varnish is present, and which may still have the original binding media, and have an eclectic palette that makes extensive use of dyes. Why are binders and dyes vulnerable? Because they are prone to irreversible degradation processes caused by radical chain reactions that may be catalysed by high intensity or high energy beam sources. Spot analysis and micro-sampling limits the possible damage to very small, well-defined areas, and mapping can bring undetected damage to a complete illumination. Thus, it is crucial to design experiments to test their safeness as exemplified in recent publications for synchrotron radiation [74].

Funding: These studies were supported by the Portuguese Science Foundation through four research projects (currently ongoing, PTDC/QUI-OUT/29925/2017) and three Ph.D. Grants, including the three awarded to Paula Nabais, Rita Araújo and Tatiana Vitorino (CORES Ph. D. programme PD/BD/105895/2014, SFRH/BD/52314/2013 and PD/BD/114412/2016), and through the Associate Laboratory for Green Chemistry- LAQV which is financed by national funds from FCT/MCTES (UID/QUI/50006/2019). Support was also given by the Calouste Gulbenkian Foundation award "Estímulo à Investigação 2016" (146301).

References

[1] Melo MJ, Castro R, Miranda A. Colour in medieval portuguese manuscripts: between beauty and meaning. In: Sgamellotti A, Brunetti BG, Miliani C, editor(s). Science and Art: the painted surface. Cambridge: Royal Society of Chemistry, 2014:170–92.

[2] Melo MJ, Miranda A, Miguel C, Castro R, Lemos A, Muralha VSF, et al. The colour of medieval Portuguese illumination: an interdisciplinary approach. Revista de História da Arte, FCSH-UNL. 2011;N°1 Série W:152–73. http://revistadehistoriadaarte.files.wordpress.com/2011/09/art11.pdf.

[3] Miguélez Cavero A, Miranda MA, Melo MJ, Castro R, Casanova C. Beatus manuscripts under the microscope: the Alcobaça Beatus and the Iberian Cistercian tradition revisited. J Mediev Iber Stud. 2016;8:217–51. DOI: 10.1080/17546559.2016.1221116.

[4] Clarke M. Anglo-Saxon manuscript pigments. Stud Conserv. 2004;49:231–44. DOI: 10.2307/25487700.

[5] Clarke M. The analysis of medieval European manuscripts. Rev Conserv. 2001;2:3–17.

[6] Coupry C. Les pigments utilisés pour l'enluminure à Fécamp aux XIe et XIIe siècles. In: Bouet P, Dosdat M, editor(s). Manuscrits et enluminures dans le monde normand (Xe-XVe siècles). Caen: Presses Universitaires de Caen, 1999:69–79.

[7] Coupry C. A la recherche des pigments. Revista de História da Arte, FCSH-UNL. 2011;N°1 Série W:127–36. http://revistadehistoriadaarte.files.wordpress.com/2011/09/art09.pdf.

[8] Muñoz Viñas S, Farrel EF. The technical analysis of renaissance illuminated manuscripts from the historical library of the University of Valencia. Valencia: Universidad Politécnica de Valencia, 1999.

[9] Mrusek R, Fuchs R, Oltrogge D. Spectral windows on the past – a new reflectographic method for analysis of book illuminations and historical manuscripts. Naturwissenschaften. 1995;82:68–79. DOI: 10.1007/s001140050143.
[10] Merian S, Mathews TF, Orna MV. The making of an Armenian manuscript. In: Mathews TF, Wieck RS, editor(s). Treasures in heaven. Armenian illuminated manuscripts. New York: Princeton University Press, 1994:124–42.
[11] Clark RJH. Raman microscopy: application to the identification of pigments on medieval manuscripts. Chem Soc Rev. 1995;24:187–96. DOI: 10.1039/CS9952400187.
[12] Melo MJ, Claro A. Bright light: microspectrofluorimetry for the characterization of lake pigments and dyes in works of art. Acc Chem Res. 2010;43:857–66. DOI: 10.1021/ar9001894.
[13] Nabais P, Melo MJ, Lopes JA, Vitorino T, Neves A, Castro R. Microspectrofluorimetry and chemometrics for the identification of medieval lake pigments. Heritage Sci. 2018;6:1–11. DOI: 10.1186/s40494-018-0178-1.
[14] Castro R, Pozzi F, Leona M, Melo MJ. Combining SERS and microspectrofluorimetry with historically accurate reconstructions for the characterization of lac dye paints in medieval manuscript illuminations. J Raman Spectrosc. 2014;45:1172–9. DOI: 10.1002/jrs.4608.
[15] Ricciardi P, Delaney JK, Facini M, Zeibel JG, Picollo M, Lomax S, et al. Near infrared reflectance imaging spectroscopy to map paint binders in situ on illuminated manuscripts. Angew Chem Int Ed. 2012;51:5607–10. DOI: 10.1002/anie.201200840.
[16] Cucci C, Delaney JK, Picollo M. Reflectance hyperspectral imaging for investigation of works of art: old master paintings and illuminated manuscripts. Acc Chem Res. 2016;49:2070–9. DOI: 10.1021/acs.accounts.6b00048.
[17] Rosi F, Miliani C, Braun R, Harig R, Sali D, Brunetti BG, et al. Noninvasive analysis of paintings by mid-infrared hyperspectral imaging. Angew Chem Int Ed. 2013;52:5258–61. DOI: 10.1002/anie.201209929.
[18] Ricciardi P, Delaney JK. Combining visible and infrared imaging spectroscopy with site specific, in-situ techniques for material identification and mapping. Revista de História da Arte, FCSH-UNL. 2011;Nº1 Série W:253–61. http://revistadehistoriadaarte.files.wordpress.com/2011/09/art19.pdf.
[19] Nabais P, Melo MJ, Castro R, Sousa L, Lopes GV. Singing with light: an interdisciplinary study on the medieval Ajuda Songbook. J Mediev Iber Stud. 2016;8:283–312. DOI: 10.1080/17546559.2016.1234061.
[20] Barreira C, Melo MJ, Araújo R, Casanova C. Through the eyes of Science and Art: a fourteenth-century winter Breviary from Alcobaça scriptorium. J Mediev Iber Stud. 2016;8:252–82. DOI: 10.1080/17546559.2016.1221119.
[21] Melo MJ, Otero V, Vitorino T, Araújo R, Muralha VSF, Lemos A, et al. A spectroscopic study of brazilwood paints in medieval books of hours. Appl Spectrosc. 2014;68:434–44. DOI: 10.1366/13-07253.
[22] Mas S, Miguel C, Melo MJ, Lopes JA, de Juan A. Screening and quantification of proteinaceous binders in medieval paints based on μ-Fourier transform infrared spectroscopy and multivariate curve resolution alternating least squares. Chemom Intell Lab Syst. 2014;134:148–57. DOI: 10.1016/j.chemolab.2014.03.012.
[23] Moura L, Melo MJ, Casanova C, Claro A. A study on portuguese manuscript illumination: the Charter of Vila Flor (Flower Town), 1512. J Cultural Heritage. 2007;8:299–306. DOI: 10.1016/j.culher.2007.02.003.
[24] Wouters J. Protecting Cultural Heritage: reflections on the position of science in multidisciplinary approaches. Chem Int. 2008;30:4–7.
[25] Melo MJ, Nevin A, Baglioni P. Special issue of pure and applied chemistry devoted to "chemistry and cultural heritage" preface. Pure Appl Chem. 2018;90:429–33. DOI: 10.1515/pac-2018-0106.

[26] Melo MJ, Castro R, Nabais P, Vitorino T. The book on how to make all the colour paints for illuminating books: unravelling a Portuguese Hebrew illuminators' manual. Heritage Sci. 2018;6:1–8. DOI: 10.1186/s40494-018-0208-z.

[27] Vitorino T, Melo MJ, Carlyle L, Otero V. New insights into brazilwood lake pigments manufacture through the use of historically accurate reconstructions. Stud Conserv. 2016;61:255–73. DOI: 10.1179/2047058415Y.0000000006.

[28] Melo MJ, Nabais P, Guimarães M, Araújo R, Castro R, Oliveira MC, et al. Organic dyes in illuminated manuscripts: an unique cultural and historic record. Phil Trans R Soc A. 2016;374:20160050:1–20. DOI: 10.1098/rsta.2016.0050.

[29] Miguel C, Pinto JV, Clarke M, Melo MJ. The alchemy of red mercury sulphide: the production of vermilion for medieval art. Dyes Pigm. 2014;102:210–17. DOI: 10.1016/j.dyepig.2013.10.041.

[30] Castro R, Miranda A, Melo MJ. Interpreting lac dye in medieval written sources: new knowledge from the reconstruction of recipes relating to illuminations in Portuguese manuscripts. In: Eyb-Green S, Townsend J, Pilz K, Kroustallis S, van Leeuwen I, editors. Sources in Art technology: back to basics. London: Archetype Publications, 2016:88–99.

[31] Blondheim S. An old Portuguese work on manuscript illumination. Jewish Quarterly Rev. 1928;19:97–135. DOI: 10.2307/1451766.

[32] Cruz AJ, Afonso L. On the date and contents of a Portuguese medieval technical book on illumination: O livro de como se fazem as cores. Medieval History J. 2008;11:1–28. DOI: 10.1177/097194580701100101.

[33] Oltrogge D. Byzantine recipes and book illumination. Revista de História da Arte, FCSH-UNL. 2011; N°1 Série W:61–71. http://revistadehistoriadaarte.files.wordpress.com/2011/09/art04.pdf.

[34] Schreiner P, Oltrogge D. Byzantinische Tinten-, Tuschen- und Farbrezepte. Wien: Verlag der Österreichischen Akademie der Wissenschften, 2011.

[35] Clarke M. Mediaeval painters' materials and techniques: the Montpellier Liber diversarum arcium. London: Archetype, 2011.

[36] Ali N. Artistic experimentation within the project "Organic paint research". Available at: http://nabil-ali.wixsite.com/artist.

[37] Roger P, Villela-Petit I, Vandroy S. Les laques de brésil dans l'enluminure médiévale: reconstitution à partir de recettes anciennes. Stud Conserv. 2003;48:155–70. DOI: 10.1179/sic.2003.48.3.155.

[38] Castro R. The book of birds in Portuguese scriptoria: preservation and access. PhD thesis, UNL, 2016. http://hdl.handle.net/10362/21481.

[39] Lemos A, Araújo R, Casanova C, Melo MJ, Muralha VSF. Regards croisés des historiens de l'art et des chimistes sur deux livres d'heures de la Bibliothèque Nationale Du Portugal, Les Mss. IL15 et IL19. In: Miranda MA, Cavero AM, editors. Portuguese studies on medieval illuminated manuscripts. Madrid: Fédération Internationale des Instituts d'Études Médiévales, 2014:145–68.

[40] Saltzman M. Identifying dyes in textiles. Am Sci. 1992;80:474–81.

[41] Cardon D. Natural dyes: sources, tradition, technology and science. London: Archetype, 2007.

[42] Orna MV, Low MJD, Baer NS. Synthetic blue pigments: ninth to sixteenth centuries. I Literature Stud Conserv. 1980;25:53–63. DOI: 10.1179/sic.1980.25.2.53.

[43] Riederer J. Egyptian blue. In: Fitzhugh EW editor. Artists' pigments – a handbook of their history and characteristics. Washington DC: National Gallery of Art, 2009:23–45. https://www.nga.gov/content/dam/ngaweb/research/publications/pdfs/artists-pigments-vol3.pdf.

[44] Accorsi G, Verri G, Bolognesi M, Armaroli N, Clementi C, Miliani C, et al. The exceptional near-infrared luminescence properties of cuprorivaite (Egyptian blue). Chem Commun. 2009;23:3392–4. DOI: 10.1039/B902563D.

[45] Bertrand L, Gervais C, Masic A, Robbiola L. Paleo-inspired systems: durability, sustainability and remarkable properties. Angew Chem Int Ed. 2018;57:7288–95. DOI: 10.1002/anie.201709303.

[46] Pradell T, Salvado N, Hatton GD, Tite MS. Physical processes involved in production of the ancient pigment, Egyptian blue. J Am Ceram Soc. 2006;89:1426–31. DOI: 10.1111/j.1551-2916.2005.00904.x.
[47] Lazzarini L, Verità M. First evidence for first century AD production of Egyptian blue frit in Roman Italy. J Archaeol Sci. 2015;53:578–85. DOI: 10.1016/j.jas.2014.11.004.
[48] Lazzarini L. The discovery of Egyptian blue in a Roman fresco of the medieval period (ninth century A.D.). Stud Conserv. 1982;27:84–6. DOI: 10.1179/sic.1982.27.2.84.
[49] Gaetani MC, Santamaria U, Seccaroni C. The use of Egyptian blue and lapis lazuli in the middle ages. The wall paintings of the San Saba church in Rome. Stud Conserv. 2004;49:13–22. DOI: 10.1179/sic.2004.49.1.13.
[50] Beeby A, Gameson R, Nicholson C. New light on old illuminations. Archives and Records. 2017;1–13. DOI: 10.1080/23257962.2017.1325729.
[51] Denoël C, Puyo PR, Brunet AM, Siloe NP. Illuminating the Carolingian era: new discoveries as a result of scientific analyses. Heritage Sci. 2018;6:1–19. DOI: 10.1186/s40494-018-0194-1.
[52] Melo MJ, Araújo R, Castro R, Casanova C. Colour degradation in medieval manuscripts. Microchem J. 2016;124:837–44. DOI: 10.1016/j.microc.2015.10.014.
[53] Miguel C, Claro A, Gonçalves AP, Muralha VSF, Melo MJ. A study on red lead degradation in the medieval manuscript, Lorvão Apocalypse (1189). J Raman Spectrosc. 2009;40:1966–73. DOI: 10.1002/jrs.2350.
[54] Muralha VSF, Miguel C, Melo MJ. Micro-Raman study of Medieval Cistercian twelve-thirteenth century manuscripts: Santa Maria de Alcobaça, Portugal. J Raman Spectrosc. 2012;43:1737–46. DOI: 10.1002/jrs.4065.
[55] Araújo R, Nabais P, Cardoso IP, Casanova C, Lemos A, Melo MJ. Silver paints in medieval manuscripts: a first molecular survey into their degradation. Heritage Sci. 2018;6:1–13. DOI: 10.1186/s40494-018-0172-7.
[56] Salvadó N, Butí S, Nicholson J, Emerich H, Labrador A, Pradell T. Identification of reaction compounds in micrometric layers from gothic paintings using combined SR-XRD and SR-FTIR. Talanta. 2009;79:419–28. DOI: 10.1016/j.talanta.2009.04.005.
[57] Miguel C. Le vert et le rouge: a study on the materials, techniques and meaning of the green and red colours in medieval Portuguese illuminations. PhD thesis, UNL, 2012. http://run.unl.pt/handle/10362/9304.
[58] Doyle BB. Infrared spectroscopy of collagen and collagen-like polypeptides. Biopolymers. 1975;14:937–57. DOI: 10.1002/bip.1975.360140505.
[59] Neevel H. The development of in-situ methods for identification of iron–gall inks. In: Kolar J, Strlič M, editors. Iron–gall inks: on manufacture, characterization, degradation and stabilization. Ljubljana: National and University Library, 2006:147–72.
[60] Quideau S, Deffieux D, Douat-Casassus C, Pouységu L. Plant polyphenols: chemical properties, biological activities, and synthesis. Angew Chem Int Ed. 2011;50:586–621. DOI: 10.1002/anie.201000044.
[61] Hidalgo RJD, Córdoba R, Nabais P, Silva V, Melo MJ, Pina F, et al. New insights into iron-gall inks through the use of historically accurate reconstructions. Heritage Sci. 2018;6:63–78.
[62] Stijnman A. Reconstructions of iron-gall ink recipes for the InkCor project. In: Stijnman A, Clarke M, Townshend JH, editors. Art of the past: sources and reconstructions. London: Archetype Publications, 2005:125–34.
[63] Ponce A, Brosto LB, Gibbons SK, Zavalij P, Viragh C, Hooper J, et al. Elucidation of the Fe(III) Gallate structure in historical iron gall inks. Anal Chem. 2016;88:5152–8. DOI: 10.1021/acs.analchem.6b00088.
[64] Leygraf C, Graedel T. Atmospheric corrosion. Wiley: New York, 2016.
[65] Larsen R (ed.). Improved damage assessment of parchment. EC Research report N° 18, 2007.

[66] Chen J, Ahn T, Colón-Bernal ID, Kim J, Banaszak Holl MM. The relationship of collagen structural and compositional heterogeneity to tissue mechanical properties: a chemical perspective. ACS Nano. 2017;11:10665–71. DOI: 10.1021/acsnano.7b06826.
[67] Schrieber R, Gareis H. Gelatine handbook. Theory and industrial practice. Weinheim: Wiley-VCH, 2007.
[68] Latour G, Robinet L, Dazzi A, Portier F, Deniset-Besseau A, Schanne-Klein MC. Correlative nonlinear optical microscopy and infrared nanoscopy reveals collagen degradation in altered parchments. Sci Rep. 2016;6:1–10. DOI: 10.1038/srep26344.
[69] Vichi A, Eliazyan G, Kazarian SG. Study of the degradation and conservation of historical leather book covers with macro attenuated total reflection–Fourier transform infrared spectroscopic imaging. ACS Omega. 2018;3:7150–7. DOI: 10.1021/acsomega.8b00773.
[70] Tempera – Teaching Emerging Methods in Palaeoproteomics for the European Research Area. Available at: https://sites.google.com/palaeome.org/tempera/home.
[71] Dinnebier RE, Fischer A, Eggert G, Runčevski T, Wahlberg N. X-ray powder diffraction in conservation science: towards routine crystal structure determination of corrosion products on heritage art objects. J Visualized Exp. 2016;112:1–17. DOI: 10.3791/54109.
[72] Panayotova S, Pereira-Pardo L, Ricciardi P. lluminators' materials and techniques in fourteenth-century English manuscripts. In: Panayotova S, Ricciardi P, editor(s). Manuscripts in the making: art and science, Volume 1. London and Turnhout: Harvey Miller Publishers, 2017:47–64.
[73] Vitorino T, Casini A, Cucci C, Melo MJ, Picollo M, Stefani L. Non-invasive identification of traditional red lake pigments in fourteenth-sixteenth centuries paintings through the use of hyperspectral imaging technique. Appl Phys A. 2015;121:891–901. DOI: 10.1007/s00339-015-9360-4.
[74] Ganio M, Pouyet ES, Webb SM, Patterson CMS, Walton MS. From lapis lazuli to ultramarine blue: investigating Cennino Cennini's recipe using sulfur K-edge XANES. Pure Appl Chem. 2018;90:463–75. DOI: 10.1515/pac-2017-0502.

Gregory D. Smith, Victor J. Chen, Kurt F. Hostettler
and Caitlyn E. Phipps

17 Disappearing ink! Unraveling the fading of a contemporary design object

Abstract: In 2009 the Indianapolis Museum of Art acquired Fernando Brizio's contemporary ceramic, *Painting a Fresco with Giotto #3*, as part of its effort to grow a collection of modern and contemporary European design objects. The artwork comprises an unglazed white faience vase that has been pierced with 30 brightly colored felt-tip markers whose dried ink stains create a whimsical polka-dot surface. The vase immediately joined a traveling exhibition, and when it returned to the museum after nearly a year of display, many of the ink spots had faded dramatically, some having nearly vanished. Technical analysis was undertaken to (1) determine the cause of the fading and the composition of its Giotto brand Turbocolor markers, (2) to determine the future fading potential of the object, and (3) to suggest new safeguards to protect this and similar objects from future damage. Non-destructive analysis of the vase using Raman spectroscopy proved challenging due to the overall application of an acrylic varnish by the artist and the intense fluorescence of many of the marker inks. Using liquid chromatography with mass spectrometry on surrogate pens acquired in 2014, the 30 markers utilized in the object were ultimately determined to contain 9 primary synthetic dyes and numerous synthetic byproducts. The inks are comprised of food colorants, which accounts for their rapid fading. In situ microfade testing showed that the vase is still extremely light sensitive and will fade further under even the most stringent lighting protocols. An artist interview provided valuable information about Brizio's thoughts concerning the life of these objects, his working methods, and the interpretation of the *Giotto* series of ceramics. The artwork has since been deaccessioned from the museum's collection but has taken on a new role in gallery didactics focused on the materials of modern design and the ephemeral and changing nature of some artworks. This project highlights the urgency of characterizing the fading rate of potentially light sensitive modern art or directly identifying the colorants used in contemporary artworks prior to their first exhibition.

Keywords: modern art, design object, dye analysis, felt-tip marker, fading, liquid chromatography, Raman spectroscopy, microfade testing

This article has previously been published in the journal *Physical Sciences Reviews*. Please cite as: Smith, G., Chen, V., Hostettler, K., Phipps, C. Disappearing Ink! Unraveling the Fading of a Contemporary Design Object. *Physical Sciences Reviews* [Online] **2019**, 4. DOI: 10.1515/psr-2018-0018.

https://doi.org/10.1515/9783110457537-017

17.1 Introduction

The Indianapolis Museum of Art (IMA), now known as Newfields, acquired *Painting a Fresco with Giotto #3* (Figure 17.1) by contemporary Portuguese designer Fernando Brizio in 2009 to add to its burgeoning collection of European design. The object comprises a white faience vase pierced with 30 Giotto brand felt-tip markers. To achieve this illusion, the wall of the vase was drilled to create tiny holes into which the pens are slotted using a small steel stickpin driven into the marker nib. The bleeding of the pen inks in contact with the unglazed, porous ceramic surface over the course of days created a playful polka-dot pattern of solid circles or concentric rings of color around each marker. The effect is described as being governed by chance, inspired by ink stains on the artists' shirt pocket from a broken pen [1], although an interview with Mr Brizio revealed the experimentation and planning that went into the execution of this work of art [2].

Figure 17.1: *Painting a Fresco with Giotto #3* from 2005, as it appeared in 2014., no. 11 of 20, by Fernando Brizio (Portuguese, b. 1968). Gift of the Alliance of the IMA. Formerly 2009.466A-B. Photo by Eric Lubrick, courtesy of Newfields. © Fernando Brizio.

Brizio is considered one of the rising stars of contemporary design, and certainly of design arts in Portugal, according to *The Wall Street Journal* [3], which called his designs "contemplative" and "cerebral," while noting their humor and playfulness.

As part of the Conceptual Design movement, Brizio's piece has been compared to earlier works by the Dutch designer Hella Jongerius where the apparent function of the object is surpassed by experimentation in color, texture, materials, and form [4]. Brizio continued this theme of whimsical color and chance in other series of bright marker-stained ceramics, but also in textiles like the *Restarted Dress* and in wooden furniture.

The vase, manufactured in 2005 as the eleventh in an edition of twenty, barely arrived at the IMA before embarking on a three venue touring exhibition, *European Design Since 1985: Shaping the New Century* [4]. Unfortunately, the object returned to the museum badly faded, with some of the ink dots, particularly the lighter pinks and purples, having nearly vanished from the surface, but with all colors showing a loss of intensity and vividness. Due to the hurried nature of the numerous acquisitions leading up to the launch of the exhibition, no professional photography of the IMA vase as it was originally received exists for comparison to the present condition. A studio photograph taken in 2014 following the vase's return from the European design show, Figure 17.1, reflects the visual state of the object today. Only a handful of snapshot images in Figure 17.2 that were captured during the exhibition of the object reveal how much color has been lost. The details below show the near total loss of the lavender ink dot and the disintegration of the pink spot in 2014 (right) compared to their appearance in a gallery snapshot from 2009 (left).

Figure 17.2: Detail image of the Giotto marker ink stains in 2009 (left) compared to the same location in 2014 following the exhibition (right). Photo left by Brittany Minton, right by Eric Lubrick. Courtesy of Newfields. © Fernando Brizio.

17.1.1 European design since 1985: Shaping the new century

The IMA co-organized the traveling exhibition *European Design Since 1985: Shaping the New Century*, which traced the evolution of western European design aesthetics from the late twentieth century to 2005, just a few years prior to the show's inauguration. The exhibition included nearly 250 objects – furniture, glass, electronics, ceramics, and metalwork – from over a hundred of the most important designers of the period [4]. The initial venue for the show was the IMA in Indianapolis (March through June, 2009) before moving to the High Museum of Art in Atlanta (June through August, 2010) and then the Milwaukee Art Museum (October 2010 to January, 2011). The total tour consisted of roughly 320 days of exhibition.

To understand the damage that occurred during that period, the loan agreements were examined to reveal the conditions under which the objects were shown. The maximum light levels for the show had been set at 20 foot-candles (215 lux). This is admittedly at the upper end of the allowable light limits for IMA objects, but these illumination levels are largely consistent with international practices for the exhibition of contemporary design objects. In fact, the museum guidelines for exhibitions at the time listed 30 foot-candles as the maximum illumination for twentieth to twenty-first century ceramics. However, the same guidelines suggest that artworks containing extremely fugitive media, specifically calling out felt-tip markers as one example, receive only 5 foot-candles of exposure. Normally, light level restrictions follow the recommendations associated with the most sensitive component of an object, but in this case a blanket lighting maximum was used for the entire exhibition. In hindsight, the short timescale between the arrival of many of these newly acquired design objects and the installation date for the first exhibition limited the amount of consideration that conservation staff could give to each individual item.

The loan agreements also contained the regular prohibitions against direct sunlight or unfiltered fluorescent lighting, and the museum conditions were specified as 70 ± 2°F (21 °C) and 50 ± 5 % relative humidity. Regular condition checks of the artworks were undertaken by museum couriers between venues, but curiously no changes were noted in the *Giotto* vase until the object was returned to the IMA in early 2011. At that point, a museum registrar noted heavy fading in multiple areas of the vase in comparison to installation snapshots taken at the start of the touring exhibition.

17.1.2 The conservation issues

The conservation science laboratory at the IMA was opened in 2011 [5], well after the acquisition and subsequent exhibition of the vase. In the lab's inaugural year, this damaged object was brought to our attention by the curator who organized the traveling exhibition. He suggested conducting a technical study to understand the conditions that led to the rapid fading of the inks and to determine a path forward for the future exhibition of the object. Three main goals were articulated. First, the lab was tasked with determining why the dyes faded so rapidly under what was considered to be typical conditions for the exhibition of contemporary design objects. This task would obviously involve chemical analysis of the marker inks to determine the colorants used in their formulation. Numerous earlier studies have warned of the particular sensitivity of dye based felt-tip markers to light exposure [6–10]. A priority was placed on causing no further damage to the object, so the analysis needed to be non-destructive and sympathetic to the condition of the artwork.

A second objective was to establish the future display conditions for the vase, including the potential for conservation treatment of the faded areas. Although exceptionally rapid color loss had occurred during the initial traveling exhibition, it was not known if the vase was still prone to fading or if it had reached something of a steady state with only slow changes expected in the future [11]. Microfade testing allows for in situ micro-destructive measurements of the current fading behavior of an artwork [12]. These accelerated fading tests are compared to those of industrial standards whose fading behavior under standard museum conditions is well known. By this method, a prediction of the future fading behavior of an artwork can be gained with no outwardly visible change to the object.

A third outcome of the proposed study was to establish a new set of guidelines to ensure better stewardship of the *Giotto* vase and similar pieces of contemporary design artworks in the future. These guidelines would reconsider light limits, exhibition duration, frequency of rotations, environmental precautions, handling instructions, and storage conditions. The results of the previous tests would be used to determine whether exhibition strategies like anoxic containers or visitor activated lighting were warranted. Finally, an artist interview was prepared with the hope of gaining insight directly from Mr Brizio regarding his knowledge of the fading issue, information on his materials and working methods, and his expectations for the life of his artworks.

17.1.3 Giotto Turbocolor markers

Painting a Fresco with Giotto #3 incorporates 30 Giotto Turbocolor brand felt-tip markers in 29 different colors arranged in four rows around the vase. Figure 17.3

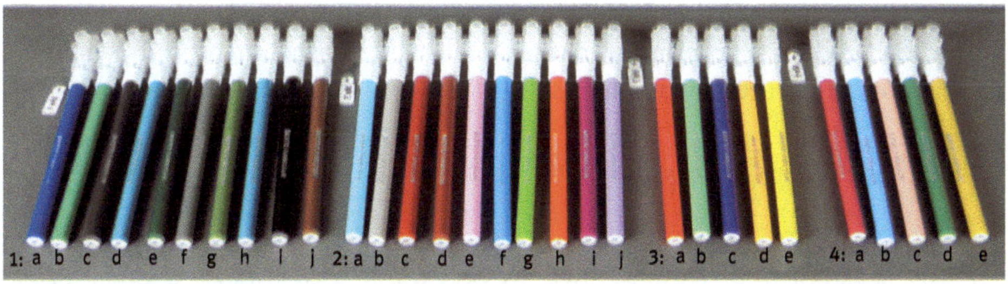

Figure 17.3: The pens used in the IMA's *Giotto* vase with unique designations based on Brizio's notations for the pens' original locations. The number indicates the row occupied by the pen, 1 being closest to the base of the ceramic and 4 being at the neck, and the letter shows the sequential location around the circumference. The white plastic holder was provided by the artist for pen storage during storage or shipping.

shows the artwork pens with unique identifiers designated by the laboratory and based on Brizio's documentation of the pen positions in the vase as described in the caption. Only 29 unique colors are used in the piece because a teal colored marker (1:d and 1:h in Figure 17.3) is used twice. The unfortunate color loss of the vase demonstrates that these markers introduce inherent vice into the long-term stability of the artwork, but at the same time, these pens are obviously important to the design concept since their name appears in the title of the piece. When the authors interviewed the artist following the study described here, Brizio [2] reported that he had tried several marker brands but liked the way the Giotto markers spread across the ceramic surface. He also commented that the markers recalled the Italian proto-Renaissance master of fresco art, Giotto di Bondone [1276–1337]. In fresco painting, the pigments are enveloped into and become one with the plaster substrate. In Brizio's conceptual design, the intercalation of the inks into the porous ceramic is like a modern day fresco.

Giotto markers are not available in the USA, but they are sold throughout Europe, where the manufacturer, F.I.L.A. Italia, is headquartered. A quick perusal of their website identifies that the brand is an old one, manufacturing art materials since 1920 [13]. The Turbocolor felt-tip pens currently come in 36 shades, 29 of which are used in the artwork. It is obvious from the website, however, that the Giotto markers are primarily marketed for school children, noting their "harmless, washable inks" and utilizing slogans like "a synonym of colour for young artists." Despite the focus on hobby use and children's art, Giotto felt-tip markers have been suggested as the possible drawing material used by the well-known filmmaker and artist Federico Fellini (Italian, 1920–1993] [14]. In fact, the bright transparent colors and fine line of these drawing implements have made felt-tip markers a common medium for the fine arts, for so called naïve or outsider art, as well as for artwork signatures, architectural drawings, and fashion drawings since their introduction in 1946 [6, 8, 10]. Because the Giotto markers are not made specifically for fine art uses, the long-term durability of their colorants was suspect. F.I.L.A. Italia politely declined to provide technical information regarding their pen ink formulations citing the product's proprietary nature. However, the company did send a 36 marker box set for these experiments.

17.1.4 Prior chemical studies of felt-tip pens

The damaged vase was brought to the laboratory in 2011, but the technical study was not initiated until January 2014. At that point, few studies concerning the composition and stability of felt-tip markers in fine arts had been undertaken [14–17], and within that small group of studies, most were concerned with so called "permanent" markers for their use and removal in graffiti. Since then, additional research has been published concerning the characterization of waterborne felt-tip markers used in art [18–21], but these largely focus on laboratory simulations or paper-based artworks and rarely consider a full line of color markers. Giotto markers have been a subject of some of these studies, but never more than six representative pens have been examined. Moreover, marker inks fall into a variety of categories, including solvent or waterborne formulations, as well as pigment or dye-based colorants. The stability of these different varieties of porous nib pens has been shown to vary significantly [7, 9, 20]. To answer the curatorial concerns regarding Brizio's compromised vase, a research plan was devised that involved the non-destructive, in situ spectroscopic identification of the marker inks on the vase and pens followed by micro-destructive, in situ microfade testing of the remaining ink stains to determine their present lightfastness. Based on these data, a set of guidelines could be decided for the storage, handling, and future exhibition of *Painting a Fresco with Giotto #3*.

17.2 Experimental

17.2.1 Samples

Reference dyestuff powders were acquired from chemical supply houses and dye manufacturers. Table 17.1 lists the materials and other pertinent information, including their structures as shown in Figure 17.4. The structures represent the neutral molecular form rather than the salts actually studied in order to facilitate discussions related to their mass spectrometry. Each dyestuff is henceforth referred to by the synonym in bold, which is chosen to reflect the name most commonly occurring in the relevant literature. These dyes were used as received with no further purification. An historic sample of Elvacite 2046, a 1:1 copolymer of iso-butyl methacrylate (iBMA) and n-butyl methacrylate (nBMA), was obtained directly from Dupont (Lot 430) in the early 1970s.

In order to avoid invasive and destructive analyses that would be particularly damaging to the artwork, a 36 marker box set of surrogate Giotto Turbocolor pens was acquired directly from the manufacturer in May of 2014. It was hoped that these surrogate markers would contain similar ink compositions to those used by Brizio in 2005, although some manufacturers' ink formulations have previously been shown to change over the course of multiple decades [14]. The style of the Giotto Turbocolor pen casings had been redesigned between 2005 and 2014, but the visible colors of 28 of the 29 original pen barrels from 2005 compared well with those produced in 2014. The one exception was a bright pink marker, labeled 2:e in Figure 17.3, whose 2005 casing was a slightly lighter hue than the one acquired in 2014, although the earlier plastic barrel might well have itself faded. Although the manufacturer lists all 36 pen colors by name in their Safety Data Sheet, the pens are not individually labeled, and it is often difficult to associate the manufacturer's

Table 17.1: Commercial dyestuffs used as reference compounds along with corresponding neutral molecular structure in Figure 17.4, chemical synonyms, and specific product information. Names in bold reflect that most commonly used in the conservation literature.

Dyestuff purchased	Structure in Figure 17.4	Synonyms	Supplier	Stated purity
Tartrazine	1	Acid Yellow 23, FD&C Yellow 5, E102, CI# 19140, CAS# 1934-21-0	Alfa Aesar #A17682	>85 %
New Coccine	2	**Acid Red 18,** Ponceau 4R, CI Food Red 7, E124, CI#16255, CAS# 2611-82-7	Acros #190230250	≥75 %
Quinoline Yellow S	3,5,6, 11,12	Acid Yellow 3, Food Yellow 13, D&C Yellow 10, E104, CI# 47005, CAS# 80583-08-0, 8004-92-0	Acros #277340250	95 %
Sunset Yellow FCF	4	Orange Yellow S, FD&C Yellow 6, E110, CI# 15985, CAS# 2783-94-0	TCI #S0141	>90 %
Chromotrope FB	7	**Acid Red 14,** Azorubine, CI Food Red 3, E122, CI# 14720, CAS# 3567-69-9	Alfa Aeser #B22328	Not given
Erioglaucine	8	Acid Blue 9, Brilliant Blue FCF, FD&C Blue 1, E133, CI# 42090, CAS# 3844-45-9	Acros #229730050	Grade "pure"
Sulforhodamine B	16	**Acid Red 52,** Food Red 106, Kiton Red 620, CI# 45100, CAS# 3520-42-1	Alfa Aeser #A14769	75 %
SERVA Violet 17	18	**Acid Violet 17,** CI Food Violet 1, CI# 42650, CAS# 4129-84-4	SERVA #35072.02	Not given
Coomassie Violet R-150			MP Biomedical #190683	Not given
Erythrosin B	19	Solvent Red 140, CI# 45430:2, CAS# 15905-32-5†	Sigma-Aldrich #200964	≥95 %
		Acid Red 51, Food Red 3, CI# 45430, CAS# 16423-68-0	Fisher #715007-A	87 %

†Solubilized through reaction with 2 eq. K_2CO_3.

fanciful color names with specific pens. Therefore, a descriptive color name was given to each marker used in the study.

17.2.2 Raman microspectroscopy

Raman spectra were acquired using a Bruker Senterra microspectrometer on a Z-axis gantry that allowed for a working distance that would accommodate the height of the ceramic. The spectrometer utilizes three selectable excitation lasers (532, 633, and 785 nm), an Andor Peltier-cooled CCD detector, and a 50 μm confocal pinhole. Laser power at the sample was always below 5 mW and often below 1 mW. The spectra are the result of 3 or 5 s integrations with 20–100 coadditions. A 50X or 100X ultra-long working distance objective was used to focus slightly below the varnish layer on the ceramic

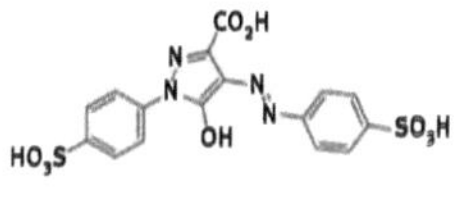
1
Tartrazine, C.I. 19140

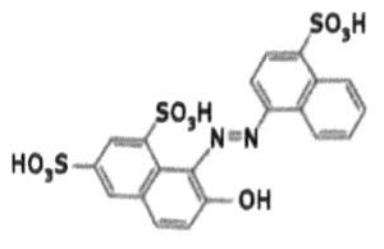
2
Acid Red 18, C.I. 16255

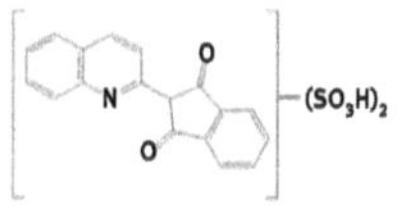
3,5,6
Quinoline Yellow S, C.I. 47005 Disulfonates I-III

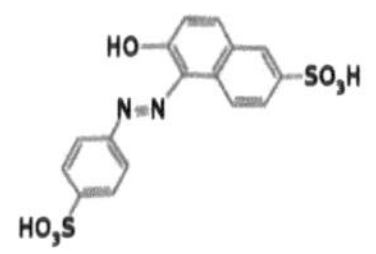
4
Sunset Yellow, C.I. 15985

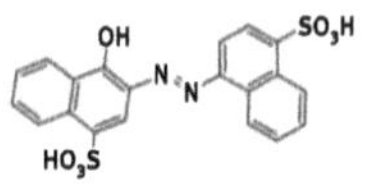
7
Acid Red 14, C.I. 14720

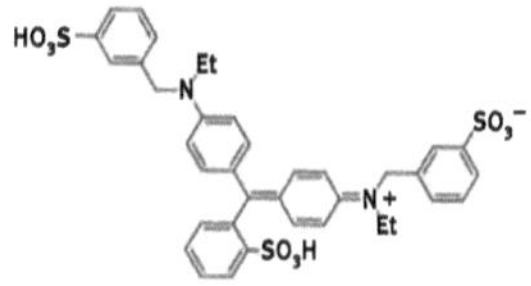
8
Erioglaucine, C.I. 42090

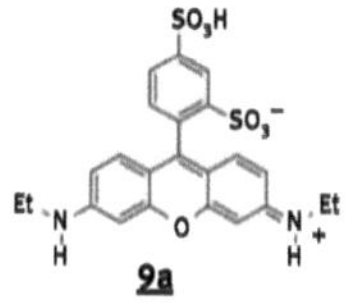
9a

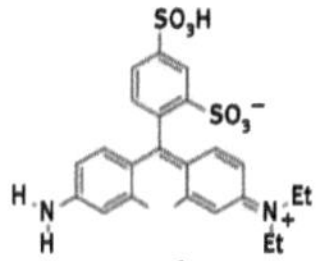
9b

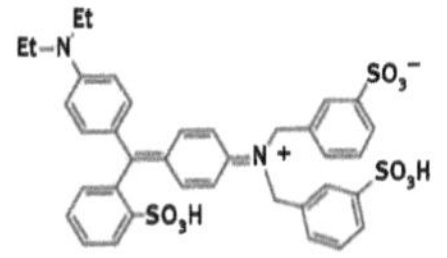
10

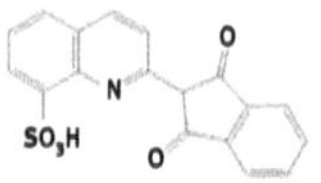
11
Quinoline Yellow S, C.I. 47005 8'-monosulfonate

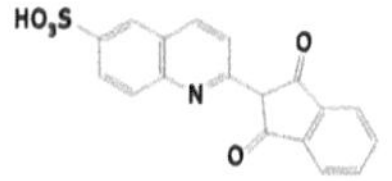
12
Quinoline Yellow S, C.I. 47005 6'-monosulfonate

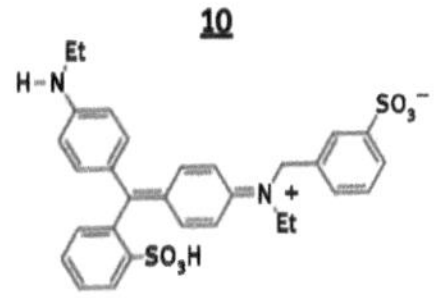
13

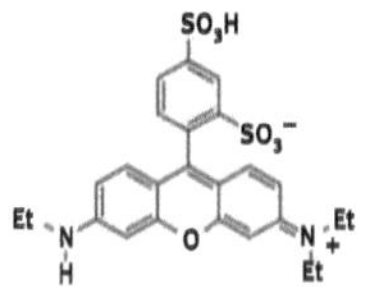
14

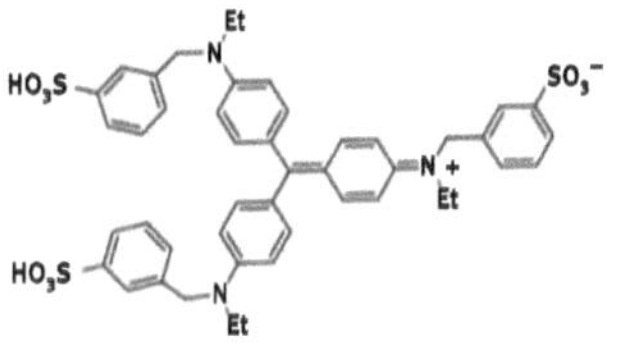
15

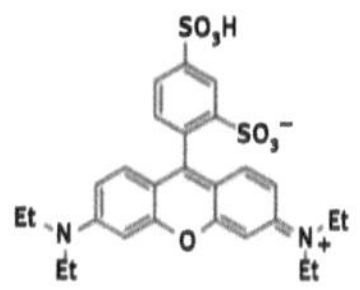
16
Acid Red 52, C.I. 45100

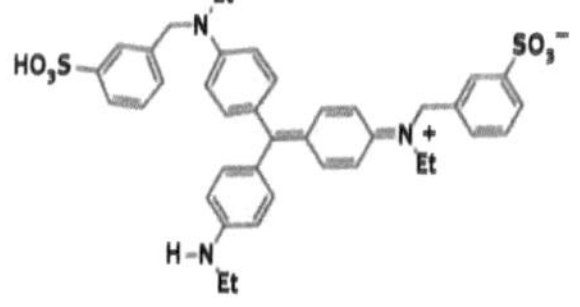
17

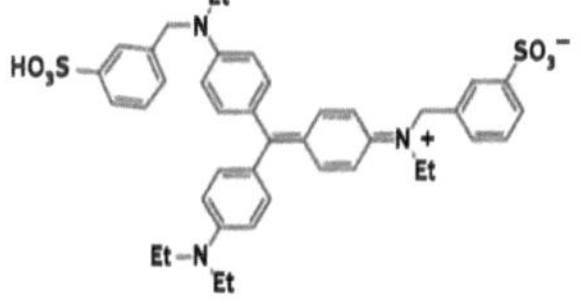
18
Acid Violet 17, C.I. 42650

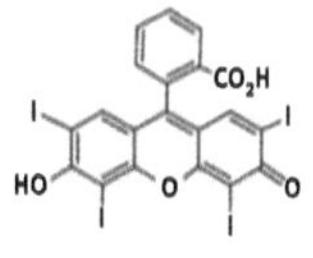
19
Erythrosin B, C.I. 45430

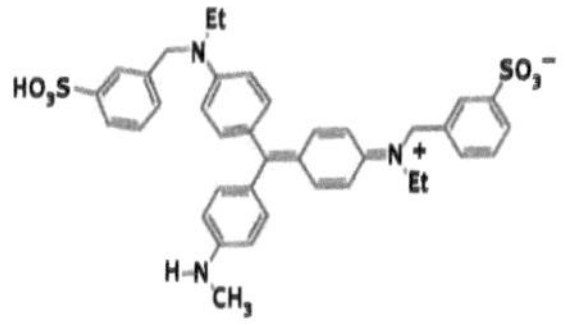
20

Figure 17.4: Chemical structures of relevant molecules shown in their neutral form to assist discussions related to mass spectrometry.

surface or directly onto the marker nibs. The analysis spot size was on the order of 1 μm, and the spectral resolution was in the range of 9–18 cm^{-1} for low spectral resolution or 3–5 cm^{-1} for high resolution spectra. OPUS software allowed for automated cosmic spike removal, peak shape correction, and spectral calibration.

17.2.3 Photography

Color images were shot using daylight balanced strobes and a Phase One digital SLR camera with a 39 megapixel P45 back and 120 mm lens. The image was made at f22 1/125th sec exposure at ISO 200. Fluorescence images were captured using a Canon 5d Mark II digital SLR camera with a 22.3 megapixel sensor and an EF100 mm F/2.8 macro lens. Camera settings were f14 with a 13 s exposure at ISO 400. The object was irradiated using two banks of four 40 W Sylvania Blacklight Blue F40 BLB bulbs. The violet emission line of the tubes was not filtered at the lamp or at the camera. The color and fluorescence images were adjusted manually in Photoshop to approximate the observed appearance of the artwork under the lighting conditions used.

17.2.4 Fourier transform infrared (FTIR) microspectroscopy

Mid-infrared spectroscopy was performed on a Continuum microscope with an MCT A detector coupled to a Nicolet 6700 FTIR spectrometer purged with dry, CO_2-free air. The spectra are the sum of 32 coadditions at 4 cm^{-1} spectral resolution. Microsamples were crushed on a diamond compression cell and held on a single diamond window during the analysis. Sample identification was performed using the Infrared and Raman Users Group (IRUG) reference spectral library or by comparison to reference materials.

17.2.5 Pyrolysis-gas chromatography-mass spectrometry (PY-GC-MS)

Pyrograms were obtained on a Frontier Lab Py-2020D double-shot pyrolyzer system with a 320 °C interface to a Thermo Trace GC and an ISQ single quadrupole MS. The method of Tsuge et al. [22] was used for the analysis. Briefly, a Thermo TG-5MS capillary column (30 m × 0.25 mm × 0.25 μm] was used for the separation with 1 mL/min of He as the carrier gas. The split injector was set to 320 °C with a split ratio of 100:1. The GC oven temperature program was 40 °C for 2 min, ramped to 320 °C at 20 °C/min, followed by a 13 min isothermal period. The MS transfer line was held at 320 °C and the EI source at 70 eV and 230 °C. The mass spectrometer (MS) was scanned from 29 to 300 amu for the first 3 min and then 45–300 amu. The electron multiplier was set to the auto-tune value. Samples were placed into a 50 μL stainless steel Eco-cup and were pyrolyzed using a single-shot method at 600ºC for 6 s. Sample identification was aided by searching the NIST08 MS library and by comparison to pyrograms of authentic samples.

17.2.6 Liquid chromatography – diode array – mass spectrometry (LC-DAD-MS)

Dyes were separated and identified using a Thermo Accela liquid chromatography system (LC) coupled sequentially to a diode array detector (DAD) and LTQ electrospray ionization (ESI) MS. The separation was achieved at 0.5 mL/min elution flow through a Restek Ultra C18 reverse phase column (150 mm × 4.6 mm, 5 μm particle) held isothermally at 30 °C. The sample injection volume was 5–25 μL. The entire system was controlled using Thermo Xcaliber v2.0 software.

The elution gradient consisted of a mixture of water and acetonitrile (ACN), each containing 0.1 % v/v formic acid. The column was initially equilibrated at 3 % ACN with a 6 min isocratic step following injection. The gradient was increased to 93 % ACN at a rate of 1.5 %/min over 60 min, followed by a final 3 min isocratic step at 93 % ACN. The column was re-equilibrated at the starting conditions for 20 min prior to the next injection. A blank was run between each analysis.

To detect eluting compounds, DAD spectra from 200 to 800 nm were collected at 1 nm step size at 20 Hz. The sequential ion trap MS was set to scan alternatingly in positive and negative ionization mode. Within each mode, a full scan in the m/z range 50–2000 was followed by two successive data dependent MS/MS scans of the collision-induced dissociation (CID) fragments generated from targeting the first and second most abundant ions detected in the full scan. In some instances the MS was scanned only in positive ion mode to increase data collection efficiency. Other MS settings include ESI source spray voltage at 5 kV, capillary temperature at 350 °C, sheath and auxiliary nitrogen gas at 80 and 10 respectively, and CID targeting of singly charged ions with isolation width 2.00, energy 45, and 0.25 activation Q for 30 ms.

From the data gathered, a search for dyes with similar electronic absorption spectra was made using a reference handbook for stains and dyes [23]. Candidate dyes were further evaluated against the LCMS data in order to select the structures that best fit the mass of the parent ion observed as well as the fragmentation profiles of the CID experiment. The identification was subsequently verified by analyzing reference samples of the likely dyes under identical conditions and by comparison to the literature.

17.2.7 Microfade testing (MFT)

Lightfastness measurements of the dried surrogate pen inks and residual color on the Brizio vase were performed using a Newport Oriel type microfade tester similar to that first described by Whitmore et al. [12]. The 75 W xenon arc lamp source was filtered to emit only visible light in the wavelength range of 400–700 nm. The luminous flux was measured at 600–1,000 millilumens using an ILT1700 radiometer with SPD024Y probe (International Light Technologies). A digital exposure controller adjusts the lamp output to maintain constancy over the course of the experiments. Spectral reflectance data were acquired from an approximately 400 μm spot every 10 s over the period of 10 min and converted into CIE L*a*b* color coordinates. Colorimetry conditions included a D65 illuminant, 45° geometry, and 2° observer. Color difference was calculated using the CIE L*a*b* color difference equation from 1976 (ΔE'76). Fading data were analyzed and plotted using GCI Spectral Viewer™ software.

Inks from the surrogate pens and reference dyestuffs prepared as 5 mM solutions in water were dotted onto the flat surface of an unglazed biscuit fired ceramic to simulate the substrate of the vase. The strengths of the applications were made to mimic the color intensities observed in photographs of other *Giotto* series vases. Duplicate analyses of fading curves for ISO Blue Wool Standards 1, 2, and 3 (BWS1-3, Talas) were collected bracketing the analysis of the samples for each day of experiments. After several days of drying, the ink spots and reference dyes were analyzed in triplicate to determine an average blue wool equivalent (BWE) fading behavior by comparison to the analyses performed on the three standards. Single in situ measurements were made at several stained areas on the ceramic surface of the artwork. Since the dried ink spots on the vase sometimes show concentric rings of color of various intensities, multiple locations were analyzed along the radius of some ink stains.

17.3 Results & discussion

17.3.1 In situ spectroscopic dye identification

Due to the already compromised condition of the *Giotto* vase, the research plan called for identifying the marker dyes directly on the artwork using a non-invasive, non-destructive spectroscopic approach. Raman spectroscopy has a long history of providing in situ, molecular specific identifications of materials used in artwork [24, 25], including early synthetic dyes in artists' inks [26, 27]. In fact, the technique had been previously used to examine a few instances of felt-tip marker inks directly on paper-based artworks [14]. The potential of the technique to identify at least the major component of the marker inks on a ceramic surface was considered high.

The Raman microspectrometer used in this study can provide a large working distance sufficient to accommodate the *Giotto* vase directly under a high powered ultra-long working distance objective. A bespoke circular soft mount was constructed to safely cradle the ceramic so that it could be rotated to make each colored ink stain accessible to the objective. Each of the ink stains was analyzed in turn, but only one spot, correlating to the position of the mauve pen 2:i, gave a reasonable Raman spectrum, albeit riding on a large fluorescence background. Figure 17.5 shows this data acquired directly from the vase compared to an in-house reference Raman spectrum of the dye Erythrosin B. This spot, along with that associated with the pink pen 2:e, shows a curious aggregation of the dye particles as seen in the inset image. These dye islands did not originally exist in the ink stains prior to exhibition.

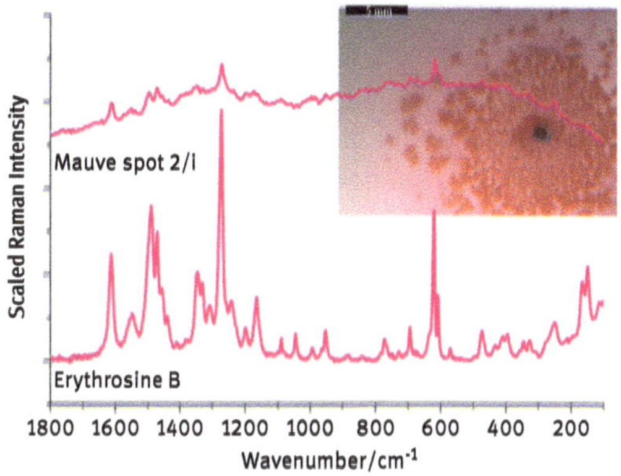

Figure 17.5: Raman spectrum taken from *Painting a Fresco with Giotto #3* at the location of the mauve ink stain, pen 2:i, compared to a reference spectrum of Erythrosin B, both collected using 785 nm excitation. Inset shows the site of the analysis.

All of the remaining attempts to acquire data directly from the ink residues on the *Giotto* series vase suffered from extreme fluorescence interference that saturated the detector or swamped the Raman signal at all excitation wavelengths.

During the course of these in situ Raman analyses, a slightly glossy surface sheen was noticed on the surface of the vase, which is somewhat visible in the inset in Figure 17.5. Suitable focus was achieved only if the Raman laser spot was positioned below this glossy medium. In some areas, small hairs or brush bristles were found to be bound to the surface, presumably adhered within a clear coating. The possibility of a surface varnish was explored using UV-induced visible fluorescence imaging. Fluorescence photography of the artwork, Figure 17.6 (right), showed a distinct appearance difference between the interior and the exterior of the vase when viewed using a black light.

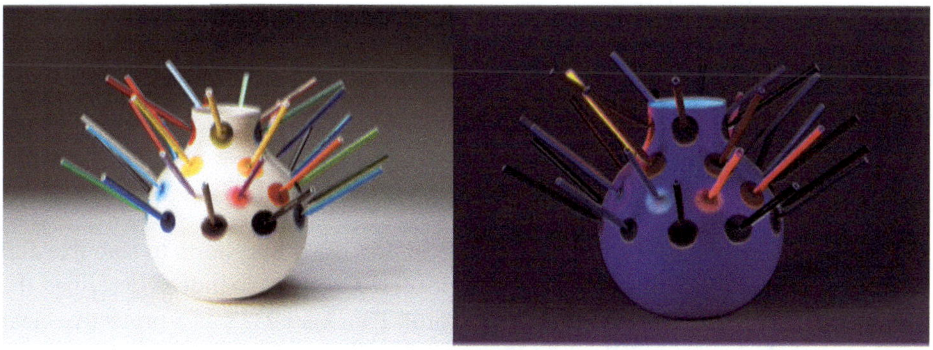

Figure 17.6: Visible (left) and UV-induced visible fluorescence (right) images of *Painting a Fresco with Giotto #3*. Photo by Eric Lubrick, courtesy of Newfields. © Fernando Brizio.

Several interesting features were noted. First, many of the ink spots, as well as some of the pen casings, are extremely fluorescent. Moreover, the surface of the object otherwise shows little luminance, although the interior fluoresced a greenish-blue. The dull violet coloration of the vase's outer surface under black light is due simply to the reflected visible Hg emission line from the bulbs. Suspecting the use of a coating on the object, a scalpel blade was gently scraped over the base of the vase. A microsample of soft, rubbery material was collected from the scalpel tip and was analyzed by FTIR microspectroscopy as well as PY-GC-MS.

The FTIR spectrum obtained from the sample (not shown) revealed strong acrylic stretching features in the carbonyl region, and the spectrum best matched a reference sample for Elvacite 2046, a 1:1 copolymer of iBMA and nBMA, which has been a popular choice for clear varnishes prepared in mineral spirits. This copolymer formulation was confirmed by a pyrogram of the same sample, which showed strong chromatographic peaks for the two acrylic monomers, Figure 17.7. In addition, a third chromatographic peak was tentatively identified as a PEGylated phenolic compound as indicated in the figure. It is not known whether this is the parent structure or a pyrolysate of a larger compound. The role of this compound could not be identified, but it is possible that it serves as a UV absorber in the varnish formulation. This might explain the non-fluorescent surface of the ceramic under a black light.

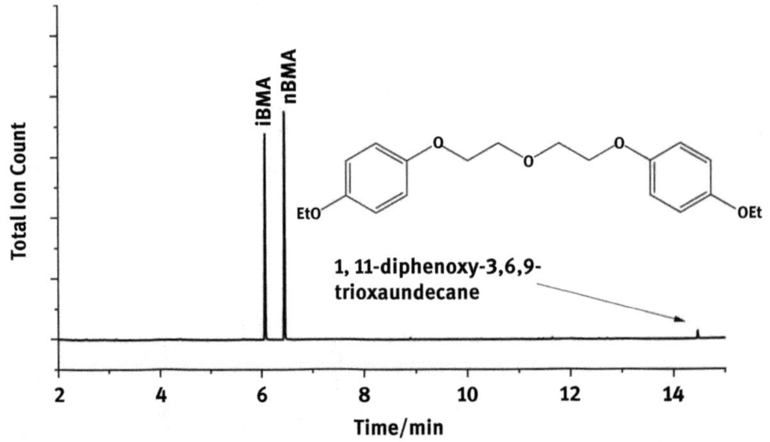

Figure 17.7: Pyrograms of the exterior varnish with major peaks identified.

In later discussions with the artist [2], the artist described his efforts to stabilize and protect the vase through the application of clear coatings. Brizio indeed described the use of an acrylic spray can varnish, Lukas Sprüh Film Matt 2322, as a protective clear coat on the vase's surface. Product literature for this solvent-borne acrylic varnish indicates that it contains a UV protectant, which Brizio thought would shield the dyes against fading, but the coating was also applied to prevent the further migration of the water soluble inks in the humid conditions found in Portugal. In addition, the artist indicated that a Spanish water-based varnish product, La Pajarita, had been brush applied to the interior of the vase before the insertion of the pens. He said this same varnish later replaced the Lukas product as the surface coating for his ceramics after he began to notice color changes in the Giotto pen inks used in his artworks. Curiously, Brizio reported that when spray applied to the surface, the La Pajarita waterborne varnish would cause some colored spots to disappear, only to reappear later. He postulated that perhaps it had to do with water solubility of the dyes and their migration into the ceramic substrate, although pH related color changes are also a possibility. La Pajarita manufactures two lines of water-based satin finish varnishes, one marketed for fine arts and another for craft use. The professional grade product is reported in their product literature to be 100 % acrylic, while the composition of the student or hobby line is not particularly specified.

A small scraping of the varnish taken from inside the vase was analyzed by FTIR microscopy, Figure 17.8, and matched a reference spectrum for Acrocryl 23822, a styrene-acrylic copolymer emulsion. The multiplet of aromatic C–H stretching bands between 3,100 and 3,000 cm^{-1} are characteristic of polystyrene polymers. Increased absorbance in the 1,120 cm^{-1} region of the spectrum from the vase interior varnish is likely due to a silica matting agent, while the bands at 3,300, 1,657, and 1,550 cm^{-1} are assignable to a protein component, perhaps an unfortunate contamination from handling the vase by placing a hand inside the mouth of the ceramic. Brizio has been photographed several times holding vases by this method [28, 29]. This styrene-acrylic composition was confirmed by PY-GC-MS where the resulting pyrogram (not shown) revealed intense peaks for styrene and nBMA, and a small peak for the same 1,11-diphenoxy-3,6,9-trioxaundecane additive seen in the Lukas surface varnish above. Seeing that the varnish has a definitive styrene component, it would appear that Brizio used the craft line of La Pajarita products to prepare the interior of the vase.

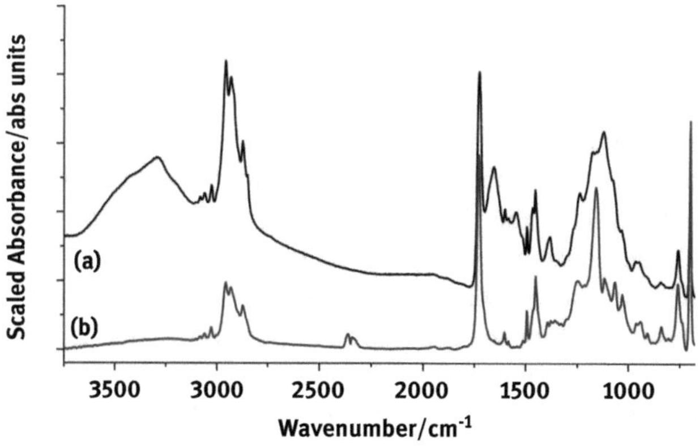

Figure 17.8: FTIR spectrum of (a) the interior varnish compared to (b) an IRUG reference spectrum for Acrocryl 23822 (IRUG, #ISR00022).

The strong fluorescence and optical complications arising from the surface varnish proved too problematic for in situ analysis of the dye stains on the ceramic vessel. To overcome the nettlesome complications of the intervening varnish, the analysis shifted to the pen nibs themselves. In each marker, the white fiber tip was occasionally stained with dried ink, but this was also mixed with rusty accretions that had formed from the steel stick pin driven into the nib to secure the pens to the vase. Unfortunately, when the nib was excited using the Raman laser, the spectrum was dominated by the intense vibrational bands of polyester, the fiber used to form the porous nib "felts," and no dye spectrum was observed. Additionally, many of the pens exhibited the strong visible and NIR-induced fluorescence that had plagued the earlier analyses of the vase's surface.

In a few instances larger globules of dried ink could be analyzed separately from the polyester substrate. In even fewer cases, fluorescence interference did not overwhelm the analytical signal from these agglomerates. The analysis of the mustard colored pen 4:e produced a strong spectrum riding on a modest fluorescence baseline. This spectrum was identified as Tartrazine by comparison to an in-house spectral library. Tartrazine is a common water soluble yellow dyestuff used in food coloring and had already been identified by Raman spectroscopy in a yellow Giotto marker [14]. Interestingly, this spectrum had long been wrongly associated with the storied and now unavailable artists' pigment Indian yellow due to confusion in the reference materials used to build a popular Raman spectral library [30]. Indian yellow is an exotic colorant produced from the urine of cows fed a diet of water and mango leaves, and has purportedly not been produced since the beginning of the twentieth century [31]. Several other red and orange pens with serendipitous dye agglomerates still on their nibs, e. g. the tangerine pen 4:a, gave spectra of reasonable quality; however matching data could not be found in our in-house reference spectral library at the time. Fortunately, the authors recognized the spectrum as Acid Red 18 based on its previous identification by Raman analyses of Giotto and other brand felt-tip markers [14].

In all of the instances reported here for the successful in situ use of Raman spectroscopy, only a single dye was detected in the ink, presumably the dominant component. In light of the many known instances of dye mixtures in porous nib pen inks, and the obvious dye segregation seen on the vase as concentric bands of color, it seemed unlikely that only one dye was present. For the majority of the artwork markers, however, the in situ analysis of the pen nibs did not produce interpretable results due to overwhelming fluorescence, poor signal strength, and/or spectral interference from the polyester nibs. Several conservation related publications report the use of thin layer chromatography (TLC) for the separation of porous nib marker inks [14, 20], and the application of surface-enhanced Raman spectroscopy (SERS) either directly on an artwork [32], directly on inks [33, 34], or to the eluted TLC spots [18] in order to overcome fluorescence and increase the Raman signal. Although these approaches were discussed for this technical study, it was decided that removal or extraction of the dye residue from the markers for TLC separation would unnecessarily endanger the artwork. The use of a SERS substrate directly on the vase was also deemed unwise considering the water solubility of the dyes, and regardless, the presence of the overall acrylic varnish on the vase would make this approach unworkable. Additionally, the application of an aqueous SERS colloid directly to the pen nibs would be dangerous considering the already corroded and fragile state of the steel pins used to hold the markers in contact with the faience vase. Instead, an alternative approach was pursued.

17.3.2 Dye analysis of marker surrogates by liquid chromatography

The initial research plan to identify the marker inks in situ on the *Giotto* vase using nondestructive Raman analysis was ineffective, and the methods for overcoming the difficulties encountered would risk further damage to the artwork. A new approach was considered that would instead use a set of surrogate markers acquired directly from the manufacturer. By examining "non-art" pens, ethical and conservation concerns regarding the already compromised artwork would be respected while allowing for a rigorous characterization of the Giotto brand inks. This approach assumes that the ink compositions would not have changed significantly in the intervening years. LC-MS was chosen as the most informative approach for the analysis of the dyes. This type of analysis has not yet been applied to artworks incorporating felt-tip marker inks. However, LC is a standard approach for the forensic analysis of dyes [35], and the technique is widely used in the conservation field to study historic objects [36]. Furthermore, mass spectrometry provides unparalleled specificity in the analysis of synthetic dyes, both modern [37] and historical [38], in terms of molecular structure elucidation.

A 36 box set of surrogate Giotto Turbocolor markers was acquired in 2014 directly from the manufacturer. A 2 cm × 1 cm section of a glass microscope slide was colored in an even layer with each marker. This inked area was then washed with 200 µL of water into a vial and further diluted fivefold if the concentration of the solution was deemed too high. A 25 µL injection of each ink was then separated using LC-DAD-MS. A list of potential color compounds was identified based on the resulting DAD and MS data, and refined through the comparative analysis of commercially acquired reference dyestuffs analyzed under identical conditions. In total, 9 dyes were identified, used alone or in mixtures, to generate the 36 shades of the Giotto Turbocolor pen palette. The individual dyes are listed in Table 17.1, and their structures can be found in Figure 17.4. From this list, Acid Red 14, Acid Red 52, Acid Violet 17, and Erythrosin B have not yet been identified in a felt-tip pen from a cultural heritage context.

All of these dyestuffs are classified as food colorants, although their regulated use in foods varies by region and country [39]. In fact, published studies using liquid chromatography to analyze colorants in gummy candy [40] and yogurt [41] were particularly helpful in this work with nearly all the dyes found here also being present in these comestibles. The use of food dyes in the Giotto pens makes sense when one considers the target market for the product is school age children rather than professional artists. Food dyes are commodity colorants that offer intense colors at a low cost, water solubility for easy clean-up, and are generally recognized as non-toxic. However, these properties unfortunately come at the price of light stability, with many of these dyes proving very fugitive in strong light [39]. For their use in food products this is not normally seen as a drawback since most food products will not be exposed to strong light for long periods and would no longer be considered edible after their expiration date anyway.

Figure 17.9 shows a set of model LC chromatograms for the nine reference dyestuffs prepared as a single solution. Table 17.2 lists the major peaks identified in the reference mixture of commercial dyes along with their associated physical, chromatographic, and spectroscopic data. The number for each peak correlates with the associated neutral molecule structure in Figure 17.4. More than nine peaks appear in the chromatogram because many of the dyes exist as heterogeneous mixtures of isomers and reaction byproducts. Because they are colored, all of the dyes could be detected by spectrophotometry, Figure 17.9a, and most gave good negative ion MS spectra, Figure 17.9b. Only a handful of the dyes yielded intense signals in positive

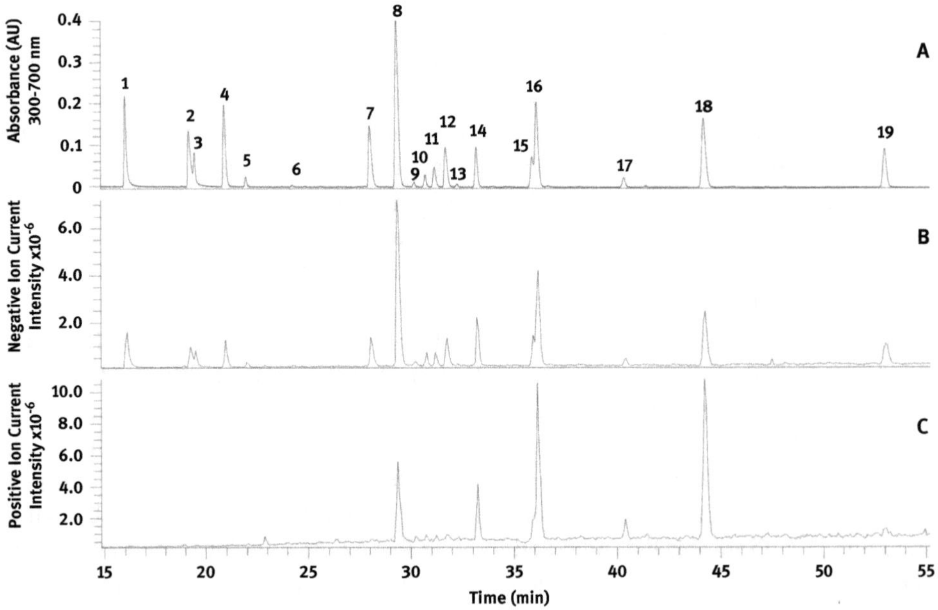

Figure 17.9: LC chromatograms based on (A) the sum of absorbance 300-700 nm, (B) negative ion full scan ESI-MS, and (C) positive ion full scan ESI-MS from a mixture having all relevant dyestuffs. Dyes were prepared as 50 fold dilutions of individual stock solutions of 5 mM assuming sample homogeneity. The data are from a 25 μL injection, and the peak numbers correspond to the structures in Figure 17.4.

Table 17.2: List of reference dyes identified in Figure 17.9 with physical, chromatographic, and spectroscopic data.

Structure Figure 17.4, Peak Figure 17.9A	Colorant	C.I. Number	MW of neutral dye	Ret. time/ min	λmax/nm	Negative ionization		Positive ionization	
						MS m/z	MS/MS m/z	MS m/z	MS/MS m/z
1	Tartrazine	19140	468.41	16.03	428	467 w 233 (M^{2-})		469	451, 407
2	Acid Red 18	16255	538.52	19.18	509	537 w 268 (M^{2-})	509, 429	539 w	ND
3	Quinoline Yellow S, disulfonate (I)	47005	433.41	19.47	430, 413, 396sh	432 w 216 (M^{2-})	368, 352	434	406, 370, 354
4	Sunset Yellow	15985	408.4	20.90	505sh, 482, 420sh	407 203 (M^{2-})	327	409 w	ND
5	Quinoline Yellow S, disulfonate (II)	47005	433.41	21.94	435, 416, 396sh	432	352	434	352
6	Quinoline Yellow S, disulfonate (III)	47005	433.41	24.25	444, 421	432	352	434	ND
7	Acid Red 14	14720	458.46	28.00	521	457 228 (M^{2-})	377	459 w	ND
8	Erioglaucine	42090	748.88	29.29	628, 560sh	747 373 (M^{2-})	667, 628, 587, 498, 306	749	628, 579, 498, 458, 306
9a, 9b	Acid Red 52, dideethylated		502.56	30.16	532, 520sh	501	486, 421	503	439, 423
10	Erioglaucine, positional isomer		748.89	30.69	628, 560sh	747 373 (M^{2-})	667, 577	749	579
11	Quinoline Yellow S, 8'-monosulfonate	47005	353.35	31.14	438, 417, 396sh	352	288	354	272
12	Quinoline Yellow S, 6'-monosulfonate	47005	353.35	31.68	433, 414, 396sh	352	288	354	336, 326, 290, 274
13	Erioglaucine, analog		578.7	32.25	613, 550sh	577	497	579	458, 409

(continued)

Table 17.2 (continued)

Structure Figure 17.4, Peak Figure 17.9A	Colorant	C.I. Number	MW of neutral dye	Ret. time/ min	λmax/nm	Negative ionization		Positive ionization	
						MS m/z	MS/MS m/z	MS m/z	MS/MS m/z
14	Acid Red 52, monodeethylated		530.61	33.18	548, 520sh	529	514, 485, 449	531	487, 467, 451
15	Acid Violet 17, analog		882.07	35.86	591, 550sh	880 440 (M^{2-})	835, 800, 708, 693, etc.	882	712, 540
16	Acid Red 52	45100	558.66	36.09	562, 520sh	557	497	559	531, 515, 479
17	Acid Violet 17, monodeethylated		711.89	40.34	588, 550sh	710	682, 665, 630, 589, 539	712	542, 526, 512
18	Acid Violet 17	42650	739.95	44.18	594, 550sh	738	710, 693, 658, 567	740	570, 554, 540
19	Erythrosin B	45430	835.9	53.00	538, 595sh	835	791, 663, 537	837	710, 583
20	Acid Violet 17, analog		697.86	38.01	582, 550sh	696	651, 616, 525, 510	698	528, 512

sh = shoulder, w = weak, sometimes not observed, (M^{2-}) = doubly negatively charged, ND = no data due to weak MS signal.

ionization mode, Figure 17.9c. Examination of the resulting data is informative regarding these principle colorants.

The commercial samples of Tartrazine, Acid Red 18, Sunset Yellow, Acid Red 14, and Erythrosin B used here were all of high purity based on the single molecular structure identified by LC-DAD-MS. However, Quinoline Yellow S, Erioglaucine, Acid Violet 17, and Acid Red 52 were heterogeneous, each to a varying degree. Interestingly, the dyes found to be pure in the reference samples were also found to be a single component in the pen inks, while those found to be heterogeneous in the reference sample were also multicomponent products in the pen inks, although not always in the same manner. An appreciation of this heterogeneity for each colorant is important for the understanding of the chemical composition of the pen inks and the possible manufacturing processes for the dyes.

Quinoline Yellow S contains the most chemical diversity with five dye components. Peaks 3, 5, and 6 in Figure 17.9 are disulfonated derivatives of 2-(2-quinolyl)-1,3-indandione while peaks 11 and 12 are monosulfonated analogs. All five compounds are synthetic variants resulting from the treatment 2-(2-quinolyl)-1,3-indandione with concentrated sulfuric acid. While the exact site of sulfo-substitution of the disulfonates I through III are not known, peaks 11 and 12 can be assigned to the 8'- and 6'-sulfonates, respectively, according to Weisz et al. [42]. Their data indicate that samples of the dyestuff obtained from different manufacturers have variable ratios of the di- to mono-sulfonates. The reference sample used here from TCI is more weighted to the monosulfonates, but still with easily detectable quantities of the disulfonates. However, the Quinoline Yellow S identified in the pen inks contains predominantly disulfonates. Interestingly, this may be a result of geography. Food colorant regulations in the USA require that Quinoline Yellow S used for this purpose contain no less than 75 % monosulphonate species, while EU restrictions dictate the dye must be at least 80 % disulphonates [39]. The Giotto pens are made in Italy, presumably with EU regulated dye, whereas the reference sample was acquired in the USA. This sort of detailed chemical analysis could help in identifying the geographic source or perhaps even the specific manufacturer of an artist's felt-tip pen.

Erioglaucine, peak 8, has two minor analogs in peaks 10 and 13, occurring in roughly 4 and 1 %, respectively. Peak 10 has the same molecular mass as Erioglaucine but a different MS/MS fragmentation. Therefore, one assumes that peak 10 a positional isomer of 8. The other minor component of Erioglaucine, peak 13, gives a parent ion 170 amu smaller than Erioglaucine and is thus assignable to structure 13 which has one less N-(3-sulfonatophenyl)methyl group. When Erioglaucine occurs in a Giotto marker, peak 8 is always the dominant species. However, the relative amounts of compounds 10 and 13 are variable with some being in the 3–5 % range and others being negligible.

Acid Violet 17 elutes as the 18th compound in the chromatogram in Figure 17.9. The reference material used in the mixture was purchased from the vendor Serva. It contains two other components, peaks 15 and 17. The former shows a molecular mass of 882 that is consistent with structure 15. This compound is an analog of structure 18 in which an N-ethyl group has been replaced with an additional equivalent of the N-(3-sulfonatophenyl)methyl substituent. The minor component eluting as peak 17 showed a molecular mass at 712, which is consistent with another analog with one lower degree of N-ethyl substitution as exemplified by structure 17. Curiously, a second reference sample from MP Biomedical has structure 18, what is commonly described as Acid Violet 17, as a minor component, being about one fifth the amount of structure 17. Also present in this reference is yet another analog, structure 20 with a MW of 698, in an amount similar to structure 18. Of the five Giotto pen inks found to contain Acid Violet 17, all had the profile of the Serva reference with structure 18 as the main component, while pen 1:a and 3:c had significant, but still relatively minor amounts of structure 15, and no marker was found to contain structure 20. Structure 17, as well as its two positional isomers, was present in variably low amounts that were less than 15 % of structure 18 in all five inks containing Acid Violet 17.

Acid Red 52 elutes as peak 16, structure 16 in Figure 17.4, and two other components eluting as the 9th and 14th peaks, corresponding respectively to the monodeethylated isomers 9a or 9b and a dideethylated analog. These could have arisen from incomplete ethylation during synthesis or from post-synthesis degradation. While peak 14 must be structure 14, the dideethylated analog in peak 9 could be one of the two positional isomers 9a or 9b. As it happens, the N,N-diethyl substituent on the aromatic ring is prone to fragment in CID via the loss of propane of mass 44 amu [43]. This diethyl structural element is present in structures 16, 14, and 9a, but not in 9b. Inspection of the MS/MS data showed that an ion due to [M-44] is prominently detected in Acid Red 52 at peak 16 and the monodeethylated analog of peak 14. However, for the compound in peak 9, the [M-44] ion is low in negative ionization and almost nonexistent in positive ionization. Therefore, the dye in peak 9 is most likely structure 9b in which the ethyl residues are bonded to separate amino nitrogens. Using selected ion extraction on the chromatogram in Figure 17.9a, a compound is in fact observed to coelute with Acid Red 14 in peak 7 that exhibits a molecular mass of 502 amu, and in MS/MS shows a prominent ion at [M-44] consistent with structure 9b. This highlights the resolving power of MS/MS in separating mixtures. Acid Red 52 was found in only two pens, and in each case it was present as the minor dye component of a mixture. In these mixtures, structure 16 was between 5 and 10 times the amount of structure 14, and the minor structures 9a and 9b were not detected.

Several points about the mass spectrometry of these dyes are worth noting. First, the disulfonated dyes like Tartrazine show a propensity in the full scan mode to form doubly charged ions (M^{2-}) in negative ionization, so much so that sometimes the singly charged species was not observed. At half the m/z of the pseudomolecular ion, the occurrence of only the M^{2-} species can confuse the identification of an unknown dye since the observed molecular weight will not correlate to that of any known dye. Fortunately, despite being very weak, the singly charged pseudomolecular ion is observable in positive ionization. In this respect, it is advantageous to have a MS capable of rapidly alternating between positive and negative ionization modes.

The fragmentation patterns in MS/MS of the sulfonated dyes are dominated by the [M-80] and/or [M-64] ions due to loss of SO_3 and SO_2, respectively. Aside from confirming that the dye is sulfonated, these offer little structural information. For the triphenylmethine class of dyes such as Erioglaucine and Acid Violet 17 and their respective analogs, MS/MS in positive ionization induces fragmentation between aryl substituents at sufficient prevalence to be helpful for structural assignment. In negative ionization MS/MS, however, the CID fragmentation produces such a large number of small fragments that it is not useful. Thus, the entirety of the DAD and MS data are needed for confident identification of pure and heterogeneous dyestuffs.

Table 17.3 shows the specific dyestuffs in each marker sorted by color from the surrogate pen set acquired in 2014 and a rough estimate of their proportions. It is interesting to note that Tartrazine, Acid Red 18, and Erioglaucine were preferentially used as the three primary colors of yellow, red, and blue. These were then blended to create other colors, such as Tartrazine and Erioglaucine to make green in pen 1:b, and all three to create black in marker 1:i. Other dyes were added in various amounts presumably to create different shades, e. g. the Powder Blue (2:a), Baby Blue (2:f), and Sky Blue (4:b) shades that are made by doping Erioglaucine with increasing amounts of Acid Violet 17. Another notable occurrence is the different shades of yellow being made from three yellow dyes: Tartrazine, Quinoline Yellow S, and Sunset Yellow. Frick [39] remarks on the importance of Quinoline Yellow S in food coloration for providing a purer yellow than the orangey-yellow generated by Tartrazine. Acid Red 52 and Erythrosin B were the least used dyes, appearing in only 2 pens each.

As mentioned earlier, a few chemical studies in the field of cultural heritage have explored the ink formulations of Giotto brand felt-tip markers, largely relying on vibrational spectroscopy either alone or to identify individual dye components following TLC separation. Sodo et al. [14] separated the dyes from three Giotto Turbocolor pens by TLC and used normal Raman spectroscopy to identify them. Although they saw two resolved spots in a yellow marker, only Tartrazine could be positively identified. The analysis here shows that all yellow markers in the Giotto line contain multiple dyes, and the unresolved component could be Sunset Yellow as seen in pen 3:d or Quinoline Yellow S in pen 3:e . The same was true of their analysis of a red marker. Two TLC spots were observed, but only Acid Red 18 could be identified. The second spot was probably the Acid Red 14 component identified here in pen 4:a, the purest shade of red. Curiously, an orange pen analyzed by their group showed only two TLC spots, but they report identifying Tartrazine, Acid Red 18, and Sunset Yellow, matching the results obtained here for pen 2:h. TLC may not have been able to resolve fully the latter two dyes, again showing the improvement possible with LC-DAD-MS.

Table 17.3: LC-DAD-MS results for the 36 Giotto Turbocolor surrogate pens from 2014 sorted by color. The pen colors that were used in the vase are identified by their location notation as in Figure 17.3, while those not used in the artwork are listed as NA1-7. Chromatographic peak intensities for each detected dye are ranked semi-quantitatively as intense (+++), strong (++), modest (+), and trace (t). The final column gives the ISO BWE fading rate for each surrogate marker.

Pen #	Color	Tart	QY	AR18	Sun	AR14	Erio	AR52	AV17	Ery B	BWE
1:a	Royal Blue						+++		+++		2
1:d/1:h	Teal	+	+				+++		++		<1
2:a	Powder Blue						+++				2-3
2:f	Baby Blue						+++		t		2-3
4:b	Sky Blue						+++		+		2
NA5	Dk. Teal	+					+++		+		--
NA6	Dk. Blue	+		+			+++		++		--
1:b	Lt. Green	+++					+++				<1
1:e	Dk. Green	++			+++		+++				<1
1:g	Forest Green	+++			+++		+++				1
2:g	Lime Green	+++			++		+				<1
3:b	Mint Green	+++					+++				<1
4:d	Green	+	++				+++				<1
NA4	Olive Green	+++		+++			+++				--
1:c	Lt. Black	+++		+++			+++		++		1-2
1:f	Gray	+++		+++	+++		+++				2-3
2:b	Lt. Gray	+++		+++	+++		+++				2-3
1:i	Black	+++		+++			+++				2-3
1:j	Brown	+++		+++			+++				2-3
2:d	Lt. Brown	+		+++			+				2-3
NA2	Lt. Tan	t	+++	+			+				--
3:a	Tangerine	+++		+++							2-3
4:a	Red			+++		+++					2-3
4:c	Peach	+++		+++							2-3
NA3	Rust	+++		+++			t				--
2:c	Dk. Red			+++		++					2-3
2:h	Orange	+++		++	+++						2-3
3:d	Dk. Yellow	+++			+++						2
3:e	Lt. Yellow	+++	+								<1
4:e	Mustard	+++		+			t				1-2
NA1	Pale Yellow	+++	+								--
2:e	Pink					+++				+++	<1
NA7	Pale Pink			+++							--
2:i	Mauve								t	+++	<1
2:j	Lavender			++				++	+++		<1
3:c	Purple							++	+++		<1

Tart = Tartrazine, AR = Acid Red, AV = Acid Violet, QY = Quinoline Yellow S, Sun = Sunset Yellow, Erio = Erioglaucine, EryB = Erythrosin B

Izzo and coworkers [20] also examined three Giotto pens using TLC, FTIR, and PY-GC-MS. For a red pen, they also report only Acid Red 18, although their published TLC plate clearly shows two resolved spots eluting in the order one would expect for normal phase chromatography if both Acid Red 18 and 14 were present, again as in pen 4:a. A similar situation exists for their analysis of a blue marker. A blue and a purple spot are seen in the TLC plate, but only Erioglaucine was identified. The purple spot should be Acid Violet 17, which was found here as the second major dye in the purest shade of blue pen 1:a. In the analysis of the black Giotto marker, FTIR identified the major components of Tartrazine, Acid Red 18, and Erioglaucine, as did the LC-DAD-MS analysis here of pen 1:i.

The final group to examine Giotto markers used a SERS reagent on the whole ink without separation [34] and in combination with reflected UV-vis spectroscopy [33]. Their SERS analysis revealed Eriogluacine and Tartrazine in a green Giotto pen. The specific shade of green was not mentioned, so it was possible pen 1:b or 3:b, what are called Lt. Green and Mint Green here, since only those shades have just the two dyes mentioned. However, it is also possible that without a prior separation of the dyes, other components of the mixture could have been missed, for instance the Quinoline Yellow S that accompanies the purest shade of green pen 4:d. Importantly, Quinoline Yellow S was found in this study not to yield a normal Raman spectrum, even though fluorescence was not overwhelming (vida infra). For a red pen, those authors again report the use of Acid Red 18, but do not observe the Acid Red 14 component. Surprisingly, they do indicate a possible mixture with the azo dye amaranth in their first paper [33] but suggest the presence of the xanthene dye Rhodamine B in their second report [34]. However, neither of these dyes occurs in the Giotto pen set studied here from 2014, and Rhodamine B seems unlikely because of its toxicity. It is possible that without prior separation that they have misinterpreted the SERS spectrum of the mixture.

A similar error could have occurred in their analysis of a blue marker. Saviello et al. report the presence of Erioglaucine and Crystal Violet based on SERS analysis [34], but they also suggest the presence of Rhadamine B based on the additional input from deconvolution of the UV-vis reflectance spectrum. However, the latter two dyes were never detected in this LC-MS study, even with selected ion monitoring. Most of the blue markers in the 2014 surrogate sample set do contain Acid Violet 17. Crystal Violet is the completely N-methyl substituted analog of Acid Violet 17, and the SERS spectra of the two could be similar; the visible absorption maxima of both are ~ 591 nm. Although one cannot rule out the use of another violet dye in the Giotto formulations sometime after 2014 when their study was undertaken, the presence of Crystal Violet in a children's marker would seem unlikely since it has medicinal properties and is not accepted as a food dye, even for animal foods. Rhodamine 6G, another dye unapproved for food use, was also detected as a minor component in their orange, yellow, and pink Giotto markers [33], but it also was not found in even trace quantities by LC-MS. The pink pen is said to contain Eosin Y as the primary component, although

LC-MS shows that this is actually the iodated analog, Erythrosin B, which shares nearly identical Raman and UV-vis spectra, but whose primary ion in mass spectrometry varies by 266 amu compared to Eosin Y, making it easily distinguishably by that technique. To resolve some of these contradictory findings, an inquiry was sent by the present authors to the quality assurance, product safety, and compliance officer at F.I.L.A. Italia regarding the use of Crystal Violet, Rhodamine B, and Rhodamine 6G in their Giotto markers. The company indicated that these dyes are not used [44].

The observations above relating to previous analyses highlight the limitations of low resolution TLC separations and the reliance on vibrational or electronic spectroscopic identification of dye components in mixtures. With these approaches, the identification of individual dyes relies on either having a reference compound to compare against, or having a reference spectrum available through spectral libraries. In many instances, the similarity between vibrational and UV-vis spectra for related compounds, especially isomers or substitutional analogs, makes positive identification of heterogeneous mixtures difficult. Even the successful Raman analyses presented in this study failed to identify more than the major dyestuff in the original artwork markers. LC-DAD-MS, however, provides a rich dataset that can identify dyes based on their color and MW or by interpreting MS/MS spectra based on known fragmentation properties. The specificity shown here for the Giotto markers demonstrates how knowing an exact dye composition for an ink used in an artwork might allow for the identification of a specific manufacturer's product, provided changes in manufacturer's formulations are known. Moreover, the technique offers excellent sensitivity as well as superb resolution of compounds through both the manipulation of chromatographic variables and the use of mass selectivity. The downside of this approach is that it requires sampling that is often at a scale undesirable for artworks, and it cannot at this point be used in situ on objects.

Having positively identified the relevant dye mixtures used in the surrogate set of Giotto markers, the innate fluorescence of the individual dye components was recorded across the UV-visible-NIR spectrum, Table 17.4. This was done to understand better the fluorescence image in Figure 17.6 and the complications encountered in the Raman spectroscopy of these dyes. UV-induced fluorescence was monitored by observing the mock-up ceramic prepared for MFT analysis while irradiating it at a distance of a few inches with a handheld black light (Spectroline ENF-280C) having longwave (365 nm) and shortwave (254 nm) lamps. The relative brightness of the fluorescence emission was gauged by eye as being intense (+++), strong (++), modest (+), or non-fluorescent (–). The UV-induced fluorescence intensities for these dyes were seen to vary if the dye concentration was lowered or if the dyes were stained onto Whatman filter paper, generally becoming more intense and more widespread across the two wavelengths studied. The impact of visible and NIR fluorescence was assessed by monitoring the Raman spectrum generated from the same reference dyes when using each excitation laser at the lowest power setting. For these wavelengths, a semi-quantitative assessment was made by noting whether the fluorescence saturated the

Table 17.4: Fluorescence of relevant dyestuffs at various excitation wavelengths on a ceramic substrate. Fluorescence emission is ranked semi-quantitatively as overwhelming (+++), strong (++), modest (+), or non-fluorescent (–).

Dyestuff	Fluorescence excitation wavelength/nm				
	254	365	532	633	785
Tartrazine	–	+	+	–	–
Acid Red 18	–	+	+++	+	–
Acid Red 14	–	++	+++	++	–
Acid Red 52	++	+++	+++	++	–
Acid Violet 17	–	–	++	++	++
Erioglaucine	–	–	+++	++	++
Sunset Yellow	+	+++	+++	+	–
Quinoline Yellow S	+	+++	+++	++	+
Erythrosin B	–	+	+++	+	–

Raman detector (+++), overwhelmed the Raman spectrum of the dye (++), competed with the Raman scatter of the sample (+), or proved a non-issue for Raman analysis of the particular dye (–).

Table 17.4 shows that many of the dyes are strongly luminescent across the UV-visible spectrum. The intense fluorescence of Erioglaucine and Acid Violet 17 also extends into the NIR excitation region. These two dyes occur in all but seven of the Giotto markers used in the artwork, explaining the difficulty in assessing the dye compositions in situ using normal Raman spectroscopy due to an intense fluorescence background even with NIR excitation. The pens most amenable to non-destructive Raman characterization should therefore be the yellow, red, and pink markers, which show little fluorescence when using the NIR laser, as observed practically in the earlier in situ analyses that identified Tartrazine, Acid Red 18, and Erythrosine B. Good quality reference Raman library spectra were obtained from all of the relevant commercial dyes using the 785 nm excitation laser with the exception of Erioglaucine, Acid Violet 17, and Quinoline Yellow S. The first two dyes were too fluorescent, while the Quinoline Yellow S appears not to be a strong Raman scatterer. Reference Raman spectra for these dyestuffs were not available in the popular IRUG database [www.irug.org] at the time of this study, but have now been submitted by the authors for inclusion.

One pen from the surrogate markers, Dk. Yellow 3:d, was further explored to determine the additives used in the ink formulation. Py-GC-MS has been applied previously to felt-tip markers used in cultural heritage to identify solvents, binders, pigments and dyes, and stabilizers [15, 18–21, 45]. In this work, LC-DAD-MS was used for the first time to assess soluble additives in a line of artists felt-tip markers. The same sample solution used in the LC-MS dye analysis was reused for the additives analysis, but this time with only DAD and full scan positive ionization MS detection.

The resulting chromatogram based on the total UV-visible absorption, Figure 17.10a, reveals the Tatrazine and Sunset Yellow colorants observed previously for this marker. However, those peaks are nearly absent in the positive ion chromatogram in Figure 17.10b, which is instead dominated by an early eluting broad peak at ~11 min followed by an envelope of evenly spaced peaks from ~15 to 26 min whose m/z data increases by 44 mass units each. The mass of the early eluting peak can be associated with triethylene glycol in its $[M + H]^+$ and $[M + NH_4]^+$ forms, whereas the latter series of peaks are the same polyethylene glycol (PEG) adducts with lengths ranging from 4 to 15 ethylene glycol units. Ammonia is not used in the LC-MS mobile phase, and so it must be present as a component of the ink formulation. PEG has been previously identified as a common additive in several brands of water-based felt-tip marker inks [20, 21], but because of the use of high temperature pyrolysis for those analyses, the exact MW distribution present in the ink is questionable. The low MW PEG detected directly here by LC-MS is presumably added to the formulation as a slowly evaporating co-solvent, a rheology modifier, binder, and/or a humectant to prevent the felt-tip from drying out if left uncapped. Pfingstag [46] refers to this as a "nonionic wetting agent" in his formulations for dyestuff based felt-tip markers. He also indicates the use of a slowly evaporating co-solvent such as diethylene glycol or propylene glycol at a concentration of 20–30 %. In concentrated injections of the marker ink, a small peak eluting around 5.6 min is observed with m/z = 107 and 124, correlating to protonated and ammoniated adducts of diethylene glycol. Palmitic and stearic acids have also been previously found in felt-tip markers including the Giotto brand using PY-GC-MS on derivatized ink samples [20]. It is

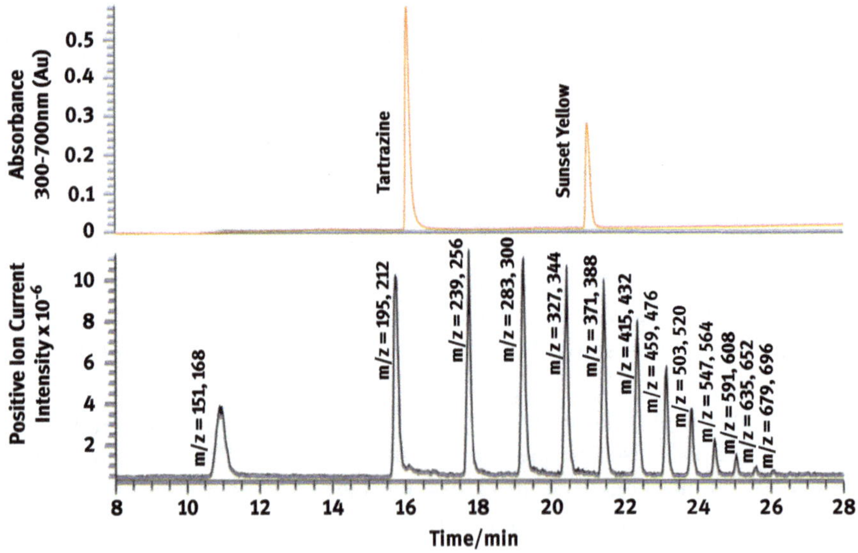

Figure 17.10: Chromatogram of Giotto Turbocolor Dk. Yellow surrogate pen 3:d acquired using (A) the sum of absorbances 300–700 nm and (B) positive ionization full scan MS.

noteworthy that minor peaks eluting at 15.88 and 16.26 min show primary m/z of 255 and 283 in negative ionization mode, which correspond to the [M-H] ions for palmitic and stearic acids, respectively. These may confirm the earlier identification of a fatty acid mixture in the inks.

17.3.3 In situ lightfastness testing

Largescale studies of the lightfastness of felt-tip and so-called "permanent" porous nib pens conducted in the second half of the twentieth century warned against the fugitive nature of most of these products [7, 9]. Feller [9] in particular argued for the creation of a lightfastness rating scale to warn artists about the incumbent fading issues. Although the *Giotto* series vase has obviously succumbed to the inherent vice of the markers employed in its creation, conservation and curatorial questions still remained regarding the anticipated future color change of the artwork. The perception of fading is often an asymptotic process, with the rate of apparent color change being greatest initially only to slow as the object loses additional color [11]. Microfade testing in the field of conservation science has allowed for an ethically acceptable, objective prediction of lightfastness directly on artworks on a laboratory timescale [12]. This in situ analysis avoids all of the inaccuracies and assumptions inherent in performing fading studies on mock-ups or surrogates. Results from MFT fading can be compared to industrial standards analyzed under identical conditions, but whose fading rates under ambient conditions of illumination have already been well established. The so-called Blue Wool Scale is well known in the conservation field, and those artworks with colorants that fall in the most susceptible range of BWS1 through 3 are considered fugitive and must be protected under the most stringent storage and exhibition conditions [11].

MFT analysis was conducted on the 29 surrogate Giotto Turbocolor pens whose colors appear in the *Giotto* series vase. The pens were drawn onto the surface of a biscuit fired ceramic, Figure 17.11, to create color swatches that approached the intensity observed in snapshot images of the vase taken before its inaugural exhibition. Similarly, 5 mM aqueous stock solutions of the reference dyestuffs were spotted dropwise onto the same surface until the color intensity was similar to that of the marker swatches. Finally, some of the colored spots on the surface of the *Giotto* series vase were also analyzed to determine their propensity for future fading. It is understood that deviations in the measured fading rates would occur between the surrogate samples and the *Giotto* series vase as a result of the strength of application of the color to the surface, the presence of the intervening acrylic varnish and the exact composition of the faience.

The analysis of the individual reference dyestuffs was informative. Erythrosin B, Acid Red 52, and Acid Violet 17 all faded at a rate much faster than even BWS1, and Acid Red 14 changed at approximately the same rate as BWS1. According to Druzik

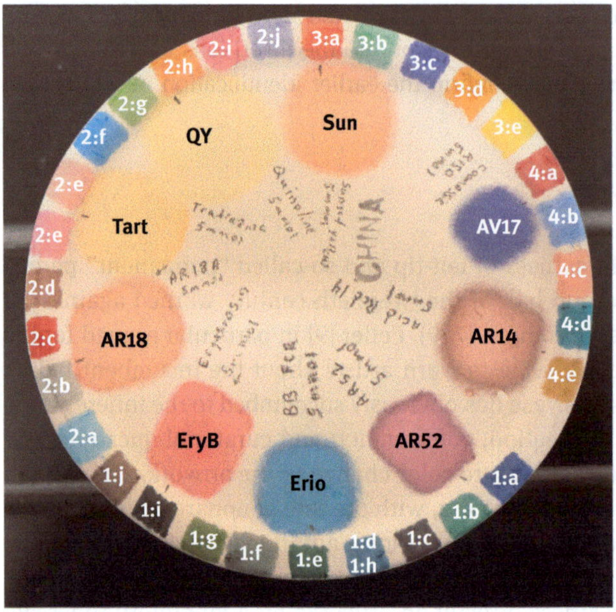

Figure 17.11: A biscuit fired ceramic cup bottom colored with the surrogate markers along the rim and stained with 5 mM solutions of the reference dyestuffs listed in Table 17.1 in the interior. Labels are as for Table 17.3.

[47], BWS1 is just noticeably faded after 300,000 lux-hrs of exposure under typical gallery lighting conditions. Assuming 3,000 hr/year exposure (about 8.2 hrs/day) at a modest 50 lux of gallery illumination, this would mean an artwork containing Acid Red 14 would be noticeably faded after just 2 years [11]. For the first three dyes mentioned, the time period would be much shorter! Erioglaucine was observed to fade at the rate of BWS2 while Tartrazine faded at a rate similar to BWS3. Under the exhibition conditions previously mentioned, fading in these dyes would be recognizable after approximately 6 and 20 years of exposure. The fading rate of Quinoline Yellow S, Sunset Yellow, and Acid Red 18 fell between BWS2 and 3.

All of the surrogate markers whose colors were used in the *Giotto* series vase were also tested for their lightfastness. The last column in Table 17.3 gives the BWS equivalent fading rate for the surrogate pens applied to the ceramic mock-up. It is interesting to note that all of these markers gave a BWS rating below 3, and many less than BWS1! None of the dyes found in the Giotto Turbocolor markers would be considered "durable" by conservation standards, and all fall into the "fugitive" class [11]. In general, the lighter shades are more susceptible to fading, a phenomenon well known with regard to translucent glazes and thin washes of color in artwork. This is compounded by the double-pass exposure that occurs when light passes through the ink only to reflect off the white substrate and pass back through the medium a second time.

Chromatically, the greens, pinks, and purples showed the worst lightfastness, while the reds, browns, and blacks proved the most stable of the Giotto Turbocolor pens. Considering the 215 lux light limits used during the initial exhibition of the *Giotto* series vase, it is understandable that the artwork returned to the IMA with significant fading. Brizio's application of a UV absorbing varnish to stabilize the inks was unfortunately misguided; the MFT fading experiments show that UV radiation is not needed to cause pronounced color change in these felt-tip pen inks. Lafontaine's [7] fading experiments on 543 felt-tip pens showed that removal of UV from the ambient illumination only extended their life on average by a modest 20 %.

Finally, a few spots on the *Giotto* series vase were explored to determine if the artwork is still prone to color change, even after the significant fading that has already occurred. In all instances measured, the fading rates still fell into the "fugitive" category. Figure 17.12 shows the fading curves for the Dk. Yellow spot (3:d) from the vase, which the inset image shows separated into a mustard yellow inner ring of color and an orange-yellow outer ring. This data is compared to the fading results from the BWS1-3 (blue lines) and the whole ink from the surrogate marker 3:d (yellow diamonds). This ink was shown by LC-MS to contain Sunset Yellow and Tartrazine, which would be expected to elute in the "normal phase" conditions of the ink bleeding out onto the faience to produce a larger migration for Tartrazine, hence the concentric rings of overlapping color. MFT testing of the individual dyes revealed Tartrazine to be more stable (BWE = 3) than Sunset Yellow (BWE = 2–3), but

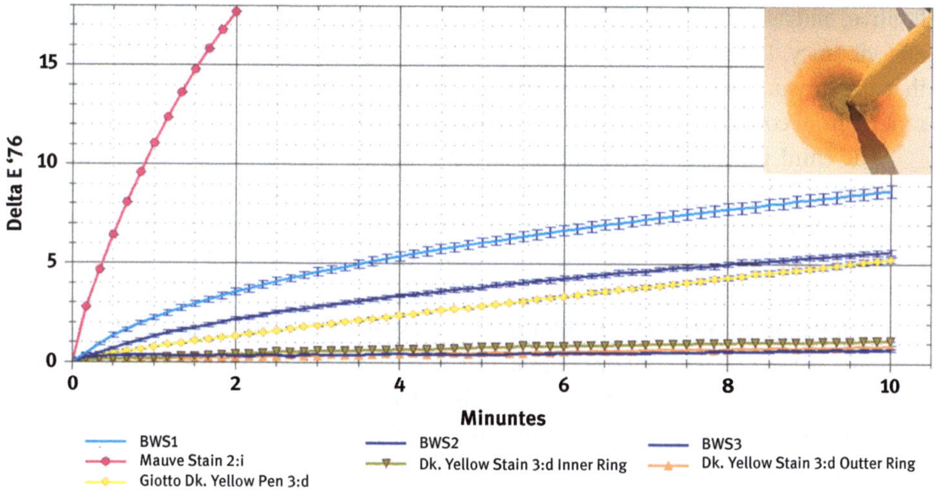

Figure 17.12: Absolute color change ΔE'76 as a function of MFT exposure for BWS1-3, surrogate Giotto Dk Yellow pen 3:d, the corresponding bands of color on the vase, and the mauve spot 2:i. Data for the BWSs and the surrogate pen ink are the average of 3 measurements, and the error bars represent 1σ. Measurements made on the vase are a single measurement. The experiment for the mauve spot was stopped after 2 min to avoid making a visually perceptible faded spot on the object.

the ink mixture in the surrogate pen faded faster than either component. Both bands of color on the vase from this marker now fade at a rate close to BWS3, suggesting that the rate of color change has slowed after the initial fading during exhibition. In contrast, a thin wash of pinkish color on the mauve pen stain on the vase (2:i) that now has a mottled surface appearance still fades much faster than BWS1, so much so that the experiment had to be ended early to avoid leaving a small white dot in the color! Based on these results, the curatorial concerns over the possibility of continued fading and color change in *Painting a Fresco with Giotto #3* are well warranted – it remains an extremely light sensitive object.

17.4 Conclusion

Traditional fresco paintings are some of the most stable forms of creative expression, lasting millennia. Ironically, *Painting a Fresco with Giotto #3* proved to be an ephemeral creation due to the inherent vice in the felt-tip markers that are so integral to the piece. In response to this incident, the resulting case study included the most comprehensive technical examination of an artwork comprised of felt-tip markers to date, and the most complete chemical characterization of a contemporary line of artists' porous nib pens. The detailed chemical analysis of surrogate pens from the Giotto Turbocolor line showed that a limited palette of food colorant dyes was used to generate a wide array of colors and shades. Importantly, LC-MS proved to be a superb technique for probing the chemical complexity of these artists' materials. Unfortunately, these dyes are incredibly fugitive, and as a result of exhibition, the artwork has suffered an overall loss of intensity and vibrancy. Lightfastness measurements made directly on the ceramic surface prove that the artwork is still very prone to fading and that its remaining color will change rapidly from continued exposure to even modest gallery lighting.

The third phase of the technical study sought to determine a path forward for *Painting a Fresco with Giotto #3*. Conservators, curators, and exhibition designers have at their disposal a limited range of options for dealing with light sensitive artworks. One immediate result of the project was the creation of a bespoke storage container for the artwork that protected the piece from light when not on view. Labels and color images attached to the enclosure ensured that museum staff knew the contents of the box without opening it and also warned of the content's lightfastness issues. Because of the use of water-soluble dyes, even with the protective acrylic varnish, additional precautions were listed to avoid inadvertent contact with water, including perspiration and condensation.

Other potential protective measures during display include limiting light levels and exhibition duration to reduce the total luminous flux experienced by the artwork. These efforts can be improved with the addition of visitor-activated lighting; there is no reason to illuminate the object if no one is actively viewing it. For materials that fade or change color due to photo-oxidation reactions, the removal of oxygen through the use of anoxic frames or vitrines has been explored for the display of watercolors [48] and photographs [49], although not for large three-dimensional objects.

Discussions among curators, conservators, and scientists weighed the various options for the future conservation and display of the artwork. Retouching to restore the lost color would be difficult without good quality photographs of the original appearance of the now faded ink stains. Moreover, the poor lightfastness of the remaining color would lead to rapid distortions in the color harmony of restored and original sections upon further exhibitions. With such extremely fugitive colors («BWS1), even exhibition at 50 lux would only prolong the loss of color by a measure of a few years without extremely limiting the duration of an exhibition. The overall gallery appearance also raised concerns since a dimly lit display case in a room that was otherwise brightly illuminated would lead to a poor aesthetic look to the gallery. The development of an anoxic display case was discarded due to the great expense of these systems and the difficulty in maintaining static anoxic conditions of such large enclosures for long periods without constant monitoring and maintenance. Furthermore, the reduction in the fading rate for some aniline dye based materials has been studied under anoxic conditions and found to benefit from it, but the particular colorants in the Giotto pens were not part of this research, and a small number of dyes were shown to actually fade faster in anoxia [50]. In 2018, a curatorial decision was made to deaccession *Painting a Fresco with Giotto #3* from the museum's collections due to the altered state of the object, the continuing instability of the remaining colors, and the harsh implications of its lightfastness issues on further display. But this was not to be the end of the story for the vase.

Since 2013, the IMA has operated the largest museum galleries in the US dedicated to modern and contemporary design. These spaces traced the development of design in the late twentieth century through modernism, postmodernism, and expressionism. In 2018, the entire design collection was reassessed, and these display spaces were renovated in response to a study of visitor experiences. The new organization of the design galleries places a larger emphasis on the design process, organizes the collection thematically rather than temporally, explores the materials of modern design, and provides numerous hands-on interactives. In consultation with the artist, *Painting a Fresco with Giotto #3* has now been reinstalled in this location as part of a didactic space that explores the materials used in contemporary art and the aesthetic consequences of those choices. The new installation of the *Giotto* series vase has delighted visitors, Figure 17.13, and has been the focus of local media surrounding the opening [51]. The artwork is described on its display case, along with a photo of an unfaded vase, as:

> *Carefully balanced felt-tip markers extend from the surface of this ceramic vase, leaving behind rings of color that seep from their tips. Designer Fernando Brizio was inspired by the unsightly ink stains left on his shirt pockets by leaking pens, and took advantage of the porous nature of faience (a type of ceramic) to execute this piece. The intense pools of color were left entirely up to chance, and in another turn of fate, they have begun to fade over time (the image shows a vase in*

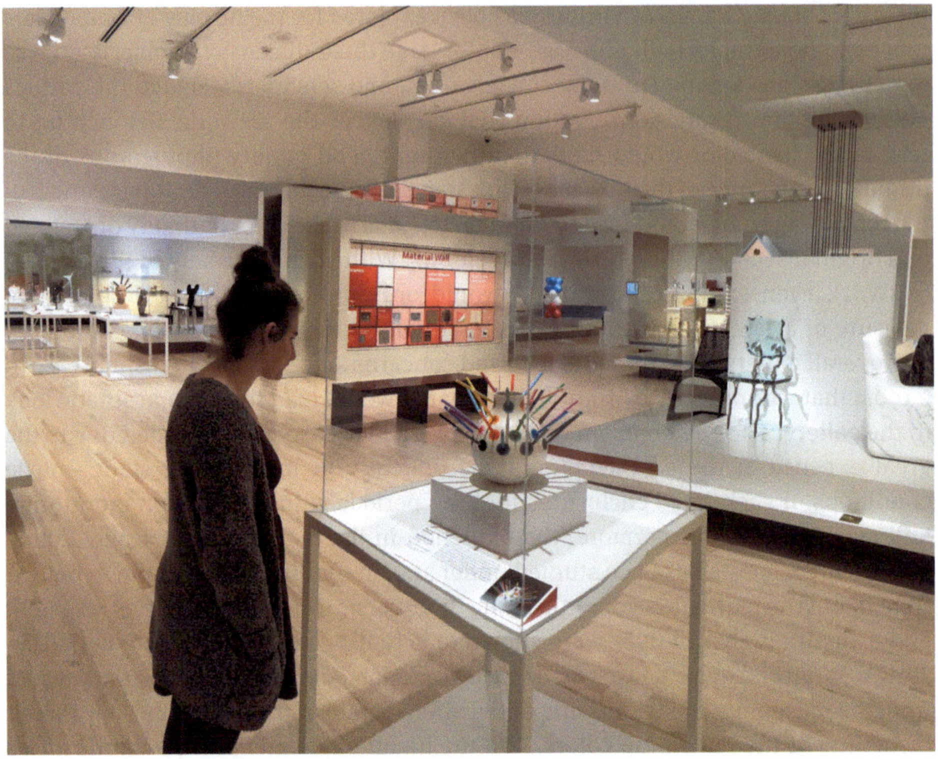

Figure 17.13: A guest admires *Painting a Fresco with Giotto #3* at Newfields during the opening week of the reinstalled design galleries. © Fernando Brizio.

> *its initial state). Through his use of fugitive materials, Brizio has created a piece that continues to change and evolve* [1].

All involved realize that the artwork will continue to change, but feel that in doing so the object makes a poignant statement about the design process and its consequences. Contemporary artists continually push the boundaries as they explore new materials and techniques for making art and design objects. Sometimes these processes incorporate unstable materials that make the long-term preservation of the piece challenging. While conservation professionals seek to educate artists about the implications of these choices, ultimately the conservation field is responsible for protecting and preserving the artist's intent as much as possible. The history of *Painting a Fresco with Giotto #3* serves as a reminder that this starts at the point of acquisition. Artworks containing very fugitive colorants can be irreversibly damaged even in their first debut. Although the *Giotto* series vase has found a new role in the museum to both delight and educate visitors, this case study highlights the urgency of characterizing the fading rate of potentially fugitive modern art or directly identifying the colorants used in contemporary artworks prior to their first exhibition.

Acknowledgements: The authors acknowledge the generous gift of the Alliance of the Indianapolis Museum of Art who funded the purchase of the artwork. Former design curator Mr R. Craig Miller is recognized for initially bringing the condition issues related to the vase to our attention. Surrogate Giotto Turbocolor markers were donated by the manufacturer F.I.L.A. Italia, and some powdered dyestuffs were generously provided by Abbey Color of Philadelphia. Eric Lubrick is thanked for his assistance in acquiring professional quality images of the artwork. Financial assistance was provided by the Carter Family Fund and the Frenzel Family Charitable Lead Trust. Mr Luke Gallion is acknowledged for his earlier work using TLC to separate the dye components of the Giotto brand markers. We appreciate the input and support of Newfields' Associate Curator of Design and Decorative Arts, Shelly Selim. Finally, Mr Fernando Brizio is to be thanked for his thoughtful replies to our inquiries regarding the *Giotto* series and his willingness to let our experiences with his vase assist in the future preservation of contemporary art and design.

References

[1] Selim S. 2018. Personal communication. February 1 2018.
[2] Brizio F. Personal communication 2014.
[3] Marcus JS. A 21st-century artisan. Wall Street J. 2011 March 25. https://www.wsj.com/articles/SB10001424052748704608504576208320312276158.
[4] Miller RC, Spark P, McDermott C. European design since 1985: shaping the new century. London: Merrell, 2009.
[5] Ahlberg P, Smith GD. Perspectives on: an art museum laboratory. Lab Manager. 2013;8:62–5.
[6] Owen A. Modern materials in drawings part 1 – media. Drawing. 1985;7:56–9.
[7] Lafontaine R. The lightfastness of felt-tip pens. J IIC Can Group. 1978;4:9–16.
[8] Ellis MH. Drawings in fibre-tipped pen – new conservation challenges. In: Richmond A, editor. Modern works, modern problems? Conference papers. Leigh, UK: Institute of Paper Conservation, 1994:114–21.
[9] Feller R. Felt-tipped markers and the need for standards of lightfastness for artists' colorants. Bulletin IIC – Amer Group. 1967;8:24–6.
[10] Ellis MH. The porous pointed pen as artistic medium. In: Fairbrass S, editor. The institute of paper conservation: conference papers, Manchester, 1992. Leigh, UK: Institute of Paper Conservation, 1992:11–8.
[11] Michalski S. 1997. The lighting decision. In Fabric of an exhibition, preprints of textile symposium 97, Canadian Conservation Institute: Ottawa, 97–104.
[12] Whitmore PM, Pan X, Bailie C. Predicting the fading of objects: identification of fugitive colorants through direct nondestructive lightfastness measurements. J Am Inst Conserv. 1999;38:395–409.
[13] Giotto. 2018. https://www.fila.it/it/en/brand/giotto/ Accessed June 27, 2018.
[14] Sodo A, Bicchieri M, Guiso M, Ricci MA, Ricci G. Raman investigations on marker pen inks. J Raman Spectrosc. 2012;43:1781–7.
[15] van der Werf ID, Andriani E, Albanese A, Daurelio G, Marano D, Sabbatini L, et al. Laser removal and chemical characterisation of graffiti paint spray and felt-tip markers on stone monuments. In: Townsend JH, Toniolo L, Cappitelli F, editors. Conservation science 2007. London: Archetype, 2008:23–32.

[16] Dupuis-Labbé D, Enshaian M-C. Les 39 dessins du “Mystère Picasso”: genèse d’une création, évolution d’un projet de conservation-restauration. Technè. 2005;22:88–95.
[17] Banks BA, Rutledge SK, Karla M, Norris MJ, Real WA, Haytas CA. Use of an atmospheric oxygen beam for restoration of defaced paintings. In: Bridgland J, editor. 12th Triennial meeting, Lyon, 29 August-3 September 1999: preprints (ICOM Committee for Conservation). vol. 1. London: Earthscan Ltd., 1999:271–5.
[18] Germinario G, Garrappa S, D’Ambrosio V, van der Werf ID, Sabbatini L. Chemical composition of felt-tip pen inks. Anal Bioanal Chem. 2018;410:1079–94.
[19] Garrappa S, Germinario G, van der Werf ID, Mirabile A, Sabbatini L. Multi-analytical study of artist felt-tip pen inks. In: 2nd IMEKO International Conference on Metrology for Archaeology and Cultural Heritage. Budapest, Hungary: IMEKO, 2016:136–9.
[20] Izzo FC, Vitale V, Fabbro C, van Keulen H. Multi-analytical investigation of felt-tip pen inks: formulation and preliminary photo-degradation study. Microchem J. 2016;124:919–28.
[21] Germinario G, Rigantea ECL, van der Werf ID, Sabbatinia L. Pyrolysis-gas chromatography–mass spectrometry of triarylmethane dyes. J Anal Appl Pyrolysis. 2017;127;229–39.
[22] Tsuge S, Ohtani H, Watanabe C. Pyrolysis-GC/MS data book of synthetic polymers. Oxford: Elsevier, 2011.
[23] Green FJ. Sigma-Aldrich® handbook of stains, dyes and indicators. Milwaukee, WI, USA: Sigma Aldrich Company. 1990.
[24] Smith GD, Clark RJH. Raman microscopy in archaeological science. J Arch Sci. 2004;31:1137–60.
[25] Smith GD, Clark RJH. Raman microscopy in art history and conservation science. Rev Conserv. 2001;2:92–106.
[26] Casadio F, Mauck K, Chefitz M, Freeman R. Direct identification of early synthetic dyes: FT-Raman study of the illustrated broadside prints of José Gaudalupe Posada (1852–1913). J Raman Spectrosc. 2010;100:885–99.
[27] Zaffino C, Passaretti A, Poldi G, Fratelli M, Tibiletti A, Bestetti R, et al. A multi-technique approach to the chemical characterization of colored inks in contemporary art: the materials of Lucio Fontana. J Cult Herit. 2017;23:87–97.
[28] Carona M. 2011a. Fernando Brizio in wonderland. http://upmagazine-tap.com/en/pt_artigos/fernando-brizio-in-wonderland/ Accessed July 24 2018.
[29] Carona M. 2011b. Fernando Brizio . . . design made in PT. http://azul-profundo-ferdos.blogspot.com/2011/10/fernando-brizio-design-made-in-pt.html Accessed July 24 2018.
[30] Smith GD. Cow urine, Indian yellow, and art forgeries: an update. For Sci Int. 2017;276:e30–4.
[31] Ploeger R, Shugar AN, Smith GD, Chen V. Late nineteenth century accounts of Indian yellow: the analysis of samples from the Royal Botanic Gardens, Kew. Dyes Pigments. 2019;160: 418–431.in press
[32] Brosseau CL, Gambardella A, Casadio F, Grzywacz CM, Wouters J, Van Duyne RP. Ad-hoc surface-enhanced Raman spectroscopy methodologies for the detection of artist dyestuffs: thin layer chromatography-surface enhanced Raman spectroscopy in situ on the fiber analysis. Anal Chem. 2009;81:3056–62.
[33] Saviello D, Trabace M, Alyami A, Mirabile A, Giorgi R, Baglioni P, et al. A combined surface enhanced Raman spectroscopy (SERS)/UV-vis approach for the investigation of dye content in commercial felt tip pens inks. Talanta. 2018;181:448–53.
[34] Saviello D, Di Gioia A, Turenne PI, Trabace M, Giorgi R, Mirabile A, et al. Handheld surface-enhanced Raman scattering identification of dye chemical composition in felt-tip pen drawings. J Raman Spectrosc. 2018;1–10. https://onlinelibrary.wiley.com/doi/abs/10.1002/jrs.5411.
[35] Griffin R, Spears L. Other methods of colour analysis: 12.1 high-performance liquid chromatography. In: Robertson J, Grieve M, editors. Forensic analysis of fibers, 2nd ed. London: Taylor & Francis, 1999: 315–22.

[36] Kirby J, ed. The diversity of dyes in history and archaeology. London: Archetype, 2017.
[37] Volná K, Holcapek M, Kolárova L, Lemr K, Cáslavsky J, Kacer P, et al. Comparison of negative ion electrospray mass spectra measured by seven tandem mass analyzers towards library formation. Rapid Comm Mass Spectrom. 2008;22:101–8.
[38] Chen VJ, Smith GD, Holden A, Paydar N, Kiefer K. The dating of an Uzbek ceremonial coat by dye analysis using liquid chromatography -mass spectrometry. Dyes Pigments. 2016;131:320–32.
[39] Frick D. The coloration of food . Rev Prog Color. 2003;33:15–32.
[40] Yang Y, Zhang J, Shao B. Quantitative analysis of fourteen synthetic dyes in jelly and gummy candy by ultra performance liquid chromatography. Anal Methods. 2014;6:5872–8.
[41] Bento WAS, Lima BP, Paim APS. Simultaneous determination of synthetic colorants in yogurt by HPLC. Food Chem. 2015;183:154–60.
[42] Weisz A, Mazzola EP, Ito Y. Preparative separation of di- and trisulfonated components of Quinoline yellow using affinity-ligand pH-zone-refining counter-current chromatography. J Chrom A. 2001;1216:4161–8.
[43] Scheifers SM, Verma S, Cooks RG. Characterization of organic dyes by secondary ion mass spectrometry. Anal Chem. 1983;55:2260–6.
[44] Cioccolani E. Personal communication. September 11, 2018.
[45] van der Werf ID, Germinario G, Palmisano F, Sabbatini L. Characterisation of permanent markers by pyrolysis gas chromatography-mass spectrometry. Anal Bioanal Chem. 2011;399:3483–90.
[46] Pfingstag G. Colorants in inks for writing, drawing and marking. J Soc Dyers Colorists. 1993;109;188–92.
[47] Druzik J. Oriel microfading tester (MFT): a brief description. In: Joel Thompson, Amanda Holden, Glenn Petersen, Sarah Stevens, editors. Textile specialty group postprints. vol. 20. Washington, DC: AIC, 2010:135–43.
[48] Townsend JH, Thomas J, Hackney S, Lerwill A. The benefits and risks of anoxic display for colorants. In: Saunders D, Townsend JH, Woodcock S, editors. Conservation and access: contributions to the London Congress 15–19 September 2008. London: IIC, 2008:76–81.
[49] Casella L, Sanderson K. Display of Alfred Stieglitz and Edward Steichen autochrome plates: anoxic sealed package and lighting conditions. Top Photographic Preserv. 2011;14:162–7.
[50] Beltran V, Druzik J, Maekawa S. Large-scale assessment of light-induced color change in air and anoxic environments. Stud Conserv. 2012;57:42–57.
[51] Bahr S. 2018. Gallery was remade with touch in mind. Indianapolis Star, July 27th, 2018, 8A.

Index

https://doi.org/10.1515/9783110457537-018